Rudenko International Conference "Methodological Problems in Reliability Study of Large Energy Systems" (RSES 2018)

E3S Web of Conferences Volume 58 (2018)

Irkutsk, Russia
2-7 July 2018

Editors:

N. Voropai
A. Michalevich
S. Senderov
H. Guliev

ISBN: 978-1-5108-7296-7

Printed from e-media with permission by:

Curran Associates, Inc.
57 Morehouse Lane
Red Hook, NY 12571

Some format issues inherent in the e-media version may also appear in this print version.

Printed by Curran Associates, Inc. (2018)

For additional information, please contact EDP Sciences – Web of Conferences at the address below.

EDP Sciences – Web of Conferences
17, Avenue du Hoggar
Parc d'Activité de Courtabœuf
BP 112
F-91944 Les Ulis Cedex A
France

Phone: +33 (0) 1 69 18 75 75
Fax: +33 (0) 1 69 28 84 91

contact-edps@webofconferences.org

Additional copies of this publication are available from:

Curran Associates, Inc.
57 Morehouse Lane
Red Hook, NY 12571 USA
Phone: 845-758-0400
Fax: 845-758-2633
Email: curran@proceedings.com
Web: www.proceedings.com

TABLE OF CONTENTS

Statement of Peer review

In submitting conference proceedings to *Web of Conferences*, the editors of the proceedings certify to the Publisher that

1. They adhere to its **Policy on Publishing Integrity** in order to safeguard good scientific practice in publishing.
2. All articles have been subjected to peer review administered by the proceedings editors.
3. Reviews have been conducted by expert referees, who have been requested to provide unbiased and constructive comments aimed, whenever possible, at improving the work.
4. Proceedings editors have taken all reasonable steps to ensure the quality of the materials they publish and their decision to accept or reject a paper for publication has been based only on the merits of the work and the relevance to the journal.

Title, date and place of the conference

Methodological Problems in Reliability of Large Energy Systems (RSES 2018)
02-07 July 2018
Irkutsk, Russia

Proceedings editor(s):

Prof. Nicolai Voropai, Melentiev Energy Systems Institute, Russia.
Prof. Sergey Senderov, Melentiev Energy Systems Institute, Russia.
Prof. Alexander Michalevich, Energy Institute of Belarus NAS, Belarus.
Prof. Husein Guliev, Azerbaijan Research and Design and Exploration Institute of Power Engineering, Azerbaijan.

Date and editor's signature

05.07.2018

Opportunities and perspectives of PMU application in power districts with distributed energy resources

Pavel Ilyshin[1*], *Alexey* Mokeev[2], and *Vladimir* Narovlyanskii[3]

[1]Petersburg power engineering institute of professional development, Aviation str. 23, Saint-Petersburg, Russia
[2]Northern (Arctic) Federal University, Severnaya Dvina emb.17, Arkhangelsk, Russia
[3]Energosetproject, Tkatskaya str. 1, Moscow, Russia

Abstract. The report examines the application of PMUs for various functional purposes of identifying emergency conditions, implement functions of relay protection and automation devices, perform a number of monitoring and diagnostic functions of the equipment. The analysis of functional requirements for control systems, protection and automation of power districts with distributed energy resources is carried out. The description of certain algorithms of emergency control and mode control is presented. Promising fields of scientific research are identified to expand the range of functions that can be implemented with the use of PMU.

1 Introduction

The development of modern telecommunication technologies and phasor management units (PMUs) enables development of distributed monitoring, protection and management systems WAMPACS [1, 2]. With PMUs emerging on the market, a device with small mass-size indicators and cost comparable with the cost of multifunction measuring transducers, the scope of the new units expands significantly [3]. One of the promising areas of PMU application is power districts with distributed energy resources (DER).

The main advantages of the PMU are related to the possibility of calculating parameters of power-district mode on the basis of current and voltage synchrophasors according to the fundamental harmonic at any control level, including the active values of the currents and voltages, active, reactive and full powers, symmetric components parameters mode etc. Based on six current and voltage synchrophasors, about one hundred parameters for the main harmonic can be calculated for each connection, and with the help of current and voltage synchrophasors obtained from various connections of the power district, the parameters of the power district in question can be calculated. This opens up wide opportunities for managing normal and emergency conditions both in power districts with distributed energy resources and in power systems to which these power districts are connected.

2 Prospects and issues of DER development

Main trends in DER development are as follows:

- Increase in gross installed power of various types of DER in the energy systems of developed countries,
- Increase in the installed power of individual DER units,
- Harmonious development of conventional generation and distributed energy resources.

As for primary equipment of power plants and power grid facilities, the key areas of development are determined by energy saving trends, development of distributed generation with a wide range of various drive motors and independent generation in large industrial enterprises using non-traditional and renewable energy resources.

An important structural feature of power supply networks of industrial enterprises with independent generation is small values of mutual resistances of generators and loads. This leads to a significant mutual influence of the operating modes of the generating units on the load operation modes and vice versa [4].

The main challenges in the implementation of the DER are conditioned by following factors:

- Positive and negative voltage deviations in all nodes of the network due to the mutual influence of power consumption schedules and power generation by DER facilities,
- Presence of reversible power flows in low and medium voltage networks,
- The added complexity of relay protection diagrams in low and medium voltage networks and the principles of applied protection and automation devices,
- Increase in the level of Isc in low and medium-voltage networks,
- The need to ensure sustainability of the power system when a large number of generating units of the DER facilities are disconnected,

* Corresponding author: ilyushin.pv@mail.ru

https://doi.org/10.1051/e3sconf/20185801001

- The need to ensure the possibility of isolated operation of all types of generating units at the DER facilities.

To effectively solve the problems mentioned above, it is necessary to introduce new technologies in control systems, relay protection and automation in power districts with distributed energy resources. Let us first consider the requirements for modern control, protection and automation systems in power districts with DER.

3 The functionality of control, protection and automation systems

The following main functions of modern automated process control systems, relay protection devices and automation of power districts with DER are identified:
- Ensuring observability and controllability of electrical modes,
- Efficient control over the choice of the composition of the operating generating equipment,
- Provision of group control over active power and voltage taking into account regulatory ranges and technological limitations,
- Implementation of the function of automatic frequency control and active power flows in an isolated power district,
- Implementation of automatic allocation for a balanced load in the event of damage to the supply network or the mode of high risks of a power failure,
- Automatic turn of generator sets (GS)/power plant from zero from cold and hot conditions,
- Automatic provision of specified network consumption parameters/power delivery to the network of the GS/power plant,
- Detection and prevention of the emergence of asynchronous modes at the GS and the links of the power district with the power system,
- Control of electromechanical transient processes in the power district at the rate of the process along a given trajectory,
- Application of high-speed devices of relay protection generators, transformers, busbars,
- The use of centralized earth-fault protection,
- Monitoring of the operation of DER equipment with control over the current mode and generation of emergency warnings,
- Removal of static and dynamic load characteristics of various technological productions,
- Calculation of technical and economic indicators.

Most of the functions listed above can be most effectively implemented using phasor management unit technology. At the same time, the functional capabilities of control, protection and automation systems using PMUs are significantly wider at the same financial costs than when applying traditional methods of building control, protection and automation systems.

4 Phasor management units

Until recently, the main deterrents for the widespread use of PMUs in distribution networks and power districts with DER were associated with the high cost of these devices and the large mass-size indicators of PMU. Therefore, the scope of application was mainly limited to the wide-area management systems (WAMS).

The experience of implementing the PMU in the WAMS and the progress in microelectronics provided the development of devices with small mass-size dimensions and low cost comparable to the cost of multifunction telemetry measurement transducers. This enables forecasting mass application of such devices at substations of various voltage classes, including 6-20 kV substations and those in power districts with DER.

At the same time, the scope of PMUs is significantly expanded: process management automated systems, automated process control systems, mode and emergency control automatics, incl. most advanced control, protection and automation systems WAMPACS. Therefore, in this article, the PMUs term refers not only to devices used in the WAMS, but also the systems listed above.

When used as part of automated control systems, devices with synchronized phasor measurements (SPM) support require measurements of the network mode parameters taking into account higher harmonics, and it can be implemented by measuring equivalent current and voltage synchrophasors [6]. This allows calculating about a hundred parameters of the network mode considering the highest harmonics at any level of control.

Expansion of the PMUs scope of application is associated with the expansion of the functionality of these devices for a number of applications, primarily for use in the system of relay protection and emergency automatics, where additional requirements for the quality of processing of input signals are imposed on the PMUs. When using synchronized vector measurements for emergency control and relay protection devices, it is important to ensure the reliability of measurements of synchrophasors in conditions of electromagnetic transients. At the same time, devices supporting SPM should provide suppression of free components of electromagnetic transients and research in this field has been actively conducted in recent years [5]. When using synchrophasors in protection and automation devices, an additional increase in the rate of data transmission is required.

For the application of the SPM technology as part of the relay protection and emergency control systems, the measurement ranges of the current synchrophasors corresponding to the emergency modes need to be extended. Similar devices are developed by both foreign and Russian manufacturers [3].

In addition, in the creation of modern control, protection and automation systems, it is expedient to integrate SPM technologies into the structure of a digital substation. Experience in implementing certain digital substation technologies, especially fast GOOSE messages, PTP time synchronization protocol and HSR and PRP backup protocols, have proved their high efficiency in substation and power plant automation systems. Taking into account the accumulated experience, as well as continuous improvement of microcontrollers and signal processors, implementation of these technologies does not have a significant effect on the final cost of devices anymore.

Current and voltage synchrophasors can be used as an alternative to sampling values which in fact represent raw digitized values of currents and voltages. When using synchrophasors instead of selective values of current and voltage, the traffic on the process bus decreases drastically, implementation of longitudinal differential protections and differential bus protectors, centralized control devices for normal and emergency operation modes is simplified.

The primary advantage of using synchrophasors for implementation of emergency control automation is the high quality of measurements of the network mode parameters on the basis of current and voltage synchrophasors for electromechanical transient processes, which is especially important in the implementation of the UFLS, under-frequency islanding system, OSPS and a number of other emergency control systems. It is also promising to use synchrophasors in the implementation of mode automation, for example, providing centralized control of reactive power compensation and maintaining the voltage in load nodes.

These data pave the way to the emergence of inexpensive and compact PMU on the market for use in distribution networks and in power districts with DER for the purpose of performing control, protection and automation functions.

Currently, the PMUs is used for simultaneous measurement of current and voltage synchrophasors. But in most practical cases, it is advisable to use separate PMUs systems for current (from the measuring and relay winding of the current measuring transformers) and for voltage (from the star and the open triangle windings of the measuring voltage transformers).

The main advantages of such an implementation of the PMUs, along with the reduction of cost and overall dimensions of these devices, are related to the need to avoid multiple duplication of voltage sync measurements in the devices of each bay or bus system taking into accountthe rare application of combined current and voltage measuring transformers in projects.

Experience in the implementation of the PTP time synchronization protocol in a number of domestic intelligent devices in the structure of a digital substation has proved its effectiveness for the PMUs. There is no need for a separate synchronization cable, as opposed to the case of time synchronization using PPS or IRIG-A/B.

In the coming years, we should expect a massive application of PMUs for solving management, protection and automation issues. But even a small number of PMU can improve the assessment of the power system condition [2, 7]. The same is true for the possibilities of implementing certain tasks of emergency control.

Consider the effectiveness of the application of PMU in the example of the implementation of a complex device for pole slip protection (OSPS) for a power facility.

5 OSPS of a power facility

OSPS of a power facility for integrated protection against the asynchronous mode (AM) of the power facility as a whole and its parts. Here, the term "power facility" means a local electric power complex which may include generation, substations, local load, power lines, distribution network feeders and other elements.

The technical solution for the complex protection of the power facility is based on the use of a distributed system of PMU connected with the central analytical unit by fibre-optic data transmission lines.

OSPS of the power facility provides protection:

- When passing the AM section along the lines of communication of the facility with the backbone network *(OSPS network function)*,
- When passing the AM section of transformers and other power elements of the power plant scheme *(OSPS of the power facility circuits-function)*,
- If there is an asynchronous move of one of the generators of the power plant *(Generator OSPS)*.

Operation of OSPS is considered using the functional diagram shown in Fig. 1, on the example of a conventional power plant with a substation through which power is delivered to the network.

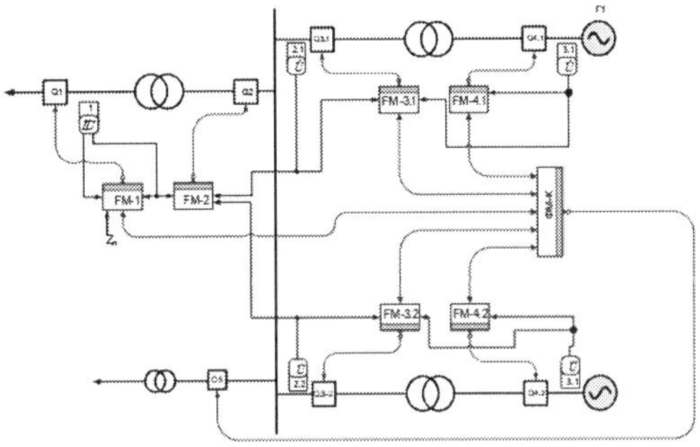

Fig. 1. Example of the OSPS functional diagram.

When implementing *the network OSPS*, the voltage and current synchrophasors of the PMU-1 are used, as well as the setting for the resistance of the monitored network section (function module FM-1). Operation of this OSPS is the same as that of the ALAR-M device [8-10]. At the same time, OSPS can be implemented both as OSPS by resistance and angular OSPS.

A more advanced version of OSPS is related to the implementation of the minimum stress points (MSP) detecting method [9]. To determine the MSP, it is necessary to know only the synchronized stress vectors in the nodes of the power district with DER. The method allows increasing the efficiency and completeness of AM detection without the need for preliminary calculation of OSPS settings, to anticipate weak links and thereby to increase the reliability of analysis and maintenance of the mode, reduce the number of failures, excessive and false alarms of emergency control devices. Fig. 2 is a diagram of control zones for possible sections in asynchronous mode.

The functional module FM-1 monitors the power system - power plant zone. If AM occurs, the section of which passes along the lines of the power output of the power plant, the power plant is separated from the power system by issuing a signal for the B1 shutdown.

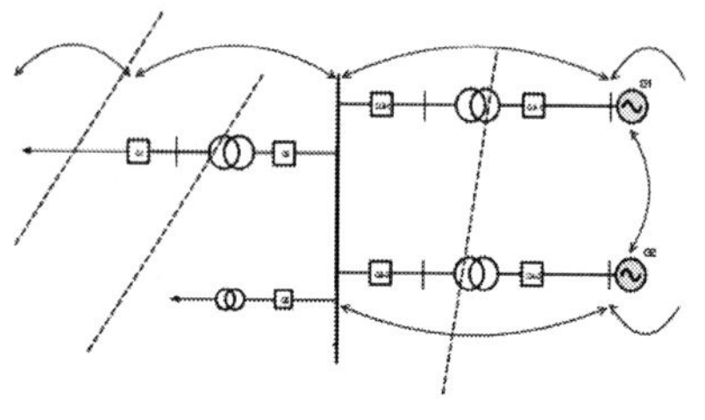

Fig. 2. Control zones of asynchronous mode sections.

OSPS in power plant circuits uses syncrovector stress measurements from PMU-1 PMU-2.1 and PMU-2.2 installed on the sides of the controlled circuit and outputs a signal of the occurrence of AM when detecting a predetermined value of the difference angle (often as a crank of 180 ^{0}). The functional module FM-2 monitors the zone of network transformers of the power plant. If AM occurs, the section of which passes through transformers of the power plant, the power plant is separated from the power system by issuing a signal to turn off B1 and in some cases to turn off B2.

OSPS in generator circuits functions similarly to *OSPS in power plant circuits*, but it is connected to the transformer generator busbar trunking measuring voltage circuits and is implemented via functional modules FM-3.1 and FM-3.2. This functionality does not require prior settings.

If AM occurs, the section of which passes through the power circuits of the generator, the generator is separated from the power plant by issuing a signal to turn off the breaker B3 (B3.1 in the circuits of generator 1, B3.2 in the circuits of generator 2) and in some cases to turn off circuit breaker B4 (B4 .1 in the circuits of the generator 1; B4.2 in the circuits of the generator 2).

OSPS of the generator is implemented by means of the functional module FM-4 and uses data from PMU-3 and information from the generator shaft run sensor. FM-4 monitors development of the asynchronous run of the generator, including accidents in the excitation system of the generator. If an asynchronous run occurs, the generator is separated from the power plant by issuing a signal to turn off B4 (B4.1 in the circuits of generator 1, B4.2 in the circuits of generator 2).

Functional module FM-K provides coordinated operation of FM-1-FM-5 modules and implements:

- Separation of the power plant from the power system when the section of AM passes through the power-circuit zones of all the generators of the power plant, i.e. when the accident is not due to technical failures of the generators' operation but is caused by the AM occurrence in the network,
- Operation of the OSPS generator functional unit by using the comparison of the angles of the voltage vectors on the power plant buses with the voltage vector angle on the stator of the generator in the absence of a generator rotation sensor or in addition to it,
- Monitoring of mutual angles of power plant generators for monitoring (and damping) of oscillations, identifying sources of low-frequency oscillations (LFO) and preventing the development of AM of generators due to tightening of stable generators in the oscillatory process.

Detection of the relative «abnormal» motion (oscillations and/or asynchronous motion) of one of the generators relative to other generators of the power plant is made by mutual comparison of the angles between the vectors of the direct sequence voltages on the stators of the generators or the values of the rotor angles of the generators.

Application of OSPS based on PMU will allow monitoring of mutual oscillation of power plant generators. The oscillation of the generator of the power plant in relation to other generators notably occurs due to inaccurate adjustment of the excitation and speed regulators, provoking the occurrence of the LFO and thereby contributing to a reduction in the stability margin.

Insignificant fluctuations of synchronous generators of power plants occur almost always due to continuous unbalances of the consumed and generated capacities. These oscillations are easily damped and do not cause complications, but in the event of significant changes in the mode, cyclic changes in frequency with amplitudes around 0.05 Hz may occur.

In this case, the behaviour of each generator (generator-turbine unit) of the plant may differ from the behaviour of adjacent generators due to differences in electromechanical characteristics (inertia, intrinsic resonance frequencies, etc.), as well as the settings of the excitation and speed regulators. This particularly applies to power plants where generators with different values of mechanical inertia constants or generators of different types and brands are installed.

In such a situation, it is important to directly monitor the behaviour of generators, and it will provide the basis for developing recommendations for tuning the excitation and speed regulators in order to reduce the mutual oscillation of the rotors of the power plant generators.

Monitoring generator oscillations allow assessing the risk of development of asynchronous move, the risk of tightening adjacent generators in the process of oscillating, it also allows ensuring generation and output of commands to the emergency control system of generators.

The frequency of formation of the voltage and current synchrophasors should not exceed 20 ms (basic value). It is useful to provide increased frequency of vector formation with a periodic sequence of 1 to 10 ms to ensure reliable operation of OSPS at the plants with low-inertia generators.

6 Conclusions

At present, there is a tendency to increase the number of

industrial power districts with distributed energy resources in the form of distributed generation facilities and various energy storage devices in which mutual influence of the operation modes of generating units and load is high.

The increase management efficiency in power districts with distributed resources of energy is largely determined by the quality of data coming from control objects.

The use of PMUs for various functional purposes can effectively enable detection of hazardous modes, implementation of functions of relay protection and automation devices, the performance of a number of functions for monitoring and diagnosing equipment in power districts with DER.

The use of PMUs allows implementation of both a centralized function of the OSPS power plant at a novel technological level and decentralized and combined one depending on the tasks assigned.

The functional capabilities of control, protection and automation systems using PMUs are significantly higher than when traditional ways of building these systems are applied, with the financial costs being equal for both options.

References

1. Wide area protection & Control technologies. CIGRE, Working Group B5.14 (Paris, 2016).
2. M. Kezunovic, S. Meliopoulos, V. Venkatasubramanian, V. Vittal. *Application of Time-Synchronized Measurements in Power System Transmission Networks* (Springer, NY, 2014).
3. A. Mokeev, Digital Substation. URL: http://digitalsubstation.com.
4. P. Ilyshin, Energoexpert, **1**, 58-62 (2015).
5. A. Mokeev, *Proceedings IEEE International Conference SIBCON* (2017).
6. A. Mokeev, V. Bovykin, A. Miklashevich, D. Ulyanov. *Proceedings International Conference Actual trends in development of Power System Relay Protection and Automation* (2015).
7. A. Gamm, A. Glazunova, Yu. Grishin, I. Kolosok, E. Korkina, Electrical Technology Russia, **6**, 2-9 (2009).
8. P. Ilyshin, P. Chusovitin, Relay Protection and Automation, **4**, 16-22 (2014).
9. V. Narovlyanskii, V. Kurmak, Power Plant, **3**, 48-51 (2012).
10. V. Narovlyanskii, *Modern Methods and Means of Prevention of Asynchronous Operation of the Power Grid* (Energoatomizdat, Moscow, 2004).

https://doi.org/10.1051/e3sconf/20185801002

Again about reliability characteristics of electrical power grids elements

Abdula Abdurakhmanov[1], *Sergey* Glushkin[1], *Vladislav* Plotnikov[1], *Andrey* Shuntov[1]*

[1]Moscow Power Engineering Institute, Department of Electric Power Systems, Krasnokazarmennaya 14, Moscow, 111250 Russia.

Abstract. The reliability characteristics of the (auto)transformers and overhead transmission lines of the unified national electric grid are subjected to statistical analysis. It is shown that their scheduled outages for scheduled or unscheduled repairs, as well as maintenance, weaken the network layout much more than emergency repairs. At the same time, there was a noticeable deterioration in the (auto)transformers and overhead transmission line reliability indexes over the last 30 years, which requires the development and adoption of appropriate organizational and technical solutions.

1 Problem description

Problem of power systems reliability performance existed before and continue to be relevant. One of its aspects is traditionally a research of reliability indexes of electrical grids. The most important are: failure intensity (precisely – mean failure intensity or failure rate); mean restoration time; medium frequency μ and duration T_{rep} of scheduled repairs.

Reliability characteristics of (auto)transformers ((A)T) and overhead power transmission lines (OHL) were analyzed in many published papers [1–6]. On it may be concluded that their fault characteristics are roughly the same, despite different time periods of monitoring: from 1980s-era till the beginning of 2000s. It is to be supposed that the reason is that equipment failures are investigated thoroughly followed by development of organizational-technical measures of their prevention.

In its turn the information about reliability characteristics of scheduled repairs is less presented and it's often controversial. Published data about characteristics of (A)T scheduled outages and single-circuit and double-circuit OHL is shown in the Table 1; Metal OHL supports are taken as a basis, as the most commonly used.

Table 1. Reliability characteristics of (auto)transformers.

Voltage, kV	μ, 1/year				T_{rep}, hrs		
110	0,5	1,0	6,3	3,2	28	30	10,5
220	0,9	1,0	6,3	3,2	28	30	12,0
330	1,0	1,0	6,3	3,2	30	30	12,8
500	1,0	1,0	6,3	3,2	50	50	14,0
750	1,0	1,0	6,3	–	50	50	–
Source	[1]	[2]	[3]	[4]	[1]	[2]	[3]

Thus, frequency of (A)T scheduled repairs in [1] and [2] is defined as 1 1/year. It clearly says in [2], that it's periodicity of only current repairs. Therefore, medium and major repairs are left out of consideration. In [3] μ values are defined at a level of 6,3 1/year with reference to [4]. However, in [4] there was talk of scheduled commutations, i.e. total bringing into and withdrawal from operation count, from where frequency of scheduled repairs μ=6,3/2≈3,2 1/year. As one can see, inconsistent information about (A)T scheduled outages is presented in considered sources.

Table 2. Reliability characteristics of single-circuit overhead lines.

Voltage, kV	μ, 1/year					T_{rep}, hrs			
110	2,9	2,1	15,0	7,2	2,7	14,9	14,5	4,2	15,2
220	5,0	2,8	13,2	7,4	2,9	25,4	17,0	5,9	12,4
330	5,7	3,0	12,0	5,2	2,4	20,1	21,0	7,9	14,3
500	12,8	3,1	10,0	5,0	1,5	16,6	18,0	12,0	16,7
750	10,0	0,17	–	–	–	9,6	–	–	–
Source	[1]	[2]	[3]	[4]	[5]	[1]	[2]	[3]	[5]

* Corresponding author sh280859@gmail.com

Works [2, 5] should be recognized as works with the most significant statistical samples. A medium frequency of scheduled repairs of OHL 220-500 kV in [5] vary (table 2) from 1.5 to 2.9 1/year; but in [2] it's within the range 2,8–3,1 1/year. Later works refers to much greater values: 10,0–13,2 [3] and 5,0–12,8 [1] 1/year. A medium frequency of scheduled OHL commutations is presented in [4], which equals 10,0–14,8 1/year, i.e. sum of their switching on/off; it is clear, that frequency of OHL scheduled repairs from [4] will be twice as little, i.e 5,0–7,4 1/year. It can be assumed, that such significant spread in presented characteristics can't give an estimate of real reliability characteristics of overhead lines in terms of scheduled outages.

Table 3. Reliability characteristics of double-circuit overhead lines.

Voltage, kV	Downtime	μ, 1/year		T_{rep}, hrs	
110	single circuit	3,8	3,2	14,8	14,8
	double circuit	0,4	0,17	19,0	21,8
220	single circuit	3,3	3,36	17,4	11,9
	double circuit	0,5	0,08	24,0	24,2
330	single circuit	7,3	7,11	15,0	9,4
	double circuit	0,3	–	14,1	–
Source		[2]	[5]	[2]	[5]

Works [2, 5] should be recognized as works with the most significant statistical samples. A medium frequency of scheduled repairs of OHL 220-500 kV in [5] vary (table 2) from 1.5 to 2.9 1/year; but in [2] it's within the range 2,8–3,1 1/year. Later works refers to much greater values: 10,0–13,2 [3] and 5,0–12,8 [1] 1/year. A medium frequency of scheduled OHL commutations is presented in [4], which equals 10,0–14,8 1/year, i.e. sum of their switching on/off; it is clear, that frequency of OHL scheduled repairs from [4] will be twice as little, i.e 5,0–7,4 1/year. It can be assumed, that such significant spread in presented characteristics can't give an estimate of real reliability characteristics of overhead lines in terms of scheduled outages.

Only in works [2,5] there were undertaken a studies of scheduled outages of double-circuit OHL. But these works refer to period of late 1970s – beginning of the 1980s.

Thus, authors set themselves the task of throwing light on reliability characteristics of (A)T and OHL in main electrical grids from the perspective of real values of frequency and duration of scheduled repairs.

2 Reliability characteristics of (auto)transformers

Monthly reports of actual carried out repairs of (A)T and OHL for a period of 2014-2016 years of Federal Grid Company (i.e. voltage of 220-750 kV) of most part of regions of the country were taken as source data. Statistical sample size for A(T) is shown in table 4.

Table 4. Statistical sample size for (auto)transformers.

Region	Amount of (auto)transformers-years in grids, kV			
	220	330	500	750
Central	615	63	111	32
North-West	213	291	6	16
Volga Region	477	–	84	–
Urals	471	–	96	–
Siberian	831	–	243	–
Southern	270	132	42	–
Total	2877	486	582	48

Classification of scheduled outages of grid elements was proposed in [6]. (A)T outages were considered by analogy with:

– for carrying out scheduled or unscheduled repairs or maintenance of (A)T (group 1);

– for carrying out scheduled or unscheduled repairs, maintenance or reconstruction of allied equipment (group 2);

Make notice that 'unscheduled repair' term was due to the fact that equipment is taken out from the service by an unscheduled dispatching form, but fundamentally it's scheduled outage. Whereby a common feature of scheduled outage is availability of certain period of time between making decision about outage and the moment of outage [6].

Number and duration of scheduled outages from table 4 is presented in table 5. As shown in table 5, scheduled outages connected with scheduled or unscheduled works on (A)T (group 1) constitute the major portion of outages (75-90%).

Table 5. Unscheduled (auto)transformers outages data.

Voltage, kV	Number of outages by groups, ea		Total duration of outages by groups, thous. hrs	
	1st group	2nd group	1st group	2nd group
220	5316	1389	902,8	124,3
330	1026	159	187,8	12,9
500	1164	276	231,4	30,4
750	108	33	31,4	0,8

There were calculated medium frequency and duration of scheduled repairs of (A)T according to table 5 data – see table 6.

Annually, (A)T 220-500 kV is scheduled taken out from the service more than twice (table 6). Thus, frequency of scheduled repairs (A)T from table 6 is close to values given in [4] about 50 years ago. But it should be noted, that actual medium durations and annual average durations of (A)T repairs turn out to be much greater than values published before. Thus, medium duration of a scheduled outage for (A)T 220 kV is 153 hours facing 30 hours shown in [1], for example. According to [1] failure rate of (A)T 220 kV

 https://doi.org/10.1051/e3sconf/20185801002

is 0,014 1/year with mean restoration time of 250 hrs. Hereof an annual average downtime of (A)T for emergency and restoration repair 0,014·250=3,5 hrs8, i.e. is in 100 times lesser than 357 hrs, shown in table 6.

Table 6. Reliability characteristics of (auto)transformers.

Voltage, kV	μ, 1/year		Total by groups 1–2	T_{rep}, hrs		Total by groups 1–2	Annual average duration of repairs, hrs
	1st	2nd		1st	2nd		
220	1,85	0,48	2,33	170	90	153	357
330	2,11	0,33	2,44	183	81	169	412
500	2,00	0,47	2,47	199	110	182	450
750	2,25	0,69	2,94	291	25	229	673

3 Reliability characteristics of overhead lines

There were carried out a research of reliability characteristics of single circuit and double circuit overhead lines. Statistical sample size for OHL is shown in table 7 and 8.

Table 7. Statistical sample size for single circuit overhead lines.

Region	Number of OHL, thous. km-year/line-year in grids, kV			
	220	330	500	750
Central	49,4/1083	5,6/123	32,6/198	7,8/30
North-West	10,6/271	15,2/255	–	3/15
Volga Region	23,2/460	–	16,4/91	–
Urals	31,5/659	–	24,4/128	–
Siberian	55,0/1422	–	45,3/271	–
Southern	13,4/313	9,9/122	12,3/39	–
Total	183,1/4208	30,7/500	131,0/727	10,8/45

There were defied following fault groups for OHL (see [6]):

– for carrying out scheduled or unscheduled repairs or maintenance of OHL (group 1);

– for carrying out scheduled or unscheduled repairs, maintenance or reconstruction of allied equipment (group 2);

– for carrying out OHL reconstruction (group 3);

– for performing construction and installation and other works in fenced-off area of OHL on requests of outside organizations (group 4).

Table 8. Statistical sample size for double circuit transmission lines.

Region	Number of OHL, thous. km-year/line-year in grids, kV	
	220	330
Central	171	15
North-West	35	29
Volga Region	85	–
Urals	169	–
Siberian	157	–
Southern	35	6
Total	652	50

Number and duration of scheduled outages of OHL from tables 7,8 are shown in table 9,10 correspondingly.

Table 10. Data of scheduled outages of double circuit transmission lines.

Voltage, kV	Number of outages by groups, ea		Total duration of outages by groups, thous. hrs	
	Single circuit downtime	Double circuit downtime	Single circuit downtime	Double circuit downtime
220	3822	107	371,4	8,2
330	296	5	37,3	0,3

Table 9. Data of scheduled outages of single circuit transmission lines.

Voltage, kV	Number of outages by groups, ea				Total duration of outages by groups, thous. hrs			
	1st	2nd	3rd	4th	1st	2nd	3rd	4th
220	9344	1318	30	48	936,3	105,5	4,1	2,4
330	1405	123	11	3	215,0	12,5	1	0,8
500	2074	251	1	11	273,5	20,7	0,4	1,7
750	101	32	0	0	18,2	2,9	0	0

Unscheduled outages connected with works, carried out on OHL (group 1), are equal 75-80% of total number of outages. Medium frequency and duration of scheduled repairs of OHL were determined based on data from tables 7-10 – see tables 11-12.

From the table 11 following that each of OHL 220-750 kV is taken out from the service from 2,6 to 3,2 1/year. This is a bit more than it was in 1980s: 1,5–2,9 1/year – see [5]. However, average duration of scheduled repair of OHL 220-500 kV was grown from 12-17 (see [5]) to 98-159 hrs, approximately by an order of magnitude, during the same period. According to [3] nonrecoverable failure for OHL 500 kV is 0,3-0,4 1/year for 100 km for mean restoration time of about 10 hrs. For average length of OHL 500 kV in Russia at a level of 200 km, its emergency and restoration downtime is 0,4(200/100)·10≈8 hrs. Whereby annual average duration of scheduled repair of OHL 500 kV is (table 11) 3,2·127=406 hrs, that almost 50 times higher its annual average emergency and restoration downtime.

Table 11. Reliability characteristics for single circuit overhead lines.

Voltage, kV	μ, 1/year				Total by groups 1–2	T_{rep}, hrs				Total by groups 1–2
	1st	2nd	3rd	4th		1st	2nd	3rd	4th	
220	2,22	0,31	0,01	0,04	2,6	100	80	139	50	98
330	2,81	0,25	0,02	0,01	3,1	153	102	88	277	149
500	2,84	0,35	0	0,02	3,2	132	83	408	152	127
750	2,24	0,72	0	0	3,0	180	91	0	0	159

Table 12. Reliability characteristics for single circuit overhead lines

Voltage, kV	μ, 1/year		T_{rep}, hrs		Annual average duration of repairs (double circuit downtime), hrs
	Single circuit downtime	Double circuit downtime	Single circuit downtime	Double circuit downtime	
220	5,8	0,16	97,2	76,2	12,1
330	5,9	0,10	126	60,9	6,1

4 Analysis of scheduled repairs duration during a year

Additionally there were plotted distributions of scheduled repairs duration of (A)T and OHL by year of period under review, i.e. proportions of the total number of repairs of given duration on time axis with accepted bin of 10 hrs. Obtained results are shown on fig. 1 and 2 as an example as applied to North-West region.

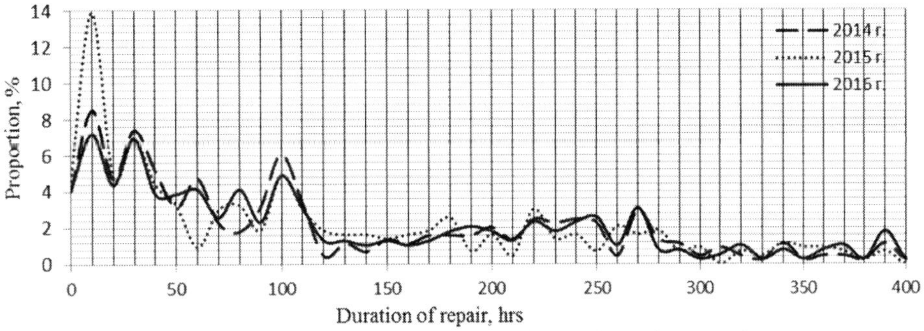

Fig. 1. Distributions of (A)T repairs by duration in North-West region.

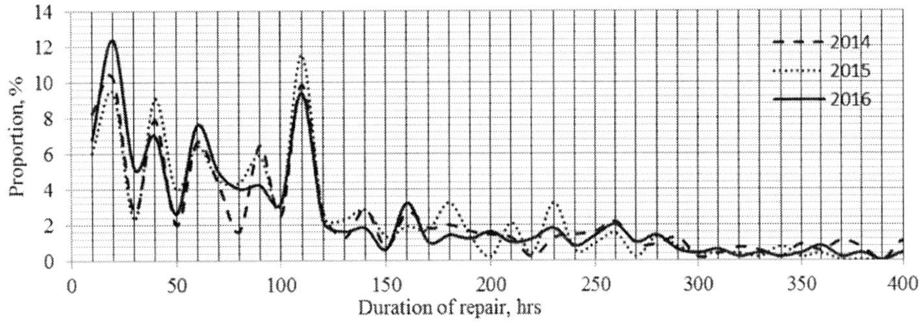

Fig. 2. Distributions of (A)T repairs by duration in North-West region.

Even visually on fig. 1 and 2 one can find similarity of distributions by year of period under review.

Additionally there was carried out correlation analysis for (A)T and OHL between regions of the country for a period of 2014-2016 yrs., where variables were: proportion of total number of outages and proportion of total duration by month without classification by voltage, i.e. relative number and duration of repairs by month. Corresponding results for OHL are shown in table 13 as an example.

 https://doi.org/10.1051/e3sconf/20185801002

Table 13. Correlation analysis for OHL by regions.

Compared regions	Pearson coefficient for	
	Number of repairs	Duration of repairs
Urals – North-West	0,93	0,92
Urals – Southern	0,89	0,92
Urals – Central	0,89	0,76
Urals – Volga Region	0,95	0,93
Urals – Siberian	0,95	0,93
North-West – Southern	0,87	0,89
North-West – Central	0,94	0,78
North-West – Volga Region	0,89	0,88
North-West – Siberian	0,96	0,92
Southern – Central	0,86	0,72
Southern – Volga Region	0,93	0,88
Southern – Siberian	0,88	0,89
Central – Volga Region	0,93	0,68
Central – Siberian	0,93	0,84
Volga Region – Siberian	0,93	0,93

Minimum value of Pearson coefficient (R) is R_{min}=0,88 for relative number of outages for comparison Southern and Siberian regions and R_{min}=0,68 for relative duration of outages for comparison Central and Volga regions (table 4). It is commonly known that correlational relationship is high for $R \geq 0,7$ and very high for $R \geq 0,9$. Thus, it is to be supposed that for accepted maintenance and repair of OHLs system for a period under review a choice of a certain year of observation is not general contributing factor.

It is commonly assumed that in reliability evaluations scheduled repairs are carried out on 'summer' periods. With that a decreasing correction factor, that is approximately K=0,5–0,6, is introduced for calculation period. Therefore, scheduled outages from table 5 and 7 were classified by months. The result is shown on fig. 3 and 4.

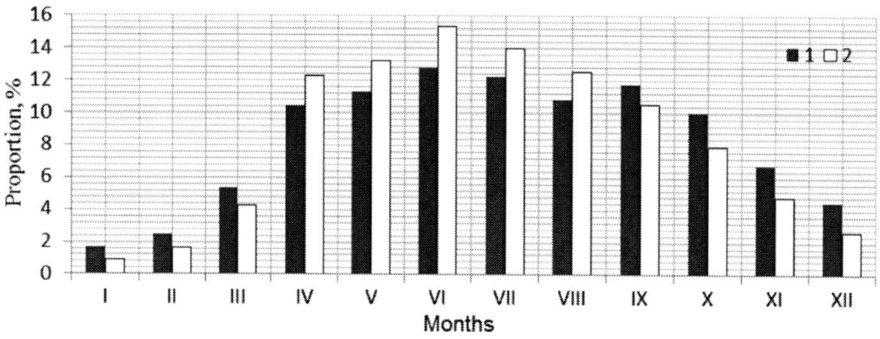

Fig, 3 Distributions of number and duration of repairs of (A)T by months: 1 - proportions of the total number of repairs; 2 – proportions of the total duration of repairs.

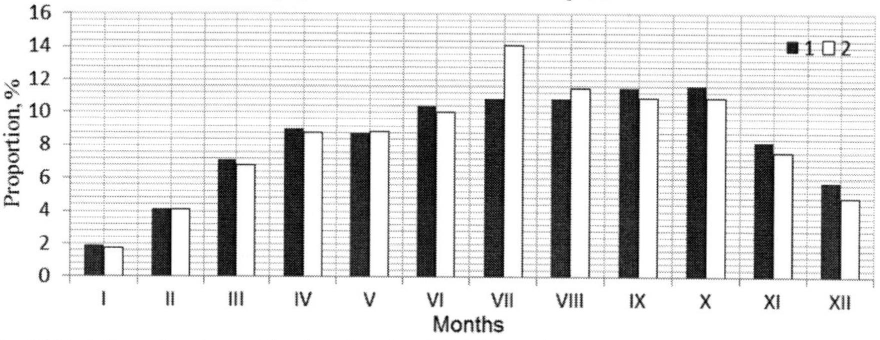

Fig, 4 Distributions of number and duration of repairs of OHL by months: 1 - proportions of the total number of repairs; 2 – proportions of the total duration of repairs.

Frequency and duration of scheduled repairs are constantly increasing from March till November; in the winter period a number of scheduled repairs is lower, than in summer period, however, they shouldn't be neglected. For example, 4-6% of annual number of repairs accrue to December, while the annual maximum is about 11-13% for both (A)T and OHL. From fig. 2 also following that histogram of total number of repairs is lower than histogram of total duration of repairs, but it's vice versa for the winter period. The reason is obvious – it's necessary to carry out more labor and time consuming repairs in warm seasons.

The maximum value of proportion of total duration of scheduled repairs of (A)T accrues to June and equals about 16%, and to July for OHL (14%). Therefore, 100/16=6,3 and 100/14=7,1 are minimum necessary number of months to carry out scheduled repairs of (A)T and OHL by an analogy with maximum load utilization time in a power system.

Where K=6,3/12=0,53 for (A)T and K=7,1/12=0,59 for OHL, which is confirm the accepted technical practice, which is used for reliability evaluation of electric circuit schemes.

5 Analysis of medium frequency of scheduled repairs of OHL dependence on their length

It was shown in [6] that frequency of scheduled outages has small dependence on a length of lines. This was explained for a variety of causes. In particular, when it's necessary to performance greater quantity of work it's accepted to increase a number of personnel (teams), but not duration of outage.

According to processed statistical data (table 9) there was analyzed dependence medium frequency of scheduled repairs on their duration:

Line length, km	to 100	100–200	200–300	Above 300
Frequency of repairs, 1/year	3,0	3,9	3,7	4,5

As we can see, some association between frequency of repairs and length appears at line lengths above 100 km. Thus, for OHL length above 300 km frequency of scheduled outages increases of 1.5 times more than short lines.

As for rather short OHLs 330 kV, which are located only in European part of the country – there's no association between given characteristics at all:

Line length, km	to 30	30–60	60–90	90–120	120–150	150–180
Frequency of repairs, 1/year	3,3	3,1	3,7	3,8	3,0	2,5

Therefore, it's confirmed once more that association between OHL length and frequency of its repairs in reliability evaluations of electric circuit schemes isn't general contributing factor and could be taken into account only for extremely long lines.

6 Conclusion

1. According to processed representative statistical data there were obtained reliability characteristics of (A)T and OHL in main electrical grids of power systems. It's found that from the perspective of medium frequency and duration of scheduled repairs they are way below than values obtained in the last century.

2. Average duration of scheduled repairs of (A)T increased of 5 times upon the average for last 30 years and annual duration of their scheduled downtimes – more than ten times. Applied in the last century normative approach to planning a periodicity of scheduled repairs of (A)T included their annual current repair and major repair once in 8-12 years. Technically, it could have been seemed more progressive than the modern approach, based on evaluation of current state of equipment, because it provided shorter periods of weakening of grid scheme during operation.

For the last 30 years average duration of scheduled repair of OHL 220-500 kV was grown from 12-17 to 98-159 hrs, i.e. almost in ten times. Upon that the main part of scheduled outages connects with repairs or maintenance of OHL itself, but not its reconstruction or other external reasons.

These facts require to be carefully thought.

3. It is found that annual duration of scheduled downtimes of (A)T and OHL exceed their annual duration of emergency and restoration repairs more than ten times. Thus, reduction of duration of repair conditions of electric grid elements could be significant reserve of power systems reliability improvement.

References

1. Karapetyan I.G Faibisovic D.L., Shapiro I.M. Spravochnik po proektirovaniyu elektricheskih setey [Reference book for the design of electrical networks]. Moscow, ENAS Publ., 2012. 320 p.

2. Guidance on the application of reliability characteristics of power system elements and the operation of power units with steam turbine units SPO Soyuztekhenergo Publ., 1985. 18 p. (in Russian)

3. Nepomnyaschiy V.A. Reliability of power systems equipment. ELECTRIC POWER. Transmission and Distribution, 2013, p.196.

4. Katson V.D., Nikitin O.V., Polovoy I.F., Halilov F.H. The annual number of switching lines and transformers of higher voltage classes Elektricheskie stancii [Power plants], 1970, **10**, pp. 42–44. (in Russian)

5. Barg I.G., Edelman V.I. Overhead power transmission lines. Moscow, Energoatomizdat Publ., 248 p.

6. Edelman V.I., Barg I.G., Korobanov S.V., Pripuskova T.P. Statistical analysis of scheduled outages of OHL 35-500 kV. Elektricheskie stancii [Power plants], 1979, **11**, pp. 49–54. (in Russian)

https://doi.org/10.1051/e3sconf/20185801003

On overhead networks transfer from 10(6) kV to 20 kV

Andrey Mayorov[1], *Kirill* Osintsev[2]*, *and Andrey* Shuntov[2]

[1]Public company «United Energy Company», 115035 Raushskaya Naberezhnaya, 8, Russian Federation
[2]National Research University «Moscow Power Engineering Institute», 111250 Krasnokazarmennaya 14, Russian Federation

Abstract. The expediency of preferential use of the rated voltage of 20 kV for the overhead networks of Russia is investigated. The feasibility study of the 20 kV voltage application areas based on the integrated parameters of the networks was performed. It is shown that due to the convergence of technical and economic characteristics of 10 and 20 kV equipment, 20 kV networks have more significant advantages in comparison with traditional 10 kV electrical installations.

1 Introduction

Feasibility of using 20 kV electric networks in our country has a long history. This voltage class was included into the list of rated voltage scale and then excluded again [1]. It was alleged that "...the introduction of 20 ... kV voltage is not necessary...". There were other opinions [2, 3, etc.]: "... the overhead networks require significantly lower costs, non-ferrous metals and especially transformer capacities... for 110/20 kV system with a direct 20/0.4 kV transformation...". When comparing alternatives of 10 and 20 kV networks it was assumed [3] that the cost parameters of overhead lines (OHL) and transformer substations (TS) of 20/0.4 kV exceed similar characteristics of 10 kV OHL and 10/0.4 kV TS in 1.1 and 1.25 times respectively.

The widespread construction of 20 kV electric networks in the country began after almost half a century, about five years ago in Moscow [4]. To date, more than 1,000 km of 20 kV cable lines have been commissioned. The construction of cable networks of this voltage class is also carried out in St. Petersburg and Yekaterinburg.

20 kV overhead networks of suburban and agricultural areas have their own specifics and have not been used all-round the country. The grounds on technical and economic efficiency of 20 kV voltage class application in overhead networks are presented below.

2 Integral network parameters

The methods for solving such optimization problems are well-known. The first one is referred to as the method of idealized structures, the second one is analytical. An example of the first approach is presented in the paper [1]. In this work, the network topology is given in the form of a lattice with a given step. TS are located in its nodes. Alternatives of the network are compared by the minimum cost through variation of initial data – load density, distances between nodes of the network, etc. In the second method, the cost indicators of OHL and TS are specified in the form of analytical (usually approximating) functions depending on their parameters (voltage, wire cross-section, line length, current, power, etc.). The extremum point is identified, for example, as in [5] – with the use of undetermined Lagrange multipliers.

The drawback of the first method is the divorcement from the actual structure and parameters of the network. The second method is even weaker, since it considers [5] the simplest configuration of the network: the power source – the load node.

We will analyse overhead networks of 10(6) kV in 18 districts of a large electric grid company from the European part of the country. In each district, typical network fragments were allocated according to the form shown in Fig. 1 [4].

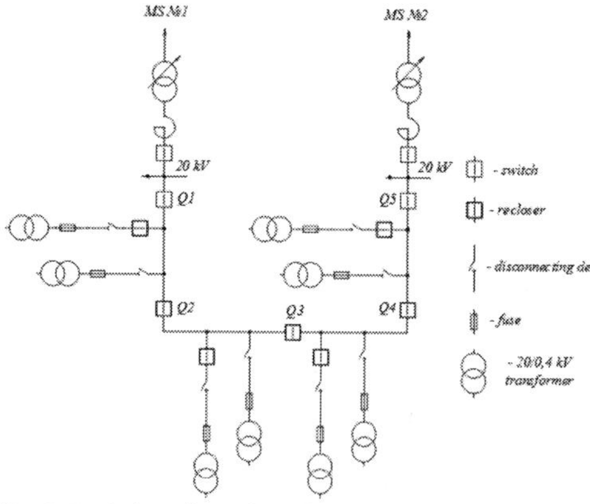

Fig. 1. Typical configuration of the overhead network.

Network fragments are connected in a loop and sourced by two geo-graphically separated main substations (MS) of 110/10(6) kV and sectionalized by reclosers, i.e. automatic posts of OHL sectionalizing. Transformer

* Corresponding author: kirmosh@mail.ru

substations of 10(6)/0.4 are connected to the backbone transmission line with a disconnector or recloser for long distance branches. Protection of 10(6)/0.4 transformers is carried out by fuses on the 10(6) kV side. In normal operation mode, the network is separated by one of the reclosers with automatic load transfer (ALT). Mainly, one-transformer TS of pole-mounted (PTS, nominal transformer power S_{nom} is 16–100 kV·A), mast (MTS, S_{nom}=160–250 kV·A) and kiosk (KTS, S_{nom}=400–1000 kV·A) types are installed.

In Table 1 the parameters relating to the linear part of the above-mentioned 18 network fragments are given: P_{max} and P_{min} – peak winter and summer load based on measurements of 2016; l_{OHL} and l_{OHLP} – the length of the line made of uninsulated wires and wires with a protective insulating sheath (OHLP); l_{CL} – the same for cable lines; l_{total} – the total length of the lines; l_{back} – the length of the backbone transmission line between the PC (without regard for the branches); σ_{ld} is the load density; σ_n is the network density.

Table 1. Parameters of the linear part of network fragments.

Frag	P_{max}/P_{min} kW/kW	l_{OHL} km	l_{OHLP} km	l_{CL} Km	l_{total} km	l_{back} Km	σ_{ld} kW/km^2	σ_n km/km^2
1	3138/ 1642	24.2	9.7	6.8	40.7	19.8	129.1	1.7
2	4751/ 3168	32.6	8.6	11.2	52.4	19.4	154.3	1.7
3	7358/ 5816	67.5	2.6	4.0	74.1	32.6	154.3	1.6
4	3609/ 1929	49.7	13.5	6.9	70.1	21.0	83.2	1.6
5	6646/ 6072	35.4	49.2	3.0	87.6	43.0	157.3	2.1
6	1781/ 989	44.7	0.6	1.6	46.9	12.9	57.3	1.5
7	1840/ 844	40.6	0.4	1.5	42.5	22.5	45.5	1.0
8	2486/ 1244	32.8	3.3	0.1	36.2	22.0	66.8	1.0
9	2992/ 1673	51.8	7.5	3.1	62.4	35.8	55.0	1.0
10	3101/ 2436	52.8	11.9	11.0	75.7	38.3	45.1	1.1
11	2716/ 1338	20.1	0.0	6.4	26.5	13.8	110.4	1.1
12	2818/ 1934	46.4	4.9	3.0	54.3	29.6	72.3	1.4
13	1640/ 1102	59.8	2.0	2.8	64.6	31.0	39.8	1.6
14	1193/ 618	49.5	10.1	2.2	61.8	41.3	24.8	1.3
15	956/ 349	43.0	7.0	0.0	50.0	30.9	26.3	1.4
16	297/ 78	35.0	0.0	0.0	35.0	30.5	8.2	1.0
17	3023/ 1872	21.0	11.4	9.9	42.3	20.3	117.6	1.6
18	1484/ 796	102.9	3.8	0.3	107.0	38.5	21.7	1.6
Aver.	2879/ 1833	45.0	8.1	4.1	57.2	28.0	76.1	1.4

The power supply area for assessing σ_{ld} and σ_n was determined on the assumption that the coverage radius of the 0.4 kV network of 10(6)/0.4 kV TS composes not more than 0.5 km. Accordingly, the width of the corridor of 10(6) kV OHL is 1 km when determining the area.

In fact, the entire overhead network (Table 1) is made with a wire of 70 mm^2, except fragments No. 5 and 17 where cross-section of 120 mm^2 is used for backbone transmission lines.

Table 2 shows the parameters relating to the TS of network fragments: S_{total} – the total installed capacity of 10(6)/0.4 kV step-down transformers; l_{TS} – the average distance between the TS; σ_{TS} – TS density; ITS – indoor TS.

Table 2. Parameters of the transformer part of network fragments.

Frag	P_{max}/P_{min} kW/kW	S_{total} kV·A	Amount, pcs				l_{TS} km	σ_{TS}	
			ITS	KTS	PTS, MTS	Total		kV·A/ km^2	pcs/ km^2
1	3138/ 1642	11700	4	36	8	48	0.85	483	2.0
2	4751/ 3168	15600	14	44	6	64	0.82	507	2.1
3	7358/ 5816	18900	11	69	7	87	0.85	396	1.8
4	3609/ 1929	12300	14	50	7	71	0.99	285	1.6
5	6646/ 6072	23600	13	82	15	110	0.80	559	2.6
6	1781/ 989	6600	2	42	10	54	0.87	213	1.7
7	1840/ 844	8200	6	31	7	44	0.97	204	1.1
8	2486/ 1244	5000	3	29	6	38	0.95	135	1.0
9	2992/ 1673	7600	0	47	0	47	1.33	140	0.9
10	3101/ 2436	11300	0	66	1	67	1.13	164	1.0
11	2716/ 1338	6300	0	20	0	20	1.33	255	0.8
12	2818/ 1934	9700	3	17	40	60	0.91	248	1.5
13	1640/ 1102	7800	6	10	41	57	1.13	189	1.4
14	1193/ 618	6300	6	6	34	46	1.34	130	1.0
15	956/ 349	9200	13	12	8	33	1.52	252	0.9
16	297/ 78	2900	4	9	2	15	2.33	81	0.4
17	3023/ 1872	10000	13	36	7	56	0.76	389	2.2
18	1484/ 796	11700	15	48	5	68	1.57	171	1.0
Aver.	2879/ 1833	10300	7	36	11	54	1.14	267	1.4

Analysis of data from Tables 1 and 2 allows to establish some statistical patterns:

1. The length of the 10(6) kV backbone line between the MS varies from 12.9 to 41.3 km with an average value of 28.0 km, which shows a relative uniformity of the MS distribution in the region.
2. The load density varies over a wide range from 8.2 to 157.3 kW/km^2 (average 76.1 kW/km^2). The range of average distances between TS in the fragments is 0.76–

2.33 km (average 1.14 km), which indicates a relatively uniform TS distribution over the territory.

3. The installed capacity of 10(6)/0,4 kV TS is 2-9 times higher than the actual maximum load, which is clearly irrational.

For comparison, the authors of [1] in their conclusions relied on the following data on the load density: in industrial areas it reaches 30–40 kW/km^2 (average 20 kW/km^2), in electrified agricultural areas – does not exceed 2–3 kW/km^2. For the 70 years passed since, the load density has multiply increased, which is an incentive to increase the nominal voltage of the network.

3 Technical and economic aspects of using 20 kV OHL

When transferring from 10 to 20 kV, the phase-to-phase distances increase slightly (from 0.2 to 0.45 m) in case of uninsulated wires [6] and there is no difference (0.4 m) when using OHLP. However, the actual phase-to-phase distances for typical reinforced concrete supports are approximately 2.5 times larger [7]. Therefore, for the same cross-section and wire type, the mechanical part of the 10 and 20 kV OHL is identical, except the number and rated voltage of the insulators in some cases, the cost of which is an insignificant part of the costs of OHL [8]. Therefore, capital investments to 10 and 20 kV overhead lines differ by not more than 1%.

The cost of constructing a 10-20 kV OHL in the European part of the country (hereinafter in 2017 prices without VAT) is determined by the simplest linear dependencies in the range of nominal cross sections from 50 to 120 mm^2 with a maximum error of 1.4%: for uninsulated steel-aluminum wires C_{OHL}=665.1+2.21s and C_{OHL}=497.5+2.18s th.RUB/km in populated and unpopulated areas respectively. Similarly, for insulated wires: C_{OHLP}=861.8+2.86s and C_{OHLP}=665.8+2.86s th.RUB/km.

When calculating discounted costs, the cost of OHL construction must be multiplied by the increasing factor $K=1+k_1k_2=1+0.0085\cdot9.4=1.8$, where k_1 stands for repairs and maintenance; $k_2=\dfrac{(1+E)^n-1}{E(1+E)^n}$; E – discount factor; n is the design lifetime of the electrical installation calculated from the moment of commissioning. Thus, at E=0.1 and n=30 years, k_2=9.4.

Moreover, it is necessary to take into account the cost of electric power losses for the design lifetime $C_{\Delta W}=\Delta P_{max}\tau C_e k_2$, where ΔP_{max} – the power losses in the peak loading conditions; τ – annual time of maximum losses; C_e – the unit cost of electric power losses. As it is known, $\tau=f(T_{max})$, where T_{max} – the number of hours of maximum load. According to the current reporting data for the region under consideration, $T_{max}\approx6000$ h (τ=3800 h) and C_e=2.18 RUB/kWh.

4 Technical and economic aspects of using 20 kV TS

Currently, domestic factories produce all the range of 20 kV TS elements. Power transformer of TMG type, used in TS of 10 and 20 kV, has the same weight and dimensions. The main difference is the height of high-voltage bushings (for 20 kV 75 mm higher than for 10 kV). For actual air insulation gaps between the fuse block and the transformer in 10 kV PTS and MTS, such an increase in bushings is not critical. The same relates to KTS transformer chambers.

The cost of 20/0.4 kV PTS is approximately 20-25% higher than of 10/0.4 kV PTS. In general, the rise in price is due to the higher price of transformers and disconnectors. The cost of TS constructing is determined by the simplest linear dependencies at the highest error of 3.6%: for PTS $C_{PTS}=273+S_{nom}+11.56U_{nom}$; for MTS $C_{MTS}=178+1.16S_{nom}+15.32U_{nom}$; for KTS $C_{KTS}=401+1.29S_{nom}+25.94U_{nom}$. For the TS, $K=1+k_1k_2=1+0.037\cdot9.4=1.35$. According to the factory data, the idling and short circuit losses of 10/0.4 and 20/0.4 kV transformers are the same and are further excluded from consideration.

The types of TS considered above have principal structural differences. Therefore, their costs are not jointly approximated with acceptable accuracy.

5 Comparison of the 10 and 20 kV network options

Table 3 shows the discounted costs (mln. RUB) for constructing networks fragments with the characteristics from Tables 1 and 2 at 10 and 20 kV: C_{TS} – costs for TS while maintaining their installed capacities; $C_{total}=C_L+C_{\Delta W}+C_{TS}$ – total expenses of the electric network; Δ – the ratio of total expenses of the electrical networks of 10 and 20 kV.

Table 3. Discounted costs for constructing network fragments.

Frag	C_L	$C_{\Delta W}$		C_{TS}		C_{total}		Δ
		10 kV	20 kV	10 kV	20 kV	10 kV	20 kV	%
1	37.8	17.6	4.1	48.8	60.3	104.2	102.2	102
2	48.6	28.7	6.3	66.3	80.6	143.3	135.5	106
3	68.9	287.8	70.2	82.9	101.8	439.6	250.9	175
4	65.1	77.8	20.8	63.1	78.2	206.0	164.1	126
5	81.4	306.5	74.7	107.1	131.5	495.0	287.6	172
6	43.6	6.0	1.3	39.5	49.0	89.1	93.9	95
7	39.5	8.9	2.0	38.7	47.8	87.1	89.3	98
8	33.6	11.5	2.6	28.9	35.9	74.0	72.1	103
9	62.4	42.6	8.1	40.6	50.7	145.6	121.2	120
10	70.4	25.1	5.2	58.1	72.2	153.6	147.8	104
11	24.6	5.2	1.4	24.8	30.2	54.6	56.2	97
12	50.4	15.3	3.1	50.8	62.4	116.5	115.9	101
13	60.0	10.2	2.3	43.4	54.1	113.6	116.4	98
14	57.4	4.5	1.0	35.2	43.5	97.1	101.9	95
15	46.5	3.1	0.8	40.5	50.0	90.1	97.3	93
16	32.6	0.2	≈0	12.2	15.3	45.0	47.9	94
17	39.3	38.4	7.9	47.1	57.6	124.8	104.8	119
18	99.4	11.9	2.7	41.5	50.5	152.8	152.6	100

The wire cross-section was taken to be equal to the actual value (mainly 70 mm^2 – see above) and the same for the backbone sections and branches. The cost of OHL construction C_L was estimated by the average value between the minimum and maximum estimates; the minimum – 650 th.RUB/km at a voltage of 10 and 20 kV for uninsulated wires in unpopulated areas; the maximum – 1070 th.RUB/km for insulated wire in the populated area; an average is 860 th.RUB/km without taking into account the factor K. When determining the cost of electric power losses in the overhead line, an annual load increase is taken to be equal to 2%.

It should be noted that the component $C_{\Delta W}$ from Table 3 differs by more than four times for 10 and 20 kV voltages: when calculating steady-state modes, no restrictions on the minimum voltage values at the network nodes were imposed. Therefore, for the 10 kV network fragments No. 1, 9, 10 from Table 3, the voltage in the network nodes decreased significantly (more than 10%) in the normal mode, while in the alternative 20 kV network it was at the rated level. For fragments No. 3–5 there was no mode convergence for 10 kV network. Here the cost of electric power losses is determined formally, by a multiple increase in losses in the 20 kV network.

The required voltage quality should be provided in accordance with State regulations GOST 32144–2013 not only in normal, but also in continuous repair and post-emergency operation modes – voltage reduction cannot exceed 10% of the rated value. In the 10 kV network this requirement is not satisfied for fragments No. 1–6, 9, 10, 12–14 at the current loading conditions and all fragments, except No. 16, in the prospective. Thus, in the foreseeable future, a deep reconstruction of practically the entire 10 kV network will be required (disaggregation of the backbone sections due to their diversion to other power centers), which is not taken into account in Table 3. In the alternative 20 kV network, the required quality is not provided only in the fragment No. 3 at the current load conditions and also in the fragment No. 10 for the perspective load growth. Thus, when using 20 kV voltage, measures to increase the network capacity will be local and significantly less expensive.

The existing approach to the comparison of diagrams of electrical networks with different levels of rated voltages is to be referred to as [9]: "The criterion for choosing a voltage system is the total costs for networks of all classes of voltage. When comparing the options of electrical networks with different voltage classes having equal costs or costs, differing up to 10%, priority is given to the option of developing networks with a higher average voltage of the distribution network". There is a certain logic in this: 10% is the generally accepted engineering accuracy of calculations, especially under conditions of great uncertainty in decision-making.

In our case, with a fixed cross-section of the wire, the capacity of the 20 kV network is twice as high as at 10 kV, and with the same transmitted power, the power losses are approximately four times less. This is a significant advantage of the electrical network of 20 kV. As can be seen from Table 3, according to the indicated criteria the transfer from 10 kV to 20 kV is reasonable practically for all the variety of values of the network parameters from Tables 1 and 2.

6 Questions of optimization of electric network parameters

For example, the cost function C_{OHLP} = 665.8+2.86s th.RUB/km is taken. Let's pass to the discounted costs C_{OHLP} = K(665.8+2.86s) = 1.08(665.8+2.86s) = 719+3.1s th.RUB/km. The discounted cost of electric power losses for 1 km of a three-phase line is $C_{\Delta W} = \Delta P_{max}\tau C_e k_2 = 3I^2R\tau C_e k_2 = 3I^2(\rho l/s)\tau C_e k_2 = 3I^2(0.0376\cdot 10^3/s)10^{-3}\cdot 3800\cdot 2.18\cdot 10^{-3}\cdot 9.4 = 8.7I^2/s$, th.RUB/km, where I – the phase current, A; R – resistance, Ohm; ρ – the specific resistance of the aluminum alloy conductor at an average temperature of 57°C (between the nominal 25 and the largest 90°C), Ohm·mm^2/m; l – the length of the line, km. Therefore, the total discounted costs on OHL are:

$$C_L = C_{OHLP} + C_{\Delta W} = 719 + 3.1s + 8.7I^2/s, \text{ th.RUB/km.} \quad (1)$$

Consequently, the optimal cross-section of the wire will be:

$$\partial C_L/\partial s = 3.1 - 8.7I^2/s^2 = 0 \quad (2)$$

$$s = \sqrt{8,7/3,1}I \quad (3)$$

From which the "economic" current density is $j_0 = I/s = 1/\sqrt{8,7/3,1} = 0.6$ A/mm^2. The authors intentionally quoted the word "economic" in order to avoid the continuation of many years of discussions [10, 11, etc.] in this field. By the way, the same value of 0.6 A/mm^2 appeared in [5]. The use of the cost characteristics of another type of OHL in (1)–(3) (see above) leads to a change of j_0 by no more than 10%.

For a cross section of conductors of 70 mm^2, the current will be 0.6·70=42 A or power of S≈730 kV·A at a voltage of 10 kV. For the configurations in Fig. 1 and Table 1, the power of a typical network fragment is estimated as 2S≈1460 kV·A. This ratio is satisfied only by case studies No. 14–16 from Table 1, which indicates an imbalance: overloaded OHL with redundant excess of transformer capacity.

When transferring to a voltage of 20 kV at the same section of the OHL wires, there is 2S≈2920 kV·A. In this case, already more than half of network fragments have satisfactory indicators from the point of j0. For the remainder of the fragments, it is advisable to further increase the cross section of the conductors to 120 mm2 (2S≈5800 kV·A). However, this is clearly not enough for fragments No. 3 and 5 from Table 1. They should be disaggregated. Accounting for the growth of loads, which was not taken into account in (1), will lead to even more lowering of j0.

Such obvious distortions in the configuration of electrical networks force to return to the well-known problem of optimizing their rated voltage. The method of undetermined Lagrange multipliers, worked out in [5] for this class of problems, will be used. Assume that condition (2) is satisfied. Then the second and third terms of (1) will be equal to each other and their sum, taking into account (3), will be:

$$2\cdot 3.1_S=2\cdot 3.1\sqrt{8.7/3.1I}=10.4I \quad (4)$$

The resulting costs for the electricity transmission over a distance l are written down as $C=C_L l+C_{TS}$. Take for example:

$$C_{TS}=C_{KTS}=K(401+1.29S+25.94U)= \\ =1.35(401+1.29\sqrt{3}\,IU+25.94U)=541+3IU+35U \quad (5)$$

Then taking into account (1), (4) and (5), it can be written:

$$C=719l+10.4Il+541+3IU+35U= \\ =541+719l+3IU+35U+10.4Il \quad (6)$$

The cost function (6) is minimized at some fixed value of the power S, i. e. we solve the problem dC=0 under the condition (the constraint equation) that $S/\sqrt{3}=IU$ by the Lagrange method.

The Lagrange function is composed as following: $C=719l+10.4Il+541+3IU+35U=541+719l+3IU+35U+ +10.4Il+\lambda(S/\sqrt{3}-IU)$.

According to the Lagrange method, there exists a coefficient λ such that:

$$\partial C/\partial U=3I+35-\lambda I=0 \quad (7)$$

$$\partial C/\partial I=3U+10.4l-\lambda U=0 \quad (8)$$

From (7) and (8) we have:

$$I(\lambda-3)=35;\ I=35/(\lambda-3) \quad (9)$$

$$U(\lambda-3)=10.4l;\ U=10.4l/(\lambda-3) \quad (10)$$

Substituting the values of I and U from (9) and (10) into the constraint equation, we obtain:

$$\lambda=25\sqrt{l/S}+3 \quad (11)$$

Taking into account (8) and (11), we have:

$$3U+10.4l-\lambda U=3U+10.4l-(25\sqrt{l/S}+3)U=0$$

From which the optimal network voltage is:

$$U=10.4l/(25\sqrt{l/S})\approx 0.4\sqrt{lS} \quad (12)$$

Condition (12) is derived for the simplest case of a load concentrated at one point, spaced from the power center by a distance l. With a load uniformly distributed along the line, the power losses in the limit will be three times less. The loss of electric power is corresponded by the term 10.4 l in the above formulas. Consequently:

$$U=0.4\sqrt{lS}/3=0.13\sqrt{lS} \quad (13)$$

Therefore, it follows from (12) and (13) that the optimum voltage for fixed power values and the distance of electric power transmission can vary over a wide range depending on the accepted design conditions:

$$U=0.13\div 0.4\sqrt{lS} \quad (14)$$

The formula (14) is used for the analysis of the preferred voltage level of the regional electric network with the characteristics from Tables 1 and 2. The average length of the backbone section of the network is 28 km (Table 1). Taking into account its sectionalizing in the normal mode, we can assume $l\approx 28/2=14$ km. The average load of the fragment in the maximum load mode is 2879 kW, whence $S\approx 2879/2=1440$ kV·A. The optimum voltage according to (14) will be 19–57 kV.

It is necessary to be cautious to the upper limit of 57 kV. With the transfer to 35 kV, the design of the 35/0.4 kV TS is significantly more complicated and heavier, but which is most importantly, the cost of OHL is 3–4 times higher compared to 10–20 kV OHL. As a consequence, with other things being equal, the cost of the 35 kV network is 1.5–2.5 times (on average 1.8) higher than for networks of 10–20 kV (especially when using transformers of relatively small, up to 100 kV·A, power). Justification of the inexpediency of the 10/0.4 transformation replacement by 35/0.4 kV (except for specific areas of oil production in the northern regions) has been considered in a large number of publications [2, 3, 11, etc.] over many years and the authors are in solidarity with this.

7 Conclusion

At this time interval, the cost characteristics of the elements of the 10 (6) and 20 kV overhead networks were either equalized (for overhead lines) or got significantly closer (for transformer substations), which allows to justify the widespread expediency of applying the nominal voltage of 20 kV in the power systems of Russia. This will ensure, at practically the same costs for ownership of the network, a doubling of its transmission capacity, which is extremely important for the implementation of new consumers technological connections.

References

1. A.A. Glazunov, S.A Gelikonskij, Electricity **11**, 24 (1948)
2. YA.M. CHervonenkis, L.M. Finger, Electricity **7**, 11 (1965)
3. A.G. Zaharin, N.S. Kanakin, Electricity **1**, 6 (1966)
4. A.V. Mayorov, K.A. Osintsev, A.V. Shuntov, Electricity **9,** 34 (2017)
5. R. Pelisier, *Energy systems* (Higher School, Moscow, 1982)
6. *Electrical Installations Code* (KNORUS, Moscow, 2010)
7. A.I. Khalyasmaa, S.A. Eroshenko, D. Bogdanov, ICSESS **November**, 75 (2017)
8. A.I. Khalyasmaa, S.A. Dmitriev, A.M. Romanov, RTUCON, (2016)

9. Company standard 34.01–21.1–001–2017, *Electrical distribution network 0.4-110 kV. Requirement for technological design* (PJSC ROSSETI, Moscow, 2017)

10. EH.N. Zuev, ELEKTRO **1**, 44 (2000)

11. V.P. Frajshteter, A.S. Mart'yanov, Oil industry **4**, 117 (2011)

12. N.A. Moguil'nittskii, B.S. Shulov, *Application of 20 kV voltage in Latvian SSR* (Gosenergoizdat, Leningrad, 1963)

Randomized algorithm of invariant zeros evaluation of electrical power system, defined in descriptor form

Magomed Gadzhiev [1], *Misrikhan* Misrikhanov[1], *Vladimir* Ryabchenko[1]*, and *Nikita* Vasilenko[1]

[1]National research university "MPEI", Institute of Electrical Power Engineering, 14 Krasnokazarmennaya St, Moscow, Russia

Abstract. A randomized algorithm for computing the invariant zeros of the electric energy system as a dynamical system with many inputs and many outputs (MIMO system), specified in the descriptor form, is proposed. Definitions of invariant zeros are carried out by randomizing the original MIMO system and it reduces to a generalized eigenvalue problem for a numerical matrix. The application of the algorithm is illustrated by the example of calculating the invariant zeros of the linear model of the United Power System.

1 Introduction

The main object of practical calculations is the linearization of nonlinear descriptor mathematical model. Thus, equations [8] are considered for nonlinear dynamic system.

$$\frac{d\mathbf{x}}{dt} = \mathbf{f}(\mathbf{x},\mathbf{z},\mathbf{u}), \qquad (1)$$

$$\mathbf{0} = \mathbf{g}(\mathbf{x},\mathbf{z},\mathbf{u}), \qquad (2)$$

$$\mathbf{y} = \mathbf{h}(\mathbf{x},\mathbf{z}), \qquad (3)$$

where $\mathbf{x} \in \mathbf{R}^{n_x}$ – state variables vector; $\mathbf{z} \in \mathbf{R}^{n_z}$ – parameter vector; $\mathbf{u}\, O\, \mathbf{R}^{r}$ – input stimulus vector; $\mathbf{u} \in \mathbf{R}^{r}$ – output quantity vector. Formula for vector functions $\mathbf{f}$, $\mathbf{g}$ and $\mathbf{h}$ is defined by system element mathematical models, measurement and correcting units.

Standard approach to llinearization (1), (2) connects with the procedure of Taylor expansion of vector functions $\mathbf{f}$ and $\mathbf{g}$ in small neighbourhood of steady-state process with $\mathbf{x}_0$, $\mathbf{z}_0$, $\mathbf{u}_0$ coordinates.

$$\mathbf{f}(\mathbf{x},\mathbf{z},\mathbf{u}) \approx \mathbf{f}(\mathbf{x}_0,\mathbf{z}_0,\mathbf{u}_0) + \frac{\partial \mathbf{f}}{\partial \mathbf{x}}_0 \Delta\mathbf{x} + \frac{\partial \mathbf{f}}{\partial \mathbf{z}}_0 \Delta\mathbf{z} + \frac{\partial \mathbf{f}}{\partial \mathbf{u}}_0 \Delta\mathbf{u}, \qquad (4)$$

$$\mathbf{g}(\mathbf{x},\mathbf{z},\mathbf{u}) \approx \mathbf{g}(\mathbf{x}_0,\mathbf{z}_0,\mathbf{u}_0) + \frac{\partial \mathbf{g}}{\partial \mathbf{x}}_0 \Delta\mathbf{x} + \frac{\partial \mathbf{g}}{\partial \mathbf{z}}_0 \Delta\mathbf{z} + \frac{\partial \mathbf{g}}{\partial \mathbf{u}}_0 \Delta\mathbf{u}, \qquad (5)$$

$$\mathbf{h}(\mathbf{x},\mathbf{z}) \approx \mathbf{h}(\mathbf{x}_0,\mathbf{z}_0) + \frac{\partial \mathbf{h}}{\partial \mathbf{x}}_0 \Delta\mathbf{x} + \frac{\partial \mathbf{h}}{\partial \mathbf{z}}_0 \Delta\mathbf{z}, \qquad (6)$$

where $\Delta\mathbf{x} = \mathbf{x} - \mathbf{x}_0 \sqcap x' \in \mathbf{R}^{n_x}$, $\Delta\mathbf{z} = \mathbf{z} - \mathbf{z}_0 \sqcap x'' \in \mathbf{R}^{n_z}$, $\Delta\mathbf{u} = \mathbf{u} - \mathbf{u}_0 \ulcorner u \in \mathbf{R}^{r}$, $\Delta\mathbf{y} = \mathbf{y} - \mathbf{y}_0 \sqcap y \in \mathbf{R}^{m}$ – deviation of corresponded vectors; $(*)_0$ – corresponded partial derivative, estimated in coordinates chosen.

Equation linearization (1) – (3) at a point of steady state motion gives linear descriptor (algebra-differential) equations

$$\frac{dx'}{dt} = \frac{\partial \mathbf{f}}{\partial \mathbf{x}}_0 x' + \frac{\partial \mathbf{f}}{\partial \mathbf{z}}_0 x'' + \frac{\partial \mathbf{f}}{\partial \mathbf{u}}_0 u, \qquad (7)$$

$$0 = \frac{\partial \mathbf{g}}{\partial \mathbf{x}}_0 x' + \frac{\partial \mathbf{g}}{\partial \mathbf{z}}_0 x'' + \frac{\partial \mathbf{g}}{\partial \mathbf{u}}_0 u, \qquad (8)$$

$$y = \frac{\partial \mathbf{h}}{\partial \mathbf{x}}_0 x' + \frac{\partial \mathbf{h}}{\partial \mathbf{z}}_0 x''. \qquad (9)$$

Hereafter for the purpose simplification of equations writing, index «0» of partial derivative is omitted.

When Jacobi matrix is invertible, equations (7), (8) can be reorganized into state space form

$$\frac{dx'}{dt} = \left(\frac{\partial \mathbf{f}}{\partial \mathbf{x}} - \frac{\partial \mathbf{f}}{\partial \mathbf{z}}\frac{\partial \mathbf{g}}{\partial \mathbf{z}}^{-1}\frac{\partial \mathbf{g}}{\partial \mathbf{x}}\right)x' + \left(\frac{\partial \mathbf{f}}{\partial \mathbf{u}} - \frac{\partial \mathbf{f}}{\partial \mathbf{z}}\frac{\partial \mathbf{g}}{\partial \mathbf{z}}^{-1}\frac{\partial \mathbf{g}}{\partial \mathbf{u}}\right)u$$
$$y = \left(\frac{\partial \mathbf{h}}{\partial \mathbf{x}} - \frac{\partial \mathbf{h}}{\partial \mathbf{z}}\frac{\partial \mathbf{g}}{\partial \mathbf{z}}^{-1}\frac{\partial \mathbf{g}}{\partial \mathbf{x}}\right)x' - \frac{\partial \mathbf{h}}{\partial \mathbf{z}}\frac{\partial \mathbf{g}}{\partial \mathbf{z}}^{-1}\frac{\partial \mathbf{g}}{\partial \mathbf{u}}u. \qquad (10)$$

If matrix $\frac{\partial \mathbf{g}}{\partial \mathbf{z}}$ is singular, then conversion to equations (10) will not exist. However, even in case of invertibility of $\frac{\partial \mathbf{g}}{\partial \mathbf{z}}$, the problem of conditionality of the matrix happens in practical calculations. For condition number

$$\text{cond}\ \frac{\partial \mathbf{g}}{\partial \mathbf{z}} >> 1 \qquad (11)$$

of matrix $\frac{\partial \mathbf{g}}{\partial \mathbf{z}}$ is close to singular matrix and mathematical model (10) becomes «stiff» in relation to minute calculation errors and variation of parameters. To put it differently, the mathematical model (10) in case of correctness of inequation (11) is ill-defined from A.N. Tihonov [1, 8].

If additional measures aren't taken, all calculations using the model (10), including controllability and observability analysis, will lead to serious errors. However, generally the method given comes across number of significant difficulties. Nonstrictly speaking, improvement of condition of matrix $\frac{\partial \mathbf{g}}{\partial \mathbf{z}}$ will be attended with deterioration in condition of matrices in (10). The alternative is the usage for calculations of original linearized descriptor model (7), (8).

* Corresponding author: ryabchenko.vn@yandex.ru

2 Invariant zeros definition

Let us introduce the following notation of block vector and matrix:

$$x=\begin{pmatrix} x' \\ x'' \end{pmatrix}\in \mathbf{R}^{n}, E=\begin{pmatrix} E_{n_x} & 0 \\ 0 & 0 \end{pmatrix}\in \mathbf{R}^{n\times n}, A=\begin{pmatrix} \frac{\partial \mathbf{f}}{\partial \mathbf{x}} & \frac{\partial \mathbf{f}}{\partial \mathbf{z}} \\ \frac{\partial \mathbf{g}}{\partial \mathbf{x}} & \frac{\partial \mathbf{g}}{\partial \mathbf{z}} \end{pmatrix}\in \mathbf{R}^{n\times n},$$

$$B=\begin{pmatrix} \frac{\partial \mathbf{f}}{\partial \mathbf{u}} \\ \frac{\partial \mathbf{g}}{\partial \mathbf{u}} \end{pmatrix}\in \mathbf{R}^{n\times r},\ C=\begin{pmatrix} \frac{\partial \mathbf{h}}{\partial \mathbf{x}} & \frac{\partial \mathbf{h}}{\partial \mathbf{z}} \end{pmatrix}\in \mathbf{R}^{m\times n},$$

where E_{n_x} – identity matrix of order $n_x \times n_x$; $n = n_x + n_z$. Then descriptor system (7)–(8) can be written over in more standard form

$$E\frac{dx}{dt}(t) = Ax(t) + Bu(t),\ y(t) = Cx(t), \qquad (12)$$

where $x(t)\in \mathbf{R}^{n}$ – state vector, $u(t)\in \mathbf{R}^{r}$ – input vector, $y(t)\in \mathbf{R}^{m}$ – output vector.

Define:

Definition [1]. *Complex frequency $\lambda = \alpha^*$, where column rank of matrix*

$$\begin{pmatrix} A-\alpha^* E & B \\ C & 0 \end{pmatrix} \qquad (13)$$

Is decreasing, is called invariant zero.

In (13) $\lambda \in \mathbf{C}$ – complex variable; $\mathbf{0}$ – zero matrix of order $m\times r$; $\mathbf{C}$ – complex plane.

Invariant zeros are defined from the condition [10]

$$\begin{pmatrix} A-\alpha^* E & B \\ C & 0 \end{pmatrix}\begin{pmatrix} x_0 \\ u_0 \end{pmatrix} = 0 \qquad (14)$$

of full «blocking» of signal transmission both eigen component and forced component of dynamic motion, i.e. for $\lambda = \alpha^*$ there are such vectors x_0 и u_0, that system output reaction equal to zero. Thus, the problem of determining of invariant zeros of the system (12) resolves oneself into eigenvalue generalized problem in the general case of rectangular pencil matrix

$$\begin{pmatrix} A-\lambda E & B \\ C & 0 \end{pmatrix}. \qquad (15)$$

The search of Invariant zeros for large dynamic systems ($n \gg 1$) comes across number of significant calculation difficulties. Calculation errors and calculation problems of zeros of generalized pencil matrices [11] play a crucial role in here. Therefore new approaches to development of computing algorithms are needed.

3 Randomized squaring

Rule of thumb calculation of invariant zeros of descriptor system (12) connected with so called dynamic system squaring up.

Suppose $m > r$ (number of outputs of a system is more that number of inputs). Let us introduce into consideration the descriptor system

$$E\frac{dx}{dt}(t) = Ax(t) + Bu(t),\ y^*(t) = T_C Cx(t), \qquad (16)$$

where $T_C \in \mathbf{R}^{r\times m}$ – full row rank matrix.

The assertion, driven by verbal proof, is valid similar to [9].

Theorem 1. Set of invariant zeros of descriptor system (12) is always a subset of a set of invariant zeros of squared descriptor system (16), but not vice versa.

Suppose hereafter $r > m$ (number of inputs of a system is more that number of outputs). Consider the descriptor system

$$E\frac{dx}{dt}(t) = Ax(t) + BT_B u^*(t),\ y(t) = Cx(t), \qquad (17)$$

where $T_B \in \mathbf{R}^{m\times r}$ – full column rank matrix.

The assertion dual to theorem 1 is valid.

Theorem 2. Set of invariant zeros of descriptor system (12) is always a subset of a set of invariant zeros of squared descriptor system (17), but not vice versa.

Lemmas 1, 2 Simply, that squaring operation generally puts into the system new zeros. This must be considered for choice of T_C and T_B matrices. The randomisation procedure, proposed by authors, is one of methods of solution for a choice of T_C and T_B matrices.

Let us introduce into consideration randomization operator $\mathrm{Rand}_{r\times m}$, which generates full rank matrices of order $r\times m$ with normal law distributed elements. Also suppose $\mathrm{Orth}_{r\times m}$ – orthogonalization operator for full rank matrix of order $r\times m$ .Then

$$\mathrm{Orth}_{r\times m}\ \mathrm{Rand}_{r\times m} \qquad (18)$$

– operators composition, which action result is a randomized orthogonalization full rank matrix.

Let us also specify Ω – set of invariant zeros of the system (12), Ω^* – set of invariant zeros of squared descriptor system. To be specific hereafter it's as if $m > r$ and и squared descriptor system is of the form of (16).

Suppose Ω_i^*, $i = 1,2,...$ – sets of invariant zeros of squared descriptor systems, where

$$T_C^{(i)} = \mathrm{Orth}_{r\times m}\ \mathrm{Rand}_{r\times m} . \qquad (19)$$

On the analogy of [12] it may be shown, that intersection of all sets of invariant zeros of squared systems Ω_i^*, $i = 1,2,...$, coincides with the set of invariant zeros of the given system Ω.

4 Evaluation of invariant zeros of large electrical power system

Let us find the solution of evaluation problem of invariant zeros of large electrical power system, composed of 25 electrical power plants. Mathematical model of electrical power system in phase coordinates «phase angle - generator slip» in form (12) has matrices of following orders:

$$A \in \mathbf{R}^{50\times 50}, B \in \mathbf{R}^{50\times 25}, C \in \mathbf{R}^{20\times 50}, E \in \mathbf{R}^{50\times 50}. \qquad (20)$$

Portrait of Jacobi matrix A is shown on fig. 1. To put it differently, there is control system on each of 25 power stations. Measurement of phase coordinates is performed on-line only on 20 power plants with synchronized vector meters included in mode parameters monitoring system.

Thus, pencil matrix (15) is rectangular and has order of 70×75 in this case. By using orthogonalization and randomization operators let us generate a set

$$T_i, = 1,\dots,k \; = \mathrm{Orth}_{25\times 20} \; \mathrm{Rand}_{25\times 20} \; , \qquad (21)$$

and then construct square pencil matrices

$$\begin{pmatrix} A-\lambda E_{50} & BT_i \\ C & \mathbf{0}_{20\times 20} \end{pmatrix}, \; i=1,\dots,k.. \qquad (22)$$

Matrix eigenvalues computational results (22) for $k=500$ (i.e. 15 000 eigenvalues) in form of their generalized portrait are shown on fig.1.

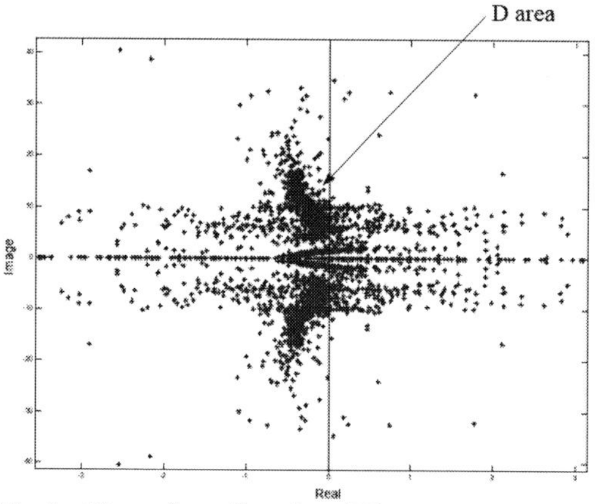

Fig. 1 – Eigenvalues of matrices (22)

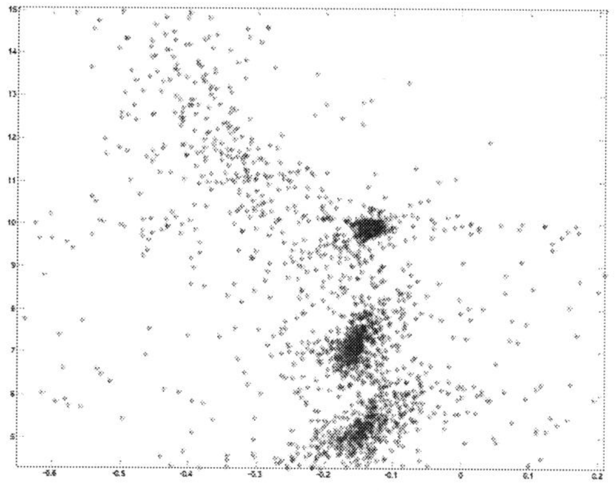

Fig. 2 – Eigenvalues of matrices (22) in D area

Research of generalized portrait shows that there are three groups of tight accumulations in D area, which could be characterized as invariant zeros candidates. Accumulation centres match up complex frequencies, on which the matrix (15) of considered electrical power system model has close to zero minimal singular values, i.e. the matrix is close to loss of its rank. This is the practical invariant zero «indicator» α^*.Thus, considered complex frequencies can be undertaken as electrical power system invariant zeros (20).

From physical standpoint estimated invariant zeros match up modal forms (mode) of oscillations, on which «blocking» of analyzed electrical energy system is happens. This is reflected in the fact that, firstly, it is impossible to reestablish conditions of other 5 power plants, by the information from phasor measurement units (PMU), installed as was shown before on 20 power plants. Secondly, the sudden loss of efficiency of stabilizing PI and PID controls, which use the given signals of mode parameters monitoring system.

5 Conclusion

Invariant zeros directly connects with controllability and observability of dynamic system, including principle possibility of PI, PD and PID control law synthesis and realisation and servocontrol as well. Analysis of Invariant zeros of large dynamic systems, to which modern electrical power systems belong, comes across number of significant calculation difficulties, where calculation errors and calculation problems of zeros of generalized pencil matrices play a crucial role. Thus, new approaches to development of computing algorithms are needed.

The randomized squaring procedure is proposed in the report, allows calculating efficiently invariant zeros of dynamic system, given in descriptor form.

Invariant zeros can be matched up with modes of oscillations, on which «blocking» of analyzed electrical energy system is happens, that is reflected in the fact that, it is impossible to reestablish conditions by the information from synchronized phasor measurement units of electrical power system elements (loss of full observability) and due to the sudden loss of efficiency of stabilizing PI and PID controls, which use given signals for electrical power system stabilization (loss of full controllability).

References

1. Brockett R.W., IEEE Trans. Automat. Control,1965, **10**. pp. 129 – 135.

2. Rosenbrock H.H. *State space and multivariable theory.* (Nelson-Wiley, 1970).

3. Zhou K. Essentials of Robust Control. Prentice Hall, 1998.

4. Zubov N.E., Mikrin E.A., Ryabchenko V.N. *Matrix approach in theory and practice of automatic flight-control system* (Bauman Moscow State Technical University Publ., Moscow 2016).

5. Zubov N.E., Mikrin E.A., Misrikhanov M, Sh, Ryabchenko V.N., *Journal of Computer and Systems Sciences International.* 2018, **9**, pp. 141 – 151.

6. Misrikhanov M, Sh, Ryabchenko V.N., Avtomatika i Telemkhanika [Automatics and teleautomatic]. 2008, **10**, pp. 31–47. (in Russian)

7. Smagina E,M. Calculation and specification of zeros of a linear multidimensional system. Avtomatika i

Telemkhanika [Automatics and teleautomatic], 1987, **12**. pp.165 – 173. (in Russian)

8. Zubov N.E., Mikrin E.A., Misrikhanov M, Sh, Ryabchenko V.N., *Journal of Computer and Systems Sciences International, 2018*. **1**, pp. 20 – 26.

9. Macfarlane A.G.J., Kareanias N. Pole and zeros of linear multivariable systems: a survey of the algebraic, geometric and complex variable theory. Int. J. Contr. 1976. Vol. 24, No 1. pp. 33–74.

10. Smagina E.M. Analysis matters of linear multidimensional objects using the concept of zero of a system. Tomsk, Tomsk Univeersity Publ., 1990.

11. Misrikhanov M, Sh, Ryabchenko V.N., Avtomatika i Telemkhanika [Automatics and teleautomatic]. 2006, **5**. 24 – 47. (in Russian)

12. Granichin O.N., Polyak B.T. Randomized algorithms for estimation and optimization under almost arbitrary disturbances. Moscow, Nauka Publ., 2003.

On the placing of synchronized vector measurements in network 110 - 330 kV for identification of the Kaliningrad region power system regime

Magomed Gadzhiev [1*], *Eugenia* Gulevich[1], *Vyacheslav* Korobka [1], *Vladimir* Ryabchenko [1], and *Yuriy* Sharov [1]

[1]National research university "MPEI", Institute of Electrical Power Engineering, 14 Krasnokazarmennaya St, Moscow, Russia

Abstract. A solution is presented for the problem of placing phasor measurement unit to identify the mathematical model of the electric power system in the state space. The criterion for complete observability of the Kalman dynamic system is taken as a basis. As a method of solution, we use the canonical genetic algorithm. The complete observability of the power system is ensured by the application of the observability rule in the form of a recursive test of observability. The proposed approach is demonstrated by the arrangement of synchronized vector meters in the Kaliningrad region power system.

1 Introduction

Identification of electric power systems (EPS) for a long time was a difficult-solvable problem because of the lack of time-allocated and time synchronized dynamic observations in space. The appearance of synchronized vector meters – phasor measurement unit (PMU) brought this problem to the practical level. Modern PMU allows measuring phase currents along transmission lines, phase voltages on buses of power stations and substations, current frequency, and (indirectly) overflows of active and reactive power. This allows the formation of a system of monitoring for transient regimes, to eliminate the lack of information on transient processes and to realize the identification of EPS [1] – [3].

Identification in the work is the task of determining the parameters of the mathematical model of EPS, which best describes the measurement data. The solution of the identification problem is based on the invariance property of the shift of the Kalman observability matrix [3]. This ensures the identification of the complete and equivalent (reduced) mathematical model of EPS and allows generating, for example, appropriate actions for controlling oscillations in EPS and ensuring a given margin of static stability.

The report shows the problem of placing PMU in the nodes of the electrical network with the purpose of real-time identification of the EPS mathematical model and solving related analysis and synthesis problems of control laws. The canonical genetic algorithm is used as a method of solution, the rule (criterion) observability of it is formulated on the basis of the recursive test analysis of a multidimensional dynamic system observability represented in the state space [4].

A practical example, considered in the paper, is the arrangement of PMU in the 110-330 kV networks of the Kaliningrad region power system.

2 EPS as an object of identification

In the present formulation of the problem, the identification of EPS is reduced to the determination of the elements of numerical matrices: A - the state matrix (Jacobi matrix), B - control effectiveness matrix, C - observation matrix and D – matrix of shunting connections, in vector linear discrete equations [3]

$$x(t+1) = Ax(t) + Bu(t),\ \ y(t) = Cx(t) + Du(t), \qquad (1)$$

describing the oscillations in EPS. Here, x (t) is the n-dimensional state vector of the EPS, u (t) is the 1-dimensional vector of control actions (input vector), and y (t) is the m-dimensional vector of the state variables (output vector), t is the discrete time.

The discrete Fourier transform (DFT) of the system (2), taking into account that the input signals are close to periodic and that an entire number of periods of EPS oscillations is observed, yields a system of linear matrix equations with complex numbers [3]:

$$\xi_k X(k) = AY(k) + BU(k),\ Y(k) = CX(k) + DU(k). \qquad (2)$$

Here ξ_k – is the shift operator, $X(k)$ – is the DFT of the state vector $x(t)$, $Y(k)$ –is the DFT of the output signal $y(t)$, $U(k)$ – is the DFT of the input signal $u(t)$.

The dependence of the second equation in (3) on the first gives the following recursion:

$$W_r(k)Y(k) = O_r X(k) + S_r W_r(k)U(k) \qquad (3)$$

where

* Corresponding author: gadzhiev_mg@interrao.ru

$$W_r(k)=\begin{bmatrix}1\\ \xi_k\\ \vdots\\ \xi_k^{r-1}\end{bmatrix},\ O_r=\begin{bmatrix}C\\ CA\\ \vdots\\ CA^{r-1}\end{bmatrix},S_r=\begin{bmatrix}D & \cdots & \cdots & 0\\ CB & D & \vdots & 0\\ \vdots & \vdots & O & \vdots\\ CA^{r-2}B & CA^{r-3}B & CB & D\end{bmatrix} \quad (4)$$

The matrix O_r in (4)is the Kalman observability matrix, and S_r – is the stabilization matrix [5]. For the observability matrix O_r (5), the following invariance property of the shift is valid [5]:

$$O_{r[1:r-1,:]}A = O_{r[2:r,:]} \quad (5)$$

Here $O_{r[1:r-1,:]}$ – is the matrix obtained from the observability matrix by choosing rows from the first to the (r-1) -th, $O_{r[2:r,:]}$ – the matrix obtained from the choice of rows from the second to the r-th. This fact is indicated by the subscripts of the matrices located in square brackets. A single «: » sign means that all columns of the matrix are selected.

Property (6) is the key and is the basis of identification algorithms. It is necessary that it be completely observable in terms of the condition of Kalman's complete observability [5] for identifying the EPS model (1)

$$\operatorname{rank} O_r = n,\ \text{при}\ r<n \quad (6)$$

2 Approach to the solution of the problem of placing PMU

Apparently, the invariance property of the shift of the observability matrix (7) is the fundamental basis of the identification algorithm used, and the Kalman complete observability condition (8) is a necessary condition for the identifiability of EPS. This criterion should be taken into account in the formation of the SMLC, including the placing of SSI.

It is proposed to use the canonical genetic algorithm [6] to search for rational variants of the SVR arrangement.

Genetic algorithms are based on the ideas of population genetics and represent methods of global search for the extremum of the objective function *g*. In this case, this problem is formulated as follows:

$$\Sigma=\sum_{i=1}^{M} z_i \rightarrow \min,\ g(Z)=0, \quad (7)$$

where z_i - is indication of the PMU installation in the i-th node (z_i =1 – SWM is installed, - there is no SWM), M – is total number of nodes in the EPS, $\Sigma=\sum_{i=1}^{M} z_i$ - the total number installed in the EPS PMU, $g(Z)$ – the objective function that determines the condition for the complete observability of the Kalman power plant.

The solution is sought on the basis of repeated cycles of the PMU arrangement options generation, further advancement and elimination of which is carried out on the basis of EPS total observability criterion according to Kalman (6) with the minimum possible number of established PMUs.

Complete observability of EPS in solving this problem is provided by applying the rule (criterion) of observability in the form of a recursive test (algorithm) [4].

The basis for using the recursive test is that applying of the rank test (6) even with low dimensionality of the state vector (more than 10) can lead to false conclusions about the uncontrollability of EPSs due to the deterioration of the numerical conditionality of the analyzed matrix O_r. In this case, a false conclusion about the loss of rank is possible, and, accordingly, the absence of the complete observability property [5].

The algorithm of the recursive test looks like this.

Step 0. An estimation of the observation matrix is formed on the basis of the variant of the PMU distribution set by the genetic algorithm C. The maximum solution is determined as $C^{\perp}$ of the matrix equation $C \cdot C^{\perp}=0$. If $C^{\perp}=0$, then the test ends. It is concluded that the system is fully observable. If $C^{\perp} \neq 0$, then the next step is taken.

Step 1. The maximum (maximum rank) solution of the matrix equation is determined.

$$\begin{bmatrix}AC^{\perp} & C^{\perp}\end{bmatrix}\begin{bmatrix}Z_1^{(1)}\\ Z_2^{(1)}\end{bmatrix}=0 \quad (8)$$

Here A – The state matrix is obtained on the basis of the EPS modeling in Matlab.

If the solution $Z_1^{(1)}, Z_2^{(1)}$ is zero, then the test ends (the system is fully observable).Otherwise, it goes to the next step.

Step 2. The maximum (maximum rank) solution of the matrix equation is determined.

$$\begin{bmatrix}Z_2^{(1)} & Z_1^{(1)}\end{bmatrix}\begin{bmatrix}Z_1^{(2)}\\ Z_2^{(2)}\end{bmatrix}=0 \quad (9)$$

If the solution $Z_1^{(2)}$, $Z_2^{(2)}$ is zero, then the test ends (the system is fully observable). Otherwise, it goes to the next step.

Step k. The maximum solution of the matrix equation is determined

$$\begin{bmatrix}Z_2^{(k-1)} & Z_1^{(k-1)}\end{bmatrix}\begin{bmatrix}Z_1^{(k)}\\ Z_2^{(k)}\end{bmatrix}=0 \quad (10)$$

If the solution $Z_1^{(k)}$, $Z_2^{(k)}$ is zero, then the test ends (the system is fully observable). Otherwise, it goes to the next step.

The criterion for stopping the test for observability is the nondecreasing rank of matrices sequence $\begin{bmatrix}Z_2^{(k-1)} & Z_1^{(k-1)}\end{bmatrix}$, $\begin{bmatrix}Z_2^{(k)} & Z_1^{(k)}\end{bmatrix}$.

At each step are sought only in the class of orthogonal matrices [7] to increase the accuracy of the computations using this recursive test of the corresponding matrix equations solutions.

3 Power system of the Kaliningrad region

The Kaliningrad region is separated from the rest of the country by the land borders of foreign states and international sea waters.

The tasks of static stability analysis and provision, evaluation and management of oscillations dominant modes, reserves of regime parameters, damper properties of EPSs acquires a high degree of relevance in conditions of a possible and rather rapid exit from the Interstate Energy Union of the Baltic States.

The initial data for the development of the Kaliningrad region energy system model were taken from the regime model for forecasting the winter maximum of the 2018 load. The total active load of EPS is 985 MW.

Presently, the power system of the Kaliningrad region has the following characteristics: the number of power stations is 5; the number of 330 kV transmission lines is 7; the number of substations is 330/110 kV is 3, the number of substations with a higher voltage of 110 kV is 86. The installed capacity of power plant generators is 923 MW.

The creation of a branched Simulink model of EPS in Matlab made it possible to determine the structure of the state matrix (the Jacobi matrix) in a continuous or discrete form. In this mode, the Jacobi matrix has dimensions of 12 × 12, 22 × 22, 42 × 42, 120 × 120, 273 × 273, 385 × 385, 450 × 450, 510 × 510, depending on the equivalence of the adjacent 110 kV network.

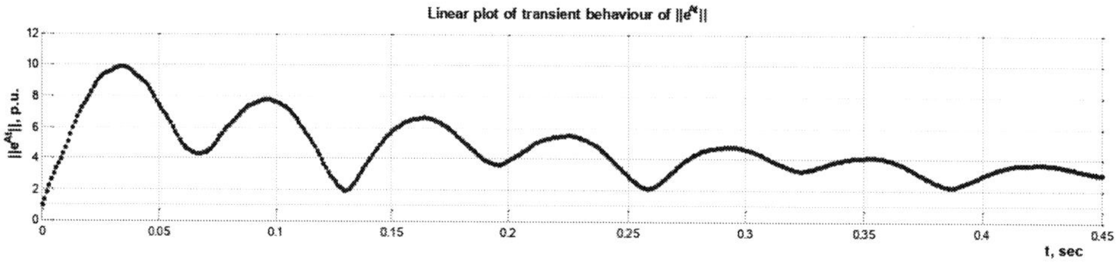

Fig. 1. Generalized transient characteristic of the power system The Kaliningrad Region

The generalized transition characteristic in the form of the norm of the exponential of the Jacobi matrix of size 42×42 $\| e^{At} \|$ [9]**Ошибка! Источник ссылки не найден.** in linear and logarithmic scales is shown in Fig. 1. It characterizes the power system of the Kaliningrad region as statically stable. High-frequency components of electromechanical processes are damped during the first seconds. Further oscillations are low-frequency with periods from one up to tens of seconds. The norms $\| e^{At} \|$ for Jacobi matrices of other sizes are qualitatively equivalent to the characteristic given in Fig. 1.

4 Research results

The sequence of the solution of the problem of placing the PMU in EPS has the form shown in Fig. 2.

The first step is the linearization procedure for the EPS model. At the initial stage of the genetic algorithm work random solutions for distribution of PMU on the on power plant and substation buses are generated randomly using the randomize number operator (sensor). Then, the "survival rate" of the solutions is calculated based on the results of a recursive observability test, which is chosen, as noted earlier, as a rule of observability. These solutions produce offspring (a population). Those decisions that are "stronger", i.e. use a smaller number of PMUs, provided that the observability of the EPS is ensured, selected for further processing, and "weak", using a larger number of PMUs, are excluded.

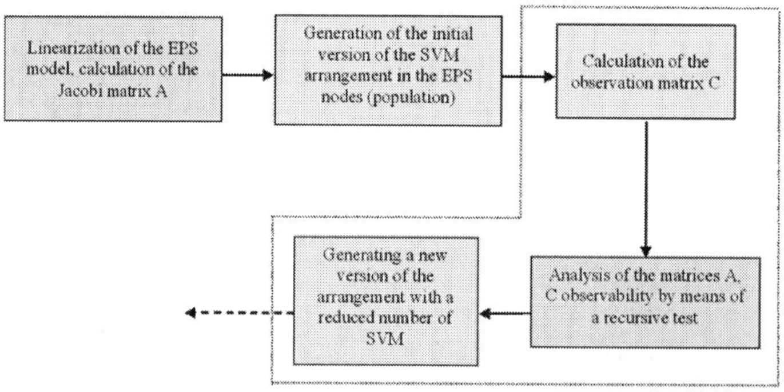

Fig. 2. Sequence of the genetic algorithm work when placing PMU

https://doi.org/10.1051/e3sconf/20185801005

The process is repeated until a solution is determined in the form of the minimum possible number of PMU that provides observability of the EPS.

At the initial stages of the search, when the number of selected and installed PMU is not less than half the dimension of the identified energy system state space, the step following the zero step of the recursive test is finite in terms of the "survival" of the solution. It consists in verifying the positive definiteness of the symmetric matrix $C^{T}A^{T}E - CC'^{T}AC' > 0$. This corresponds to the simplified criterion of observability introduced. If this condition is not met, the decision is also considered unacceptable.

The required number of PMUs (fig. 3) and the locations of their arrangement in the power system of the Kaliningrad region were determined, as a result of the genetic algorithm tests in conjunction with the recursive test.

In the fig. 3 a relatively small number of PMUs for high values of the dimensionality of the state space of the mathematical model is explained by the existing multiply connected energy system. At the same time, the growth of the installed PMUs number, depending on the dimensionality of the state space of the mathematical model, is nonpolynomial, which makes it difficult to predict solutions in such problems.

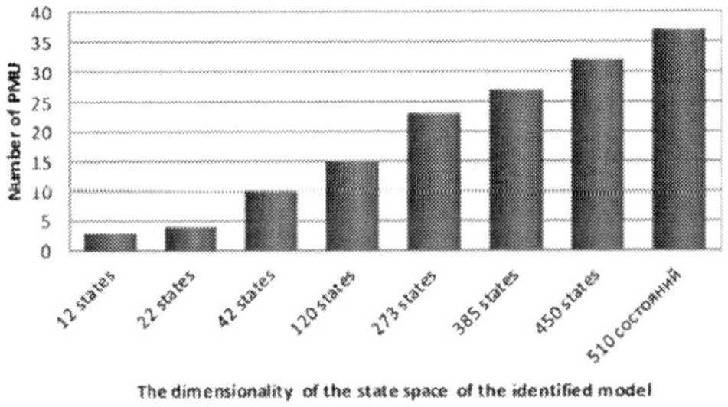

Fig. 3. Distribution of the number of PMUs required for the Kaliningrad region power system

In all the calculated versions of the locations of the PMU, these devices were located in the generator nodes and the most loaded load nodes.

Note that in the post-emergency calculation modes, it is necessary to provide for the appropriate reservation of PMU, which can lead to a twofold increase in the number of meters.

Real-time identification of the simplest power system model, which has the 12th order and takes into account the basic modes of electromechanical oscillations, proves to be sufficient only for 3 PMU, one of which is located at Kaliningradskaya Thermal power station (TPS) – 2, and others at substations 330 kV "Sovetsk" and "Centralnaya".

The general relationship between the practical tasks being solved in the interests of the power systems, and the conditions for the arrangement of the SWMs is shown in Table 1.

We will make the necessary explanations regarding Table1. The authors used models with Jacobi A matrices of 42 × 42 and 450 × 450 sizes for the analysis of the static stability of the power system of the Kaliningrad Region (Task No. 1 in Table 1). In this case, a smaller Jacobi matrix was obtained by reducing the 450 × 450 matrix A based on Krylov subspaces [7] and containing relatively "slow" oscillation modes located near the imaginary axis in the complex plane.

When estimating the dominant oscillation modes (task No. 2 in Table 1), a reduced model with a Jacobi matrix of size 42 × 42 was used, containing, as noted earlier, "slow" oscillation modes. These modes that determine the electromechanical transients and make the main contribution to the fluctuations in the power system of the Kaliningrad region are obtained from the results of mathematical modeling. To estimate the reserves of the regime parameters (Task No. 3 in Table 1), we used the Jacobi A matrices of the entire ruler of the obtained dimensions, i.e. 12 × 12, 22 × 22, 42 × 42, 120 × 120, 273 × 273, 385 × 385, 450 × 450, 510 × 510, in order to determine sufficient conditions for the required dimensionality of the power system model. It was revealed that, starting from the size 42 × 42 and above, the stocks of the regime parameters do not change significantly. Models of smaller dimensions give overestimated (too optimistic) values of the regime parameters reserves.

Table1. Conditions for the arrangement of the PMU

	Class of tasks to be solved	Mathematical Problem	Power system mode	Kind of model and dimension
1	Static stability analysis	The complete problem of the eigenvalues of matrix A	Steady mode	Full high-dimensional model
2	Estimation of dominant modes of oscillation	A partial problem of the matrix A eigenvalues	Electromechanical transients	Reduced modelof smalldimension
3	Estimation of reserves of mode parameters	Analysis of the matrix A pseudospectrum [9]	Steady mode, electromechanical transients	Complete or reduced model
4	Evaluation of damping properties	Calculation of the damping decrement of the vibration modes on the basis of the matrix A eigenvalues	Electromechanical transients	Reduced model
5	Centralized, including emergency control	Synthesis of state observers and laws of centralized control based on matrices A, B, C, D [11]	The steady state, electrical and electromechanical transients	Complete model
6	Control of system stabilizers and reactive power compensation devices	Synthesis of decentralized control laws based on matrices A, B, C, D [12]	The steady state, electrical and electromechanical transients	Complete and reduced model

Models with Jacobi matrices A 12 × 12 and 22 × 22 were used to estimate the damper properties (Task No. 4 in Table 1). These models represented oscillation modes with eigenvalues located in close proximity to the imaginary axis of the complex plane. The degree of damping of these vibration modes was estimated as determining in the damping properties.

The 510 dimension model (the 510 × 510 Jacobi matrix) was used to synthesize the laws of centralized emergency control (Task No. 5 in Table 1). This model allows taking into account the dynamic properties of the power system to the maximum extent and synthesizing the laws of centralized emergency control. These laws provide both specified reserves of static stability, and other important properties: the quality of transient processes, robustness (coarseness, invariance) with respect to perturbations and available information, etc.

When solving the problems of control of system stabilizers and static reactive power compensation devices (Task No. 6 in Table 1), they were used as high-dimensional models (Jacobi A matrixes of sizes 273 × 273, 385 × 385, 450 × 450, 510 × 510) and Reduced models (Jacobi matrices of sizes 12 × 12, 22 × 22, 42 × 42). Such a variety of models makes it possible to use different approaches and synthesis methods (centralized and decentralized synthesis of control, local control, etc.) and to perform a comparative analysis of the results with the choice of the preferred option.

5 Conclusion

The creation of the system of monitoring for transient regimes brought the problem of identifying the energy system to a level where the success of its solution depends on the methods and algorithms used.

The application of discrete algorithms for anenergy system linear mathematical model identification with realization in state space is based on the property of invariance of the observability matrix shift. Ensuring the full observability of the Kalman power system is possible due to the installation of PMU in certain nodes. The number of such nodes is much smaller than the state space dimension of the identified mathematical model.

Determination of the PMUs location and their number allows a genetic algorithm that works together with a recursive test of the dynamic system observability analysis. It is considered as the chosen rule (criterion) of observability.

These devices should be placed on the buses of power stations and substations in all calculated versions of the PMU locations. With the growth of the dimensionality of the state vector by non-linear law, the required amount of PMU also increases.

In all calculated versions of PMU locations, these devices should be placed on power plant buses and power system substations in the Kaliningrad region. At the same time, when the dimension of the state vector increases, the required amount of PMU also increases according to a non-linear (NP) law. To solve the entire range of problems considered in the work, it is sufficient not more than 40 PMUs.

The introduction of a system for managing power systems in real time inevitably leads to high-dimensional dynamic models and the complication of the PMU arrangement task. The nonpolynomial nature of the growth in the number of installed PMU, depending on the dimensionality of the mathematical model state space, makes it difficult to predict the solutions in these kinds of problems. However, the pace of development of mathematical methods and methods of artificial intelligence provides good chances for overcoming this problem.

References

1. A.V. Zhukov, D.M. Dubinin, D.N. Utkin, F.N. Gaydamakin, A.V. Danilin, D.N. Toporkov Development and implementation of an automatic system for collecting information from the monitoring system of transient regimes registers in the UES of Russia. Relyschik, **3**, 18 – 23 (2013).

2. V. Terzija, G. Valverde, C. Deyu, P. Regulski, V. Madani, J. Fitch, S. Skok, M.M. Begovic, A. Phadke Wide-area monitoring, protection, and control of future electric power networks, Proc. IEEE, Vol. 99, **1**, 80 – 93 (2011).

3. M.G. Gadzhiev, E.A. Gulevich, Y.V. Sharov Identification of the linear model of the electric power system in the state space, Electricity, **7**, 30 – 38 (2017).

4. E.Y. Zybin, M.S. Misrikhanov, V.N. Ryabchenko. Recursive tests on controllability and observability of large dynamical systems, Automation and Telemechanics, **5**, 119 – 132 (2006).

5. T. Kailath Linear Systems. Prentice Hall. Englewood Cliffs. N.J, 682, (1980).

6. L.A. Gladkov, V.V. Kureichik, V.M. Kureichik *Genetic algorithms*. (FIZMATLIT, 2006).

7. N.E. Zubov, E.A. Mikrin, V.N. Ryabchenko. *Matrix methods of automatic control systems for aircraft in the theory and practice,* 666 (Publishing house of the Moscow State Technical University Bauman, 2016).

8. M.S. Misrikhanov, V.N. Ryabchenko Placing poles in large dynamic systems with many inputs and outputs. Reports of the Academy of Sciences, T. 434, **4**, 1 – 3 (2011).

9. Y.V. Sharov On the development of methods for analyzing the static stability of electric power systems. Electricity, **1**, 12 – 18 (2017).

10. M.G. Gadzhiev, Y.V. Sharov Pseudo-modal estimation of static stability stocks of electric power systems. Electricity, **8**, 12 – 17 (2017).

11. M.G. Gadzhiev, M. Misrikhanov, Y.V. Sharov. Modal synthesis of control laws in electric power systems based on multi-step decomposition. Electricity, 11, 4 – 11 (2017).

12. Y.V. Sharov Application of the modal approach to solve the problem of ensuring the static stability of electric power systems. Izvestiya RAN. Power engineering, 2, 13 – 29 (2017).

Enhancing reliability of power supply systems with distributed generation plants

Bulatov Yuri [1,*], *Kryukov* Andrey [2]

[1]Bratsk State University, Faculty of Power Engineering and Automation, Bratsk, Russia
[2] Irkutsk State Transport University, Irkutsk National Research Technical University, Power Supply and Electrical Equipment Department, Irkutsk, Russia

Abstract. The article present results of studies aimed at emergency control methods development in power supply systems equipped with distributed generation plants. The main purpose of emergency control is ensuring postemergency mode dead-beat stability and high quality of dynamic transitions for electrical network load-shedding. Control actions were formed by changing the vector of controlled parameters along a path corresponding to the shortest distance to the hypersurface of the limit modes. The dynamic processes quality when implementing the control actions was ensured based on the special settings of excitation and frequency automatic regulators. Based on computer-aided modelling, it is shown that the effective in-troduction of a postemergency mode to the stability region can be performed based on generalized equations of the limit loads while using the start algorithm ensuring mode's entering the stability region's neighboring boundary. The application of the fuzzy control algorithms for tuning regulators allows formation of a qualitative dynamic transition while load-shedding distributed generation plants in the postemergency mode.

1 Introduction

One of the goals of emergency control (EC) [1-3] systems application is to ensure the static stability of postemergency modes (PEM) of electrical power systems (EPS). The control actions in this case, are used to introduce the PEM into the permissible region (stability region) along a certain trajectory $D\mathbf{Y}$ in the space of the controlled parameters $\mathbf{Y}$. In practice, the trajectory $D\mathbf{Y}$ is assumed to be linear and can be defined proceeding from the following assumptions:

- to be set based on the preliminary calculations;
- to be determined based on the shortest distance to the limiting hypersurface [2], or by minimal damage related to cutting power sources and consumers.

Due to introduction of distributed generation (DG) plants in the immediate future, the modes introduction to the stability region becomes urgent for distribution networks and power supply grids and systems (PSS). It acquires special importance in networks equipped with DG plants, implemented on the basis of non-traditional renewable energy sources (NTRES) [4-9]. Such installations, as mini hydropower plants and offshore windmill farms can be removed from the consumption centers, which leads to a 'narrowing' of the areas of static dead-beat stability.

2 Formulation of the problem

The EC task is illustrated in Fig. 1, where the cross section of the stability region is shown with the coordinate plane of the active generating powers P_i, P_j is shown; here and further in the text it is assumed that the stability and transmitted power limits coincide [2]. Curve 1 corresponds to the boundary of the stability region for the complete network diagram, curve 2 corresponds to a similar boundary when one of the main transmission lines is disconnected, and curve 3 corresponds to condition $\Im$ = const, where $\Im$ – the required value of the stability margin of the PEM.

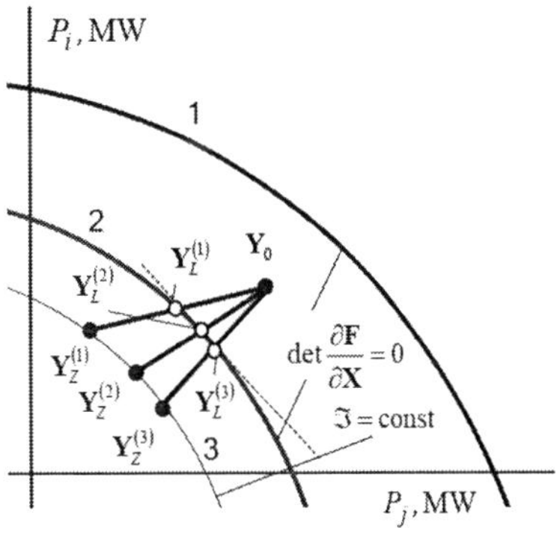

Fig. 1. To the sum of PEM introduction to the stability region: $\frac{\partial \mathbf{F}}{\partial \mathbf{X}}$ – matrix of Jacobi for steady-state equations (SSE); $\Im$ – stability reserve

The task of EC application is to reach one of the points $\mathbf{Y}_Z^{(1)}, \mathbf{Y}_Z^{(2)}, \mathbf{Y}_Z^{(3)}$:

* Corresponding author: bulatovyura@yandex.ru

$$\mathbf{Y}_Z^{(1)} = \mathbf{Y}_0 + D\mathbf{Y}^{(1)} = \mathbf{Y}_0 + t_1\Delta\mathbf{Y}^{(1)};$$
$$\mathbf{Y}_Z^{(2)} = \mathbf{Y}_0 + D\mathbf{Y}^{(2)} = \mathbf{Y}_0 + t_2\Delta\mathbf{Y}^{(2)};$$
$$\mathbf{Y}_Z^{(3)} = \mathbf{Y}_0 + D\mathbf{Y}^{(3)} = \mathbf{Y}_0 + t_3\Delta\mathbf{Y}^{(3)},$$

where $D\mathbf{Y}^{(k)}, k = 1..3$ – the trajectories of regime change in space **Y**; $\Delta\mathbf{Y}^{(k)}, k = 1..3$ – direction of regime change; $t_k, k = 1..3$ – scalar parameters that determine the amount of load-shedding in the relevant direction; index k = 1 corresponds to direction set antecedently, index k = 2 corresponds to load shedding the direction of the normal to the limiting hypersurface, while k = 3 – to load shedding, ensuring minimal possible damage from cutoffs of power sources and consumers.

Load-shedding in the direction chosen $\Delta\mathbf{Y}^{(k)}$ should be effected for an acceptable quality of the dynamic processes, which can be achieved through the use of automatic excitation regulators (AER) and automatic regulator of rotor speed (ARRS) of DG plants synchronous generators. The relatively small power of DG plants and the low value of rotors inertia constant of their generators require taking into account the mutual influence of AER and ARRS when tuning them. It should also be noted, that optimal control requires adjusting the tuning of AER and ARRS when significant changes are introduced in the operation modes of both DG plants and PSS. These requirements can be met by using the intelligent control algorithms [10-15].

Below are the results of research aimed at developing methods for introduction of the mode into the stability region by the set trajectories, as well as algorithms for the functioning and adjusting of AER and ARRS.

3 Introduction of a postemergency mode to the stability area

The introduction of PEM to the stability region can be effected based on limit loads equations (LLE) [2]:

$$\left.\begin{aligned} &\mathbf{F}[\mathbf{X}, \mathbf{Y}(t)] = \mathbf{0}; \\ &\mathbf{V}[\mathbf{X}, \mathbf{R}, \mathbf{Y}(t)] = \left(\frac{\partial \mathbf{F}}{\partial \mathbf{X}}\right)^T \mathbf{R} = \mathbf{0}; \\ &U(\mathbf{R}) = \mathbf{R}^T\mathbf{R} - 1 = 0, \end{aligned}\right\} \qquad (1)$$

where **F** – ℓ – dimensional function corresponding SSE; **X** – ℓ – uncontrolled parameters dimensional vector; **Y** – m – controlled parameters dimensional vector; $\mathbf{Y}(t) = \mathbf{Y}_0 + t\Delta\mathbf{Y}$; **R** – matrix own vector $\left(\frac{\partial \mathbf{F}}{\partial \mathbf{X}}\right)^T$, corresponding to zero eigenvalue.

Multiple calculation experiments [2] showed that on the basis of equations (1) the EPS regime can be introduced to the boundary of the stability region: the point $\mathbf{Y}_L^{(1)}$ (Fig. 1). To achieve a stability reserve necessary, an additional load shedding should be effected in direction $\Delta\mathbf{Y}^{(1)}$ (point $\mathbf{Y}_Z^{(1)}$) or in the direction of **R** vector, which coincides with the normal to the limiting hypersurface [2].

Though, in some cases the outreach to a 'distant boundary' of stability region [16] is possible: the point $\mathbf{Y}_L^{(db)}$ in Fig. 2; in this case, the solution obtained differing in the nodal powers injections symbols reverse, cannot be used in practice.

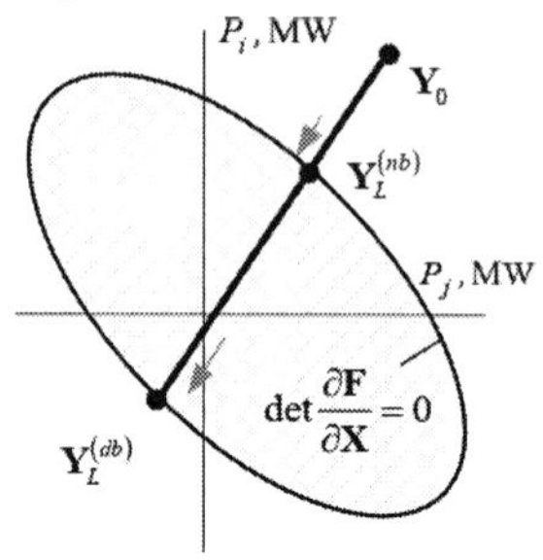

Fig. 2. To the problem of the 'distant boundary'

An effective and generalized way to solve the 'distant boundary' problem can be implemented based on starting algorithms using 'high order' methods [2, 17] or residual functional minimizing of the Steady-State Equations (SSE) [17]:

$$\aleph(\mathbf{X}) = \mathbf{F}^T(\mathbf{X}, \mathbf{Y}_0) \cdot \mathbf{F}(\mathbf{X}, \mathbf{Y}_0). \qquad (2)$$

The use of **X** parameters as first approximations calculated as a result of functional (2) minimizing when solving limit loads equations (LLE) and their modifications, ensures a reliable convergence to the required points $\mathbf{Y}_L^{(nb)}$, found on the 'neighboring' boundaries of the stability region.

4 The formation of quality dynamic transition when load-shedding generators

Due to relatively small power of DG plants generators and their rotors low inertia constant, harmonized configuring of AER and ARRS becomes especially urgent. The harmonized configuring principle is defining optimal coefficients of AER and ARRS configuring ensuring minimal voltage and frequency deviations from the set values, high damping properties for electromechanical transition processes and the required stability reserve. Optimization of settings is performed based on genetic algorithm while using quadratic criterion [14, 15]:

$$J = \int_0^{\Omega} e^2(j\omega)d\omega \to \min, \qquad (3)$$

where $e(j\omega) = D^D(j\omega) - D^M(j\omega)$ – error between the desired and model sets of characteristic polynomials factors; ω – frequency current value from range [0; Ω] that determines the system 'bandwidth'. Butterworth or other polynomials can be used as desired polynomials.

The method of harmonized AER and ARRS configuring allows defining the optimal coefficients of regulators setting for different modes of PSS operation. For this purpose the use of auto-tuning unit is suggested

 https://doi.org/10.1051/e3sconf/20185801006

with mode identification modules and regulators harmonized setting. Fig. 3 shows a block diagram of the suggested fuzzy control system. The input parameters are the current values of voltage U_g, rotor rotational frequency ω_g and DG plant power values P_g, Q_g. AER and ARRS tuning factors optimal for a current mode are determined based on the system performance. A more detailed description of the fuzzy control of AER and ARRS tuning is represented in works [14]. During the emergency control mode, the generators powers are varied in the postemergency mode ($\mathbf{Y}_Z$ vector components) defined by the above method.

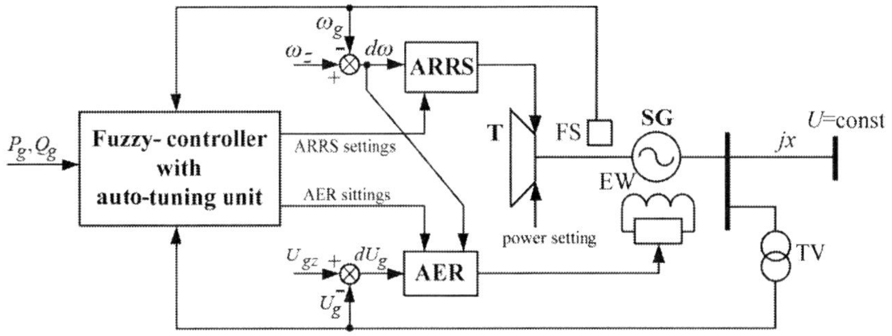

Fig. 3. Structural scheme of fuzzy system for control AER and ARRS of the DG plant:
FS - frequency sensor; EW - excitation winding; SG - synchronous generator; T - turbine; TV - voltage transformer

When identifying the mode based on experimental data, the closed-loop control system model for DG plants is formed. For this purpose, nonparametric identification is used, as a result of which, on the basis of a priori information about the process, the numerical values of the complex activation factor are determined as the ratio of the spectra of the output and input signals [18]. The 'turbine-generator' closed-loop system is demonstrated by the diagram shown in Fig. 4.

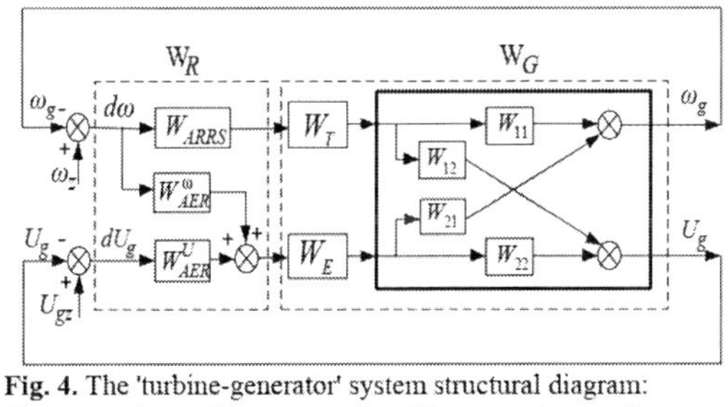

Fig. 4. The 'turbine-generator' system structural diagram:
W_T – complex turbine gear ratio; W_E – complex exciter gear ratio

Characteristic polynomial of the system under consideration is determined using the expression:

$$D^M(j\omega) = \det\left[\mathbf{E} + \mathbf{W}_G(j\omega)\cdot \mathbf{W}_R(j\omega)\right], \qquad (4)$$

where $\mathbf{E}$ – a single matrix; $\mathbf{W}_G(j\omega)$ – matrix activation function of the tunned object (of 'turbine-generator' system) determined experimentally; $\mathbf{W}_R(j\omega)$ – matrix activation function of the controller considering interconnection AER and ARRS:

$$\mathbf{W}_R(j\omega) = \begin{bmatrix} W_{ARRS}(j\omega) & W^{\omega}_{AER}(j\omega) \\ 0 & W^{U}_{AER}(j\omega) \end{bmatrix};$$

$W_{ARRS}(j\omega)$ – ARRS complex activation factor; $W^{\omega}_{AER}(j\omega)$ – AER complex activation factor of the frequency-tuning channel; $W^{U}_{AER}(j\omega)$ – AER complex activation factor of the voltage-tuning channel.

The following ratios can be recorded for activation factors:

$$W_{\text{ARRS}}(j\omega) = \left(k_p + \frac{k_i}{0{,}1 j\omega} + \frac{jk_d\omega}{j\omega+1}\right)\cdot\frac{1}{0{,}01 j\omega+1},$$

$$W^{\omega}_{\text{AER}}(j\omega) = \frac{1+0{,}5j\omega}{0{,}5j\omega}\cdot\left[\frac{2k_{0\omega}j\omega}{(2j\omega+1)(0{,}02j\omega+1)} + \frac{0{,}05k_{1\omega}j\omega}{0{,}05j\omega+1}\right],$$

$$W^{\text{U}}_{\text{AER}}(j\omega) = \frac{1+0{,}5j\omega}{0{,}5j\omega}\cdot\left(k_{0u} - \frac{0{,}02k_{1u}j\omega}{0{,}06j\omega+1}\right),$$

where k_p, k_i, k_d – ARRS tuning factors; k_{0u}, k_{1u}, $k_{0\omega}$ and $k_{1\omega}$ – AER adjusting channels tuning factors.

5 The modelling results

The point defining $\mathbf{Y}^{(1)}_L$ based on limit loads equations (LLE) (1) was carried out with regard to the PSS diagram showed in Fig. 5.

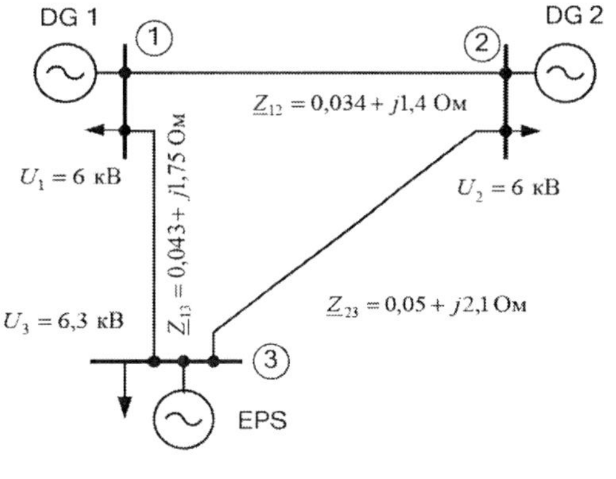

Fig. 5. Network diagram

In this diagram, two mini hydropower plants with a power of 24 MW each, operate at an industrial point load (timber processing facilities) connected at nodes 1 and 2. Network voltage choice is determined by the presence of a large amount of high-voltage motors with 6 kV voltage. The enterprises operate in one shift and in the hours of the evening maximum of the EPS, a power output of 15 MW from each generator to the receiving system (node 3) is carried out. The network is implemented via flexible symmetrical electrical pathways [19]. As an emergency mode, the line 1-3 switching off was considered. One of the options for PEM entry into the stability region by the shortest trajectory on the basis of equations (1) is illustrated in Fig. 6.

The computer-aided modelling showed that when using LLE (1) and the starting algorithm, based on residual functional minimizing (2) a reliable outreach to the required boundary of the stability area is ensured.

In addition, the modelling was carried out in the MATLAB system, taking into account the AER and ARRS models. In this case, using the method of concordant setting, regulators parameters were identified for different modes and the base of fuzzy control system rules was generated.

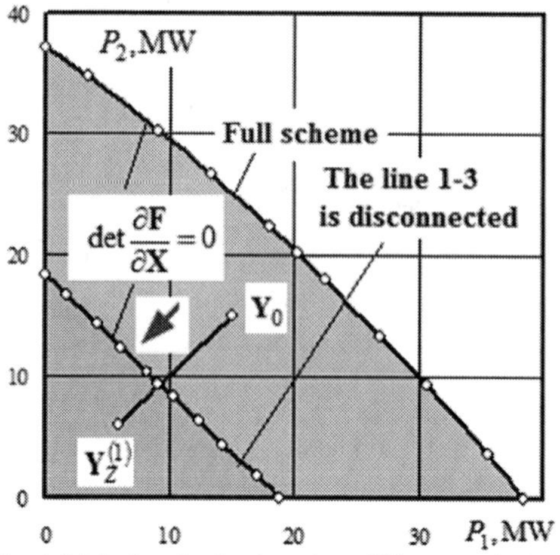

Fig. 6. Mode introduction into the stability area via the set trajectory while using equations (1): $\mathbf{Y}_0 = [15 \quad 15]^T$; $\mathbf{Y}_Z^{(1)} = [6 \quad 6]^T$

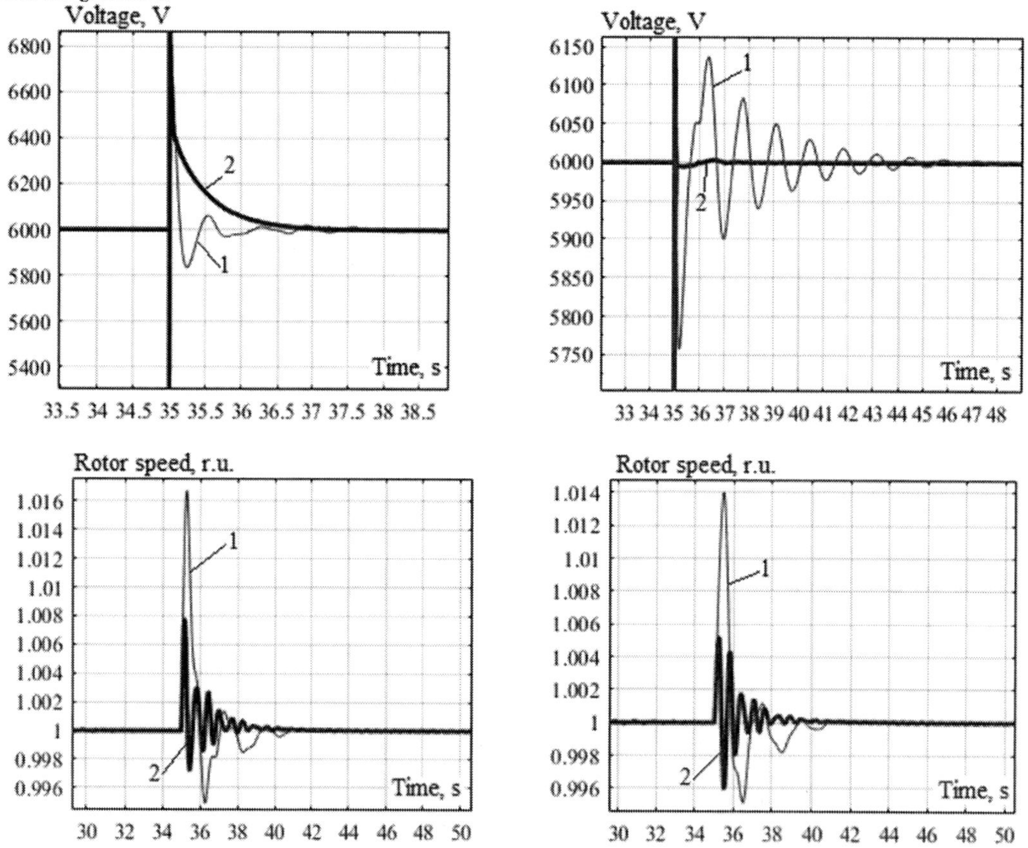

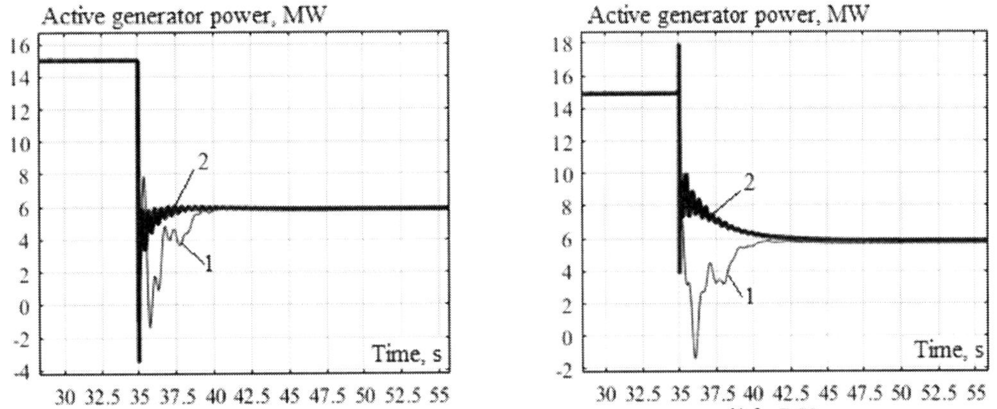

a) for DG1 b) for DG2

Fig. 7. Oscillograms for voltage, frequency and power of DG plants when lines 1-3 are cut off: 1 – without change in AER and ARRS tuning factors; 2 – fuzzy control of AER and ARRS settings

When DG plants operate without regulators, the system loses stability, in case one of connections with infinite buses is cut off. To ensure PEM stability, DG plants generators should be load-shedded. In this case, the regulators harmonized tuning and its change in generators different operation modes allows to significantly enhance the transient processes quality parameters. The corresponding oscillograms of the voltage, frequency, and power of the DG plants when the lines 1-3 are cut off are shown in Fig. 7.

6 Conclusion

The following conclusions can be formulated based on the performed calculations and computer-aided modelling:

1. When DG plants operate without regulators, the stability of DG generators parallel operation is lost, in case one of connections is cut off from EPS receiving system.

2. Based on computer-aided modelling, it is shown that the effective introduction of a postemergency mode to the stability region can be performed based on generalized equations of the limit loads while using the start algorithm ensuring mode's entering the stability region's neighboring boundary.

3. The application of the fuzzy control algorithms for tuning AER and ARRS allows formation of a qualitative dynamic transition while load-shedding DG plants in the postemergency mode.

References

1. *Innovative Power Engineering - 21* / Ed. V.M. Batenin, V.V. Bushuev, N.I. Voropai, Moscow: Information Center "Energia", 584 p. (2017)
2. A.V. Kryukov *Limiting regimes of electric power systems*, Irkutsk: IrGUPS, 236 p. (2016)
3. N.I. Voropai, V.G. Kurbatsky, N.V. Tomin [and others], *Complex of intellectual means for prevention of major accidents in power systems*, Novosibirsk: Science, 332 p. (2016)
4. N.I. Voropai, Z.A. Stychinsky, *Renewable energy sources: theoretical foundations, technologies, technical characteristics, economics*, Magdeburg: Otto-von-Guericke-Universität, 223 p. (2010)
5. O. Ellabban, H. Abu-Rub, F. Blaabjerg, Renewable energy resources: Current status, future prospects and their enabling technology, Renewable and Sustainable Energy Reviews, **39**, pp. 748-764 (2014)
6. A.Q. Huang, M.L. Crow, G.T. Heydt, J.P. Zheng, *The Future Renewable Electric Energy Delivery and Management (FREEDM) System: The Energy Internet*, S. J. N. I. S. (2010). Proceedings of the IEEE. DOI: 10.1109/JPROC.2010.2081330
7. D. Egre and J.C Milewski, The diversity of hydropower projects, Energy Policy, **30** (**14**), pp. 1225-1230 (2002)
8. S.C.E. Jupe, A. Michiorri, P.C. Taylor, Increasing the energy yield of generation from new and renewable energy sources, Renewable energy, **14** (**2**), pp. 37-62 (2007)
9. Torriti J. Demand, Side Management for the European Supergrid: Occupancy variances of European single-person households, Energy Policy, **44**, pp. 199-206 (2012)
10. K. Suslov, N. Solonina, V. Stepanov, *A principle of power quality control in the intelligent distribution networks*, International symposium on smart electric distribution systems and technologies EDST 2015 Proceedings, pp. 260-264 (2015)
11. Y. Chen, Y. Ma, W. Yun, Application of Improved Genetic Algorithm in PID Controller Parameters Optimization, Telkomnika, Vol. **11**, No. **3**, pp. 1524-1530 (2013)
12. A.Y. Jaen-Cuellar, R. de J. Romero-Troncoso, L. Morales-Velazquez, R.A. Osornio-Rios, PID-Controller Tuning Optimization with Genetic Algorithms in Servo Systems, International Journal

of Advanced Robotic Systems, **10**, pp. 324-337 (2013).

13. Yu.N. Bulatov, A.V. Kryukov, K.V. Suslov, Multi-agent technologies for control of distributed generation plants in the isolated power systems, Far East Journal of Electronics and Communications, T. **17**, № **5**, pp. 1197-1212 (2017)
14. A.V. Kryukov, S.K. Kargapol'cev, Yu.N. Bulatov, O.N. Skrypnik, B.F. Kuznetsov, Intelligent control of the regulators adjustment of the distributed generation installation, Far East Journal of Electronics and Communications, T.**17**, № **5**, pp. 1127-1140 (2017)
15. Yu.N. Bulatov, A.V. Kryukov, *Optimization of automatic regulator settings of the distributed generation plants on the basis of genetic algorithm*. 2nd International Conference on Industrial Engineering, Applications and Manufacturing (ICIEAM). IEEE Conference Publications, pp. 1-6 (2016). doi: 10.1109/ICIEAM.2016.7911456
16. B.I. Ayuyev, V.V. Davydov, P.M. Erokhin, Optimization model of limiting modes of electrical systems, Electricity, **11**, pp.3-11 (2010)
17. V.I. Tarasov, *Theoretical bases of the analysis of the steady-state regimes of electric power systems*, Novosibirsk: Science, 344 p. (2002)
18. Yu.N. Bulatov, A.V. Kryukov, Application of wavelet transform and genetic algorithms to tune automatic regulators of distributed generation plants, Sci. Bull. Novosibirsk State Tech. Univ., No. **2**. T.**63**, pp. 7-22 (2016)
19. A.M. Semchinov, *Current conductors of industrial enterprises*, Energia, 200 p. (1972)

Recovery time reduction of a damaged 6-35 kV transmission line after double earth fault applying the iteration fault location method

A.L. Kulikov[1*], *V.J. Osokin*[1], *and M.D. Obalin*[1]

[1]NSTU n.a. R.E. Alekseev, Nizhniy Novgorod, Russian Federation

Abstract. The task of determining the fault location on a power line is difficult and the longest operations to restore damaged areas of the network. Determining fault location with double earth faults is a priority due to the high wear of medium voltage networks and the absent of currently accurate algorithms. The usage of the phase coordinates method allows to obtain the calculated relations for determining the distances to the nearest and furthest point of damage. A significant increase in the accuracy is achieved by using iterative recalculation, the essence of which is to rectify such parameters as the transient resistance, the voltage at the neutral point of the load and the distance to the damage. The proposed algorithms allow reducing the calculation error and reducing the bypass zone of the damaged line, which in turn determines the power transmission line recovery time and increases the reliability of the power supply network.

1 Introduction

The task of determining the fault location on a power line is difficult and the longest operations to restore damaged areas of the network. Accurate and efficient fault location on power lines affects amount of damage caused by power supply interruptions and reduction of the transmitted electricity quality. In medium voltage electrical networks, there are mainly used fault location methods based on unilateral measurement of the emergency mode parameters [1-3, 6-10], which allow to determine distances to phase-to-phase faults. A phase-to-earth fault can be determined using expensive specialized indicators (for example, indicating the direction of flow of the capacitive fault current, active sensing methods [4], and wave method of fault location [13]. Determining fault location with double earth faults is a priority due to the high wear of medium voltage networks and the absent of currently accurate algorithms. Despite the variety of software and hardware offered, the problem of fault location at double earth fault does not to date have a generally recognized solution [5-10], and modern methods that allow calculating the distance to the fault locations have a large error.

2 The fault location algorithm for double earth faults on different lines.

The developed method of fault location at double earth fault on different lines is based on the simulation modeling and following statistical processing of the results [9-15]. The parameters of the simulation model and the developed methods of fault location are below. Calculation of the distances to the damage sites with double earth fault on different lines is fulfilled according to the schemes shown in Figure 1.

The components of the transient resistance R_{t1} are resistance of foreign objects between the wire and ground or support, ground resistance of the support.

During the calculation, the following assumptions were made:

1) the three-phase elements of the system are assumed to be symmetric;

2) capacity admittance of transmission lines is not taken into account;

3) the transition resistance is purely active;

4) backfeed of the short-circuit point by the load current is absent.

The use of the scheme shown in Fig. 1, in phase coordinates [7], allows obtaining expressions to determine distances to the near and far points of damage with double earth fault on different lines.

* Corresponding author: inventor61@mail.ru

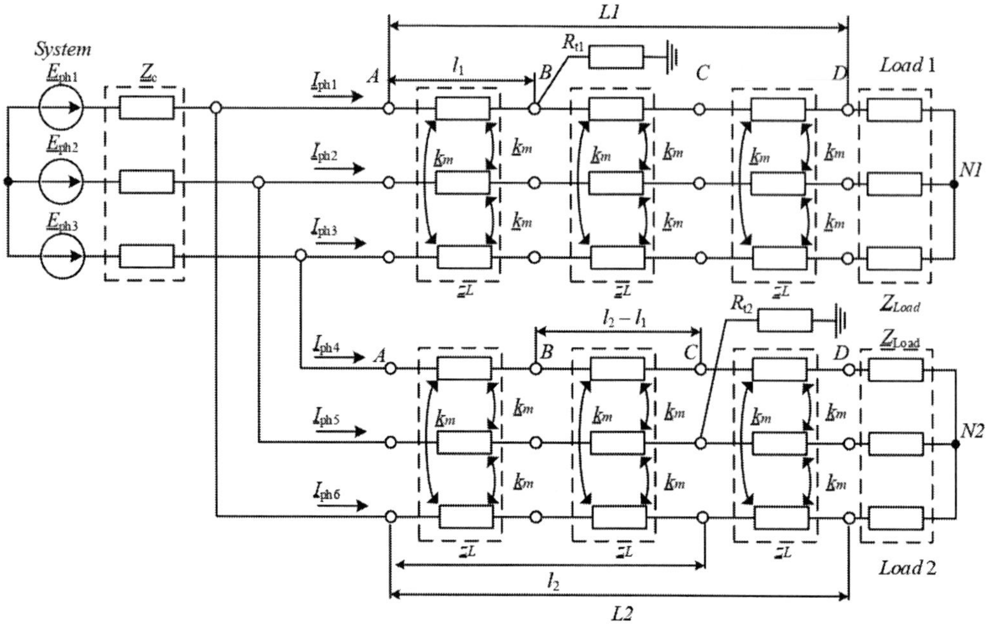

Fig. 1. The scheme for replacing the network in the phase «Ph1» in double fault mode on different lines at a distance *l*1 and phase «Ph2» at a distance *l*2

In figure 1, the following designations are accepted: $\underline{E}_{\text{ph1}}, \underline{E}_{\text{ph2}}, \underline{E}_{\text{ph3}}$ – is the equivalent of system electromotive force; $\underline{Z}_{\text{S}}$ – equivalent system resistance; $\underline{z}_L$ – line resistivity; $\underline{k}_m$ - ratio of mutual induction; $\underline{z}_{Load}$ – equivalent load resistance; $R_{\text{t1}}, R_{\text{t2}}$ - transient resistance in the places of short circuit; l_1, l_2 – distances to near and far ground faults.

In general terms, the calculation expressions look like this:

$$l_1 = \frac{X_{ph1}}{(x_L + x_{m12} + x_{m13})}, \quad (1)$$

$$l_2 = \frac{X_{ph2}}{(x_L + x_{m21} + x_{m23})}, \quad (2)$$

where

$$x_{m12} = Im\left(\underline{z}_m \cdot \frac{\underline{I}_{ph11}}{\underline{I}_{ph21}}\right); x_{m13} = Im\left(\underline{z}_m \cdot \frac{\underline{I}_{ph13}}{\underline{I}_{ph11}}\right);$$

$$x_{m21} = Im\left(\underline{z}_m \cdot \frac{\underline{I}_{ph21}}{\underline{I}_{ph22}}\right); x_{m23} = Im\left(\underline{z}_m \cdot \frac{\underline{I}_{ph23}}{\underline{I}_{ph22}}\right)$$

– resistivity of mutual inductance of the nearby phases regarding the damaged one; X_{ph1}, X_{ph2} – calculative inductances proportional to the distances to the faults; $\underline{I}_{ph11}$ – the current of the damaged phase of the first line; $\underline{I}_{ph22}$ – current of the damaged phase of the second line; $\underline{I}_{ph12}, \underline{I}_{ph13}$ – currents of the undamaged phase of the first line; $\underline{I}_{ph21}, \underline{I}_{ph23}$ – currents of the undamaged phase of the second line.

The verification of the proposed fault location method for double earth fault on the different lines was carried out according to the scheme presented in Figure 2, by simulation modeling of the emergency mode in question in the PSCad and Visual Basic software.

 https://doi.org/10.1051/e3sconf/20185801007

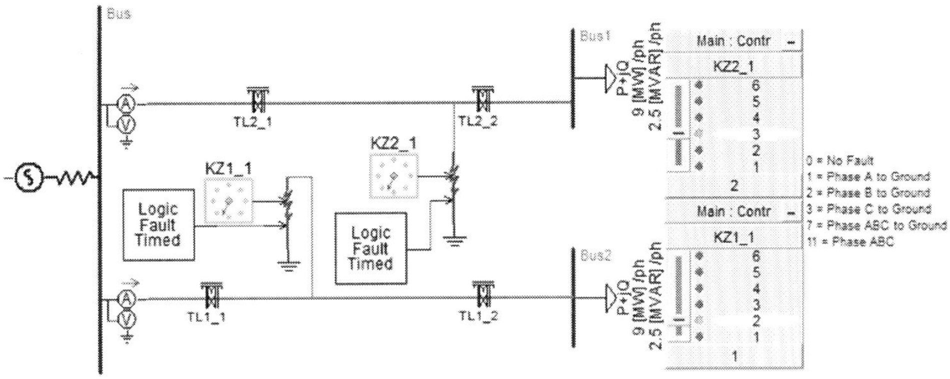

Fig. 2. Simulation modelling of the transmission line 35 kv with a double earth fault on different lines in the PSCad software

The considered scheme has the following parameters:

- voltage of 35 kV network;
- length of the line $L1 = 10$ km, $L2 = 10$ km;
- specific resistance of the phase:

$\underline{z}_L = 0.079 + j0.697$ Ohm / km;- resistivity of mutual induction: $\underline{z}_m = 0.0451 + j0.353$ Ohm / km;- transient resistances in the places of closures are determined by a random quantity distributed with uniform law in the range from 0 to 10 Ohm;

- load power consumption: $\underline{S}_{Load} = 9 + j2.5$ MBA.

The simulation results were processed in Mathcad and the dependencies of the estimated distances obtained from formulas 1 and 2 were constructed from the actual values (Fig. 3).

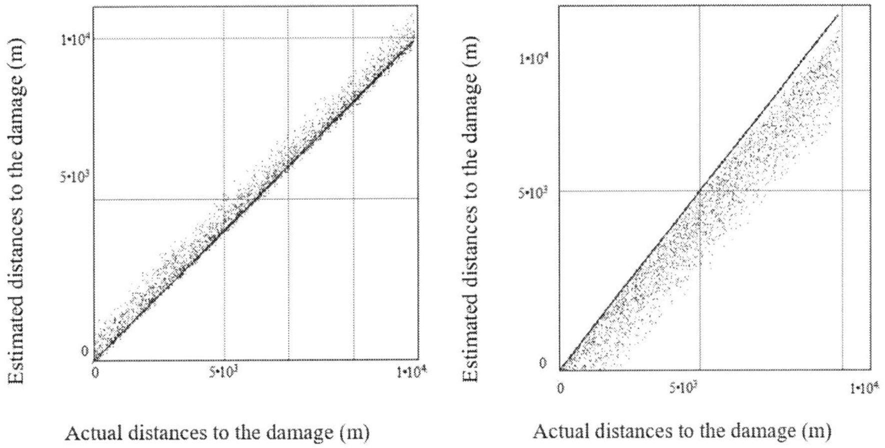

a) b)

Fig. 3. Dependence of the estimated distances to the damage points of the transmission line on actual ones

Analysis of the obtained results showed that the distance to the first fault point (fig 3a) is determined with a relative error up to 12%, and the mathematical expectation of the obtained error was 3%. The distance to the second fault point (fig 3b) is determined with a relative error up to 26%, and the mathematical expectation of the obtained error was 7%.

Thus, the above method makes it possible to determine the region of double earth fault location for different transmission lines by installing a resistance measuring device connected to the phase current and the phase voltage of the outgoing lines.

3 Fault location rectification applying the iteration recalculation

For increasing in the accuracy of fault location algorithms it is proposed to use the iterative rectification of the calculation results of the distance to the fault. As the initial data, currents and voltages oscillograms in the steady-state short-circuit mode are used, also the values of distances to the faults obtained in the first approximation by formulas (1) and (2).

The algorithm rectifying the distances to the point of the fault on the first power line is below.

Let's find the value of the voltage at the neutral point N1 (Fig. 1), using the parameters of undamaged phase:

$$\underline{U}_{n1} = \left(\underline{U}_{u.ph1} - \underline{I}_{ph12} \cdot \left(\underline{z}_L + \underline{z}_{Load1}\right)\right) - -\left(\underline{I}_{ph13} \cdot \underline{z}_m + \underline{I}_{ph11} \cdot \underline{z}_m \cdot l_{11}\right), \quad (3)$$

where, $\underline{U}_{u.ph}$ – the voltage of the undamaged phase of the first line; $\underline{U}_{n1}$ – the voltage in the neutral point of the load; $\underline{I}_{ph11}$ – the current of the damaged phase in the first transmission line;; $\underline{I}_{ph12}, \underline{I}_{ph13}$ – currents of the undamaged phase of the first transmission line; $\underline{z}_L$– specific resistance of the transmission line; $\underline{z}_{Load1}$ – load resistance per phase; l_{11} – distance to the nearest point of damage, obtained by formula (1) regarding the full length of the line.

Consider the damaged phase on the first power line (Fig. 1), taking into account the mutual induction of neighboring phases (Fig. 4).

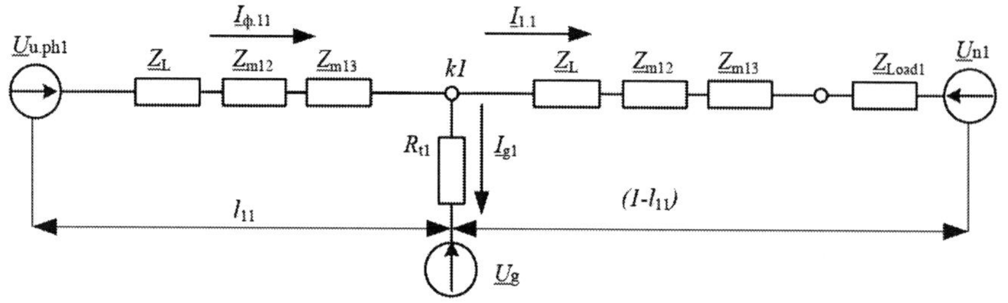

Fig. 4. The replacement scheme of the damaged phase on the first line, with a double earth fault.

Taking into account the fact that the resistance of the equivalent section obtained by Ohm's law is equal to the sum of the resistance of the equivalent circuit, the following expression can be written:

$$\frac{\underline{U}_{u.ph1} - \dfrac{\dfrac{\underline{U}_g}{R_{t1}} + \dfrac{\underline{U}_{n1}}{(1-l_{11}) \cdot \left(\underline{z}_{L1} + \underline{z}_{m12} + \underline{z}_{m13}\right)}}{\dfrac{1}{R_{t1}} + \dfrac{1}{(1-l_{11}) \cdot \left(\underline{z}_{L1} + \underline{z}_{m12} + \underline{z}_{m13}\right)}}}{\underline{I}_{ph.11}} = \quad (4)$$

$$= \left(\underline{z}_{L1} + \underline{z}_{m12} + \underline{z}_{m13}\right) l_{11} + + \frac{R_{t1}\left(\left(\underline{z}_{L1} + \underline{z}_{m12} + \underline{z}_{m13}\right) \cdot (1-l_{11}) + \underline{z}_{Load1}\right)}{R_{t1} + \left(\underline{z}_{L1} + \underline{z}_{m12} + \underline{z}_{m13}\right) \cdot (1-l_{11}) + \underline{z}_{Load1}}$$

If $\underline{z}_{L1} + \underline{z}_{m12} + \underline{z}_{m13} = \underline{z}_1$,and ground voltage $\underline{U}_g = 0$, then, taking into account the transformations, the resulting expression looks like this:

$$\frac{\underline{U}_{u.ph1} - \dfrac{\underline{U}_{n1} \cdot R_{t1}}{(1-l_{11}) \cdot \underline{z}_1 + R_{t1} + \underline{z}_{Load1}}}{\underline{I}_{ph.11}} = \underline{z}_1 l_{11} + \frac{R_{t1} \cdot \left(\underline{z}_1 \cdot (1-l_{11}) + \underline{z}_{Load1}\right)}{R_{t1} + \underline{z}_1 \cdot (1-l_{11}) + \underline{z}_{Load1}} \quad (5)$$

Having expressed R_{t1} from this equation, and taking its absolute value, we obtain a formula for determining the transient resistance:

$$R_{t1} = \left|\frac{\underline{U}_{u.ph1} \cdot \left(\underline{z}_1 + \underline{z}_{Load1} - \underline{z}_1 \cdot l_{11}\right)}{\underline{I}_{ph.11} \cdot \left(\underline{z}_1 + \underline{z}_{Load1}\right) - \underline{U}_{u.ph1} + \underline{U}_{n1}}\right| - -\left|\frac{\underline{I}_{ph.11} \cdot \underline{z}_1 \cdot l_{11}\left(\underline{z}_1 - \underline{z}_1 \cdot l_{11} + \underline{z}_{Load1}\right)}{\underline{I}_{ph.11} \cdot \left(\underline{z}_1 + \underline{z}_{Load1}\right) - \underline{U}_{u.ph1} + \underline{U}_{n1}}\right| \quad (6)$$

Knowing the value of the transition resistance in the place where the arc arises, it is possible to obtain the value of the current through the transient resistance:

$$I_{g1} = \frac{\underline{U}_{u.ph1} - \underline{I}_{ph.11} \cdot l_{11}\left(\underline{z}_1 + \underline{z}_{m12} + \underline{z}_{m13}\right)}{R_{t1}} \quad (7)$$

where $\underline{z}_{m12} = \underline{z}_m \cdot \frac{\underline{I}_{ph12}}{\underline{I}_{ph11}}$; $\underline{z}_{m13} = \underline{z}_m \cdot \frac{\underline{I}_{ph13}}{\underline{I}_{ph11}}$ – mutual inductance of the first line regarding the damaged phase; $\underline{I}_{ph11}$ – the current of the damaged phase on the first line; R_{n1} – transient resistance in the places of the first short circuit.

Then, according to the first Kirchhoff's law, it is possible to determine the supply current of the short-circuit point from the load:

$$I_{1.1} = \underline{I}_{ph.11} - I_{g1}. \quad (8)$$

Then, taking into account the above, recalculate the voltage at the neutral point N1 (Fig. 1):

$$\underline{U}_{n1} = \left(\underline{U}_{u.ph1} - \underline{z}_{L1}\underline{I}_{ph12} - \underline{z}_{Load1}\underline{I}_{ph12}\right) - -\underline{I}_{ph13}\underline{z}_m - \left(\underline{I}_{ph11}\underline{z}_m l_{11} - I_{1.1}\underline{z}_m(1-l_{11})\right) \quad (9)$$

The solution of the expression (4) regarding to l_{11} allows to determine the distance to the short-circuit point.

The following expression for determining l_{11} can be obtained:

$$l_{11} = \left|\frac{\underline{I}_{ph.11} \cdot \left(\underline{z}_1 + \underline{z}_{Load}\right) + \sqrt{a_1 + b_1 + c_1} + \underline{U}_{u.ph1}}{2 \cdot \underline{I}_{ph.11} \cdot \underline{z}_1}\right|$$
$$l_{11} = \left|\frac{\underline{I}_{ph.11} \cdot \left(\underline{z}_1 + \underline{z}_{Load}\right) - \sqrt{a_1 + b_1 + c_1} + \underline{U}_{u.ph1}}{2 \cdot \underline{I}_{ph.11} \cdot \underline{z}_1}\right| \quad (10)$$

The coefficients a, b, c formulated from the following expressions:

$a_1 = \left(\underline{I}_{ph.11} \cdot \underline{z}_{Load}\right) \cdot \left(\underline{I}_{ph.11} \cdot \underline{z}_{Load} + 2\underline{I}_{ph.11} \cdot \underline{z}_1 + +4R_{t1} \cdot \underline{I}_{ph.11} - 2\underline{U}_{u.ph1}\right)$;

$b_1 = \left(\underline{I}_{ph.11} \cdot \underline{z}_1\right) \cdot \left(\underline{I}_{ph.11} \cdot \underline{z}_1 + 4R_{t1} \cdot \underline{I}_{ph.11} - -2\underline{U}_{u.ph1}\right)$;

$c_1 = 4R_{t1} \cdot \underline{I}_{ph.11} \cdot \left(\underline{U}_{n1} - \underline{U}_{u.ph1}\right) + \underline{U}_{u.ph1}{}^2$.

The obtained dependences (expression 10) allow to rectify the distance to the point of damage. To choose the only correct solution, it is necessary to compare the obtained values of l_{11} with 1 (the values were obtained in p.u. regarding the real length of the line).

On the next iteration, taking into account the new value of the distance to the fault according to the formula (6), the value of the transition resistance is rectified. Onward, according to to the expressions (8) and (9), the current of the load-point loading and the voltage at the points N1 are determined. Onward, taking into account the updated values, the distance to the point of damage is determined from expression (10). The iterative process continues until the changes in the values obtained in two consecutive iterations become insignificant (no more than the required fault distance calculation error).

The distance to the damage point located on the second line is determined similarly.

Verification of the proposed method was tested using the scheme shown in Fig. 2 taking into account the simulation results in the PSCad and Visual Basic.

Figures 5a and 5b show the estimated distances to the nearest and the furthest point of damage on the transmission line dependence on the actual ones.

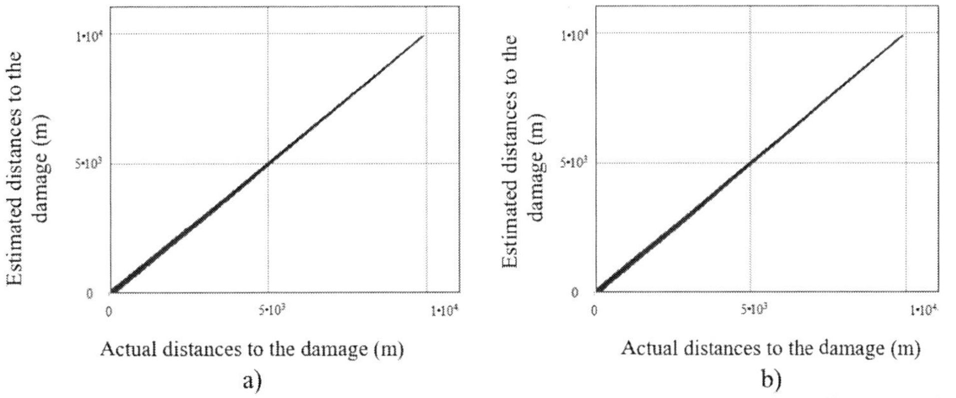

Figure 5. Dependence of the estimated distances to the damage points of the transmission line on actual ones $1 \cdot 10^4 \cdot 5 \cdot 10^3$ 0

The analysis of figure 5 confirms increasing of the accuracy of the fault location and the reduction of the calculation error in comparison to the initial results. The maximum value of the relative error of the calculation in the near point of damage is 2.2 %, and in the far point of damage 2.3%, which is approximately 38 meters.

4 Formation of the bypass zone under the assumption of the normal law of error distribution of the fault location.

At the core of the organization of the search for the damage location there is the concept of the bypass zone. When solving fault location problems, it is important to determine not only the location of the closure, but also the bypass zone of the line. The size of the zone assesses the period of time required to carry out repair-and-renewal operations [8-10].

Having a sufficiently large amount of statistical data of fault location errors, it is possible to determine the probability density of the error distribution relative to the estimated damage site and, by specifying the confidence probability, to obtain a confidence interval that will specify the bypass zone of the transmission lines.

In order to form a bypass zone in the case of a double earth fault on different lines, the probability density of the distribution of absolute errors calculation with respect to the estimated fault location was determined for each of the possible fault locations (Fig. 6).

Figure 6a shows the probability density of the absolute error distribution, relative to the estimated value of the distance to the nearest closure point (solid line - before the application of the correction factor, dashed line after it). According to the rule of three sigma (3σ), there was calculated a confidence interval corresponding to the bypass zone. For the point of damage obtained before the application of the iteration recalculation, the bypass area was 1.5 km which is 15% of the total length of the line and after application of the iteration recalculation became 0.2 km which is 2% of the total length of the line.

 https://doi.org/10.1051/e3sconf/20185801007

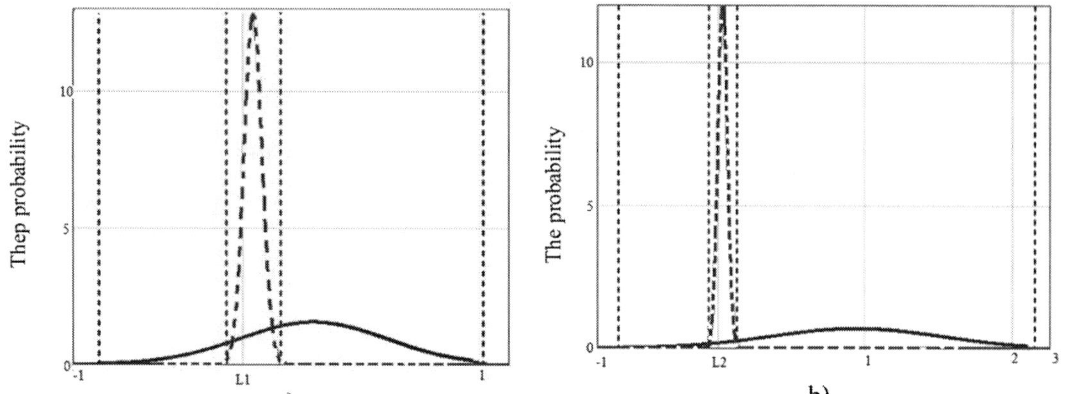

a) Length of the line   b) Length of the line

Figure 6 - Example of the probability density distribution of the absolute error relative to the mathematical expectation

The results of bypass zones calculation for the example (Figure 6a) are shown in Table 1.

Table 1. Results of bypass zones calculation for double earth fault on different lines (near point of damage)

	Left border of the bypass zone, km	Right border of the bypass zone, km	Total bypass areas length, km	Total bypass areas length.,%
Before the application of iteration recalculation	-0,5	1	1,5	15
After the application of iteration recalculation	-0,06	0,13	0,2	2

The iteration recalculation use clarifying the near point of damage, allowed to reduce the bypass area from 15% to 2% and, therefore, to shorten the time required to eliminate the damage.

Figure 6b shows the probability density distribution of the absolute error, relative to the estimated value of the distance to the far point of closure (solid line - before the application of the correction factor, dashed after it).

For the short circuit point (figure 6b) obtained before the application of the iteration recalculation, the bypass area was 3.5 km - 35% of the total length of the line and after application of the iteration recalculation was 1.3 km - 13% of the total length of the line.

The results of bypass zones calculation for the example (Figure 6b) are shown in Table 2.

Table 2. Results of bypass zones calculation for double earth fault on different lines (far point of damage)

	Left border of the bypass zone, km	Right border of the bypass zone, km	Total bypass areas length., km	Total bypass areas length.,%
Before the application of iteration recalculation	-0,8	2,7	35	35
After the application of iteration recalculation	-0,07	0,13	0,2	2

The use of the iteration recalculation to clarify the far point of damage, allowed to reduce the bypass zone from 35% to 2%.

4 Conclusion

It is proposed a method to increase the accuracy of the finding fault location in the case of a double earth fault in a 6-35 kV network using simulation modeling, accumulating statistical information, and error compensation in real calculations based on simulation results. The application of iteration recalculation made it possible to reduce the bypass area of the damaged power line from 15% to 2% for the near point of damage and from 35 to 2% for the long settlement point.

Proposed algorithms provide a significant increase in the accuracy of fault location and a reduction of the damaged bypass zone, as well as cost reduction for the 6-35 kV transmission lines repair-and-renewal operations.

This work was supported by the Ministry of Education and Science of the Russian Federation (agreement No. 14.577.21.0244 for provision of support of September 26, 2017. Unique project identifier RFMEFI57717X0244).

References

1 V.G. Glovatskiy, A.G. Khalidov, "Opredelenie rasstoyaniya do mest dvoynykh za-mykaniy na zemlyu v elektricheskikh setyakh 35 kV", Energetik, vol.9., P.31, 1985.

2 E.A.Vorob'eva, D.I. Gandzhaev, G.A. Filatova, V.A. Shuin, "Informatsionnye parametry elektricheskikh velichin perekhod-nogo protsessa dlya opredeleniya mesta zamykaniya na zemlyu v raspredelitel'nykh kabel'nykh setyakh napryazheniem 6-10 kV", Vestnik IGEU, vol.2., pp.34-42, 2017.

3 E.F. Khakimzyanov, R.G. Mustafin, A.I. Fedotov, "Opredelenie rasstoyaniy do mest dvoynykh zamykaniy na zemlyu na linii elektroperedachi raspredelitel'noy seti srednego napryazheniya", Izvestiya vuzov, Problemy energetiki, vol.3-4, pp.132-137, 2015.

4 A. L. Kulikov, M. Sh. Misrikhanov, A. A. Petrukhin, "Opredelenie mest povrezhdeniy LEP 6 – 35 kV metodami aktivnogo zondirovaniya", – M.: Energoatomizdat, P.162, 2009.

5 A.N. Visyashchev, "Diagnostika sostoyaniya vozdushnykh liniy elektroperedachi 10-110 kV v normal'-nykh i avariynykh rezhimakh", IrGTU, Uchebnoe posobie, Irkutsk. P. 270, 2012.

6 Geoinformatsionnaya sistema OMP 6-35 kV. URL: www.relematika.ru (data obrashcheniya: 17.07.2017) www.relematika.ru/produkty/6-35_kv/sistema_omp_6-35_kv.

7 B.V. Papkov "Toki korotkogo zamykaniya v elektricheskikh sistemakh" NGTU , Ucheb. Posobie, P.275, 2005.

8 Yu. S. Belyakov, "Aktual'nye vopro-sy opredeleniya mest povrezhdeniya vozdushnykh liniy elektroperedachi", Bibliotechka energetika, prilozhenie k zhurnalu «Energetik», vol. №11.,P.80, 2010.

9 A.L. Kulikov, M.D. Obalin, P.A. Kolobanov, "Kompleksnye algo-ritmy OMP LEP na baze statisticheskikh metodov", Energetik, vol.№1, 2012.

10 A. L. Kulikov, M. D. Obalin, P. A. Kolobanov, "Analiz i povyshenie tochnosti pri opredelenii mesta povrezhdeniya liniy elektroperedachi", Izvestiya vuzov. Elektromekhanika, vol. №5, pp.57-62, 2013.

11 H. Diaz, M. López, "Fault location techniques for electrical distribution networks: a literature survey," in Proc. the Fifth IASTED International Conference, pp. 311-318, June 15-17, 2005.

12 M. Lehtonen, "Novel techniques for fault location in distribution networks," in Proc. Power Quality and Supply Reliability Conference, pp. 1-6, Aug. 27-29, 2008.

13 A.L. Kulikov, V.V. Anan'ev,. Adaptivnoe volnovoe opredelenie mesta povrezhdeniya linii ehlektroperedach // Vestnik IGEU, vol.4., pp.21-25,2014.

14 V.F Lachugin., D.I. Panfilov, A.N. Smirnov, "Realizatsiya volnovogo metoda opredeleniya mesta povrezhdeniya na liniyakh elektroperedachi s ispol'zovaniem statisticheskikh metodov analiza dannykh", Izvestiya Rossiyskoy akademii nauk, Energetika, vol.6, pp.137-146, 2013.

15 A.L. Kulikov., M. D. Obalin, V.J. Osokin, T.R. Sharafeev "Primenenie imitacionnogo modelirovaniya LEP 6–35 kV dlya povysheniya tochnosti opredeleniya rasstoyaniya do dvojnyh zamykanij na zemlyu", // Vestnik IGEU, vol.1., pp.40-49,2018.

Strategy of energy efficiency enhancement for sustainable and environmentally balanced development of gold deposits at its final stage

Marina Ryl'nikova[1*]

[1]IPKON RAS, 111020 Moscow, Russia

Abstract. The article presents the rationale of the concept and defines the principles of keeping the production potential of a mining engineering system at the final stage of development to prevent a mine closure accompanied by a range of social, economic and environmental problems resulting from the disturbance of habitat, as well as lithosphere and hydrosphere parameters, etc. This concept can be implemented due to radically new geotechnology solutions aimed at the alteration of gold ore mining and processing technologies with the eventual disposal of wastes to mined-out areas. At the current stage of mining science advancement, energy efficiency and resource saving are inseparably connected with the sustainable use and conservation of non-renewable natural energy sources and transition to the use of technology-related energy sources regenerated in the course of mining. In this case, energy efficiency is viewed as economically feasible electricity consumption achieved at the current level of mining machinery and technology development and mandatory compliance with environmental safety requirements. Enhancement of energy efficiency envisages the minimization of energy consumption for mining processes while maintaining the scale of production and tackling the accompanying non-production challenges.

For the first time, for gold deposit conditions the author proposes general methodological approaches, in combination with a certain level of details, to the problem of systematization and estimation of opportunities of renewable energy sources use in the course of implementation of mining processes. The importance of the mentioned problem solution for mining sciences is in finding the evidence of the fact that in the course of implementation of mining processes within the framework of a mine various technology-related renewable energy sources are formed, which were not taken into account earlier, while the discovered objective laws of the formation of these energy sources, which, depending on the design and technology parameters of mining operations, provide a choice of preferable technological solutions to minimize the consumption of external energy supply by a mine. Development of new technical and technological solutions on the conversion of technology-related energy sources into electricity, its accumulation and consumption for in-mine needs will significantly minimize the consumption of external power supply, and in general it will enhance energy efficiency of mining operations and involve deposits located in remote and hard-to-reach areas in production.

1 Introduction

In the latest decades the problem of mining energy efficiency enhancement and resource saving, along with the issues of mine technical modernization and innovative technological development are viewed as top priorities [1-4].

In some countries and regions with a well-developed mining sector mines consume 10-55% of the total electricity generation. For instance, in SAR the share of mines accounts for 6%, in Brazil – for 4%, in Colorado – for 18%, in the USA – for 3% as an average, in Germany – for 46%, while in Russia – for 55% [5]. Moreover, in the pattern of energy consumption in Russia the share of mineral mining and processing industry prevails (Fig. 1).

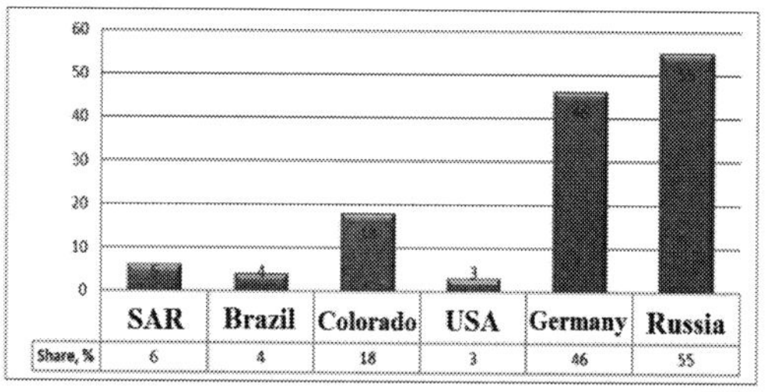

Fig. 1.1 Share of mining sectors in the total electricity consumption by country

It is important to note that the natural growth of mining depth results in the increase of energy consumption. It is often the case, when electricity demand of a large mine is higher than that of the municipal infrastructure providing services to it. Therefore, in case of continuous current trends there will

* Corresponding author: rylnikova@mail.ru

inevitably come a moment, when energy costs of mineral mining at many Russian deposits in terms of energy source global prices will be comparable with the recoverable value of mined minerals.

In this context, in the latest years in particular, the global mining science and practice attach special attention to the substantiation of methodological and engineering solutions providing the enhancement of mining energy efficiency [6-8].

Energy efficiency is viewed as economically feasible electricity consumption achieved at the current level of mining machinery and technology development and mandatory compliance with environmental safety requirements. Enhancement of energy efficiency envisages the minimization of energy consumption for mining processes while maintaining the scale of production and tackling the accompanying non-production challenges.

At the current stage of mining science advancement, energy efficiency and resource saving are inseparably connected with the sustainable use and conservation of non-renewable natural energy sources and transition to the use of technology-related energy sources regenerated in the course of mining operations. These particular solutions form the philosophy of the full environmentally balanced cycle of comprehensive development of mineral deposits and of the transition period for the formation of Industry 4.0 innovation wave of a mine.

Long-term benefits of the research into the use of renewable energy sources are determined by the fact that their creation and application on the basis of innovative process flow sheets of energy saving and energy regeneration in mine planning both "from scratch", and in the course of modernization and reconstruction provide the enhancement of mine operation energy efficiency within the shortest possible period. For this purpose the operation of each mineral deposit site under development must envisage an opportunity of energy cost increase compensation in case of depletion of reserves of high- and medium-grade ore, deterioration of run-of-mine ore quality, strengthening of product quality requirements, particularly in conditions of growing mining depth.

2 Basic approach to the rationale of gold deposit development at the final stage

Application of modern energy saving and energy regeneration systems is rather important for operating gold mines run by Yuzhuralzoloto Group of Companies (YuGK AO) specializing in gold ore mining and saleable gold production at refineries.

Development of vein gold deposits of the South Ural by YuGK AO is currently characterized by a significant decrease of their resource portfolio reproduction, deterioration of quality characteristics of mined ore, more and more unfavorable geomechanical conditions, increasing production costs [9]. Most deposits are currently at the final stage of the development process, i.e., at the stage of finalizing the development of on-balance (economically mineable) reserves in accordance with the baseline plan. A mining company with ever increasing frequency has to bring into development the sites of a deposit with unfavorable conditions, low and rather variable gold ore grades and to transfer mining to deeper horizons, where "conventional" technologies" become less economically feasible, and entailing risks associated with rock bumps or even loss-making that predetermines significant operation costs.

In the above described conditions it is very important to enhance productivity of ore underground mining due to the most efficient use of the total resource potential of deposits, namely: standard- and low-grade ores, associated minerals, middlings, technology-related formations of the previous years, mineralized water, underground space, as well as already disposed and current ore mining and processing wastes, with the concurrent reduction of saleable production costs and improvement of industrial and environmental safety of production [10].

Generalization of gold deposit development practice has made possible the grounding of the concept and defining of the principles of keeping the production potential of a mining engineering system at the final stage of development to prevent a mine closure accompanied by a range of social, economic and environmental problems resulting from the disturbance of habitat, as well as lithosphere and hydrosphere parameters, etc. This concept can be implemented due to radically new geotechnology solutions aimed at the alteration of gold ore mining and processing technologies with the eventual disposal of wastes to mined-out areas.

The main subject-matter of the concept comes down to an idea that at the final stage of deposit development the complete unlocking of mine reserves and resources potential and provision of sustainable development are inseparably tied with comprehensive multipurpose use of all kinds of georesources, which can be efficiently brought into development on a particular site of lithosphere in the period of deposit development specified by the plan. Opportunity of phased use of georesources is based on the properties of the Earth underground resources – their multifaceted nature, multifunctionality and non-depletability. It is these particular properties of underground resources that determine the fact that the final stage of mine evolution can and preferably must become a stage of stepwise evolution rather than its closure, and at every stage envisage the involvement of a new kind of georesources in advantageous and efficient use in an arbitrary indefinitely long period.

To provide the conditions for mine evolution at the final stage of development it is possible and desirable to perform the restructuring of georesources in the course of reforming, reorganizing and technical upgrading of production. In this case, it is possible to create new kinds of underground resources or convert the potential resources to actual ones due to the application of resource reproduction technologies [11, 12].

 https://doi.org/10.1051/e3sconf/20185801008

Along with the issues of technical modernization of mining YuGK AO currently deals with a problem of energy efficiency enhancement as a top priority at stages of mineral mining, processing and metallurgical extraction. This problem can be solved by way of energy sources consumption cutting and more efficient use of natural energy sources, explosives, fuel and gas, cancellation of centralized electricity supply and transition to in-mine power generation that provide nearly two-fold cutting of energy costs in the structure of ore production costs [10]. The problem solution acquires even greater importance, as most gold deposits are located in the regions, where local power supply sources are not available.

3 Evolution of energy regenerating geotechnologies

The growth of energy efficiency of ore mining at deep horizons of deposits is based on the search for and advantageous use of electricity regenerated in mining processes due to technology-related renewable energy sources for in-mine energy consumption.

Studies performed by IPKON RAS researchers have proved the idea that implementation of geotechnology processes in the course of technology-related transformation of underground resources there exists a real opportunity of compensation of some portion of energy consumed in the process of mining. It is realizable by way of energy regeneration due to energy sources formed in geotechnology processes, such as, water drainage, ventilation, backfilling of mined-out areas, ore drawing, transportation (recuperation of energy of downhill movement of transport machinery and vessels), downward force of heavy machinery, use of energy of rock pressure, natural and induced rock mass motion [13-15]. These sources are not obvious outside a mining engineering system, and can be reproduced and converted in electricity only in technology-related transformation of underground resources. The principal condition determining the materializing of the above listed kinds of regenerated energy is a typical feature of ore deposit mining, namely the difference in elevations of horizons, at which mineral is mined. It is important to note that in the earlier years, the world practice did not take into account these resources even in the pattern of the known renewable energy sources.

For instance, in the process of ore mining and processing within the boundaries of mining and processing enterprises the multi-million volumes of run-of-mine material, fluids and slurry are transferred. They have energy, which is equivalent to dozens of MW.h per year. For example, a backfilling mixture pumped from the surface of Gubkin Mine (KMA) to mined-out areas to a depth of 250 m at a rate of 4.5 million t/year has potential energy, which is equivalent to 2.5 million kW.h [13].

Research into opportunities of generation, collection and storage of electricity from renewable energy sources of natural and technology-related origin performed in cooperation with the special division of the Laboratory for Environmentally Balanced Mineral Deposit Development located in the town of Plast and Yuzhuralzoloto Group of Companies has shown that in conditions of underground mines of the Urals developing steeply dipping vein gold ore bodies at great depths of Kochkarskoe, Svetlinskoe, Berezovskoe and Novotroitskoe deposits the greatest energy potential belongs to the formed in mines and moving downward gravity flows of hydraulic mixture circulating on a year-round basis. These mixtures are contaminated waters containing high concentrations of solids within a water drainage system, as well as consolidated and hydraulic backfilling mixtures formed on the basis of ore mining and processing wastes. These flows are formed at horizons collecting surface and ground waters, as well as in surface backfilling systems, ore processing plants, tailing storage facilities, and they move predominantly by gravity downward to main horizons of water drainage or zones of backfilling operations of underground mines [14].

Energy potential of these flows is rather high. As it has been proved by the performed calculations, the collection and conversion of hydraulic flow energy to electricity is able to make good up to 40-50% of energy consumed by underground mines.

In the course of research, the main parameters of geotechnology of electricity regeneration within the framework of the mining and engineering system incorporating small-capacity hydropower plants (HPP) (up to 100 kW) have been designed with due account for the actual mining depth at underground mines and specified values of level and sublevel height: intermediate – from 10 m and more, sublevel – at least 15 m, and level – within 30-60 m [15].

For the materialization of the energy potential of hydraulic mixture flows in the course of pilot run of the novel technology of mine flow energy regeneration some HPP operational models with small-capacity electricity generators having different types of blades and wheels have been designed. Estimates have been made of the efficiency, conditions of operation and performance of hydro turbines for the conversion of energy of falling downward hydraulic mixtures varying in composition, viscosity and density depending on the head pressure and angle of feeding to the turbine, its location in a mining engineering system have been found.

Historic analysis of the world hydropower achievements and comparison of design varieties of hydro turbines used both in distributed power generation, and in modern dam hydropower plants of extra high capacity have revealed the main trends of hydro turbine design development, made possible the selection of the best type of bucket-, rotor- and axial-type hydro turbine design, which is most suitable for underground mine conditions of energy regeneration with the use of small capacity HPP (up to 100 kW), as well as the right choice of the receiving nozzle design with due account for the feeding direction and material composition of mixtures, as well as the development and manufacture of hydro turbine models at Uralenergoresurs OOO (in Magnitogorsk) for recovery of energy of water drainage

https://doi.org/10.1051/e3sconf/20185801008

and backfilling systems contaminated hydro mixtures in underground mine conditions (Fig. 2).

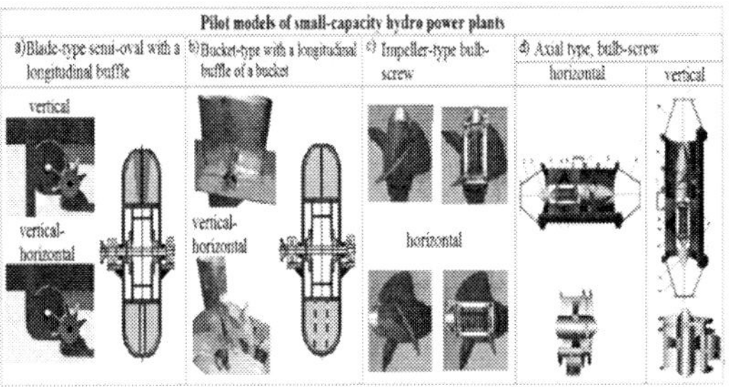

Fig. 2. Hydro turbine design approved for pilot-scale tests

With due account for constant input and variable design parameters of the mining engineering system the parameter characteristics of mixture feeding to a turbine have been calculated to provide the highest efficiency and minimized wear of the hydro turbine for the HPP being designed: height of a mixture single-stage transfer – 30-100 m; angle of mixture falling on a turbine – 65-90o.

Design calculations and full-scale experiments have proved in practice that in underground mine conditions application of the electric energy regeneration system based on mine water flows of a water drainage system and hydraulic flows of backfilling mixtures on the basis of thickened ore dressing tailings on upper and lower horizons of underground mines of the Urals provides a 20% decrease of mine electricity consumption and in general enhances energy efficiency of mineral deposit development.

Irregular loading of mine transport machinery at different sections of a mine, repair, maintenance and other auxiliary processes can also be viewed as reserves of the air flow residual energy application for the formation of renewable sources of electricity. A schematic circuit diagram of electricity generation due to the air flow residual energy with the use of small-capacity generators is shown in Fig. 3.

In underground mine conditions, the downward force of heavy machinery, recuperation of brake energy of lifting and transport equipment are additional sources of renewable energy. Recuperation systems have got widespread use throughout the world. In this area work is under way in some of the known Russian companies using recuperation systems with stationary hoist and winder installations (for instance, Kazsk Branch of Evrazruda OAO, Primorsky GOK, Vorkutaugol OAO, Osinnikovskaya Mine of Yuzhkuzbassugol OAO).

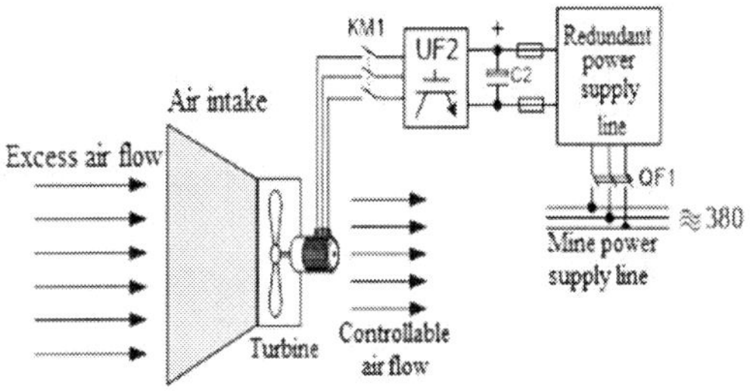

Fig. 3. Generator schematic circuit diagram for electricity generation based on the air flow residual energy as a renewable energy source.

Solutions have been developed for some conditions of cage hoist operation providing energy generation both for man and material roundtrip running, and for electricity return to the mine power supply line with its eventual use for other purposes. Energy saving is estimated at up to 75% as compared to conventional two-speed hoists.

Besides, rather promising is the conversion of heavy machinery downward force energy into electric power. For this purpose in mine transport operational stop points (discharge point, ore or rock passes, weighing point, drift entry, etc.) specially-designed gyro-platforms can be installed. Heavy transport machinery downward force energy is converted into electricity with the use of systems transferring power fluid mechanical energy to mechanical energy and then to electric power, or, in case of ferromagnetic fluid, immediately to electric power.

Another challenging process solution, particularly for deeply occurring mineral deposit development with the application of hot air systems, is the use of water in air cooling systems. Air-heated mine water is used in hydro lift systems for pressure handling of mined material to the surface. Formation of a common hydro network in a mine provides comprehensive minimization of ventilation and mine water drainage costs, as well as mined material transportation costs.

Cutting of natural sources energy consumption and involvement of technology-related renewable energy sources in mining are also realizable due to technological advancement. Energy of virgin rock failure under the force of gravity; energy of rock mass straining under rock pressure forces; energy of heat exchange processes in virgin rock caused by endogenous processes of drillhole and shaft in-situ leaching, geothermal energy and technology-related thermal energy resulting from the implementation of physical and chemical geotechnologies are also the areas of research into technology-related renewable energy sources, formed in the course of geotechnology processes.

For the creation of energy-efficient geotechnologies at the final stage of gold deposit development it is necessary to implement the following system of measures:

a) assessment of energy potential of mine flows transferred in the course of implementation of gold-ore underground mining processes, namely: hydraulic

mixtures, mine air flows, downward force of heavy machinery, temperature gradient of the environment;
b) process audit and systematization of opportunities of the enhancement of process efficiency based of novel geotechnology application for the formation and use of renewable energy sources;
c) understanding of the objective laws governing the influence of geological and mining factors on energy efficiency of geotechnology processes;
d) elaboration of methods and procedures for grounding of the parameters of main, auxiliary and return mine flows to enhance energy efficiency of the application of geotechnology process energy potential;
e) elaboration of methods and procedures to provide the design rationale of the parameters of energy-efficient mining engineering systems with a full cycle of comprehensive mineral deposit development.

4 Conclusions

Tests and trials of the performed design and engineering solutions at mines of YuGK AO based on the results of the performed studies for the preparation of the rationale of the new concept and geotechnology of vein gold deposit development at the final stage have proved conclusively that for the finalizing of the development of non-placer gold deposit on-balance (economically mineable) reserves the multifaceted nature, multifunctionality and non-depletability of underground mineral resources determine the opportunity of multi-stage and long-term use of these diverse resources and properties.

In-mine consumption of electricity regenerated in the processes of gold deposit development based on technology-related renewable energy sources and exploring new non-conventional electrical and mechanical aids is one of the areas of ore mining energy efficiency growth at gold mines.

The studies are performed with the support of the Russian Science Foundation (Project No14-37-00050)

References

1. K.N. Trubetskoi, Yu.P. Galchenko, Geoecology of subsoil development and ecogeotechnology of mineral resources exploitation. Publishing house "Nauchtekhlitizdat", 360 (2015)
2. I.A. Tangayev, Energy intensity of minerals mining and processing. Nedra, 231 (1986)
3. V.I. Golik, S.V. Leonov, E.V. Shevchenko, Nature management rationalization in the strategy of industrial enterprises development: an academic project. Culture, 380 (2012)
4. K.K. Ilkovsky, D.I. Timofeev, Mining Journal, 11, 73-73 (2011)
5. A. McIvor, Cleantech, 5, 16-19 (2010)
6. K.N. Trubetskoy, D.R. Kaplunov, M.V. Ryl'nikova, D.N. Radchenko, Journal of mining science, 3, 58-66 (2011), DOI: 10.1134/S1062739147030087
7. D.R. Kaplunov, D.N. Radchenko, Mining Information and Analytical Bulletin (Scientific and Technical Journal), 1, 447-455 (2011)
8. V.N. Kalmykov, M.V. Rylnikova, O.V. Petrova, T.P. Plesovskikh, The "Mine Surveying Bulletin" magazine, 1, 59-64 (2012)
9. K.I. Strukov, Gornyi Zhurnal (Mining Journal), 9, 21-25 (2017), DOI: 10.17580/gzh.2017.09.04.
10. K.I. Strukov, M.V. Rylnikova, I.L. Nikiforova, Mining Industry Journal, 2, 46-49 (2018), №2, DOI: 10.30686/1609-9192-2018-2-138-46-48
11. K.I. Strukov, R.V. Berger, V.V. Fedoseev, M.V. Rylnikova, Gornyi Zhurnal (Mining Journal), 6, 5-8 (2017), DOI: 10.17580/gzh.2017.06.01
12. M.V. Rylnikova, V.A. Ezhov, I.L. Nikiforova, S.N. Plotnikov, Gornyi Zhurnal (Mining Journal), 9, 35-40 (2017), DOI: 10.17580/gzh.2017.09.07
13. D.R. Kaplunov, M.V. Rylnikova, Gornyi Zhurnal (Mining Journal), 9, 72-75 (2015), DOI: 10.17580/gzh.2015.09.16
14. D.R. Kaplunov, S.G. Leizerovich, V.K. Tomaev, Gornyi zhurnal (Mining Journal), 4, 62-66 (2013)
15. M.V. Rylnikova, K.I. Strukov, V.V. Olizarenko, I.S. Turkin, Gornyi Zhurnal (Mining Journal), 11, 71-76 (2017), DOI: 10.17580/gzh.2017.11.13

 https://doi.org/10.1051/e3sconf/20185801009

Mining engineering system as an energy asset in industry 4.0

Dmitriy Radchenko[1]*, *Alina* Bondarenko[1]

[1]IPKON RAS, 111020 Moscow, Russia

Abstract. Mines now absorbing Industry 4.0 technologies face the challenges of production energy efficiency enhancement and energy safety improvement. Following the environmental standards involves the requirement of turning from diesel to electric drive of mining machinery and equipment. This significantly improves miners' working environment in ore mines. At the same time, the share of energy consumption increases in the situation of low energy security of mines, particularly at peak loads. The article discusses the conditions, and proposes a scenario of mine energy supply system reliability enhancement. For this purpose, a mining engineering system is viewed as an energy asset, with identification of its expanding capabilities.

1 Introduction

Global trends of all without exception large production corporations joining Industry 4.0 give a conclusive interpretation of the prospects of mining industry development – transition to intelligent control systems with the alteration of the role of the human factor in design and development, as well as operation of mining engineering systems and their health monitoring.

Industry 4.0 is not only and not so much new technologies; rather, it is a new approach to production and consumption. It is based on the principles of sustainable development, implementation of "green" technologies, collection of big data, their processing and use for coordinated actions and operation timing performed without human intervention. The base trend of Industry 4.0 development is the attribution of such functions as optimization and setting of autonomous operation to machines and mechanisms.

As a result, the country's mining complex, which is already an energy-intensive sector (6% of the total power generation of Russia [1]; as an average, 1 tonne of coal output requires 50.5 kWh; that of gold – 250 kWh; and that of bauxite – 13 kWh), on a mid-term horizon it is expected to increase energy consumption for process needs. It is explained by the fact that Industry 4.0 base technologies, such as robotization, digitization of deposit sites and production processes, positioning of machinery and mine personnel, collection, storage and processing of big data, cybersecurity provision and some other are rather energy-intensive.

The main obstacle preventing the application of Industry 4.0 principles and technologies is the current high share of diesel-driven machinery and mechanisms both in terms of environmental safety (restrictions and consequences of the personnel attendance in toxic zones and ICE-driven machinery gas emissions), and technology features (mines have to spend up to 40% of the total energy for ventilation to remove exhaust gases of mine transport machinery). Even in technologically developed mining countries the share of energy generated as a result of diesel fuel combustion and consumed in technological processes of mining reaches 45% [2]. In Russia, it is impartially higher and accounts for 75% as an average. On the one hand, this can be explained by the fact that a significant share of mines operates in the areas with poor energy infrastructure, and by the widespread use of heavy load-haul-dump machinery and dump trucks with diesel engines in underground mines, on the other hand.

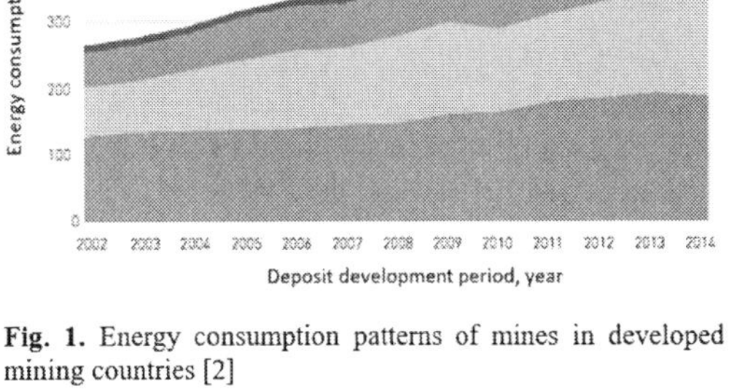

Fig. 1. Energy consumption patterns of mines in developed mining countries [2]

In this context, a cross-cutting priority of mining and energy engineering is the identification of expanding capabilities of energy in mining engineering systems for the enhancement of energy efficiency, industrial, energy and environmental safety of production with the minimization of the share of diesel fuel-fired power plants and internal combustion engines in the total pattern of energy consumption of mines.

* Corresponding author: mining_expert@mail.ru

2 Research methods

Analysis of mine behavioural and investment models aimed at the enhancement of energy security of processes and perfection of energy safety is the main research method.

Research has been performed into the world practice of mines' approaches to the application of renewable energy.

Expandable capabilities of a mining engineering system have been identified, namely, energy sources, which can be specified as technology-related, i.e., caused by and existing as a result of human intervention underground [1]. Analysis of the following expandable capabilities has been made:

- renewable energy sources (solar and wind energy, geothermal energy);
- technology-related sources (energy of in-mine air flow, energy of in-mine hydraulic flows).

Design and development of mining engineering systems have been performed and 4 most probable generic design patterns have been identified, in which the application of technology-related sources of energy of a mining engineering system is feasible.

3 Key research findings

Transition to Industry 4.0 in the mining sector must be based on the phase-out of the application of load-haul and transport machinery with diesel engines.

Transition to load-haul-dump machinery and underground dump trucks with electric drives will minimize operation costs and provide environmental advantages in transition to mineral mining at deep horizons with high temperatures, as well as in conditions, where ventilation costs are of critical importance. Main concepts reflecting the advantages and shortcomings of mine underground transport machinery application options are provided in Table 1. For the first time in mining industry RDH's heavy 3 TEB Muckmaster LHD was demonstrated in May 2016. The payload capacity of this model is 3 t, its bucket capacity is 1.15 m3. The machine is powered by an Artisan LiFePO4 battery. Currently, the only electrically-powered LHD manufactured in Russia based on power-storage unit goes through testing (Rosatom).

Table 1. Alternative options of mine transport machinery

No	Notations	Key features	
		Advantages	Shortcomings
1	Diesel 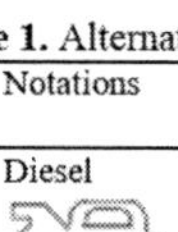 	Low cost, large-scale production, tried and tested operation practice.	High environmental hazard, inability to operate in confined space because of temperature and oxygen content requirements, inability to operate as robotized equipment.
2	Electric cable 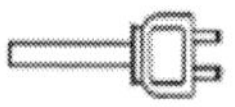 	Low heat emission; advantageous requirements to ventilation of underground workings and volume of air supply; Low operation costs.	Limited mobility, and thus inability to adapt to robotized working conditions; problematic turnout in underground workings; high wear of cables; high capital costs; risk of short circuit with disablement of a power supply system section, high safety requirements.
3	Storage battery  	High mobility, robotization opportunity, 4-fold cutting of energy costs as compared to diesel-driven analogs, free from hydrostatic oil and filters; hydraulic oil change at 2-year intervals; favorable requirements to ventilation of underground workings; low heat emission.	High cost, need of organizational and engineering solutions for battery charging in conditions of underground mines, higher energy consumption of mines; compulsory retraining of mine workers with the issue of work permits for them.

The analysis of data from Table 1 shows that in addition to high costs the main obstacle preventing the application of mining machinery with electric drives based on power storage units (storage batteries) are high loads on a mine power supply system.

Studying of the approaches to mine power supply and energy security problem solution has shown that use is made of numerous business models, which may help mining companies enhance their energy efficiency, and get profit due to application of renewable energy sources [4]. It has been found that some large mining companies, such as MSPL Limited or Essel Mining & Industries Limited, expanded the area of their activity and launched wind parks. Their activity in the area of renewable energy sources is not directly connected with their mining business. They generate electricity both for their own consumption, and for supply to outside consumers. There exist many other business models, in which electricity generated by windpower plants and solar-cell panels is consumed directly by mines. As many mining companies do not view electricity generation as a component of their scope of main competences, there appear broad opportunities for outside investors and power plant leasing models. Particularly an area of renewable energy development in remote regions, where diesel-generator substations are the only energy source, is being shaped up.

In Russia, where conditions for the implementation of technologies of energy regeneration due to natural sources are rather limited because of unfavourable weather and climate, as well as the current level of technological development, it is necessary to seek alternative scenarios. It is proved by well-known research monographs on energy [5-6].

For the enhancement of operation energy efficiency of mining engineering systems equipped with robotized machinery the IPKON Department of the Theory of Mineral Deposit Development Planning has proposed an approach to consumed electric energy compensation due to the application of in-mine electricity regeneration plants in mineral deposit mining. [3, 7]. Energy conversion of mine flows, which volume in underground mines is considerable, is the most promising area (Table 2, [8]). Analysis of Table 2 shows that even at not very deep mines with the mining depth of not more than 400 m, the water inflow may reach 500 m3/h, while at deeper mines it exceeds 1000 m3/h. It is important to note that the total volume of mine water goes by gravity to lower horizons (Fig. 2, a), from where it is pumped to the surface. At mines of 4,000 m depth the total length of underground workings may reach 900 km, i.e., underground mines, as a rule, have an extensive network of vertical, inclined and horizontal workings, boreholes for water transfer from one horizon to another (vertical boreholes for mine water) providing the migration of mine water gravity flows.

Table 2. Classification of deposits by hydrogeological conditions of mining

Group	Mining depth, m	Expected inflow of mine water, m^3/h
I	200-400	below 500
II	500-600	500 – 1000
III	Свыше 600	over 1000
IV	Свыше 1000	the same

Moreover, according to the requirements to underground mine planning, at each underground mine its horizontal workings are driven with a 2-3-degree pitch towards the mine shaft. As a result, the total volume of water is accumulated in the main drain sump and then pumped to the surface and disposed to the environment with or without preliminary treatment.

It is evident that energy efficiency of mining engineering systems, in which so much energy is spent for water drainage (and ventilation, as can be seen above), is rather low.

For the enhancement of energy efficiency of mining engineering systems it is proposed to alter the requirements to planning and design of underground working drivage. The new philosophy of mine water disposal is shown in Fig. 2-b. According to the proposed philosophy, water is accumulated at each horizon due to the workings incline turning from the shaft to the periphery. As a result, at each horizon there will be an opportunity of water accumulation for its controlled transfer to the lower horizon, and so on from overlying to underlying horizons. Energy of mine flows formed this way may be and must be converted to electricity. With the current depth of modern underground mines and rather impressive opportunities of water transfer from one horizon to another (any structure can be built) the proposed mining and engineering system planning and design philosophy seems rather promising.

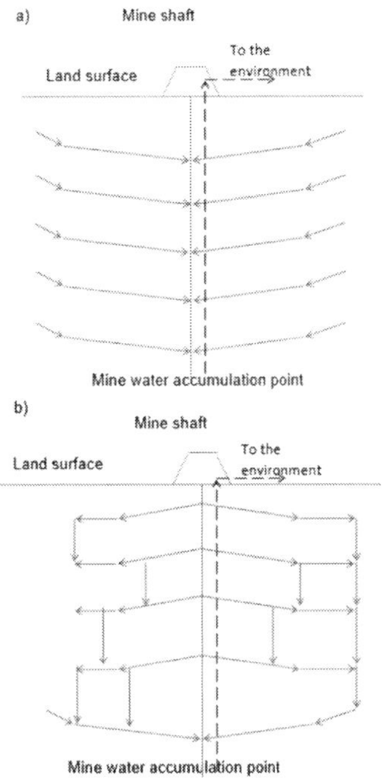

Fig. 2. Philosophy of mine water drainage design applicable globally (a) and proposed for regeneration of energy of mine flows (b). Direction of water outflow and transfer from one horizon to another is shown by arrows.

For grounding of electricity regeneration plant parameters (small HPP, underground hydroelectric pumped storage power plants) it is important to rely on the estimates of geological, hydrogeological and mining conditions of ore deposit development determining the volume of water inflow to a mine due to continuous inflow of ground waters and peak water inflows resulting from precipitations and snow melting. Such estimates must be made as early as at the stage of mine planning on the basis of comprehensive research of geological, hydrogeological and mining conditions of deposit development. In this case, geological conditions cover: strata pressure, chemical composition of surface and ground waters, quantity and thickness of aquifers, ground water flow connection with surface water bodies and watercourse, nature of ground water migration and filtration properties of rock in the large, as well as concentration of suspended particles and solid impurities [9].

Furthermore, the value of energy flows will depend on the volumes of hydraulic mixture (Q_{gm},m^3/h), its density, difference in elevations and equipment used for energy conversion (Fig. 3). The results of preliminary estimation show that in heavily-watered mines during the transfer of water at a rate of 5 m^3/s from a horizon lying 70 m higher than underlying horizon may generate up to 70 MW of electrical power. It is evident that this amount of energy is sufficient for energy supply of a mine, especially in view of the fact that for the achievement of the objective of mine transfer from mine machinery with diesel engines to those equipped with environmentally safe electric drives.

Yet, mine water flows of horizons are not of permanent volume, i.e. there may be peak inflows, but in some periods water does not actually flows to a mine or some of the horizons. Therefore, design, manufacture and use of energy storage installations, including storage battery systems for operation of transport machinery in mines are very important tasks.

It is a forward-looking idea to install these systems at each horizon, where at the same time:

- work is under way on mineral mining with the use of robotized LHD units with storage batteries. Judging by the available data, such storage batteries must be charged after 4-hour operation, with the charging time of 1.5-2 hours;
- water flow transfer is provided;
- energy regeneration complex including a generator-inverter (transformer) – energy-storage system is in operation.

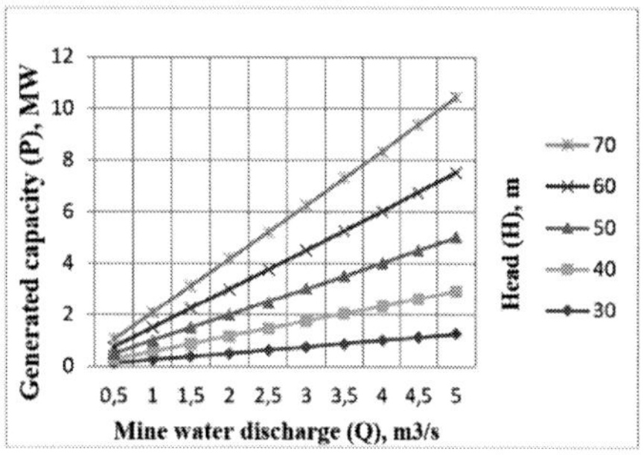

Fig.3. Dependence of regenerated capacity on the volume of mine water transferred from one mine horizon to another and head (distance between horizons).

The principle of charging plant location for the replacement of robotized equipment storage batteries at an underground mine horizon is shown in Fig. 4. The analysis of Fig. 3 shows that the scheme includes 2 energy generation and storage systems; moreover, the system on the overlying horizon converts energy of a mine flow transferred from horizons 1 and 2 ($Q_1 + Q_2$) to the third underlying horizon. The underlying system of energy generation and storage converts the energy of mine water flows from three horizons $Q_1 + Q_2 + Q_3$ and it has the highest capacity.

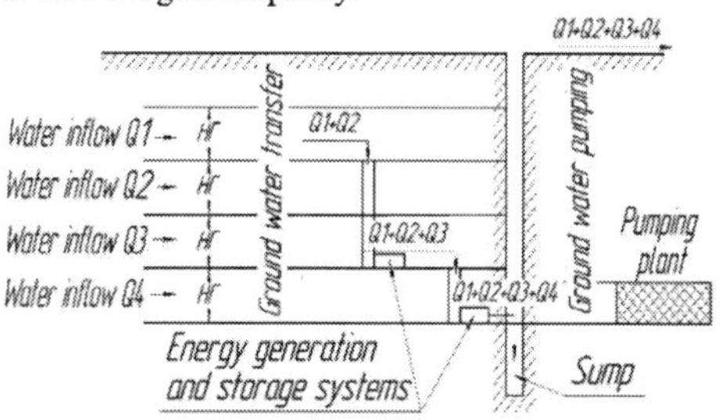

Fig. 4. Principle of installation of a charging plant for mine battery-powered transportation machinery in an underground mine.

It is envisaged that robotized equipment with battery-powered electric drives, when necessary, goes to energy storage systems for battery replacement and then back to ore mining operation.

The analysis of the schematic diagrams shown in Fig. 2 and Fig. 4 proves the fact that the mine consumes electric energy for drainage of the total water inflow, ventilation, etc., and it also regenerates it. This suggests the conclusion that the energy balance of the mining engineering system as an energy asset of Industry 4.0 can be defined as:

$$E_{mes} = E_r - E_c \quad (1)$$

where E_{mes} is the total energy of the mining engineering system, MW/h; E_r is regenerated energy, MW/h; E_c is consumed energy, MW/h.

It is evident that if the total water inflow is pumped, such energy balance is negative, i.e., $E_r < E_c$. Such standard mining engineering systems are shown in Fig. 5-a and Fig. 5-b. However, it has been found that two fundamentally different mining situations are possible, when the energy balance of the mining engineering system described by formula 1 will be positive. Such standard design options of mining engineering systems are shown in Fig. 5.

According to the data of Fig. 5-c, deposit development takes place in the upland area. The total water inflow is transferred stepwise from upper to lower horizons, while energy generation and storage systems are installed on the third, second and first horizons. After water transfer to the lower horizon the total water inflow goes by gravity to the environment, i.e., the mine spends nothing for forced water drainage.

According to the diagram shown in Fig. 5-d the ore body is mined in the wall of a depleted open pit. In this case, after electric energy generation and accumulation the total water inflow goes by gravity to the depleted open pit. Thus, no water pumping from the underground mine takes place. It suggest a conclusion that the energy balance of the mining engineering system looks as follows: $E_r > E_c$, i.e., in these conditions (Fig. 5-c and Fig. 5-d) the mine is able to generate more energy than it consumes.

With regard to the balance equation 1 it is important to note that this is the case of implementation of Industry 4.0 environmentally safe technologies in the mine, i.e. ore mining technologies without application of mine LHD and transport machinery with diesel engines. In other words, mining engineering system with the positive balance ($E_r > E_c$ condition) means the situation, when there are no 40% ventilation costs due to the application of battery-powered equipment, and no water drainage costs.

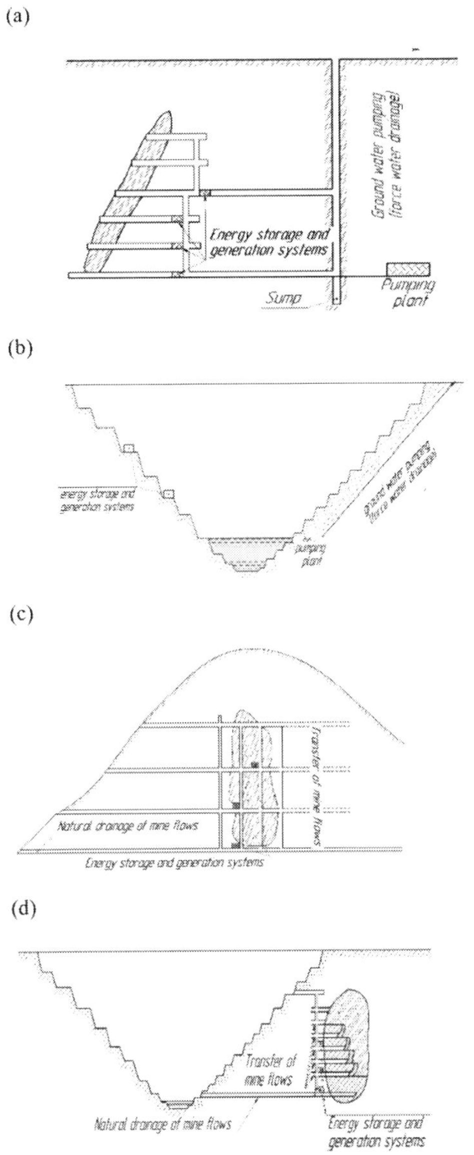

Fig. 5. Typical mining engineering systems and opportunities for energy production with forced pumping out of mine waters (a, b) and natural water disposal (c, d)

4 Conclusions

A mining engineering system, which was designed earlier based on conditions and restrictions concerning work environment of people, in the future, with the development of remote and robotized geotechnologies, autonomous equipment for rock mass monitoring, and combination of these technologies, will be based on grounded parameters of mine structures and geotechnologies aimed only at the provision of conditions for operation of robotized and autonomously controlled machinery in zones of mining operations. Elimination of the necessity of health and safety requirements concerning people attendance in these zones, as well as restrictions in terms of industrial safety determines new approaches to the formation of mining engineering systems and grounding of the living environment parameters in a mining engineering system and in the area of its influence.

The situation with launching of Industry 4.0 technologies based on the application of robotized environmentally safe equipment with electric drives, implementation of measures on the enhancement of energy efficiency on the basis of mining technology-related energy sources is either poorly developed or even has not got any start in the international practice. Therefore, setting a target of mine re-equipment on the basis of energy cost compensation due to conversion of mine flow energy within a mining engineering system seems a very important and timely solution for the enhancement of energy security and minimization of negative environmental effect.

The studies are performed with the support of the Russian Science Foundation (Project No14-37-00050).

References

1. Data of Rosstat, Technological development of industries economy, Energy efficiency, URL: http://www.gks.ru/wps/wcm/connect/rosstat_main/rosstat/en/statistics/economydevelopment/ (circulation date 12.02.2018).
2. Energy in Australian mining sector. SunSHIFT Pty Ltd, 20 (2017)
3. D.R. Kaplunov, M.V. Ryl'nikova, D.N. Radchenko, J. Min. Sci, **51**, 1, 111–117 (2015), DOI: 10.1134/S1062739115010147
4. Dr. Thomas Hillig Energy Consulting - THEnergy. Business Models for renewable energy applications at mines [Electronic resource]. URL: https://www.th-energy.net/english/platform-renewable-energy-and-mining / business-models / (circulation date 03/03/2018).
5. S.M. Senderov, V.I. Rabchuk, N.I. Pyatkova, S.V. Vorobyov, Ensuring Russia's energy security: choice priorities. Sib. Otd-Institute of Energy Systems. L.A. Melentieva. Novosibirsk: Science, 116 (2017)
6. Pyatkova N.I. [and etc.]. Energy Security Russia: problems and solutions. Russian Acad. Sci., Sib. Branch, Institute of Systems power them. L.A. Melentieva., 146 (2014)
7. M.V. Rylnikova, K.I. Strukov, V.V. Olizarenko etc, Gornyi Zhurnal, **12**, 71–76 (2017), DOI:10.17580/gzh.2017.11.13
8. VNTP 13-2-93. Norms of technological design mining enterprises of metallurgy with underground method of development, by the Committee of the Russian Federation on metallurgy (1993)
9. Rylnikova M.V., Radchenko D.N., Gadzhieva L.A., Materials Jubilee scientific and practical conference dedicated to the 100-anniversary of the RSGRUU. S. Ordzhonikidze "New in geological study of mineral resources ", 111-113 (2018)

Economic aspects of ensuring the capacity adequacy of electric power systems

Nikolay Belyaev[1]*, *Andrey* Egorov[1], *Nikolay* Korovkin[2] and *Vladimir* Chudny[2]

[1] Russian Energy Agency Ministry of Energy of the Russian Federation, Russia
[2] Peter the Great St.Petersburg Polytechnic University, Russia

Abstract. The present report deals with economic issues of selecting means ensuring the capacity adequacy of electric power systems. Cost analysis of various engineering measures ensuring the capacity adequacy has been performed, they are as follows: construction of margin generating capacities, erection of new power transmission lines or increase of transmission capacity of existing power lines. The cost-based analysis has been conducted following investment programs and regulatory acts in place. Recommendations on developing computational models of power systems have been given based on the results obtained in order to assess the capacity adequacy indices with account made for the cost of various engineering measures to be taken for their improvement.

1 Introduction

The capacity adequacy (CA) studies represent as traditional as ever a highly topical problem in terms of planning the electric power system (EPS) development to which an interest is expressed not only by scientific community or organizations operating in the industry and being in charge for EPS operation and development, but by consumers of electricity and capacity markets as a whole. This is due, first of all, to the problem of generating capacity availability, as the capacity margin technologically required to meet the consumers' need in reliable power supply to be paid on capacity market may be well-grounded only by calculations of capacity adequacy indices (CAI) [1–3].

Issues connected with the analysis of capacity adequacy have been given a great attention in national and foreign literature. Except for various models and techniques of CAI evaluation, a range of specialized software packages has been developed as of today both in Russia and abroad to perform relevant calculations (GE MARS, GridView, Orion, Yantari etc.) [4–6]. However, some issues related to the evaluation of power grid CAI do not have a unique solution and in view of this present an area of special scientific interest. These issues include the simulation of transmission line and accounting for capacity transmission constraints when evaluating CAI of meshed systems with weak tie lines.

The paper examines economic aspects of transmission line simulation when analyzing the EPS capacity adequacy. First, common issues of EPS simulation have been studied to evaluate CAI and to optimize the capacity margin, then a feasibility study of taking into account various power transmission constraints has been carried out on the basis of comparing the cost of margin capacity and that one of various engineering solutions aimed at ensuring the transmission capacity increase in power grid.

2 EPS transport models

In prevailing practice of CAI calculations the structure of EPS transmission network is presented as a transport (multi-node) model which makes an aggregate of reliability zones (concentrated nodes) and tie-lines between them of given transmission capacity [3,5]. Each reliability zone is an aggregate of nodes of basic grid in EPS where there are no transmission constraints or where they do not affect a mutual reserve capability of given nodes under eventual capacity settlement within the system. Ties between reliability zones are determined based on power grid topology, they may represent one or several overhead electric lines (OHL) whose power flow is limited by some criteria or others. When evaluating CAI, their values are defined for each reliability zone, thus the correctness of multi-zone model formation defines directly the correctness of results obtained on its basis.

The main task of EPS multi-zone model formation is to define transmission constraints between zones. The formation of reliability zones and determination of tie-line transmission capabilities are to be performed with due account made for these constraints. Saying that we know that each power grid component has its limit transmission capability, which is specified by one or several factors. However, not each of these limitations may affect the level of CA, that is why it makes no sense to take into account all transmission constraints identified for EPS, as calculations become very complicated, but they do not affect their results. In this

* Corresponding author: Belyaev.NA@yandex.ru

context, when forming multi-zone EPS models it should be reasonable to consider only the constraints having an impact on the possibility to utilize capacity margins in order to meet all power demands, in other words to identify the determinant levels of CA in different points of electric power system.

An analysis of worldwide studies of CA given in [5] shows that in spite of large-scale application of software packages for CAI assessment, there are no completely formalized approaches with regard to the formation of transport models in EPS. This is due, first, to the fact that in developed countries the studies of EPS CA are performed on a regular basis and the gained calculation and operation experience makes it possible to set margins of reliability zones without applying a formalized analysis of the structure and eventual operating states of EPS. Moreover, a rather simple structure of EPS allows defining in some cases the reliability zone margins at the expert level.

At the same time in UES of Russia the common practice describing the domains of admissible operational states for EPS has been developed on the basis of real power with the use of maximum allowed power flow (MAPF) values in controlled sections [7]. The formalized methods applied for the formation of EPS multi-zone models and CAI assessment with the use of MAPF values in controllable sections have been developed and presented in [8,9]. Let us consider the possibility to optimize the approaches proposed in cited papers with due account made for economic aspects of the choice of facilities ensuring CA.

3 Cost-correlation of facilities ensuring CA

As specified above, the influence of a given constraint on EPS CA levels is one of the criteria for transmission constraint satisfaction while forming a multi-zone transport model. On the other hand, as it was mentioned, the main objective of CAI calculation is to define the required level of generating capacity margin, i.e. its volume and location within EPS in order to ensure a required level of CA. This being the case, when evaluating and accordingly when forming multi-zone transport models it should be useful to take into account the transmission constraints only for those power grid segments whose cost of transmission capacity increase correlates with that one of new generating capacity construction.

Major factors constraining the transmission capacity of power grid are as follows:

1. Continuous current carrying load of power equipment (under normal and post-fault operational conditions) which is, as a rule, a determinant transmission constraint for 220 kV and above power grid;

2. Minimum accessible margin in terms of steady-state stability accord-ing to [10] (under normal and post-fault operational conditions) which is, as a rule, a determinant factor for long-distance extra- and ultra-high voltage transmission lines.

Let us evaluate the cost of power grid transmission capacity growth as compared to that one of new generating capacity. To do that, data from [11] have been taken to estimate the cost of new generating capacity construction while data from [12] have been taken to evaluate the cost of engineering measures aimed at increasing the power grid transmission capacity (in average for territorial entities of the Russian Federation). Relevant data are given in Table 1.

Table 1. Cost of generating facility and measures providing the increase of power grid transmission capacity.

Measures	Cost, mln. RUR
New generating facility, per 1 MW of installed capacity	28,8–41,9 (fuel – gas) 49,2–53,5 (fuel – coal)
Construction of single-circuit overhead line, per 1 km at rated voltage of:	
110 kV	7,9–8,7
220 kV	12,0–12,2
330 kV	12,8–13,4
500 kV	18,2–20,1
Controlled shunt reactor, per за 1 MVAr	2,0–4,1
Static capacitor bank, per 1 MVAr	0,3–0,6

The calculation of the cost of power grid transmission capacity has been performed on the basis of data given in Table 1 for 220 kV and 500 kV transmission lines. The results for 1 kW of transmission capacity against transmission route length are presented on Fig. 1. Charges due to the construction of overhead lines (wire type AC-500) and provision of substations with equipment (bank of transformers, breaker cells) of relevant voltage class have been taken into account.

The line transmission capacity was taken for calculations according to the following expression:

$$P_{\max} = \min\left(P_S;\ 0{,}8\frac{U_{nom}^2}{x_0 \ell} \right), \quad (1)$$

where P_S – maximum transmission capacity in terms of continuous current-carrying load accepted upon ata from [13] (for 220 kV- 342 MW OHL, for 500 kV – 1700 MW OHL), U_{nom} – OHL rated voltage, x_0 – OHL inductance per unit length, ℓ – OHL length. A critical OHL length (400 km and 1200 km for 220 kV OHL and 500 kV OHL respectively) has also been taken from [13].

On the basis of calculation data given on Fig. 1 one may state that for OHL of short or average length the cost of 1 kW of its transmission capacity will be considerably lower than that one of installed capacity of new generating facility and only in the case when it is close to its critical length (400 km for 220 kV OHL and 800 km for 500 kV OHL) the above costs may correlate. However, it should be noted that for 250-350 km long OHL there is a linear relationship between its cost and length. The transmission capacity of this OHL is limited by continuous current-carrying load and is independent of its length. The transmission capacity of long-distance OHL is specified by operating transmission limit in terms of steady-state stability which is, in its turn, inversely proportional to OHL inductance and accordingly to its length. In this context, the cost of 1 kW of transmission capacity of such an OHL is in quadratic dependence of its length.

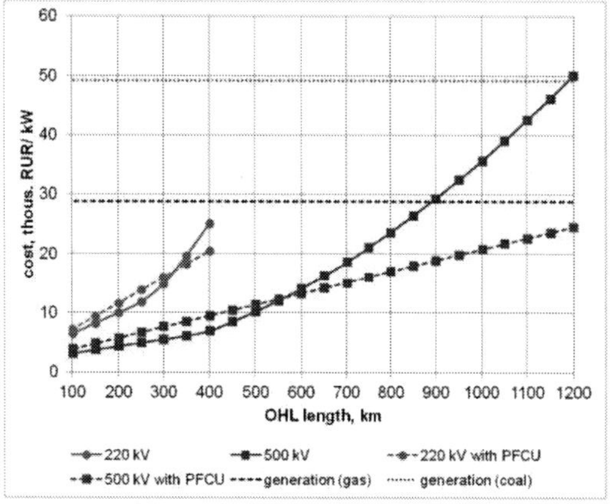

Fig. 1. Cost of 220 kV and 500 kV OHL transmission capacity.

In addition, the cost of OHL transmission capacity with due account made for power factor compensation has been considered. Figure 1 shows these relationships as dashed lines with relevant markers. The installation of power factor correction units (PFCU) [14] at intermediate substations requires additional expenses and results in OHL cost rise, but it allows regulating voltage levels and increasing OHL transmission capacity up to the value P_S. As it is shown on Fig. 1, this involves the reduction of 1 kW transmission capacity cost of long OHL as the effect due to the transmission capacity increase exceeds additional expenses connected with PFCU. Hence, the cost of 1 kW transmission capacity of long OHL is beyond that one of installed capacity of new generating facility even at OHL critical lengths.

The following expression has been applied to estimate a PFCU overall capacity in order to fully compensate the reactive power losses of OHL:

$$Q_{PFCU} = \frac{P_S^2}{U_{nom}^2} x_0 \ell . \quad (2)$$

With reference to the foregoing, a general conclusion may be drawn, i.e. to eliminate a local power system deficiency it would be economically feasible to consider, first, the development of power grid, including the increase of transmission capacity of existing OHL by using PFCU and to build new power transmission capacities when required. Consequently, while estimating and forming multi-zone transport models of EPS it would be useful to consider only transmission constraints for long OHL under conditions of full compensation of reactive power losses, because the increase of transmission capacity of these power grid segments correlates in cost with the construction of new generating capacities.

4 Correlation between facilities ensuring CA in terms of their effect

The conclusion made at the end of previous section does not consider the constraints available at selecting the OHL voltage class and respectively their transmission capacity.

It should be noted that if the cost of new generating facility is practically pro rately dependent on its installed capacity, the OHL cost is mostly determined by its voltage class. The choice of wire type and other equipment affecting the transmission capacity of OHL has a minimum effect on its overall cost. Meanwhile, the choice of voltage class for long OHL is limited by OHL critical length, which is specified, among other things, by losses due to capacity transmission [13].

In order to evaluate the effect of the above factor on the choice of facilities ensuring CA there has been carried out the correlation between alike in cost engineering measures aimed at increasing the installed generating capacity and the transmission capability of power grid on the basis of data given in Table 1. Calculation results are shown on Fig. 2. The OHL length is laid off as abscissa while the installed power of new generating facility is laid off as ordinate. The curves exhibit the conformance between the installed power of new generating facility and the length of OHL alike in cost. Calculations have been performed for OHL with rated voltage of 220, 380 kV (wire type AC-300) and 500 kV (wire type AC-400).

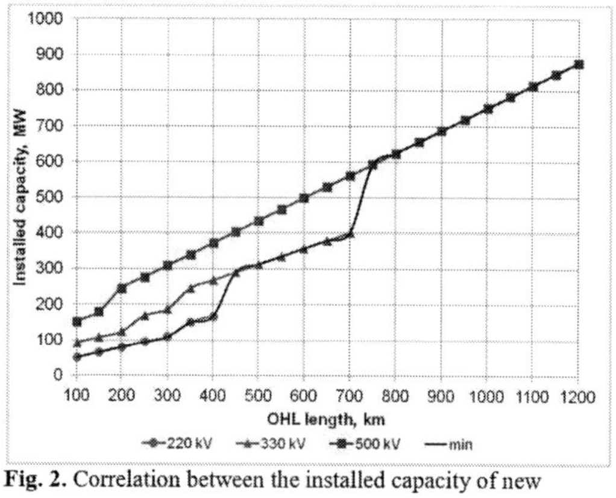

Fig. 2. Correlation between the installed capacity of new generating facility and the length of OHL alike in cost.

The curves on figure 2 are limited to the right by OHL critical length upon data of [13]. Thus, if at OHL length up to 400 km, its minimal cost (voltage class 220 kV) correlates with the cost of 100-150 MW of new generation, then in case of longer distance the construction of new OHL may be justified if the power need makes 300 MW and more. If the required need is below a "minimum" curve shown on figure 2, the construction of new OHL makes no economic sense and it should be reasonable to consider the construction of a new power-producing facility of required capacity.

In this context, when forming multi-zone transport models of EPS it is useful to take into account not only the distance between EPS nodes, but their potential need of margin capacity which may be estimated as being that one for concentrated power system [15]. As far as it concerns remote nodes in case of their small need in margin capacity, it should be reasonable to separate them as specific reliability areas because the increase of power grid transmission capacity towards these nodes may become more expensive as compared to the installation of additional generating facilities therein.

5 Concluding remarks

Based on performed calculations the following recommendations may be proposed to form EPS transport models in order to estimate CAI with account made for the cost of engineering measures ensuring CA:

1. When forming EPS transport models and performing further CAI calculations, one should take into account the transmission capacity of OHL with PFCU use allowing increasing the transmission capacity of power grid and adjusting the load of some its components, as the cost of PFCU is by an order of magnitude lower than that one of new generating facility or new OHL. The rated power capacity and places of PFCU installation may be specified on the basis of calculations of power system states.
2. It should be useful to integrate EPS nodes located within up to 400 km distance (critical length for 220 kV OHL) in one zone of reliability because the cost of transmission capacity growth per 1 kW is greatly lower than that one of new generating facility and the minimum cost of new OHL is comparable to that one of generating facility of average capacity (up to 150 MW).
3. It should be reasonable to integrate EPS nodes (group of nodes) located within 400 – 1000 km in one area of reliability only when their potential need in margin capacity makes more than 300-750 MW (with respect to distance). Otherwise, it makes no sense to consider the construction of new OHL and the transmission constraints between these nodes (groups of nodes) should be considered when estimating CAI.
4. The EPS nodes (groups of nodes) within more than 1000 km are considered as individual reliability areas because the cost of new OHL construction over such distances is comparable to that one of new generating facility of relevant capacity.

References

1. R. Billinton, D. Huang, *Canadian Conference on Electrical and Computer Engineering*, 611 (IEEE, 2005)
2. N.A. Manov, Yu.Ya. Chukreev, N.V. Bobyliova, L.P. Solovieva, Proc. of the RAS. Energ. E **5**, 74 (2009)
3. D.S. Krupenev, L.M. Lebedeva, G.F.Kovalev, N.A. Belyaev, A.Ye. Yegorov, Pv.E. Gromov, Energy policy,E **1**, 33 (2018)
4. W. Li, *Risk Assessment of Power Systems: Models, Methods and Applications* (IEEE Press, New Jork, 2005)
5. Yu.Ya. Chukreev, M.Yu.Chukreev, *Models of evaluation of capacity adequacy indices when managing the development of electric power systems* (Komi scientific centre, Syktyvkar, 2014)
6. G.F.Kovaliov, L.M.Lebedeva; N.I.Voropay, *Reliability of electric power systems* (Nauka, Novossibirsk, 2015)
7. Yu.M. Sharguin, A.L. Koviazin, Ye.Ye. Popov, L.S. Smirnova, Proc. of HVDC Power Transmission Research Institute, E **65**, 66 (2011)
8. N.A. Belyaev, A.Ye. Yegorov, N.V.Korovkin, V.S.Chudny, Power plants, E **11**, 47 (2015)
9. N.A. Belyaev, N.V. Korovkin, V.S. Chudny, Electricity, E **5**, 11 (2018)
10. Methodical provisions with regard to power system stability. Approved by the Order of the Ministry of Energy of Russia dated June 30, 2003, No. 277
11. Regulations concerning power pricing sold upon capacity supply agreements. Approved by the Decision of the Government of the Russian Federation dated April 13, 2010, No. 238
12. Aggregate standards of pricing typical process solutions for major construction works of electricity generation facilities with regard to power supply facilities. Approved by the Order of the Ministry of Energy of Russia dated February 8, 2016, No. 75
13. D.L.Faibisovitch, *Handbook for power grids designing* (NC ENAS, Moskow, 2012)
14. N.G. Hingorani, L. Gyugyi , *Understanding FACTS. Concepts and Technology of Flexible AC Transmission Systems* (IEEE press, New Jork, 2006)
15. I.M. Markovitch, *Power system states* (Energy, Moskow, 1969)

 https://doi.org/10.1051/e3sconf/20185801011

Analysis of additional factors for determining heat network pipelines failure rates

Irina Akhmetova[1,*], and *Timur* Akhmetov[2]

[1]Federal State State-subsidized Educational Institution of Higher Professional Education «KSPEU», Kazan, Russia
[2]JSC «Kazenergo», Kazan, Russia

Abstract. The paper suggests the methodology to assess the reliability of both the heat network schemes under design and the existing heat network schemes, that allows to determine availability factor and failure-free operation probability on the basis of the input data including the length, diameter and exploitation time of pipeline segments of the heat networks. As a result of the research, the additional factors that have a significant impact on the reliability of heat supply to consumers have been determined. The impact of these factors on the heat network reliability relying on the experimental data provided by regional heat supply organizations has been assessed. Additional factors were taken into account when developing a new methodology and algorithm for calculating heat supply reliability indices. The program for calculating the consumer heat supply reliability is worked out.

1 Introduction

The input data for determining the reliability of consumer heat supply systems include failure rates and the average restoration time of heat networks and equipment.

The actual level of reliability in the given heat supply system should be assessed on the basis of the processing of statistical data of the system components failures. Given that statistical samples should meet certain requirements of homogeneity, completeness, and significance [2], the collection of input failure data in each system should be organized in accordance with the form recommended in [3].

If the failure statistics are not used, heat networks failure rate λ considering their exploitation time is calculated on the basis of Weibull distribution [4] at an initial failure rate of 1 km of a single-line heat pipe λ^{in} equal to $5.7 \cdot 10^{-6}$ 1 / (km h) or 0.05 1 / (km year) [1]. The initial failure rate corresponds to the period of normal operation of the new heat network pipelines after the early failure period (shake-down period).

The average failure rate of one shut-off control valve is set equal to $2.28 \cdot 10^{-7}$ 1 / h or 0.002 1/ year [1].

For heat supply schemes designed for cities and urban districts with a total population exceeding 100 thousand people, reliability indices are calculated referring to the generalized consumers nodes. The thermal accumulation rate of the buildings used in this case is that referring to the buildings representative for this buildings category or for the building presenting the worst heat stability.

To assess the reliability of the heat network schemes under design and the existing heat network schemes, PJSC «Gazprom Promgaz» has presented a methodology [6], which allows to determine heat network availability factor and failure-free operation probability rate on the basis of the input data including the length, diameter and exploitation time of heat network pipeline segments.

The disadvantage of this approach is that the calculations do not take into account a number of factors directly affecting the heat network reliability.

The purpose of the given research was to identify and take into consideration some additional factors for determining the heat network failures intensity, as well as suggesting a new methodology for calculating consumers heat supply reliability indices.

The goal is achieved by solving the following tasks:
- reveal the additional factors that significantly affect the consumers heat supply reliability, but are not considered in the approved methodology and algorithm for heat network reliability calculating;
- assess the additional factors impact on the heat network reliability based on the experimental data provided by regional heat supply organizations;
- consider the additional factors impact in the development of a new methodology and algorithm for calculating consumers heat supply reliability indices;
- present a program for calculating the consumers heat supply reliability indices based on the proposed methodology and calculation algorithm.

2 Revealing additional factors not considered in the approved methodology

*Corresponding author: irina_akhmetova@mail.ru

It was assumed that the following factors affect the heat network segment failure rate, besides their lifespan: the residual pipeline wall metal thickness, the presence of other (preceding) failures in the segment, the soil corrosiveness, the presence of channel flooding (traces of flooding), the heat network pipeline material, the presence of intersections with communications, the percentage of pipeline lot failure.

To solve the issue of considering certain factors in a new methodology and algorithm for calculating the consumers heat supply reliability rates, an analysis of failure statistics in various parts of Kazan city heat network was carried out [7].

Failure statistics were grouped according to the presence of certain factors (a combination of factors). A direct impact on the heat network component failure rates was revealed with regard to the following factors: the residual pipeline wall metal thickness (K1), the presence of other (preceding) failures in the segment (K2), the soil corrosiveness (K3), the presence of channel flooding (traces of flooding) (K4), the presence of intersections with communications (K5).

These factors impact on the consumers heat supply reliability was assessed in [5].

Such factor, as pipeline material, was excluded from consideration because up to date failure statistics have been accumulated referring to the steel pipelines only. The present-day construction materials for designing heat network systems have been coming to stay relatively recently and therefore failure statistics concerning polymer pipelines are not present.

During the data analysis, a direct relationship between the segment exploitation period and the heat network lot failure percentage was revealed (Table 1). Since the network segment lifespan is already taken into consideration as the main parameter of heat network segment failure rate calculation, such additional parameter as the heat network lot failure percentage had to be eliminated.

Table 1. Year of heat network segment commissioning and pipelines lot failure percentage

Commissioning year	Pipelines lot failure percentage (average value), %
before 1989	85
1990	85
1994	75
1999	75
2001	60
2002	50
2004	50
2010	10
2012	5

3 Assessment of the additional factors impact on the heat network components failure rates

According to the methodology [6], the heat network components failure rates are calculated as follows:

$$\lambda = \lambda^{init} \cdot (0.1 \ \tau^{expl})^{(\alpha-1)}, 1 / (km\ h) \quad (1)$$

where: λ^{init} - initial failure rate of 1 km of a single-line heat pipeline; t ^{expl} – network segment exploitation period, years; α – coefficient considering the network segment installation and exploitation features:

$$\alpha = \begin{pmatrix} 0{,}8 \text{ for } 0 < \tau^{expl} \leq 3 \\ 1 \text{ for } 3 < \tau^{expl} \leq 17 \\ 0{,}5 \cdot e^{\left(\frac{\tau^{expl}}{20}\right)} \text{при } \tau^{expl} > 17 \end{pmatrix} \quad (2)$$

In order to consider the factors revealed in [1], a special coefficient for additional factors was introduced into the existing methodology:

$$\mathrm{K}i = f(\mathrm{K1; K2; K3; K4; K5}) \quad (3)$$

Heat network components failure rate, taking into account this coefficient, is calculated according to the formula:

$$\lambda = \lambda^{st} \cdot (\tau^{expl})^{\alpha-1} \quad (4)$$

where α - coefficient that takes into consideration the service lifespan of the network segment:

$$\alpha = 0{,}5 \cdot e^{\mathrm{K}_i} = 0{,}5 \cdot e^{f(\mathrm{K1;K2;K3;K4;K5})}. \quad (5)$$

To assess the impact of each coefficient in formula (5), the calculated values of Ki were determined:

$$K_i = \ln(2\alpha) \quad (6)$$

For this purpose, the failure rates for each (i-th) emergency pipeline segment were calculated:

$$\lambda_i = \frac{1}{8760 \cdot \tau_i^{экспл} \cdot L_i}, \quad (7)$$

then, according to the expression obtained by converting formula (4), a special coefficient was determined, considering the heat network segment installation and exploitation features:

$$\alpha = \frac{\log(\lambda / \lambda_{нач})}{\log(\tau_{экспл})} + 1 \quad (8)$$

On the basis of the precalculated values of the coefficient Ki, one can assess the impact of additional factors on the heat network components failure rate.

Assessment of any parameter impact was carried out considering failures groups, united by some common factor (a group of factors), as follows:

1. K1 parameter impact.

At the first stage the failures sample analysis was worked out considering a single parameter - the residual pipeline wall metal thickness. The analysis was carried out by comparing the wall thinning with the

 https://doi.org/10.1051/e3sconf/20185801011

experimental value of the coefficient Ki (Table 2). Wall thinning was calculated according the formula (W.T.):

$$W.T. = \left(1 - \frac{res.thickness}{tot.thickness}\right) \times 100\% \quad (9)$$

Table 2. Comparison of wall thinning with the value of Ki

Wall thinning, %	Coefficient for additional factors *Ki*	Present parameter
11,1	1,017314159	K1
12,5	1,050576511	K1
20,0	1,081818332	K1
27,5	1,133125107	K1
34,3	1,198338195	K1
40,0	1,233266444	K1
52,5	1,291050063	K1
54,3	1,327608694	K1
62,5	1,372850791	K1

While comparing the parameters, a linear dependence of the additional factors concerning coefficient of Ki on the residual pipeline metal wall thickness was revealed (Fig. 1):

$$Ki = 0.00673 \cdot K1 + 0.954 \quad (10)$$

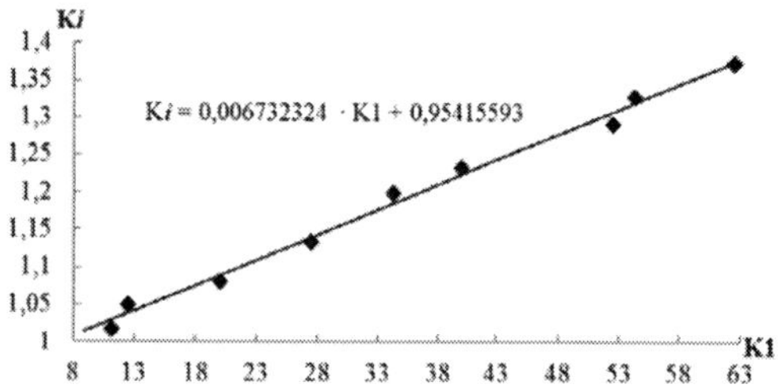

Fig. 1 - The impact of the residual pipeline metal wall thickness on coefficient accounting for additional factors Ki

Since all the remaining cases of heat network component failures include the K1 parameter, the impact of the other factors is determined only by the inclination angle and the shift of this linear correlation line.

2. K1 and K3 parameters impact

At the next stage the failures samples analysis was conducted, which considered two additional factors - the pipeline metal wall thinning and the soil corrosiveness (Table 3).

Table 3. Comparison of wall thinning with the value of Ki

Wall thinning,%	Coefficient Ki	Participating parameter
24,7	1,126785915	K1, K2, K3
25,0	1,149985359	K1, K3
37,5	1,187818695	K1, K3
60,0	1,372850791	K1, K3

In the total statistics volume, a single case of pipeline segment failure had occured, which resulted from a combination of three parameters - K1, K2 and K3. It was decided to supplement the sample with this given single case (Figure 2).

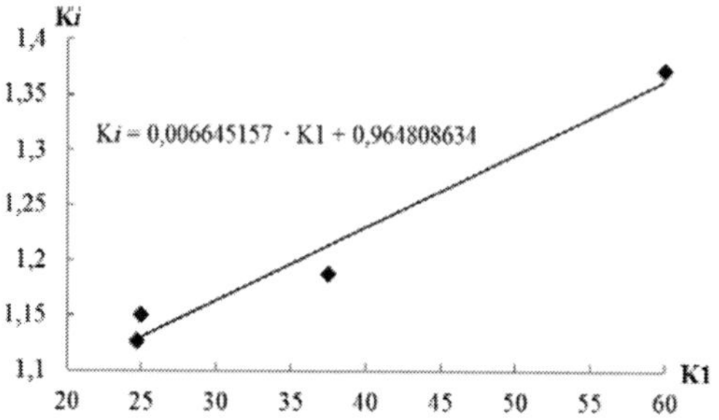

Fig. 2 - K1 and K3 parameters impact on coefficient accounting for additional factors Ki

The functional dependence of the coefficient Ki on the K1 and K3 parameters:

$$Ki = 0.00664 \cdot K1 + 0.964 \quad (11)$$

3. K1 and K4parameters impact.

Further, the analysis of the failure samples was carried out, which includes two additional factors - pipeline wall metal thinning and the presence of channel flooding (traces of flooding) (Table 4).

Table 4. Comparison of wall thinning with the value of Ki

Wall thinning,%	Coefficient Ki	Participating parameter
8,9	0,952114679	K1, K4
11,1	1,007054898	K1, K4
13,3	1,050576511	K1, K4
16,5	1,039035538	K1, K4
17,5	1,039035538	K1, K4
22,5	1,062634197	K1, K4
32,5	1,102538978	K1, K4
40,0	1,133125107	K1, K4
42,5	1,168133183	K1, K4

The functional dependence of the coefficient Ki on the parameters K1 and K4 (Figure 3):

$$Ki = 0.00494 \cdot K1 + 0.949 \quad (12)$$

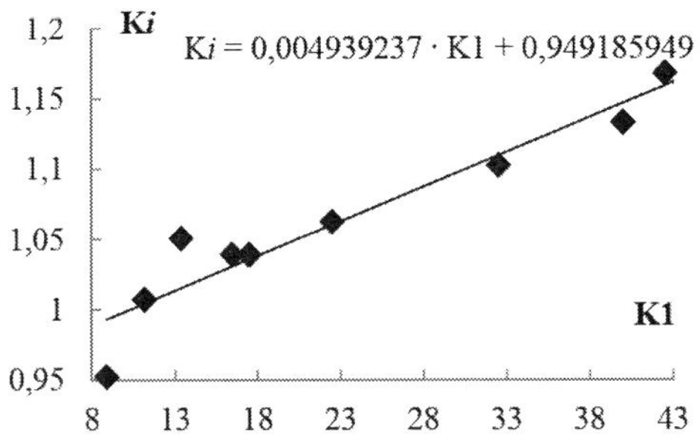

Fig. 3 - K1 and K4 parameters impact on the coefficient accounting for additional factors Ki.

4. K1 and K5 parameters impact

The next sample included the following parameters: the pipeline wall metal thinning and the presence of intersections with communications (Table 5).

Table 5. Comparison of wall thinning with the value of Ki

Wall thinning,%	Coefficient Ki	Participating parameter
15,6	1,050576511	K1, K5
30,0	1,187818695	K1, K5
35,0	1,220994033	K1, K5
47,5	1,260154364	K1, K5
62,9	1,372850791	K1, K5

The functional dependence of the coefficient Ki on the parameters K1 and K5 (Fig. 4):

$$Ki = 0.00641 \cdot K1 + 0.973 \quad (13)$$

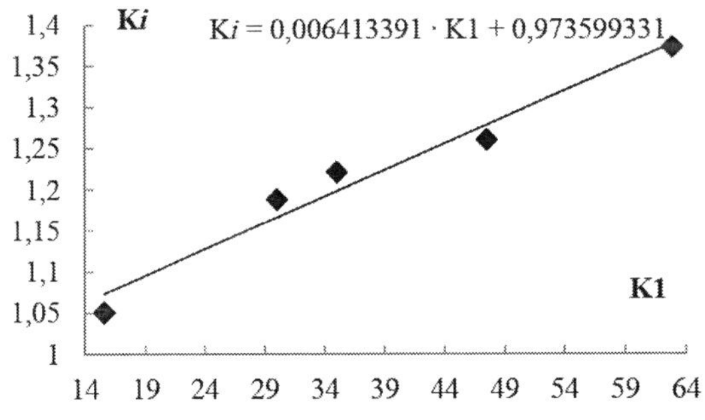

Fig. 4 - K1 and K5 parameters impact on the coefficient accounting for additional factors Ki.

5. K1, K2, K3, K4, K5 parameters impact

The final failures sample considered five parameters at once: the thinning of the pipeline wall metal, the presence of other (preceding) failures in the segment, the soil corrosiveness, the presence of channel flooding (traces of flooding and the presence of intersections with communications (Table 6):

Table 6. Comparison of wall thinning with the value of Ki

Wall thinning,%	Coefficient Ki	Participating parameter
7,5	0,943862731	K1, K2, K3, K4, K5
11,1	1,007054898	K1, K2, K3, K4, K5
35,0	1,134155834	K1, K2, K3, K4, K5
53,3	1,243020534	K1, K2, K3, K4, K5
62,5	1,364564407	K1, K2, K3, K4, K5

The functional dependence of the coefficient Ki on the parameters K1, K2, K3, K4 and K5 (Figure 5):

$$Ki = 0.006889744 \cdot K1 + 0.905045926 \quad (14)$$

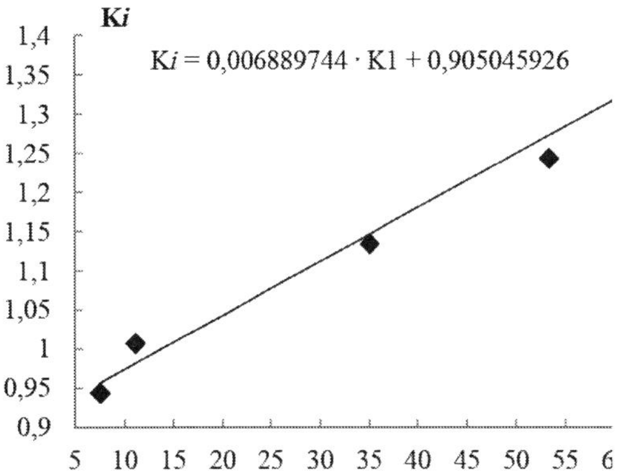

Fig. 5 – the K1, K2, K3, K4 and K5 parameters impact on the coefficient accounting for additional factors Ki

6. Considering the impact of parameters for non-sampled failures.

The remaining single cases of heat network components failure (for example, combinations K1-K2, K1-K3-K5, etc.) that were not included in any sample were taken into account by means of the general functional dependence, plotted on the basis of all experimental values of the coefficient Ki (Figure 6):

$$Ki = 0.00704 \cdot K1 + 0.918 \quad (15)$$

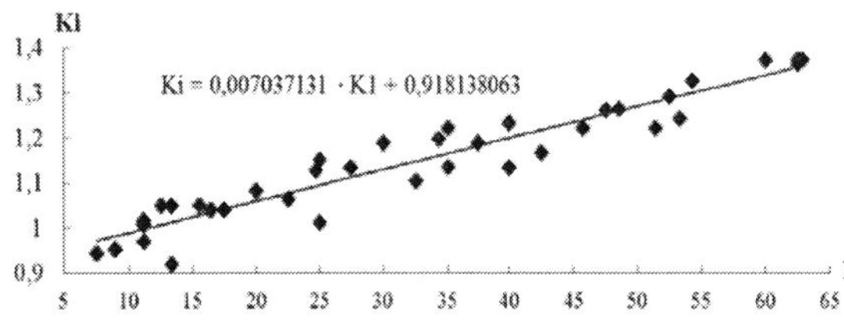

Fig. 6 - General functional dependence for all heat network components failure cases

On the basis of the functional relationships, combined in Table 7, the theoretical values of the coefficient Ki were calculated. Calculated theoretical

https://doi.org/10.1051/e3sconf/20185801011

values of Ki for all cases of heat network segments failure are given in [5].

Table 7. Functions of additional parameters impact

№	Presence of factors	Functional dependence
1	K1	Ki= 0,00673·K1+0,954
2	K1, K3	Ki= 0,00664·K1+0,964
3	K1, K4	Ki= 0,00494·K1+0,949
4	K1, K5	Ki= 0,00641·K1+0,973
5	K1, K2, K3, K4, K5	Ki = 0,00689·K1+0,905
6	Combinations of factors other than options 1-5	Ki= 0,00704·K1+0,918

In Fig. 7 the comparison of the theoretical and experimental values of the factor of additional factors Ki is presented.

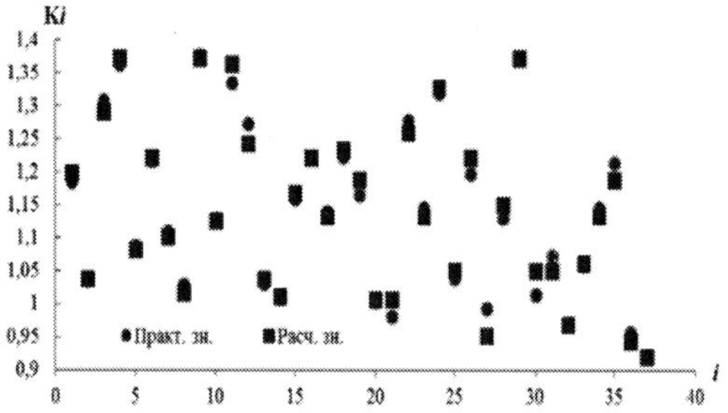

Fig. 7 - Theoretical and experimental values of the coefficient accounting for additional factors Ki.

As can be seen from the graph in Fig. 7, the deviation of the theoretical values from the experimental data is not so significant - the average error does not exceed 1.13%. Thus, the calculations and experimental studies presented in this work have confirmed that the functional dependencies (1) - (15) can be used to calculate the consumers heat supply reliability.

References

1. Reliability of energy systems and equipment: Reference edition in 4 volumes, edited by acad. Yu.N. Rudenko. V. 4 Reliability of heat supply systems / E.V. Sennova, Novosibirsk: Nauka, 2000 - 351 p.
2. Barlow R., Proshan F. Statistical Reliability Theory and Testing for Reliability. - Moscow: Nauka. - 1984. - 328 p.
3. MDK 4-01.2001 "Methodological recommendations for technical investigation and accounting of technological violations in public energy supply systems and the work of energy organizations of the housing and communal services complex" (approved by the Order of the Minister of Construction of Russian Federation dated 20.08.01 No. 191).
4. Gnedenko V.V., Kovalenko I.N. Introduction to the theory of queuing. - Moscow: Science. - 1987. - 336 p.
5. Akhmetova I.G .System of integrated assessment and centralized heat supply efficiency increase for housing and communal services and industrial enterprises: Thesis for doctor's degree. - Kazan: Kazan State University, 2017. - 32 p.
6. Sennova E.V., Kiryukhin S.N., Shimanskaya A.O. Methodology and algorithm for calculating heat supply reliability indicators for consumers and reserving heat networks for the development of heat supply schemes // Heat supply news. - 2013. - No. 12 (160).
7. Gaidyshev, I.P. Solution of scientific and engineering tasks by means of Excel, VBA and C/C++. SPb.: BHV-Petersburg, 2004. - 512 p.
8. Ivan V. Postnikov, Valery A. Stennikov, Ekaterina E. Mednikova, Andrey V. Penkovskii. A Methodology for Optimization of Component Reliability of Heat Supply Systems // Energy Procedia Volume 105, May 2017, Pages 3083-3088
9. Bożena Babiarz. Reliability analysis in subsystem of heat supply // International Conference on Information and Digital, 2016
10. Z. J. Guo, Z. L. Chen. Case investigation in safety and reliability of a family heat supply system // Materials Science and Engineering Technology. September 2017, pages 905-911

Assessment of Power System Adequacy with Renewable Energy Sources and Energy Storage Systems

Dmitry Krupenev[1,*]

[1] Melentiev Energy Systems Institute of Siberian Branch of the Russian Academy of Sciences, ESI SB RAS, Irkutsk, Russia

Abstract. The paper deals with the problem of the accounting of renewable energy sources and energy storage systems in assessment of power system adequacy. Development of renewable energy sources and energy storage systems in the present day power systems is one of the main focuses. In power systems of some countries the share of electric energy generated by renewable energy sources is above 50 % in the energy balance. Therefore, the plans on development of the present day power systems must be elaborated with the proper accounting of operation of renewable energy sources and energy storage systems and the sound capacity reserves in terms of these facilities. The paper presents the algorithms for the accounting of renewable energy sources and energy storage systems. The experimental studies performed illustrate feasibility of the suggested algorithms.

1 Introduction

The present day power systems are characterized by the large-scale integration of renewable energy sources (RESs) and energy storage systems (ESSs). Solar and wind power plants (SPPs and WPPs, respectively) are most widely used among RESs. An extensive nomenclature of ESSs allows the use of this or that energy storage in power systems depending on the external conditions.

The key feature of RESs is their random electricity production, which depends on atmospheric and climatic conditions. Random electricity production at RESs is often weakly connected with its consumption and, therefore, has an adverse effect on electricity supply reliability. Note that sometimes electricity production at WPPs and SPPs will approach zero. As concerns ESSs, they are characterized by some specific features, which also influence electricity supply reliability within power systems.

As far as RESs and ESSs influence electricity supply reliability, they must properly be taken into account in assessment of power system reliability. This work deals with the accounting of the features of RES and ESS operation in adequacy assessment, which is a power system property to satisfy consumer requirements within the specified values and limitations on supplies of energy resources in terms of the planned and unforeseen interruptions in the work of its elements and operating constraints [1]. The accounting of RESs and ESSs in the adequacy assessment will make possible the resolution of such important, practical and critical tasks facing the electric power industry as:

- validation of RES and ESS integration into power systems;
- validation of transmission lines installation to transmit power and energy from RESs to ESSs;
- validation of generating capacity reserve levels of the power systems comprising RESs and ESSs;
- support of the required electricity supply reliability in the systems with RESs and ESSs, and other tasks.

As a rule, the power system adequacy is assessed for prospective power system schemes. This fact imposes certain limitations on the initial data on power systems, and rather, on their completeness. Really, it is not always possible to obtain accurate parameters of the commissioned equipment to be taken into account for power system adequacy assessment for a long term. The situation with the initial data on the renewable energy resource activity is the same: the models of power system adequacy assessment require special algorithms to be applied to obtain data on its activity.

2 Power system adequacy assessment

The current programs for power system adequacy assessment involve the estimation technique based on the Monte Carlo method [2-5]. Moreover, all active programs apply the mechanism of random events, which means independence of all generated power system states. The general scheme of the adequacy assessment technique can be presented in the following form:

1. Generation of independent calculated power system states for the considered period (as a rule, the power system adequacy is assessed for a year). A set of the calculated states is generated for each hour of the considered period with the analysis of the relative error value in the obtained result. The error value may range within 5%, which is acceptable for similar calculations (some uncertainty of the initial information, discreteness

* Corresponding author: krupenev@isem.irk.ru

of the initial information, etc.). The states are generated on the basis of the Monte Carlo method. Each state contains a set of random events: failures of power system facilities, random load fluctuations, as well as scheduled repairs of facilities for the considered period. One random event K_i, $i = 1,...,n$, whose probability is equal to P_i, $i = 1,...,n$, is modeled in generation of one random number x_i, $i = 1,...,n$ from the uniformly distributed set in the interval [0,1] [6]. If x_i is in the interval [0, P_i], the event K is considered to take place, if not, then:

$$K_i = \begin{cases} 1, & если\ x_i \in [0, P_i] \\ 0, & если\ x_i \in (P_i, 1] \end{cases}, \quad i = 1,...,n. \quad (1)$$

Similarly, this procedure can be extended to a greater number of incompatible events, for example, in the analysis of irregular load fluctuations, where the rule of 3σ is used for discrete approximation of the normal distribution law of load deviations and hence, the distribution series consists of 7 steps.

2. Calculation of the conditions in the generated system states. This stage involves the calculation of the power system condition with minimum power shortage for each generated system state. To assess the power system adequacy the mathematical formulation of the problem is.

for assessment of the power shortage in the k-th power system state, $k = 1, \dots, N$, it is necessary to determine:

$$\sum_{i=1}^{n}(\bar{y}_i^k - y_i) \to \min \quad (2)$$

subject to the balance constraints

$$x_i - y_i + \sum_{j=1}^{n}(1 - z_{ji}a_{ji})z_{ji} - \sum_{j=1}^{n} z_{ij} = 0, \ i = 1,..., n, i \neq j, \quad (3)$$

$$y_i \leq \bar{y}_i^k, \ i = 1, ..., n, \quad (4)$$

$$x_i \leq \bar{x}_i^k, \ i = 1, ..., n, \quad (5)$$

$$z_{ij} \leq \bar{z}_{ij}^k, \ i = 1, ..., n, \ j = 1, ..., n, \ i \neq j, \quad (6)$$

$$y_i \geq 0, \ x_i \geq 0, \ z_{ij} \geq 0, \quad (7)$$

where: x_i is the usable capacity at node i, $\bar{x}_i^k$ is the available generating capacity at node i, y_i is the served load at node i, $\bar{y}_i^k$ is the load value at node i, z_{ij} - is the power flow from node i to node j, $\bar{z}_{ij}^k$ is the transfer capability of transmission lines between nodes i and j, a_{ij} - are the given positive coefficients of specific power losses at its transmission from node i to node j, $i \neq j$, $i = 1, \dots, n$, $j = 1, \dots, n$, $i \neq j$, $k = 1, \dots, N$.

3. Calculation of power system adequacy indices. After calculation of all generated states the calculated power system parameters are processed by the statistical methods, yielding a spectrum of adequacy indices. The main calculated adequacy indices are:
- the probability of failure-free operation;
- the expectation of electricity undersupply;
- the expectation of power shortage;
- the energy index of reliability;
- the probability of operating parameter deviation beyond the maximum admissible values.

3 Modeling of RESs for power system adequacy assessment

Wind energy is most rapidly developing among RESs to day. In 2017 the total generating capacity of WPPs in the world amounted to more than 539 GW [7]. In many countries the wind energy share in the energy balance exceeds 20 %. In 2018 it is planned to put into operation above 3 GW of wind generation in Russia by 2023 [8]. WPPs can operate both independently of the power system and in parallel with it. The central task for the latter existing and designed WPPs is their modelling for power system adequacy assessment.

WPPs are modelled using the approach described in [9]. In this approach, at the stage of forming the calculated power system states the possible failures of wind turbines were modelled at the first step and the random value of the wind speed that determines the generating capacity of wind turbines remained in operation was generated at the second step.

This study suggests a single-step procedure for determination of WPP capacity for generation of a random power system state. For this purpose prior to start the procedure of generating a random state, the distribution series of the state (operable and inoperable) of wind turbines and the wind speed are composed [2]. Then, the power generated by WPP at the specific time point is determined on the basis of the resulting series of WPP generation.

The WPP modeling algorithm for the power system adequacy assessment is as follows:

1. Generation of the series of operable and inoperable states of wind turbines at the considered WPP as a function of the emergency rate g_a of the considered wind turbines:

$$(p_a[X] + g_a[0])^o = 1, \quad (8)$$

where p_a is the probability of failure-free operation of a wind turbine; o is the number of homogeneous wind turbines at the considered WPP; X is the rated capacity of a turbine, kW.

If the considered WPP has several types of wind turbines (turbines of different capacity or with different reliability and engineering characteristics), then one series is obtained by multiplication of the series of operable state of turbines.

2. Composition of the hourly series of wind activity in the region with the installed or planned WPPs. The adequacy can be assessed by generation of the distribution series of wind speed in the considered region on the basis of the mean multi-year series, taking into account the global cyclical climate change in the considered regions, if possible.

The wind activity series for the considered region is generated by processing the hourly data on wind speeds at least for 10 previous years, yielding the distribution series of the wind speed $p_v[V]$ for the considered region in the form:

$$p_v[V_1]+p_v[V_2]+...+p_v[V_{10}]=1. \qquad (9)$$

We will limit to 10 wind speed steps in the considered territory.

3. Transformation of the obtained wind activity series into the power generation series, replacing the corresponding wind speed values by generated power at this speed (dependence of the generated power on the wind speed is given in the nameplate data of the wind turbine):

$$p_v[X_1]+p_v[X_2]+...+p_v[X_{10}]=1. \qquad (10)$$

4. Multiplication of distribution series (8) and (10).

5. Formation of the random generating capacity state for WPP and other power system equipment by the random number generator.

The coefficient of wind speed increase depends on the wind turbine height. One more important factor to be taken into account in modeling of the hourly wind intensity is the variation of wind activity in different regions during a day, and the hourly statistics will consider these variations.

4 Modeling of SPPs for power system adequacy assessment

Solar energy similarly to wind energy is dynamically developing worldwide. In 2017 the installed capacity of SPPs on photovoltaic cells in the world amounted to 402 GW [6]. Another type of SPPs is the so called plants concentrating solar thermal power (CSP), whose installed capacity in 2017 made up about 5 GW in the world [6]. In about 150 MW to be installed before 2022 [8].

In modelling of SPPs for power system adequacy assessment it is also necessary to take into account the following two main factors: non-delivery of power by the SPPs due to technical problems caused by failures of their elements and non-delivery of power due to insufficient intensity of solar radiation.

We suggest the following algorithm for SPP modelling:

1. Determination of the distribution series of operable and inoperable states of photovoltaic and concentrated solar cells comprising a SPP:

$$(p_s[X]+g_s[0])^r=1. \qquad (11)$$

where p_s is the probability of failure-free operation of one solar cell; g_s is the probability of failure of one solar cell; r is the number of homogeneous solar cells.

If the considered SPP consists of several types of solar cells (panels) with different capacities and probabilities of failures, the common distribution series of SPP operability is obtained by multiplication of the series.

2. Determination of the hourly series of solar radiation intensity. As far as the adequacy is assessed for a long-term period and sometimes for a very long-term period (above 20 years), it is valid to apply the data on the mean multi-year values of solar radiation in the considered territory. In this case the distribution series of solar radiation for an individual hour can be of the form:

$$p_s[R_1]+p_s[R_2]+...+p_s[R_{10}]=1. \qquad (12)$$

where $p_s[R]$ is the probability of solar radiation intensity.

3. Transformation of the obtained series of solar radiation intensity into the distribution series of power generation, replacing the corresponding solar radiation intensities by the power generated at this intensity:

$$p_s[X_1]+p_s[X_2]+...+p_s[X_{10}]=1. \qquad (13)$$

In generation of series (13) account is taken of both the solar radiation intensity and the ambient temperature. Dependence of the power generated by the solar panel on the ambient temperature is presented in the nameplate data of solar panels.

3. Multiplication of the distribution series of operable and inoperable states of photovoltaic and concentrated solar panels and solar radiation intensity (11) and (13).

4. Generation of the random generating capacity state of SPP using the random number generator.

5 Modeling of ESSs for power system adequacy assessment

The nomenclature of ESSs integrated into the present day power systems is sufficiently extensive. Traditionally, the energy storage practice is based on the diverse methods and materials intended basically for storage of thermal and mechanical energy and subsequent conversion into electric energy. For example, the superheated steam storage tanks between the boiler and turbine are used at thermal power plants to level off the steam flow rate, when the turbine generator load is highly non-uniform over time; hot water storage tanks at cogeneration plants to provide their more uniform load during daily heat consumption fluctuations, etc.

There exist a fairly large number of diverse electricity storage units and they are steadily developing in both technology and scale.

This work deals with large power interconnections, where it is expedient to consider operation of large

energy storage systems, participating in maintaining the energy and power balances and hence, in ensuring the required level of power system adequacy. The main characteristics of an energy storage unit from the viewpoint of its participation in these balances and the power shortage coverage are:
- the maximum capacity of an energy storage unit;
- the time of energy storage unit charge;
- the characteristic of power supply to the system by the energy storage unit depending on the residual charge;
- the energy storage unit availability to supply some part of power to the system at any time instant.

The power system adequacy is assessed in the work using the mechanism of random events that is based on the assumption of independence of all considered events. Hence, in the energy storage unit modelling to assess the power system adequacy it is necessary to take into account that:
1. In the power shortage conditions the energy storage unit must supply a maximum possible power to the power system to cover the shortage, i.e. participate as an electric energy generator;
2. In the shortage-free conditions the energy storage unit must accumulate energy, i.e. participate as an electric energy consumer;
3. Possible failures of energy storage equipment must be taken into consideration in generation of random power system states.

Thus, for power system adequacy assessment the energy storage system is specified by an additional generating unit or a group of generating units, if the energy storage system, for example, the pumped storage plant (PSP), consists of several units. It is regarded that the energy storage system involves all available resources to cover the shortage occurred in the system.

6 Experimental studies

The work of the suggested models is illustrated by the model of the interconnected power system (IPS) of Siberia with changed parameters, which is presented in Fig. 1. IPS of Siberia is clustered into the reliability zones in accordance with the division of the Russian Federation into its entities, except for the Bodaibo load center that is referred formally to the Irkutsk region and factually is connected with the Buryatia power system.

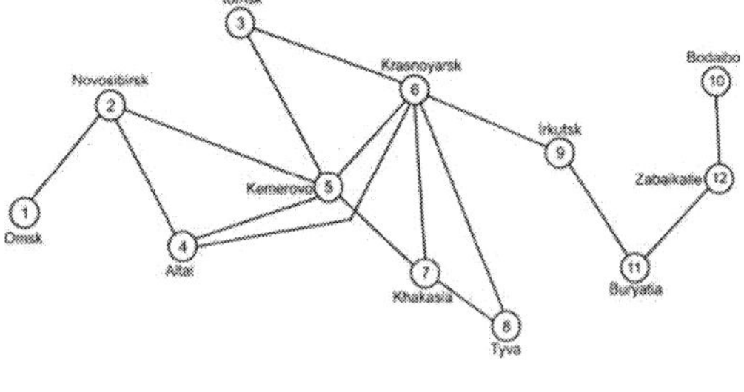

Fig. 1. Model of IPS of Siberia for adequacy assessment.

The results of operation of WPPs, SPPs and ESSs are illustrated by the characteristics of IPS of Siberia to determine the excess generating capacity, i.e. the generating capacity in the reliability zones is taken to be not real, but reduced. Table 1 presents the characteristics of the reliability zones in the calculation model of IPS of Siberia.

Table 1. Characteristics of the reliability zones of IPS of Siberia.

№ zone	Zone name	Annual peak load, MW	Available capacity, MW
1	Omsk	1782	1479
2	Novosibirsk	2690	2730
3	Tomsk	1302	918
4	Altai	1884	1444
5	Kemerovo	4535	4462
6	Krasnoyarsk	6235	6848
7	Khakasia	2155	5025
8	Tyva	152	40
9	Irkutsk	7570	9048
10	Bodaibo	90	20
11	Buryatia	945	898
12	Zabaikalie	1260	1156

Table 2 demonstrates the transfer capabilities of the interzonal tie lines in the calculation model of IPS of Siberia.

Table 2. Transfer capabilities of the interzonal tie lines of IPS of Siberia.

№ tie line	Connected zones	Capability of tie line, MW
1	1. Omsk – 2. Novosibirsk	1305
2	2. Novosibirsk – 4. Altai	1440
3	2. Novosibirsk – 5. Kemerovo	950
4	3. Tomsk – 5. Kemerovo	1170

5	3. Tomsk – 6. Krasnoyarsk	780
6	4. Altai – 5. Kemerovo	950
7	4. Altai – 6. Krasnoyarsk	850
8	5. Kemerovo – 6. Krasnoyarsk	1560
9	5. Kemerovo – 7. Khakass	1650
10	6. Krasnoyarsk – 7. Khakass	3400
11	6. Krasnoyarsk – 8. Tyva	135
12	6. Krasnoyarsk – 9. Irkutsk	3630
13	7. Khakasia – 8. Tyva	135
14	9. Irkutsk – 11. Buryatia	885
15	10. Bodaibo – 11. Buraytia	66
16	11. Burayta – 12. Zabaikalie	410

Let us locate SPPs in the Omsk power system, WPPs in the Novosibirsk power system and ESSs as PSPs in the Altai power system. The main parameters of the located objects are presented in Table 3.

Table 3. Main parameters of WPPs, SPPs and ESs.

Reliability zone	Type of located object	Model x pcs..	Total capacity, MW
Omsk	SPPs	Eagle 72P 340 W x 500000	170
Novosibirsk	WPPs	ENERCON E-138 EP3 / 3,5 MW x 50 [10]	175
Altai	PSPs	POHT-115 x 2	230

The results of a series of calculations: without commissioning of SPPs, WPPs and PSPs; with commissioning of only SPPs; with commissioning of only WPPs; with commissioning of SPPs, WPPs and PSPs are presented in Table 4.

Table 4. Calculation results.

Zone name	Without commissioning	
	Probability of failure-free operation	Expectation of generation reliability, MWh
Omsk	0.99874	1826
Novosibirsk	0.99994	64
Tomsk	0.9996	455
Altai	0.9985	3167
Kemerovo	0.99999	0.0
Krasnoyarsk	0.99978	440
Khakasia	0.99999	0.0
Tyva	0.99742	2075
Irkutsk	0.99904	2321
Bodaibo	0.99525	1050
Buryatia	0.99714	2669
Zabaikalie	0.98324	14994
System	0,98171	29061
	SPPs	
	Probability of failure-free operation	Expectation of electricity undersupply, MWh
Omsk	0.99946	550
Novosibirsk	0.99993	60
Tomsk	0.99974	239
Altai	0.99924	1501
Kemerovo	0.99999	0.0
Krasnoyarsk	0.99978	368

https://doi.org/10.1051/e3sconf/20185801012

Khakasia	0.99999	0.0
Tyva	0.99866	1099
Irkutsk	0.99937	1074
Bodaibo	0.99656	674
Buryatia	0.9982	1712
Zabaikalie	0.98521	12792
System	0.98386	20069
WPPs		
	Probability of failure-free operation	**Expectation of electricity undersupply, MWh**
Omsk	0.99894	1201
Novosibirsk	0.99995	14
Tomsk	0.99956	726
Altai	0.99861	2907
Kemerovo	0.99999	0.0
Krasnoyarsk	0.99962	697
Khakasia	0.99999	0.0
Tyva	0.99768	1902
Irkutsk	0.99891	2245
Bodaibo	0.99567	983
Buryatia	0.99745	2403
Zabaikalie	0.98286	14727
System	0.98138	27805
SPPs, WPPs, PSPs		
	Probability of failure-free operation	**Expectation of electricity undersupply, MWh**
Omsk	0.99974	227
Novosibirsk	0.99996	10
Tomsk	0.99961	541
Altai	0.99965	542
Kemerovo	0.99998	7
Krasnoyarsk	0.99977	410
Khakasia	0.99999	0.0
Tyva	0.99913	732
Irkutsk	0.99947	739
Bodaibo	0.99752	474
Buryatia	0.99897	924
Zasbaikalie	0.98568	11634
System	0.98449	16240

Table 4 shows that commissioning of SPPs, WPPs, PSPs in the power systems had a positive effect on the adequacy indices.

7 Conclusion

The paper studies a topical problem of the modern electric power industry, namely the modelling of SPPs, WPPs and ESSs for assessment of power system adequacy. Development of modelling principles of these objects is important, since the power generation at WPPs and SPPs depends on the random character of natural phenomena, namely wind activity and solar radiation in the region. As to ESSs, the paper studies the principles of ESS accounting for adequacy assessment by the Monte Carlo method using the mechanism of random events. The testing of the suggested models and principles showed their efficiency.

References

1. *Handbook on terminology in electric power industry*. (M.: Printing house "KEM", 2008) (in Russian)
2. Kovalev G.F., Lebedeva L.M. *Reliability of electric power systems*. (Novosibirsk: Nauka, 2015) (in Russian)
3. Chukreev Yu.Ya. *Models for electric power system reliability support*. (Syktyvkar, 1995) (in Russian)
4. Wenyuan Li, *Probabilistic Transmission System Planning* (Wiley-IEEE Press, 2011)
5. https://www.geenergyconsulting.com/practice-area/software-products/mars.
6. Krupenev D.S., Boyarkin D.A., Iakubovskii D.V. Generation of random states of electric power systems for their reliability assessment by the Monte Carlo method. Reliability and safety in energy. V.10. No. 1. (2017) (in Russian)
7. *Renewables 2018 global status report* (http://www.ren21.net/wp-content/uploads/2018/06/17-8652_GSR2018_FullReport_web_-1.pdf)
8. *List of the selected RES projects* (https://www.atsenergo.ru/vie/proresults)
9. Krupenev D.S., Perzhabinsky S.M. Reliability assessment of electric power systems with wind power plants , Bulletin of the Russian Academy of Sciences. Energy. No. 2. (2017) (in Russian)
10. https://www.jinkosolar.com/index.html?lan=en
11. https://www.enercon.de/en/home/

 https://doi.org/10.1051/e3sconf/20185801013

An analysis of dynamics of a change in the electricity intensity of the economy of the country and eastern regions and the forecasting of electricity consumption in the long term

Anatoliy Korneev[1,*]

[1] Energy Systems Institute of Siberian Branch of the Russian Academy of Sciences, 664033 Lermontov str., 130, Irkutsk, Russia

Abstract. The paper presents a retrospective analysis of the trends towards changes in the electricity intensity of the economy of Russia and its eastern regions and assesses the dynamics of its change in the long term. The macroeconomic forecast of Russia's economic development is given and the forecast of electricity consumption is made for the country and its eastern regions.

1 Introduction

The eastern regions of Russia, including the Entities of the Russian Federation in the Siberian Federal District (SFD), with focus on the regions of Eastern Siberia, and the Entities of the Russian Federation in the Far Eastern Federal District (FEFD), have a rich natural resource potential for economic development. This is, first of all, mining and processing industry, that requires advanced development of production and social infrastructure. The development of such an infrastructural industry as electric power industry occupies a special place in terms of power supply to electricity intensive productions. In this regard, it is essential for the eastern regions to forecast their electricity needs. The level of electricity consumption will depend on the growth rates of their economy and on the electricity intensity of the production.

The electricity intensity of the products (specific electricity consumption) is the most important parameter for all prospective forecasts of electricity consumption: short-term, medium-term and long-term. Depending on the forecasting period and the quantitative composition of industries, the electricity intensity may have a different degree of aggregation. In the models of forecasting the economy and power consumption with a large number of activities, which is typical of short- and medium-term periods, the prospective electricity intensity is estimated according to the types of activity. This allows us to more specifically consider the influence of various factors on the level of the prospective electricity intensity of a particular activity. For the long-term forecasting, however, and especially for the periods of 30 or more years, the economic activities can be aggregated up to the production of GRP due to the high uncertainty of the impact of various factors on the electricity intensity [1]. In this case, electricity intensity *(e)* is calculated as a ratio of the total electricity consumption *(E)* by the economy to the total volume of the produced gross regional product *(P)* in year *t*:

$$e(t)=E(t)/P(t)$$

The macroforecast in comparison with the detailed multi-sectorial forecast of gross regional product has both advantages and disadvantages. However, many of the macroforecast disadvantages associated with the high aggregation of the economy and its electricity intensity are largely overcome by constructing the scenarios for the future economic development based on retrospective trends and their prospective assessment.

2 Scenarios for the national economic development for the long term future

Based on the state and trends in the national economic development over the past 15 years, we can consider two main scenarios for the long-term economic development: a 'no-change' scenario (I) and a moderately optimistic scenario (II). In the 'no-change' scenario, the economy of Russia will develop under the following conditions:

- orientation to the external raw-material market;
- various types of sanctions associated with geopolitical and economic competition with the leading developed countries;
- world financial and economic crises, recurring approximately every 8-10 years, caused, as a rule, by the interaction of various money and credit factors [2, 3];
- limited internal and external investment and their unstable growth rates due to recurring world crises;
- a slow, sluggish transition to effective technologies and innovative implementations that require large investment.

The main difference between the moderately optimistic scenario and the 'no-change' scenario is that the former is oriented to real structural and production shifts towards an increase in the amount of products for the domestic market, which provides import substitution.

* Corresponding author: korn@isem.irk.ru

 https://doi.org/10.1051/e3sconf/20185801013

Another important condition for this scenario is the development of a stable and accessible domestic investment market that ensures intensive innovative development of the real sector of the economy. However, based on past experience, especially in 2007-2015, when, at high world prices for energy resources, favorable investment conditions for internal innovation development and restructuring of the real economy were missed, which led to the significant economic stagnation in Russia, which has been observed in recent years, the most likely scenario for the future economic development, at least in the period before 2035, is the 'no change' scenario. This scenario is of the greatest interest from the point of view of not optimistic (desirable), but minimum possible social and economic expectations, therefore, here the forecasting of the economic development is focused on this scenario.

3 The preconditions for the macroeconomic forecast of the electricity consumption in the country and its eastern regions according to the 'no-change' scenario

The basic condition for the macroeconomic forecast of the gross regional product of the country according to the 'no change' scenario is the unstable nature of the dynamics of the investment growth rates associated with the cyclicality of crises. This, accordingly, leads to an unstable growth rates of GRP production and to their relatively low average annual values.

The base year for the forecast is 2015. A general scheme of macro-forecasting of the economy and power consumption is shown in Fig. 1.

INVESTMENT IN THE CAPITAL ASSET OF RUSSIA: $I(t)=j(t)*I(t\text{-}1)$, ($t = \overline{2015,...,2050}$), where:
$I(t)$ – investment in the fixed assets in the year t; $j(t)$ – investment growth index in the year t

↓

GROSS REGIONAL PRODUCT OF RUSSIA: $P(t)=k*I(t)$, where:
$P(t)$ – gross regional product of the country in the year t; k – coefficient of rate of return in terms of GRP ($k=P/I$),rub.
GROSS REGIONAL PRODUCT BY FEDERAL DISTRICT: $P_{fd}(t)=s_{fd}*P(t)$, where:
$P_{fd}(t)$ – *gross regional product of federal district in the year t;*
s_{fd} – share of the federal district in Russia's GRP, %

↓

RUSSIA'S GRP ELECTRICITY INTENSITY: $e(t)=j_e(t)*e(t\text{-}1)$, kWh/rub, where:
$e(t)$ – GRP electricity intensity in the year t; $j_e(t)$ – *index of decline in electricity intensity of Russia's GRP*

↓

ELECTRICITY CONSUMPTION IN RUSSIA: $E(t)=e(t)*P(t)$, where:
$E(t))$ – electricity consumption of the country in the year t, *kWh*

↓

ELECTRICITY CONSUMPTION OF RUSSIA'S EASTERN REGIONS: $E_{fd}(t)= s_{fde}*E(t)$ *where:*
$E_{fd}(t)$ – electricity consumption of the Federal District in the year t, *kWh;*
s_{fde} – share of the Federal District in electricity consumption of the country (%)

Fig. 1. A General scheme of macroforecasting of the economy and electricity consumption in the country and its eastern regions

The calculation of prospective indicators of the national GRP production is linked to the investment through the investment return coefficients. Their value in dynamics has a small spread: in 2010 - 2016 it was within a range of 5.7-6.2 rub. [4].This makes it possible to use its average value for the GRP forecasting (Fig. 2).

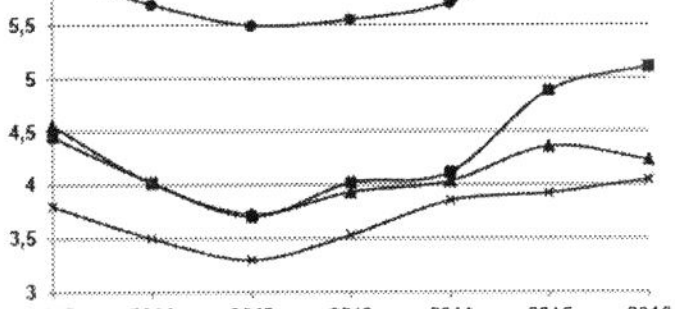

Fig. 2. The dynamics of investment return for GRP

Fig.3 demonstrates the dynamics of Russia's GRP production until 2050 which, like investment, is affected by crises. Despite the high growth rates of GRP in the phases of "recovery" during the crisis period (104-108%), their average annual value from 2015 to 2050 is approximately 102.5%.

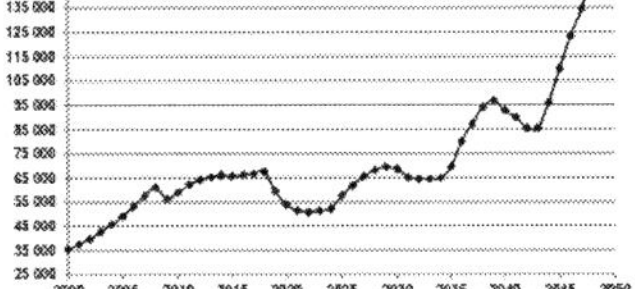

Fig. 3. The national GRP forecast

The observed stable structure of Russia's GRP by Federal District of the eastern regions allows us to assess the prospective levels of regional GRPs by means of a

 https://doi.org/10.1051/e3sconf/20185801013

share-based disaggregation of the total GRP produced in the country (Figures 4, 5).

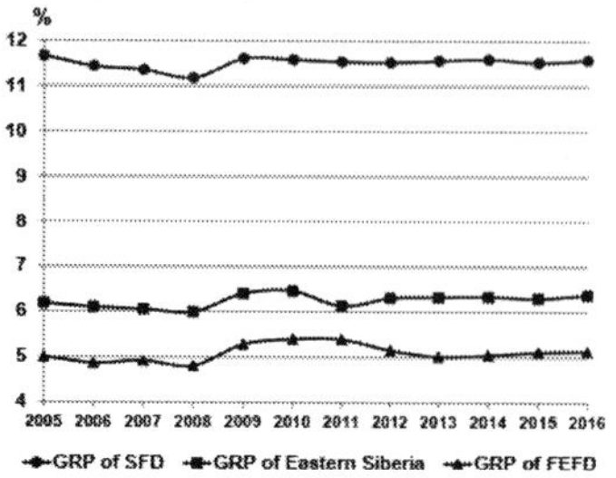

Fig. 4. Share of eastern regions in the GRP of the country

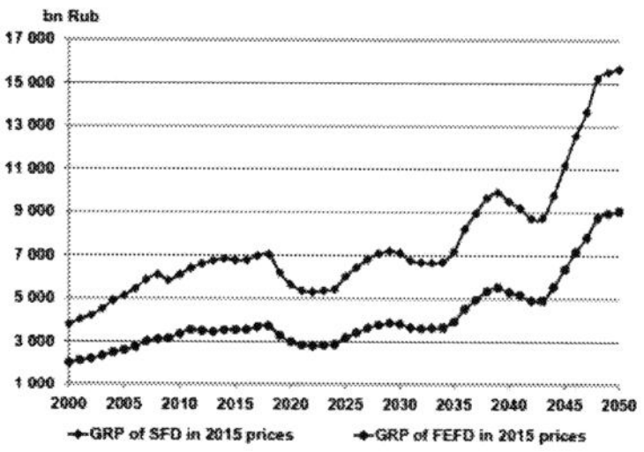

Fig. 5. GRP of the eastern regions

4 The dynamics of GRP electricity intensity and the factors that affect it

The level and dynamics of indicators of the electricity intensity of the production of various products and, accordingly, GRP depend on many factors. The main of them are:

- The improvement in production technologies (for example, according to some estimates, a 1% increase in spending on purchase of new equipment and advanced technologies by individual enterprises contributes to a decrease in the electricity intensity of production by 0.24% [5]);
- The organizational and technological measures for energy conservation both in production, and in the service sector;
- The change in the structure of production in terms of an increase in the share of commodities produced by manufacturing industries, and especially the production of machinery, equipment, instruments, space and military equipment, etc., that are characterized by high added value and low energy intensity of production, or vice versa – an increase in the share of energy intensive products (metallurgy, mining and beneficiation of minerals, chemical productions, etc.);
- The dynamics of and relationship between the prices for energy resources (it has an indirect effect associated with the transition to alternative energy resources or the need for energy-saving measures);
- The level of loading of production capacities and degree of wear and tear of fixed assets;
- The investment resources for upgrading of production, for technological innovations and energy saving, etc.

The electricity intensity of the national and regional GRP is a complex macro-indicator that reflects the cumulative effect of many social and production factors. Whereas for an individual enterprise we can quite accurately determine the influence of each factor on the electricity intensity of the products produced by it, for the aggregate GRP electricity intensity, this is problematic, and, firstly, because of the incompleteness of all the necessary information. However, on the basis of available information and analysis of the most significant factors affecting the electricity intensity, their approximate contribution to the formation of the GRP electricity intensity and the trends in its changes, is assessed both in retrospect and in prospect.

The most important period for the analysis of the dynamics of GRP electricity intensity of Russia and its eastern regions is the last 15 years that reflect the market development of the economy, when economic entities have had an investment resource for the development and the sales markets have largely been determined. In these years, high world prices for energy resources and the flow of investment, both from the private and the budget sectors, have become an important factor in the development. The revival of production and economic activities in the country has also led to an increase in its effectiveness. An important indicator of this is the dynamics of the GRP electricity intensity in the country as a whole. The period of 8 years from 2000 to 2008 saw its decrease by 32%, i.e. on average by 4.7% for each year. In 2009-2016, that is, from the year of the global financial crisis, the rate of decline slowed sharply and amounted to only 6.9%, and on average 1.2% yearly. Similar dynamics of the change in the GRP electricity intensity for the first 8-year period was typical of the Far East and Eastern Siberia. In the second period, in the Far Eastern Federal District, there was an increase instead of a decrease in the electricity intensity: by 9.5% - over the period, and by 1.3% - on average yearly (Fig. 6).

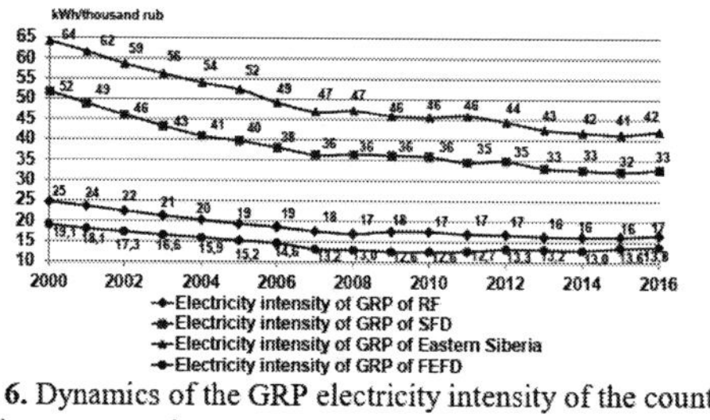

Fig. 6. Dynamics of the GRP electricity intensity of the country and its eastern regions

In 2008, the electricity intensity of the national GRP was 16.9 kWh/rub which is by 7.8 kWh/rub lower compared to 2000. In the crisis year 2009, it rose to 17.5 kWh/rub, by 2013 it amounted to 16.3 kWh/rub, and has remained the same for all past years. This is explained by the crisis state of the economy and the influence of structural and production factors: a decrease in the capacity utilization, a decrease in the investment in the energy saving measures, the development of the energy intensive industries in the eastern regions, etc. In the emerging internal and external conditions, the cardinal changes towards favorable conditions are not expected until 2025-2030. After 2030-2035 only, there can be a moderate revival of the investment activity in the country, which will have a positive effect on the dynamics of GRP electricity intensity. According to optimistic estimates, the GRP electricity intensity by 2050 will decrease to about 10 kWh/rub, and according to more cautious forecasts - up to 12.5 kWh/rub.

5 Forecast of electricity consumption in the country and its eastern regions

The estimates of the electricity intensity dynamics were used to make a long-term forecast of the national electricity consumption up to 2050. In 2050, its value may reach 1575 billion kWh, which is approximately 1.5 times the volume of electricity consumption in 2015. Electricity consumption in the eastern regions was estimated in terms of the unchanged shares of electricity consumption in the country (Fig. 7).

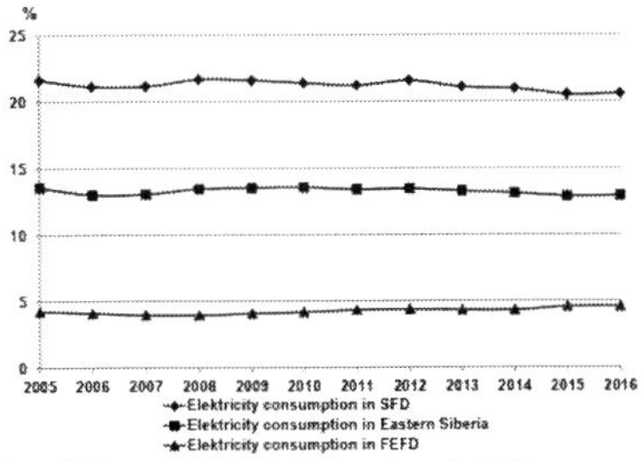

Fig. 7. Electricity consumption structure in RF

In the Siberian Federal District, the electric power consumption will amount to 315 billion kWh by 2050, which is 1.4 times the volume of 2015, in Eastern Siberia - 215 billion kWh (1.6 times), and in the Far Eastern Federal District - 90 billion kWh (by a factor of 1.9) (Fig. 8).

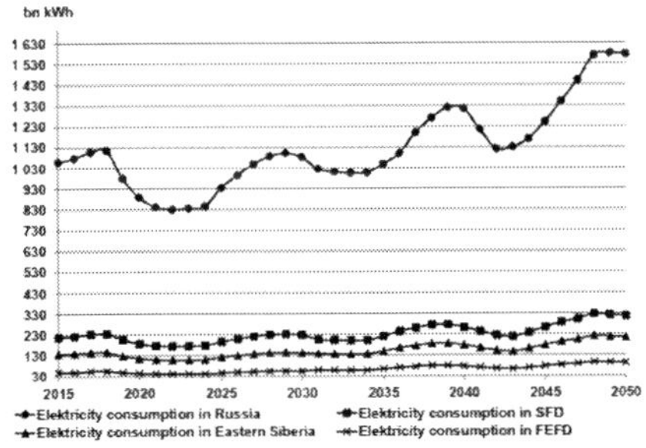

Fig. 8. Electricity consumption of RF and its eastern regions

Fig. 9 presents the data of the official forecast from the draft Energy Strategy of 2017 to compare the indicators of the electricity consumption forecast for the country and the eastern regions, given cyclical crises [6].

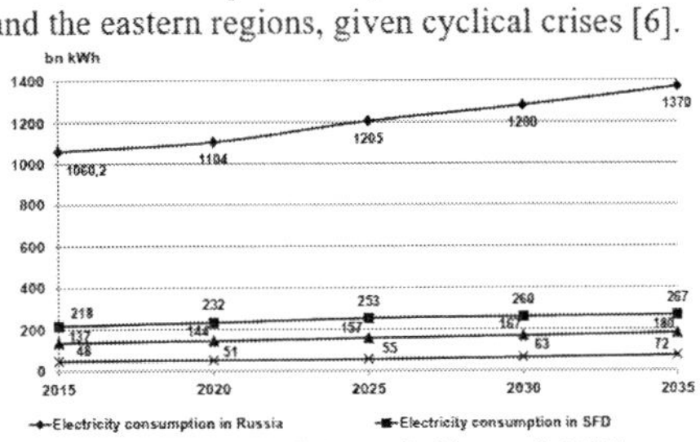

Fig. 9. Forecast of electricity consumption of the country and its eastern regions from the draft Energy Strategy of 2017

As evidenced from a comparison, the official forecast gives very optimistic levels of electricity consumption, which is poorly related to the current and forthcoming unfavorable conditions for the economic development of the country. According to the macroforecast from the 'no change' scenario presented in this paper, the level of official electricity consumption in the country in 2035 is reached only by 2045.

6 Conclusion

The forecast level of electricity consumption in Russia and its eastern regions according to the 'no change' scenario can be used as a guide for the development of State and non-State Strategies and Programs for the long-term development of the national and regional energy sector, and its individual industries.

The work is done in the framework of the project III.174.2.2 program of fundamental research SB RAS, registration number AAAA-A17-117030310435-0

References

1. Kononov Yu.D. Ways to increase the validity of long-term forecasts of the energy development. Novosibirsk: Nauka, 2015, p. 34-36.
2. When to wait for a new world crisis. [Online]. Available: https://storm100.livejournal.com/3541359. html (Accessed June 27, 2018).
3. Types of cycles. Economic crises and their classification. [Online]. Available: https://storm100.livejournal.com/3541359.html (Accessed June 27, 2018). – p. 5.
4. Korneev A.G. Monitoring of the current state of long-term forecasts of electricity consumption in the eastern regions of Russia and methodological issues of their improvement // Methodological problems in reliability study of large energy systems: Collected papers/ Ed. by N.I. Voropai, M.A. Korotkevich, A.A. Mikhalevich. Minsk: Publishing house BNTU, 2015. Issue 66, p. 125-133.
5. Anisimova T.Yu. Influence of various factors on the achievement of strategic goals to reduce energy intensity of the domestic economy. [online] Available: https://kpfu.ru/docs/F929450089/2_Anis2.pdf (Accessed June 28, 2013), p. 4.
6. Draft Energy Strategy of the Russian Federation for the period up to 2035 (revised as of 01.02.2017). [online]. Available: https://minenergo.gov.ru/node/1920 (Accessed June 28, 2013).

Optimization of joint operation of district and distributed heat sources for effective and reliable heat supply to consumers

Ivan Postnikov[1,*], *Valery* Stennikov[1], *Andrey* Penkovskii[1]

[1]Melentiev Energy Systems Institute of Siberian Branch of the Russian Academy of Sciences, Laboratory of Heat Supply Systems, 664033 Irkutsk, Russia

Abstract. The problem of optimal load distribution between district and distributed heat sources (heat sources of prosumers) in the district heating systems are considered. The methodical approach and mathematical model for solving this problem are proposed. They are based on methods of the theory of system energy researches, theory of industrial organization, theory of hydraulic circuits, and basic laws of cogeneration. The case study results obtained by using the developed methodology are presented. The potential economic effect of the prosumer adoption under specified conditions is demonstrated. The conclusions and directions for further research are formulated.

1 Introduction

Modern energy systems are developing according to an integration paradigm. Its main idea is to integrate various energy technologies into single multi-resource super-systems with intelligent control [1–3]. The issues of establishing integrated intelligent power systems (Smart Grid) are the subject of a large number of studies that form a theoretical framework for their successful practical implementation. A rather extensive review of the literature on this subject is given in [4].

Optimization of the generation structure when combining various sources, including distributed and localized sources of *prosumers*, into a single power supply network is one of the key tasks in control and operation of integrated energy systems. A prosumers that have their sources and/or energy storage systems, basing on the energy consumption and production balance, regulate their consumption to improve the efficiency and reliability of both their own energy supply and the entire energy system. As evidenced by the analysis of publications on prosumers, most of the studies are devoted to their operation within electric power systems [5–11]. The prosumer concept is however no less relevant for district heating systems (DHS) that are the largest fuel consumers, especially in the countries with a cold climate (in Russia, for example, more than 45% of total fossil fuel consumption goes for heating). In [12–14], the authors present methodological and practical studies on the participation of prosumers in district heating in Sweden and Norway.

In the previous studies [15, 16] we proposed a model for optimal control of the DHS that involve prosumers with owned heat sources (HS). The model is based on the bi-level programming method. This model demonstrated good performance in calculations, but it does not include the cost components related to the heat network (HN) operation, which is its main disadvantage. This paper presents an alternative approach to the load management of the prosumer HS. The approach takes into account network conditions and constraints, as well as some other factors that are significant for the studied problem.

2 Mathematical formalization of optimal management of the prosumer HS load in DHS, given network conditions

Calculated scheme of DHS is represented by a network structure with a corresponding hydraulic circuit, which consists of m nodes and n branches [17]. This structure is defined by ordered sets of nodes $J=\{j:j=1,\ldots,m\}$, including HS nodes $J_{\mathrm{hs}}\subset J$, heat consumer nodes $J_{\mathrm{c}}\subset J$ and branching nodes $J_{\mathrm{o}}\subset J$; and a set of branches (HN sections) – $I=\{i:i=1,\ldots,n\}$ that reflect the specified pair-wise ties between the nodes. The operation process of such a system is modeled with a time interval of 1 hour starting at time τ_0, that corresponds to the calculated heat load, and finishing at the final (calculated) time τ_Σ, h. The entire set of time instants is denoted by T. A set of heat consumers connected to the DHS will be divided into 2 subsets: $J_{\mathrm{c}}=J_{\mathrm{c}}^{\mathrm{ord}}\bigcup J_{\mathrm{c}}^{\mathrm{pro}}$, where $J_{\mathrm{c}}^{\mathrm{ord}}$ – ordinary consumers, and $J_{\mathrm{c}}^{\mathrm{pro}}$ – prosumers with their own HS. The total heat demand at node j at time τ is expressed as follows:

*Corresponding author: postnikov@isem.irk.ru

$$q_{j\tau}^{\mathrm{c}} = q_{j\tau}^{\mathrm{ord}} + q_{j\tau}^{\mathrm{pro}},\ j \in J_{\mathrm{c}}^{\mathrm{ord}} \cap J_{\mathrm{c}}^{\mathrm{pro}}, \quad (1)$$

where $q_{j\tau}^{\mathrm{ord}}$ and $q_{j\tau}^{\mathrm{pro}}$ – heat demand of ordinary consumers and prosumers at node j, respectively, GJ/h.

On the other hand, heat demand of consumers at node j is determined by the Rossander formula [18], according to which heat load at time τ is determined by the following expressions:

$$q_{j\tau}^{\mathrm{c}} = q_j^{\mathrm{h}}[1-(1-\omega_j)(\tau/\tau_\Sigma)^{\sigma_j}] + q_j^{\mathrm{hw}},\ j \in J, \tau \in T; \quad (2)$$

$$\omega_j = (1-\varphi_j)[(t_1-t_3)/(t_1-t_2)],\ j \in J; \quad (3)$$

$$\upsilon_j = (1-\varphi_j)[(t_1-t_4)/(t_1-t_2)],\ j \in J; \quad (4)$$

$$\sigma_j = (\upsilon_j - \omega_j)/(1-\upsilon_j),\ j \in J; \quad (5)$$

where q_j^{h} – calculated heating load, GJ/h; q_j^{hw} – calculated hot water supply load, GJ/h; ω_j, υ_j and σ_j – heat load curve irregularity factors; φ_j – proportion of hot water supply load; t_1 – calculated indoor temperature, °C; t_2, t_3, t_4, – ambient air temperature: calculated, corresponding to the beginning of the heating season and average throughout the heating season, respectively, °C.

Management of the operation of DHS with prosumers implies the distribution of heat load between the district HS and of prosumer HS according to some criteria that provide the required (expected) parameters of the system operation. Given the above-mentioned conditions, the problem of heat supply management can be mathematically solved using the objective function of the total prosumer costs that represent a sum of the costs of heat purchase from the DHS and the costs of heat self-production over the considered time interval:

$$z_\Sigma^{\mathrm{pro}} = \sum_{\tau \in T} \sum_{j \in J_{\mathrm{c}}^{\mathrm{pro}}} [c_\tau^{\mathrm{h}} \cdot (q_{j\tau}^{\mathrm{pro}} - q_{j\tau}^{\mathrm{pro(hs)}}) + z_{j\tau}^{\mathrm{pro(hs)}}], \quad (6)$$

where c_τ^{h} – tariff for the heat supplied from the DHS, rub/GJ; $q_{j\tau}^{\mathrm{pro(hs)}}$ – heat self-production volume at the HS of the j-th prosumer, GJ/h; $z_{j\tau}^{\mathrm{pro(hs)}}$ – costs of heat self-production at the HS of the j-th prosumer, rub.

Modern DHS are regulated natural monopolies where heat tariff for consumers is calculated by summing the specific total costs of heat production and transportation in the DHS and standard profitability of the unified heat supply organization (UHSO). Depending on the individual characteristics of the system and its operating conditions, the profitability level of the UHSO lies in a range of 4–7% [19]. Thus, the heat tariff within the UHSO is calculated by the following formulas:

$$c_\tau^{\mathrm{h}} = AC(1+R/100\%), \quad (7)$$

$$AC = \left(\sum_{j \in J_{\mathrm{hs}}} z_{j\tau}^{\mathrm{hs}} + z_\tau^{\mathrm{hnw}} \right) \Big/ \sum_{j \in J_{\mathrm{hs}}} q_{j\tau}^{\mathrm{hs}}, \quad (8)$$

where AC – specific total heat supply costs, rub/GJ; R – profitability, %; $z_{j\tau}^{\mathrm{hs}}$ – heat production costs of the j-th heat source, rub; z_τ^{hnw} – heat transportation costs (HN operating costs), rub; $q_{j\tau}^{\mathrm{hs}}$ – volume of heat produced by the j-th HS, GJ/h.

The costs of heat production by district and distributed HS of prosumers for any time instant τ can be represented as a quadratic function [20, 21], rub:

$$z_{j\tau}^{\mathrm{hs}} = \alpha_j (q_{j\tau}^{\mathrm{hs}})^2 + \beta_j q_{j\tau}^{\mathrm{hs}} + \gamma_j,\quad j \in J_{\mathrm{hs}}; \quad (9)$$

$$z_{j\tau}^{\mathrm{pro(hs)}} = \alpha_j (q_{j\tau}^{\mathrm{pro(hs)}})^2 + \beta_j q_{j\tau}^{\mathrm{pro(hs)}} + \gamma_j,\ j \in J_{\mathrm{pro(hs)}}; \quad (10)$$

where α_j, rub/(GJ/h)², β_j, rub/(GJ/h), γ_j, rub – approximation coefficients of cost characteristic of respective heat sources; J_{hs} – a subset of the DHS heat sources; $J_{\mathrm{pro(hs)}}$ – a subset of the prosumer heat sources.

The costs of HN include operational (constant) costs and the costs of pumping a heat carrier through the HN (variable). They are determined by the following analytical dependence [18]:

$$z_\tau^{\mathrm{hnw}} = \sum_{i \in I} z_{i\tau}^{\mathrm{hnw}} = F_1 + F_2 \sum_{i \in I} x_{i\tau}^2 |x_{i\tau}| s_i, \quad (11)$$

$$F_1 = \frac{f_{\mathrm{c}}}{N_{\mathrm{pump}}} \sum_{i \in I} \ell_i (a_i + b_i \ell_i^{0.19u_i} \chi_i^{0.19u_i} s_i^{-0.19u_i}), \quad (12)$$

$$F_2 = c^{\mathrm{e}} / (367 \eta_{\mathrm{pump}}), \quad (13)$$

where F_1 – semi-fixed costs, rub; F_2 – coefficient of semi-fixed costs of HN, rub; $x_{i\tau}$ – heat carrier flow rate in the i-th branch of HN at time τ, t/h; s_i – coefficient of hydraulic resistance of the i-th branch, mh²/t²; f_{c} – share of semi-fixed and operational costs of HN (normally taken equal to 0.075); N_{pump} – the number of pump unit operation hours, h/year; ℓ_i – length of the i-th network section, m; a_i, b_i – approximation coefficients of costs for the i-th network sections with various diameters; χ_i – coefficient depending on the pipeline roughness of the i-th network section; c^{e} – electricity price, rub/kWh; η_{pump} – pump unit efficiency, %.

The cost of heat transportation is determined by the optimal flow distribution in HN. The mathematical model of flow distribution for each time point τ in has the following matrix form [17]:

$$\mathbf{A}\mathbf{x}_\tau = \mathbf{g}_\tau, \quad (14)$$

$$\overline{\mathbf{A}}^{\mathrm{T}}\mathbf{p}_\tau = \mathbf{h}_\tau - \mathbf{H}_\tau, \quad (15)$$

$$\mathbf{S}\mathbf{X}_\tau\mathbf{x}_\tau = \mathbf{h}_\tau, \quad (16)$$

where $\mathbf{A}$ – matrix of couplings of linearly independent nodes of HN; $\overline{\mathbf{A}}^{\mathrm{T}}$ – complete transposed branch and node incidence matrix; $\mathbf{x}_\tau$ – vector of heat carrier flow rates in the branches of HN at time τ, t/h; $\mathbf{g}_\tau$ – vector of flow rates at network nodes at time τ, t/h; $\mathbf{p}_\tau$ – vector of nodal pressure in HN at time τ, mm wc; $\mathbf{h}_\tau$ – vector of head losses in the network branches at time τ, mm wc; $\mathbf{H}_\tau$ – vector of acting heads at HS at time τ, mm wc; $\mathbf{S}, \mathbf{X}_\tau$ – diagonal matrices of coefficients of hydraulic resistance in the branches, m/(h²t²), and absolute values of flow rates in them, t/h, at time τ.

Heat flow rates at each node j at time τ are determined depending on mass flow rates by the known relationship [22], GJ/h:

$$q_{j\tau} = \nu g_{j\tau}\Delta t, \quad (17)$$

where ν – heating capacity of the heat carrier, GJ/°Ckg; $g_{j\tau}$ – mass flow rates of a heat carrier at the j-th node, t/h; Δt – network water temperature difference in the supply and return pipelines, °C.

Thus, according to the above-presented mathematical expressions, the problem of optimal load distribution between district HS (of DHS) and prosumer HS is formulated as follows:

– find the minimum of function (6) that determines the total costs of all prosumers by using the following mathematical expressions:

1) equations (2)–(5), to determine the schedule of heat consumption during the heating season (demand for heat);

2) equations (7) and (8), to determine heat tariff for each calculated time point;

3) empirical dependencies (9) and (10), to calculate the operational costs of district HS and distributed HS of prosumers at each time point, respectively;

4) formulas (11)–(13), to calculate the operational costs of HS used to determine the heat tariff in DHS;

5) model (14)–(16) and dependence (17), to calculate the thermal and hydraulic conditions in HN.

Search for optimal solutions to the formulated problem is based on the application of the methods of univariate relaxation (the method of coordinate descent) and iterative approximation. Within the computation cycle, the multidimensional optimization problem is reduced to a one-dimensional one with a step-by-step procedure for improving solutions on the heat production volumes at all HS in the system.

3 Case study and discussion

The DHS under consideration is presented in Fig. 1 and consists of one district HS(0) with an output of 132 GJ/h, 9 consumers (nodes 1–3, 7–10 and 12, 13) with total load equal 130 GJ/h and a circuit HN consisting of 14 sections (branches) with their characteristics indicated in the Fig. 1. The summary load of consumers is 131.6 GJ/h and total length of the HN is 1150 m.

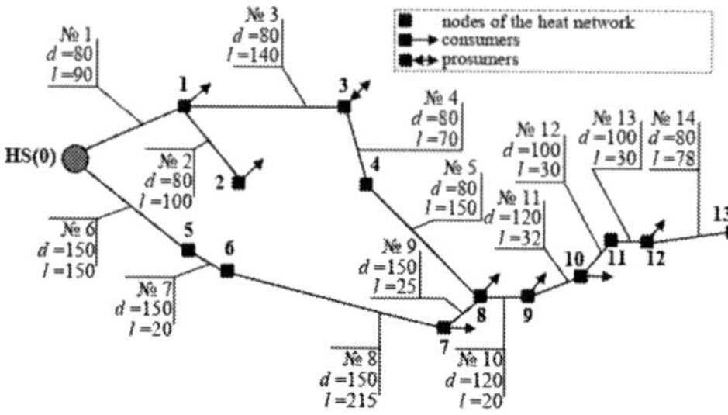

Fig. 1. A calculated scheme of DHS: in the callouts d – diameter of a network section, mm; l –length of a network section, m

It is assumed that consumer 3 is a prosumer (hereinafter referred to as P3) and has its own HS running on fossil fuel with a capacity of 17 GJ/h, which completely covers designed load of P3. We specify the quadratic cost function (10) for HS P3 with the following values of the approximation coefficients: α_j – 0.45 rub/GJ²; β_j = 980 rub/GJ; γ_j = 15864 rub. The production costs of HS(0) are also described by a quadratic function with the following values of the approximation coefficients: α_j = 0.25 rub/GJ²; β_j = 380 rub/GJ; γ_j = 12047 rub.

Fig. 2 demonstrates a diagram describing the total (production + transportation) costs of heat supply to P3, depending on the load of its HS throughout the entire heating season assumed to be 5000 h. For each time instant, there is a minimum point that corresponds to the optimal loading of the HS P3 (these points lie on line A–B). In this case, its value is 25% for all considered conditions.

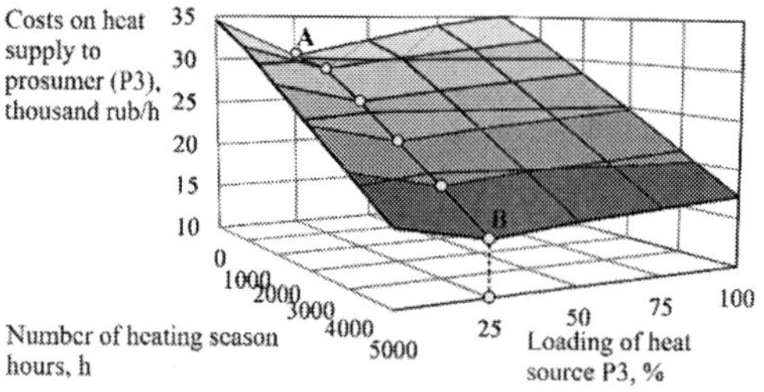

Fig. 2. Change in the costs of heat supply to prosumer 3 (P3) depending on loading of its HS during the heating season

Table 1 presents the indices of optimal load distribution between HS(0) and HS P3 for heat supply to P3 during the heating season. These data are clearly demonstrated in Fig. 3 where diagram 1 shows the heat load curves of P3 with highlighted heat volume generated by HS(0) and HS P3; diagram 2 illustrates the total (production + transportation) costs of heat supply to P3 that are calculated both with and without the consideration of the prosumer heat source operation.

As evidenced by Figures 3-1 and Table 1, the heat source of prosumer 3 operates during the entire heating season with an output of 1.6–4.1 GJ/h depending on the prosumer load.

Table 1. Results of an optimal load distribution between HS to cover heat load of P3 during the heating season*

Indices	Number of heating season hours, h					
	1	1000	2000	3000	4000	5000
Load of P3, GJ/h	16,8	11,6	9,9	8,5	7,3	6,5
Loading of the district HS (0), GJ/h	12,6	8,7	7,3	6,3	5,5	4,9
Loading of HS P3, GJ/h	4,1	3,0	2,6	2,2	1,8	1,6
Costs of heat supply to P3 without HS P3, thousand rub/h	34,5	31,8	28,0	24,5	20,9	17,2
Costs of heat supply to P3 with HS P3, thousand rub/h	30,2	29,0	26,0	22,5	18,9	15,4
Cost saving owing to the use of HS P3, thousand rub/h	4,3	2,8	2,0	2,0	2.0	1,8

*calculations were performed using the "GAMS" software

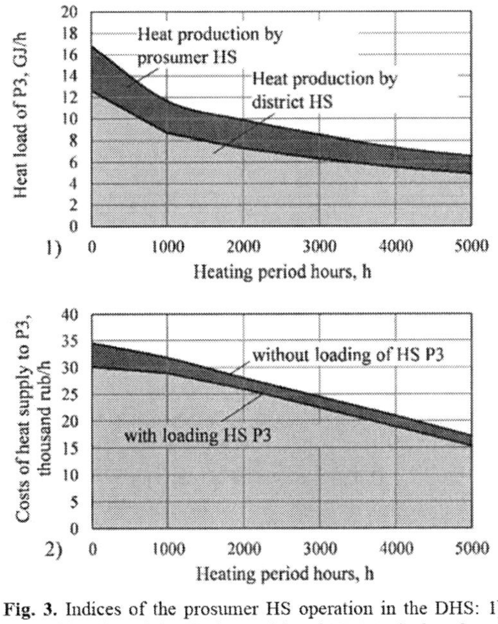

Fig. 3. Indices of the prosumer HS operation in the DHS: 1) optimal loading of the district HS(0) and HS P3; 2) plot of total costs of heat supply to P3 (the area of a dark figure corresponds to an economic benefit owing to the HS P3 loading)

The total heat consumption of P3 over the considered period is about 50.5 thousand GJ, including about 12.6 thousand GJ (25%) covered by the prosumer HS. According to the calculations, it is exactly this ratio of the HS loading that corresponds to the minimum total costs of heat supply to P3. Thus, the specific total costs of heat supply to P3 at the maximum load are 34.5 thousand rub/h without the use of the HS P3, and 30.2 thousand rub/h when the HS P3 is used. The total costs of heat supply to P3 from the HS(0) for the entire heating season is 130.7 million rub. The involvement of the HS P3 reduces this value to 118.3 million rub. Consequently, the economic effect of the HS P3 operation over the considered period is about 12.3 million rubles or 9.4%.

4 Conclusions

In this study, we state the problem of managing district heating system with a prosumer, which means an economically-effective load distribution between the heat sources to cover the heat load of the prosumer by using both district heat sources and prosumer heat sources. Solving this problem is based on the search for the conditions that correspond to the minimum production costs of heat supply to the prosumer, including operation of the heat source and heat network. The optimization is performed according to the load curve of consumers, which makes it possible to obtain an integral cost saving estimate for the whole heating period.

The results obtained in the calculation of the test district heating system show the effectiveness of the developed methodology and the possibility of gaining an economic effect when the prosumer heat source is used. An analysis of the results obtained in the test calculations provides the basis for further research in this area.

The research was performed at Melentiev Energy Systems Institute SB RAS under the support of Russian Science Foundation (Grant №17-19-01209).

References

1. Pierluigi Mancarella. MES (multi-energy systems): An overview of concepts and evaluation models // Energy, 2014, vol. 65, p. 1–17.
2. Pierluigi Mancarella, Göran Andersson, J.A. Peças-Lopes, Keith R.W. Bell. Modelling of integrated multi-energy systems: Drivers, requirements, and opportunities // Power Systems Computation Conference (PSCC), 2016, Genoa, Italy. DOI: 10.1109/PSCC.2016.7541031.
3. Nikolai I. Voropai, Valery A. Stennikov, Evgeny A. Barakhtenko, Oleg N. Voitov, Ivan V. Postnikov. A Model for Control of Steady State ofIntelligent Integrated Energy System // Energy System Research, 2018, vol. 1(1), p. 57–66.
4. Henrik Lund, Poul Alberg Østergaard, David Connolly, Brian Vad Mathiesen. Smart energy and smart energy systems // Energy, 2017, vol. 137, p. 556–565.
5. Ni Zhang, Yu Yan, Wencong Su. A game-theoretic economic operation of residential distribution system with high participation of distributed electricity

prosumers // Applied Energy, 2015, vol. 154, p. 471–479.
6. Luka Perkovic, Hrvoje Mikulcic, Neven Duic. Multi-objective optimization of a simplified factory model acting as a prosumer on the electricity market // Journal of Cleaner Production, 2017, vol. 167, p. 1438–1449.
7. Hongming Yang, Tonglin Xiong, Jing Qiu et al. Optimal operation of DES/CCHP based regional multi-energy prosumer with demand response // Appled Energy, 2016, vol. 167, p. 353–365.
8. Rehman Zafar, Anzar Mahmood, Sohail Razzaq et al. Prosumer based energy management and sharing in smart grid // Renewable and Sustainable Energy Reviews, 2018, vol. 82, p. 1675–1684.
9. Stig Ødegaard Ottesen, Asgeir Tomasgard, Stein-Erik Fleten. Prosumer bidding and scheduling in electricity markets // Energy, 2016, vol. 94, p. 828–843.
10. Dimitrios J. Vergados, Ioannis Mamounakis, Prodromos Makris, Emmanouel Varvarigos. Prosumer clustering into virtual microgrids for cost reduction in renewable energy trading markets // Sustainable Energy, Grids and Networks, 2016, vol. 7, p. 90–103.
11. Lakshmi Prakash, Sugatha Kumari P.R, Sharanya Chandran et al. Self-sufficient Smart Prosumers of Tomorrow // Procedia Technology, 2015, vol. 21, p. 338–344.
12. Lisa Brange, Jessica Englund, Patrick Lauenburg. Prosumers in district heating networks – A Swedish case study // Applied Energy, 2016, vol. 164, p. 492–500.
13. Lisa Brand, Alexandra Calvén, Jessica Englund et al. Smart district heating networks – A simulation study of prosumers' impact on technical parameters in distribution networks // Applied Energy, 2014, vol. 129, p. 39–48.
14. Hanne Kauko, Karoline Husevåg Kvalsvik, Daniel Rohdeb et al. Dynamic modeling of local district heating grids with prosumers: A case study for Norway // Energy, 2018, vol. 151, p. 261–271.
15. Stennikov V.A., Postnikov I.V., Penkovskii A.V. Methods and models of optimal managing of district heating systems with prosumers // E3S Web of Conferences "Methodological Problems in Reliability Study of Large Energy Systems", 2017, vol. 25, 5 p.
16. Postnikov I.V., Stennikov V.A. Methods and models of ensuring reliability of district heating systems with prosumers // E3S Web of Conferences "Methodological Problems in Reliability Study of Large Energy Systems", 2017, vol. 25, 6 p.
17. Merenkov A.P., Khaselev V.Ya. Theory of hydraulic circuits / Moscow. Nauka, 1985, 278 p.
18. Sennova E.V., Sidler V.G. Mathematical modeling and optimization of developing district heating systems / Novosibirsk: Nauka, 1985, 222 p.
19. Gimadi V.I. The reform that warms: what is expected from reforms in heat supply // Industrial Organization, 2014, No. 4 (46), p. 49-64.
20. Stennikov V.A., Khamisov O.V., Pen`kovskii A.V. Optimizing the Heat Market on the Basis of a Two-Level Approach // Thermal Engineering, 2011, vol. 58(12), p. 1043–1048.
21. Penkovskii A.V., Stennikov V.A., Khamisov O.V., Mednikova E.E., Postnikov, I.V. Search for a Market Equilibrium in the Oligopoly Heat Market // Energy Procedia, 2017, vol. 105, p. 3158–3163.
22. Sokolov V.Ya. Heating and heating networks / M.: Publishing house of MEI, 1999, 472 p.

The analysis of dispersed generation influence on power system automatics settings and function algorithms

Pavel Ilyushin[1*]

[1] Petersburg power engineering institute of professional development, Aviation str. 23, Saint-Petersburg, Russia

Abstract. Dispersed generation (DG) objects are usually connected to distribution grid. This-type of grid supposes to have power system automatics (PSA) that comprise emergency control system (ECS). ECS serves for normal power supply scheme recovery after fault elimination, fault development prevention and system parameters maintenance in acceptable limits within fault and post-fault conditions. Generating power connection to the grid causes significant power flow and current distribution change. This might become a reason of ECS automatics misoperation. This paper presents the results of DG objects influence analysis on PSA settings and function algorithms. Implemented analysis has allowed to define post-fault conditions envelopes that have to be analysed within DG objects grid connection design and provide the recommendations for PSA settings and function algorithms change.

1 Introduction

At present, the tendency of distributed generation (DG) objects commissioning increase is resuming in Russia. They are usually built by the owners of industrial enterprises in the oil and gas, mining, metallurgical, pulp and paper and chemical industries. They are connected to distribution electric grids or to internal power supply networks of these enterprises.

At the same time, power system automatics are widely used in these grids: automatic switching of backup power supply (ABS), automatic frequency discharge (AFD) and automatic control of voltage reduction (ACVR). These devices are used to restore the normal circuit of power supply, prevent the development of accidents, as well as normalize the parameters of the electrical mode in emergency and post-emergency circuit-mode situations. They are necessary to ensure the normal operation of the most part of consumers' electric receivers.

The connection of small and medium power generating units (GUs) of DG objects inevitably leads to a change in the circuit-mode situations in the grid, significantly affecting, in some cases, the automation devices of the power systems in operation. This applies both to the correctness of their operation, in accordance with the specified algorithms, and to the correct tripping, in accordance with the setpoints. It is important to avoid failures in the operation, as well as excessive and false trips of power automation devices.

Therefore, when developing the project of technological connection (TC) of the DG facility, it is necessary to conduct an analysis of the correctness of the technical solutions for the operation algorithms and the adjustment of the previously installed units of power systems automation in the adjacent grid. For this, it is necessary to make calculations of electrical modes, including calculations of stable modes and electromechanical transient processes. In addition, it is necessary to check the consistency of the operation algorithms and the settings of the automation devices of the power systems in operation with the algorithms of operation and the settings of the new automation devices that must be put into operation at the TC of the DG object.

2 Characteristics of ABS operation in the grids with DG

Automatic backup power and equipment switching devices (ABS) are used to restore power of power receiving installations. This is done by automatically connecting a backup power source when the consumer's electrical installations are de-energized and switching on the backup equipment when the working equipment is disconnected with a violation of the normal process. ABS devices can be installed on transformers, power lines, sectional and bus breaker, electric motors, etc. [1].

If the circuit-mode situation and the disturbance are such that the ABS can not trip due to the operational input disconnection (the input does not turn off) and therefore trips due to the voltage drop, it is clear that the presence of a GU on the redundant part of the network increases the residual voltage. This will interfere with the normal operation of the ABS device. Let us consider the case when the GUs are equipped with a system of independent excitation, which provides for the maximum forcing value independent of the voltage at the GU terminals, and are not de-energized due to the voltage drop at short-circuit.

* Corresponding author: ilyushin.pv@mail.ru

 https://doi.org/10.1051/e3sconf/20185802001

The main parameter determining the critical time for the dynamic stability of the GU is the magnitude of the mechanical constant of inertia of the GU, T_J, s. T_J values close to 1 s are typical for gas piston units (GPU) and three-shaft gas turbine units (GTU), about 3 seconds for high-power GPUs and twin-shaft gas turbines, 6-8 seconds for small and medium-sized steam turbine generators, 10-12 c - for single-shaft gas turbines. The value of T_J in the calculations is assumed to be 2 s, which is typical for the GU of the DG objects.

Calculations show that in many cases, with different combinations of load and generation ($0<d<100\%$, $0<p<100\%$), the operation of electric receivers is violated. The areas of possible post-accidental modes (PAM) are shown in Fig. 1.

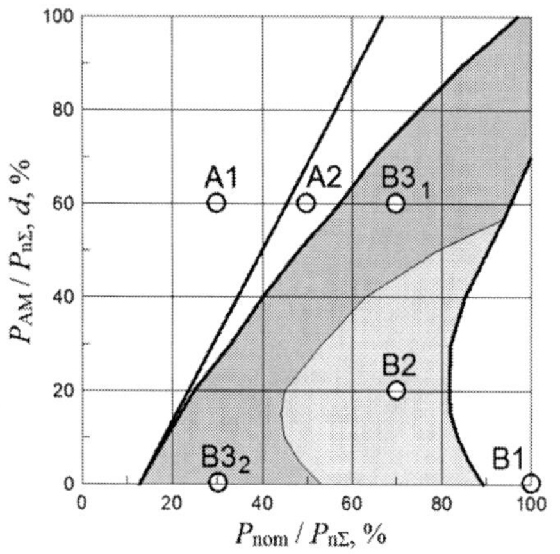

Fig. 1. Post-emergency modes when ABS is triggered by a voltage drop.

In areas where ABS devices operate, the following are possible:

- Normal trip (the initial parameters of the mode correspond to the area **A1**),
- Increase in the response time of the ABS due to the fact that the GUs increase the voltage and the voltage starting point (VSP) does not trip until the braking of the asynchronous motors (AM) causes a sufficient voltage drop (area **A2**).

The failure of the ABS device can result in:

- The modes where the initial parameters correspond to the area **B1**, which can be considered as short-time permissible voltage (above the critical values) and frequency (conditionally accepted f > 45 Hz). These modes can appear due to a decrease in the power consumed by the AM and synchronous motors (SM) with a decrease in frequency, as well as by static electric receivers with a decrease in voltage. They are limited by the permissible duration of the forced GU excitation (20-30 s) and the triggering of the AFD devices,
- The modes corresponding to area **B2**, where the frequency decreases, but the voltage, due to the forced excitation of the GU, remains above the critical one,
- The occurrence of a voltage collapse - area **B3**.

In the **B3** area, when there is a high probability of a voltage collapse, triggering of the ABS is necessary in this case, but:

- There may be a failure in the operation of the VSP, since in the case of a voltage collapse, it decreases in different ways, and its effective values may contain variable components,
- May be unacceptable, since the closure of the link between substations (SS), where a voltage collapse has just passed and normally operating SS, can cause the spread of a voltage collapse to these substations and lead to the disconnection of an additional number of consumers [2].

The transient process in area **B3** (point **B3$_1$**), where d = 60%, p = 70%, is characterized by the fact that the voltage after the elimination of short-circuit faults does not recover above 48% of U_{nom}, which leads to the disconnection of a significant number of electric receivers, while the frequency in the power district rises to 52 Hz.

The transition process in the **B3** area (point **B3$_2$**), where d = 0, p = 30%, is characterized by the fact that the voltage after the elimination of short-circuit is restored to 70% of U_{nom}, which leads to a smooth decrease in frequency, considering the significant shortage of active power, of a power district from 50 to 24 Hz for 4 s from the beginning of the fault (without taking into account the operation of the AFD).

Remote and incomplete phase short-circuits result in the failure of the ABS to such PAM, which are more favorable, since at the time of such a short-circuit the motors are braked less than in the case of a near three-phase short-circuit.

Taking into account the above said, the operation of ABS devices due to the voltage decrease, adjusted as delayed from the trip of relay protection devices (RP), ABS devices located closer to the power supply and automatic reclosing devices does not meet the requirements for the speed required in the presence of DG facilities in the power center.

The increase in the speed of the ABS is possible due to the use of launching units that monitor the parameters of the previous mode, detecting the trip of the RP devices and the current state of the switching devices in the grid, including GU breakers. At the stage of the technological connection project development, on the DG object it is required to verify the correct operation of all ABS devices in the grid of external and internal power supply and, if necessary, provide for the implementation of the necessary organizational and technical measures.

3 Features of AFD operation in grids with DG

Automatic frequency discharge (AFD) is designed to prevent unacceptable frequency reduction according to the standards of stable operation of generating equipment and power-receiving units and its subsequent recovery [3]. The operation of the ABS devices and the automatic reclosure must be consistent with the operation of the AFD devices in such a way that their action does not restore the power to the consumer load, disconnected by the AFD devices, from the same or other electrically connected power supplies. The tripping of AFD devices in transient modes characterized by frequency reduction, not associated with an emergency shortage of active power or with an interruption in power supply should be avoided [4].

Obviously, the consequences of a power district outage from the power system, in which the GU facilities of the DG can not cover the entire load, are determined by a combination of two factors: a shortage of active power, which leads to frequency decrease, and a shortage of reactive power, which leads to voltage decrease and can trigger a voltage collapse. The latter, if it covers the entire energy district, will make it impossible for all electric receivers to work.

Prior to the commissioning of the DG objects the consumers' frequent cases were:

- Violation of power supply to the power district quickly led to deep voltage failures and the normal operation of the ABS,
- The disconnection of a small part of the generation in the power system or its island part was accompanied by a frequency decrease while maintaining the voltage at an acceptable level with the AFD effect [5].

In the intermediate case, when the energy district is not large and contains the DG object, during the separation from the power system, there can be an arbitrarily large shortage of active and reactive powers. Currently, there are likely qualitative differences between the processes under consideration from the same processes to the construction of DG objects with modern GUs having small T_J values, as indicated earlier. Differences are due to two circumstances:

- The rate of frequency reduction increases, which may require more speed of the AFD devices,
- If the power shortage is significant, then the probability of voltage collapse occurrence substantially increases.

In the latter case, if the voltage fails deep and fast, a very significant load drop is obtained and the balance of active power can be restored, which will lead to normalization of the frequency, while the AFD will not work. In addition, the voltage collapse can pass so quickly that it is impossible to prevent it with ACVR devices.

Consider a small power district with an industrial load, which, in addition to the static load, includes the AM and the DG object. The power district is connected to the power system by a transmission line, while the voltage on the buses of the power system does not depend on the processes in this power district. In the initial mode, the share of AM in the total load varies and is $d = P_{AM} / P_{n\Sigma}$,%. The value of the generation varies and is set relative to the value of the total load ($P_{nom} / P_{n\Sigma}$,%); T_J = 2 s; loading of GU in the initial mode is nominal, short-term available power is 110%; the GU excitation is independent, the GU parameters are the same as for the 2 MW GPU. In the initial mode, the reactive power is set equal to 0.5 P_{nom}, the initial voltage on the load buses is 10.2 kV.

If the connection between the power district and the power system is broken, the calculations show that the primary factors determining the course of the process with loss of external power are the values of the initial active power shortage $D_P = 1 - P_n / P_{n\Sigma}$ (%) and the share of AM d (%). In these coordinates, the process can in principle be represented by four areas **A**, **B**, **C**, **D**, as shown in Fig. 2.

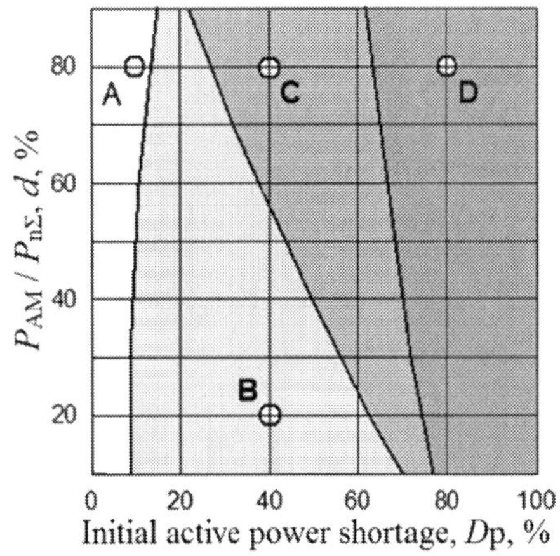

Fig. 2. Features of the processes in the islanding of the energy district with the DG object.

In area **A**, successful processes: in stable modes U > 0.8 U_{nom}, f > 49 Hz, which is due to the presence of short-term reserves of the GU (overload of the drive gas piston engines and the excitation forcing) and the regulating effects of the load, linearly dependent on d (for d = 0, the control effects $K_U \approx 2$, $K_f = 0$; if d = 100%, then $K_U \approx 0$, $K_f \approx 1.8$).

Area **B** corresponds to the conditions when the GU of the DG object keeps the voltage above the critical value, the AM continues to operate, the available power of the GU is insufficient and the frequency decreases; the share of AM is not large and they can not provoke a voltage collapse. This is the circuit-mode situation for which AFD devices are designed.

Area **C** is the same as **B**, but the share of AM is larger, so the increase in their reactive power consumption (as the frequency decreases) leads to a significant decrease in voltage.

Area **D** corresponds to a very large initial power shortage and quickly causes the occurrence of a voltage collapse, the AM stop. When the voltage is about 10% of the rated voltage, the total load is small and the GU speed controllers provide for a frequency increase in the power district to the nominal value.

The described processes refer to cases when at the beginning of the process there is no significant voltage drop. If the emergency mode starts with a short-circuit, the rapid braking of the motors increases the total reactive load and reduces the voltage faster, in comparison with the case of absence of short-circuit. The consequence is a decrease in the active load and a lower frequency decrease in the power district.

The same effect - the expansion of the **D** area - is manifested if the GUs on the DG object have not an independent excitation system, as in the previous cases, but a self-excitation system in which the maximum possible excitation degree decreases in proportion to the reduced voltage at the GU terminals. At the initial moment, the short-circuit voltage decreases after a few seconds to almost zero, and a full load drop occurs, which leads to a frequency increase. Thus, in the **D** area, the size of which depends on the parameters of the power district and the initial disturbance, the AFD devices can not serve as the main means of normalizing the mode in the power district [6].

Taking into account the above said, at the stage of technological connection project development of the DG object it is required to conduct an analysis of the AFD efficiency in the power district when the communication with the power system is broken and, if necessary, provide for the installation of additional emergency control system (ECS) for normalizing the parameters of the mode and keeping the largest part of the consumer electric receivers operational.

4 Features of ACVR operation in the grids with DG

Automation of voltage reduction limitation (ACVR) is designed to prevent the violation of load stability and the occurrence of a voltage collapse in post-emergency modes of the power system. It should effect the excitation forcing of synchronous machines, compensation devices forcing, reactors shutdown and as an exception, and in case of insufficient network activities and the existence of a justification - the consumers switch off [1].

The initial conditions for selecting the ACVR devices setting parameters that act to disconnect consumers' electric receivers or their groups are as follows:

- The criterion by which the control action (CA) to disconnect the electric receivers is formed is the voltage after the RP devices of the grid eliminate the emergency mode by disabling the damaged element (controlled voltage),
- ACVR trip should occur if the reference voltage is below the critical voltage[a] for a given load (with an appropriate redundancy),
- ACVR trip should not occur with direct motor starts and other disturbances in normal production processes at the customer site,
- There should be no unnecessary disconnections of consumers' electrical receivers.

It follows from item 1 that the realization of the recovery should be made after the occurrence of a voltage failure with a time delay, the value of which determines the slowest RP device, on which the recovery of the monitored voltage depends. It is obvious that the possibility of electric motors self-start and, correspondingly, the necessary amounts of their shutdown depend on many factors, and not only on the value of the controlled voltage at the time when the ACVR device trips [7].

The need to increase the efficiency of ACVR devices is dictated by the load properties, the value of the damage from violations of the production process and statistical data on the frequency of occurrence of violations in electrical grids. The use of complex and high-precision technological lines of foreign manufacturers in domestic industrial enterprises and the presentation of high quality requirements for products to enter the world markets necessitates the provision of quality in power supply systems in accordance with the requirements of GOST 32144-2013 [8]. And this, in turn, entails the need to improve the principles of emergency control.

The ACVR device, which enables the recovery only by load shedding (LS) with a time delay of 6-22 s, may not be very effective or completely ineffective if a group of electric motors of considerable power is decelerated. In this case, the longer the voltage is reduced, the higher the total starting current of the motors, the lower the voltage, and the greater the probability will be that the process will end with a voltage collapse. Obviously, the earlier the shutdown of a part of the electric motors is performed, the smaller the amount of load disconnection is required.

Consider an example of calculations performed for a real industrial enterprise with large number of electric motors in the load. The total power consumption of the unit is about 10 MW, about 90% of it is by AM. The load is supplied from a powerful TPP through a 6 kV distribution grid (external to the unit under consideration); the total equivalent resistance of the external grid, reduced to a voltage of 6 kV, $Z_{eq} \approx 0.6$ Ohm. The load unit contains a DG object with a 2 MW GPU. In the calculated processes, due to a three-phase short-circuit with a duration of 0.2 s, the connection between the power system and the specified unit is loosened (the resistance Z_{eq} is multiplied by 1.5).

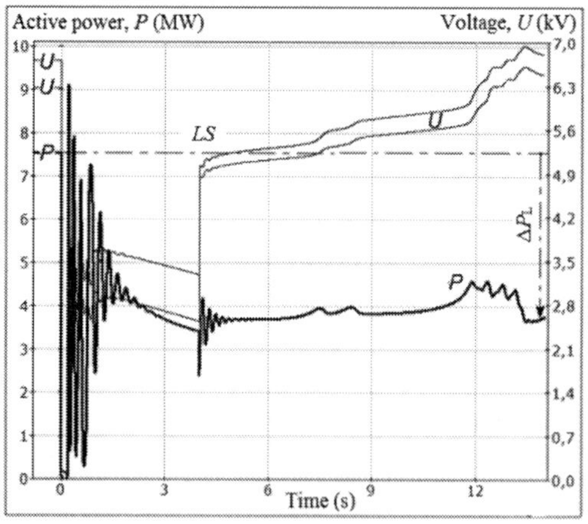

a)

[a] The critical voltage is the minimum voltage at which the electrical receivers can continue to perform their functions.

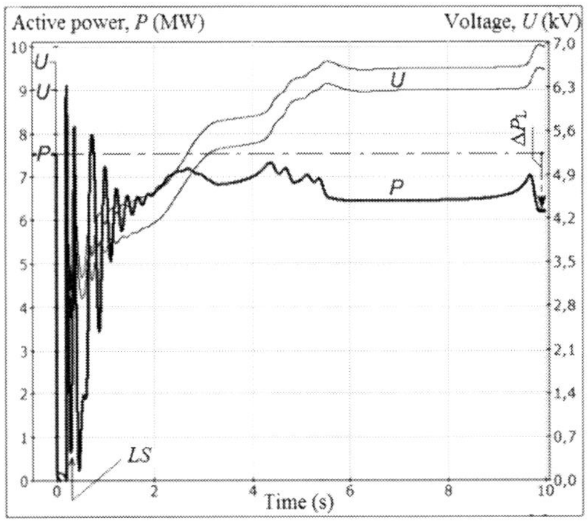

b)

Fig. 3. Comparison of processes that differ in ACVR response speed: a) $\Delta T = 4$ s; b) $\Delta T = 0.3$ s.

The delay in unloading from the beginning of the SC ΔT was set at 4 s and 0.3 s, as shown in Fig. 3a, 3b, where U is the voltage at two node points of the object, P is the active power coming to the object from the power system, LS is the load shedding, ΔP_{n} is the reduction of the post-emergency unloading compared to the pre-emergency one. Volumes of unloading from ACVR in both calculations are set by the limiting ones to prevent the occurrence of a voltage collapse.

The values of the minimum required ΔP_{L} shutdowns depend on all load and grid parameters, as well as on the type and location of the SC. The Fig. 4 shows the averaged dependence ΔP_{L} of the discharge delay ΔT, for the same object as above.

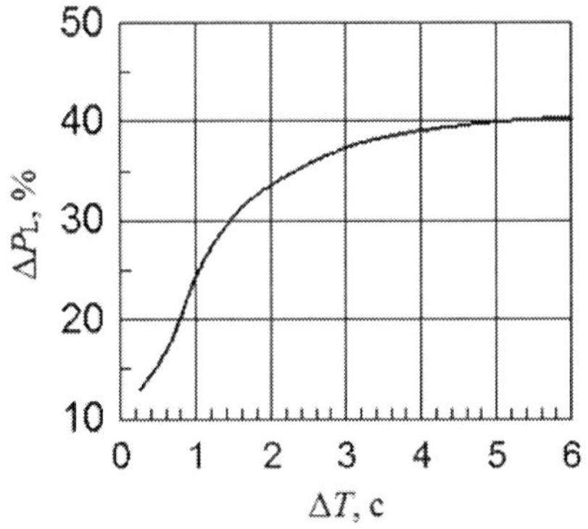

Fig. 4. The effect of the unloading delay on its required value at three-phase short-circuits.

If ΔT is large, unloading is required, allowing all non-disconnected motors to start from a stationary state. If the value of ΔT is small, the self-start starts with the slip achieved during the short-circuit, in this case the total self-starting currents are much smaller and the possibility of restoring the normal power supply to the electric receivers is more favorable.

Considering the foregoing, in order to prevent the voltage collapse and not to allow excessive unloading, which causes significant damage to consumers, it is necessary to use signals from start-up and interlock devices of other ECS devices so that the set of all signals indicates the need for high-speed unloading in this circuit-mode situation with sufficient probability. With such a formulation of the problem, it is not possible to consider a high-speed ACVR a local ECS.

5 Conclusions

Technological connection of DG objects to distribution electric grids or to internal power supply grids of enterprises inevitably leads to a change in the circuit-mode situations in the grid, significantly affecting, in some cases, the power system automatics in operation.

In order to prevent trip failures, as well as unnecessary and false trips of power system automatics, when commissioning the DG objects, it is absolutely necessary to analyze the correctness of the algorithms and the correct selection of the settings of the previously installed automation devices in the adjacent grid.

In addition, it is necessary to check the consistency of the operation algorithms and the settings of the power system automatics in operation with the algorithms of operation and the settings of the new automation devices that must be put into operation during technological connection of the DG object.

In some cases, based on the analysis of circuit-mode situations, it may be necessary to install additional ECS to normalize the parameters of the mode and to keep the most of consumers' receivers in operation, as well as a fundamental change in the algorithms of operation of ECS.

References

1. Rules for the designing of electrical installations. 7th ed. (Publisjer NTs ENAS, M, 2004).
2. P. Ilyushin, Relay protection and automation, **4**, 28-36 (2017).
3. GOST R 55105-2012 (Standartinform, M, 2013).
4. Ya. Artsishevskiy, B. Gieev, Energetik, **5**, 18-21 (2017).
5. Ya. Artsishevskiy, B. Gieev, University News. «Electromechanics», **6**, 37-44 (2017).
6. P. Ilyushin, Energetik, **12**, 21-25 (2016).
7. Yu. Gurevich, K. Kabikov, *Peculiarities of power supply, aimed at failure-free operation of industrial consumers* (ELEKS-KM, M, 2005).
8. GOST 32144-2013 (Standartinform, M, 2014).

The substantiation of regulatory requirements to the reliability of consumer support in the conditions of the perspective planning of the UPS of Russia

Yuri Chukreyev[1,*], *Mikhail* Chukreyev[1], and *Alexey* Stepskov[1]

[1]Komi Scientific Center of the Ural Branch of the Russian Academy of Sciences, ISE and EPN, 167000, Russia

Abstract. The issues of the influence of the information content of the balance reliability problem on the decisions taken to justify the operational power reserve of the UPS of Russia are considered. The calculated indicators of adequacy are presented with different representation of the power consumption and generating capacity for the conditions of the optimal distribution of the power reserve.

1 Concepts of the regulatory reserve of capacity in the balance of capacity of the power system

When planning the development of the industry, the so-called balance of power for the UPS of the country as a whole and its unified electric power system (UPS) [1, 2] was developed earlier, is being developed today and will be developed in the future. Its form includes two positions (Figure 1): the expenditure part "demand" (maximum load, export and standard power reserve); the incoming "cover" (installed power, unused power for the period of passage of the maximum, power inputs after passing the maximum, underutilization of power).

The issue that concerns both consumers and suppliers of electrical energy is to justify the regulatory power reserve, which is one of the components of the expenditure part of the balance sheet. In the tasks of long-term planning of EPS it is called the full (normative) reserve, and conditionally divided into three components: repair, strategic and operational, intended to compensate for unscheduled (emergency) conclusions of the main generating and network equipment for repair. Therefore, in order not to confuse the concept of operational reserve in the current planning and short-term (long-term), this component of the full reserve has recently been called the compensation reserve.

Repairing power reserve is made to compensate power generating equipment of power plants that is going to be maintained according to the schedule (average, current and capital). Compensatory reserve is used to compensate: decrease of power generators at electric power stations in emergency situations, occasional excess load of electricity consumption. This might be caused by the forecast errors and temperature variations. Therefore, the compensation reserve is divided into emergency and loaded. These components can be considered independent that simplifies the solution of the problem of the justification of the normalized power reserve a lot.

Market relations in the electric power industry exacerbate the issues of justifying the power reserve. With short-term planning (from one year to 4 years), the price for capacity during long-term (competitive) power

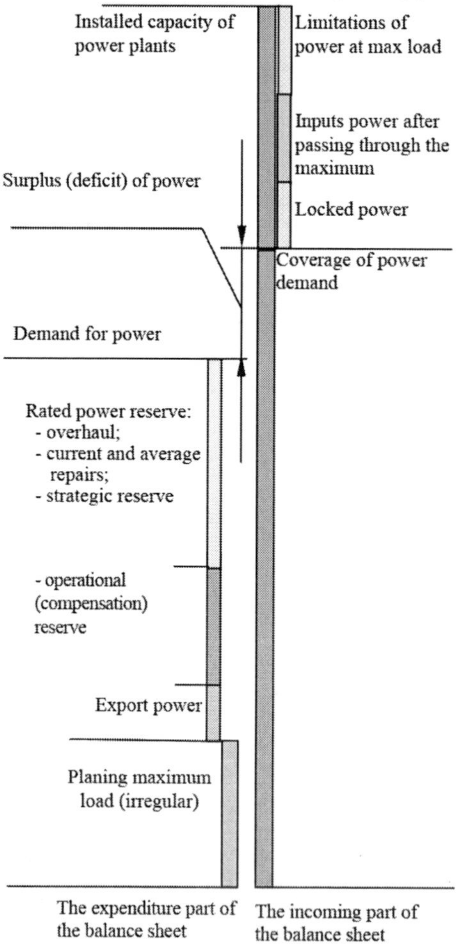

Fig. 1 - Structure of the projected balance of power.

* Corresponding author: chukreev@iespn.komisc.ru

take-off depends on the requirements for a full (normative) reserve in the conditions of market relations. There is a conflict of interests between power supply companies and consumers of electricity.

If the repairing power reserve is determined on the basis of the analysis of statistical information on the operation of the main equipment or the rules for their organization[1], strategic - expert, then the definition of the operational (compensation)power reserve requires the solution of the task of assessing the indicators of adequacy (IA). The regulatory definition of reserve generating power is difficult mainly due to the need to find compensatory power reserve [3 - 6, etc.].

The basis for determining the requirements for power reserve in the development of prospective balances was, is and will be the task of assessing the indicators of balance reliability of development options of the UPS of Russia, the so-called reliability analysis problem. Regardless of the management principles of the electric power industry (centralized or market), the methodology for its solution involves the implementation of two phases. On the first, by some other methods random states of generating power are generated, caused by emergency outputs of equipment and load due to its random changes due to forecast errors. At the second stage, these states are evaluated from the standpoint of a possible restriction of consumers due to the impossibility of achieving a power balance.

The shortage of power to the consumer and the damage caused by this phenomenon are usually more expensive than purchasing power regardless of the form of management (market or centralized management) by the industry. This leads to the following. At each state of the system, a power shortage in the UPS of Russia is possible in cases of (emergency exit of the main generating and network equipment, random load changes): a shortage of generating capacity; lack of stocks of communication capacities; one and the other together. Therefore, we can say that the problems characteristic for solving the problem of estimating the indicators of balance reliability in the conditions of market relations practically do not differ from those in the conditions of the centralized management principle of the UPS of Russia. Differences can only concern the issues of distribution of system power deficiency between the territorial zones of the model of the settlement scheme of the UPS of Russia. Thus, under the conditions of liberalization, the traditional formulation of the task of ensuring balance reliability remains. Balance reliability is provided due to the optimal development of generating capacities and reserves of the capacity of system-forming links.

2 Indicators of adequacy and their normative values in our country and abroad

At the present stage, the most acceptable IA from the standpoint of justifying the means of ensuring reliability (compensation capacity reserve) are considered [5-7]:

- In the domestic practice, the probability of deficit-free operation of territorial zones of the EPS ($\rho = 1 - J_д$), where $J_д$ is the integral probability of power shortage;
- the average number of days of power shortage (loss of the anticipated load – *LOLE* [2]);
- the average number of hours per year of power shortages (loss of load hours – *LOLH*).

At us in practice of designing till now the standard to probabilistic indicators of balance reliability $\bar{\rho}$ =0,996 is applied. In developed countries the minimum value of power reserve from unplanned changes must also correspond to some pre-selected probability targets. Some of the normative values of *LOLE* and *LOLH* exist have at the expert level in North America and Western Europe have. They serve as indicators in decision-making. In the United States and Canada the minimum amount of reserve power in these countries should correspond to the index of adequacy is not higher values of *LOLE* = 0,1 d/year. Common standards for some European countries [12]: France *LOLH* = 3 h/year, the UK and the Netherlands – *LOLH* = 4 h/year, Ireland – *LOLH* = 8 h/year.

3 Problematic issues of justification of the standard reserve capacity in relation to modern conditions for planning the development of power systems

The justification of the reserves of capacity of the territorial zones of the UPS of Russia when managing their development, in modern conditions, should be directed to the application of methodological approaches aimed at minimizing the functional of the resulted costs $З_\Sigma$ (П) with the obligatory performance of the normative indicators of balance reliability.

$$З_\Sigma(П) = З_R(П) + З_L(П) \to min,\ npu\ \rho(П) > \bar{\rho}, \quad (1)$$

where $З_R(П) = \sum_{j=1}^{n} з_{R_j}^{уд.} R_j$, $З_L(П) = \sum_{l=1}^{m} з_{L_l}^{уд.} P_l^L$ – respectively, the reported costs for the maintenance of reserves of generating capacity (R_j) *j*-th territorial zones and throughput (P_l^L) *l*-th communication; $з_{R_j}^{уд.}$, $з_{L_l}^{уд.}$ - unit costs (rub / kW), respectively, to create a reserve for generation of capacity in the *j*-th territorial area of EPS and to strengthen ties; $\rho(П)$, $\bar{\rho}$ – accordingly, the indicators of adequacy given the means to ensure reliability (П) and their normative significance.

The use of standards for the indicators of balance reliability is due to the difficulty of obtaining information on compensation costs from unreliability (damages). At the same time, the question arises of

[1] SO 34.04.181-2003. Rules for the organization of maintenance and repair of equipment, buildings and structures of power plants and networks (http://regulation.gov.ru)

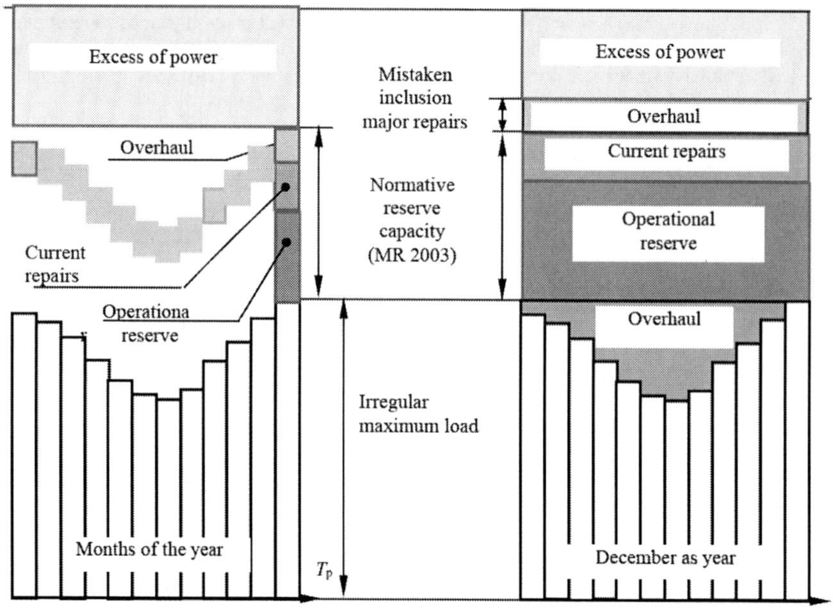

Fig. 2. Effect of simplifications in the representation of the power consumption mode in the justification of the operational (compensation) power reserve.
on the left - carrying out major repairs evenly for the seasons of the year;
right - the inscription of capital repairs in seasonal failure.

determining the numerical value of the standard for estimating the balance sheet reliability indicators $\overline{p}$ for making decisions on the justification of power reserves in the territorial UPS of Russia. In the domestic practice of planning the development of the country's UPS [1, 2] for these purposes, the reliability factor was used in the form of the probability of a deficit-free work $\overline{p}$ = 0.996. This corresponds to the integral probability of power shortage of a power deficit in the *j*-th territorial zone of the UPS of Russia $J_{д}^{опт.} = 0.004$.

The connection of the indicator $J_{д}^{опт.}$ used in Russia since the 1980s with the LOLH load loss indicators in hours and especially LOLE in the day is quite complicated. The article considers only the issues of comparing the domestic indicator with the LOLH index. Important from the standpoint of justifying the operational reserve of capacity is the understanding of two mutually affecting each other moments. The first is that in our country when planning the development of the UPS of Russia, information on the years of the prospective period is reduced to the period of maximum loads of December[2]. That is why in the practice of calculating the indices of balance reliability of the territorial zones of the UPS of Russia various simplifications are applied. One of them is the submission of the settlement year by one daily schedule of December lasting all 250 working days.

The second moment of justifying the power reserve, without which the application of the first one would be impossible, consists in using the principle of capital repairs of generating equipment during periods of seasonal load reduction (Figure 2, right). This, with a certain degree of reliability, allows us to assume that the average December daily load schedule lasts the whole year.

By the third moment, which is not directly related to the first two, the composition of the generating equipment used to justify the operational (compensating) power reserve should be included. Unfortunately, in all software complexes used in national practice, estimates of the balance reliability indicators use the principle of independence in determining the components of the operational (compensation) and repair capacity reserves. This leads to the fact that when building a generation model that takes into account the composition of the generating equipment and possible its withdrawal into emergency unscheduled repair, equipment and equipment that is not used in the balance sheet and put out for scheduled repairs (even in December) are present. Not taking into account these features adds

[2] Decree of the Government of the Russian Federation of December 27, 2010, No. 1172 (Edited on 19.01.2018) "On Approval of the Rules for the Wholesale Electricity and Capacity Market and on Amending Certain Acts of the Government of the Russian Federation on the Organization of the Wholesale Electricity and Capacity Market".

Order of the Ministry of Energy of Russia dated 07.09.2010, No. 431 (Ed., Dated August 17, 2017) "On approval of the Regulation on the procedure for determining the demand for capacity for long-term power take-off on a competitive basis in the wholesale electricity (capacity) market and the procedure for determining the planned coefficients reserve capacity in zones (groups of zones) of free flow of electrical energy (power) "(Registered in the Ministry of Justice of Russia on 29.09.2010, No. 18578) inaccuracy in the solution of the problem of justifying the reservation in the UPS of Russia.

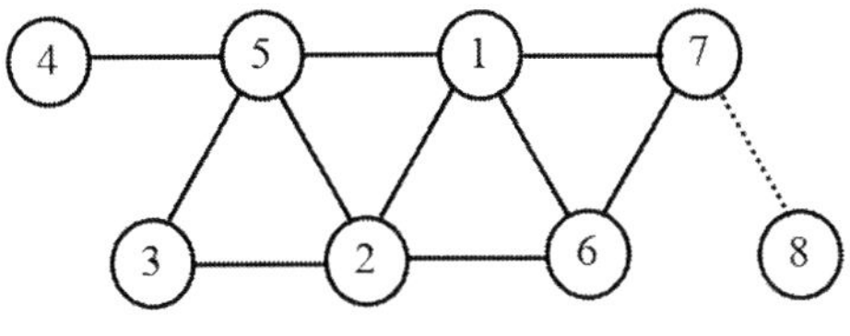

Fig. 3. Model of the settlement scheme of the UPS of Russia.
1 - UPS Urala; 2 – UPS of the Middle Volga;
3 - UPS of the South; 4 - UPS of the North-West;
5 - UPS of the Center; 6 - Kazakhstan;
7 - UPS of Siberia; 8 - UPS East.

In Table 1, the results of the influence of the above-mentioned moments on the amount of the operational (compensation) power reserve are presented. The results are given for the model of the settlement scheme of the UPS of Russia, represented as a unified power plant (Figure 3). In the first line, for clarity, information is given on the irregular maximum of the load, applied to the year 2022 considered from the work of the Russian Unified Power System of Russia for 2016-2022. Lines 2 and 5 show the results of the distribution of the operational (compensating) power reserve when the norm $\bar{p}$ =0.996, for two variants of the power consumption mode representation.

The first (line 2) consisted in optimizing the distribution of the operational power reserve when presenting the power consumption mode in the form of a single average load schedule for December, lasting all 250 working days of the year (Figure 2, right). Probabilistically determined information about the operation of the generating equipment and random deviations of the power consumption mode for this option was obtained by the specialists of SO UPS JSC and is presented in the work[3] and [8]. In the calculations of the justification for the means of reserving the territorial zones of the UPS of Russia, the provisions of the 2003 LR were fully implemented in this version [1]. At the same time, this and the second version did not take into account the impact of the actual generation equipment, due to the non-use of excess capacity and the withdrawal of part of the generating equipment for scheduled repairs. This will be discussed separately below.

In the second variant (line 5), optimization of the distribution of the operational power reserve taking into account compliance with the normative indicator of adequacy $\bar{p}$ = 0.996 was performed when the electricity consumption mode was presented taking into account its seasonal decline by the month of the year (Figure 2, left). In this case, almost the same probabilistically determined information about the operation of the generating equipment and the random deviations of the power consumption regime was used, as for the first variant. It is clear that according to the power consumption regimes this information was to undergo certain changes (in the summer months, there are less random deviations, the throughput capacity will change, etc.), but it is adopted unchanged. Seasonal (monthly) reduction of the load relative to the December maximum is also assumed conditionally, based on the data available to the authors on the UPS of the North-West.

The probabilistic indicators of the balance reliability of the territorial zones of the UPS of Russia shown in lines 3 and 6 for the optimal variants for the placement of operational (compensating) power reserves with different representations of the electric power consumption regime are of no practical interest. It can be seen that these indicators are close to the normative value adopted in our country for *j*-th territorial zones $J_{\text{д}j}^{\text{опт}}$ = 0.00418. At the same time, the *LOLH* load loss indicator obtained in the course of examining 8760 steps of the annual load schedule is determined by simply multiplying the indicator $J_{\text{д}j}^{\text{опт}}$ by 8760 h, which leads to its approximate value of 36.5 h / year (lines 3 and 6 Table 1).

Attention is drawn to the significant decrease in the optimal value of the operational reserve of power when the mode of power consumption is changed from those shown in Figure. 2, right to left. According to UPS of Russia as a whole, the decrease in the size of the operative (compensating) power reserve was 2.33% (from 10.02% to 7.69% - lines 2 and 5), in territorial zones about the same figures. Their comparison of these two lines shows how much the impact of the submission of the electricity consumption regime on the decisions taken to justify the magnitude of the operational (compensation) power reserve in the UPS of Russia.

Consider now the 4th line of Table 1, in which the mode of power consumption is changed for the received using the average daily schedule of December, lasting 250 working days of the year (Figure 2, on the right), of the power reserve allocation. Electric power consumption is represented by 12 months with characteristic daily hourly graphs (Fig. 2, left). A kind of this approach with a certain degree of reliability is the simulation of 8760 hourly changes in the load during the year. It is seen how significant in relation to row 3 of Table 1, the probabilistic indicators of balance reliability of the territorial zones of the UPS of Russia have changed.

[3] Report on R & D "Substantiation of the normative values of the components of the total reserve capacity in the context of the UPS and UPS of Russia as a whole in the planning of their development". - Syktyvkar: ISEiEPS Komi Scientific Center of the Ural Branch of the Russian Academy of Sciences (on behalf of JSC "SO UPS"). 2016. - 66 p.

Table 1. Probabilistic indicators of balance reliability for the optimal distribution of the value of the operational (compensating) capacity reserve in UPS of Russia for various views of the electricity consumption regime

Indicators	UPS Russia without UPS East	Name of UPS					
		Ural	Middle Volga	South	Northwest	Centre	Siberia
1. Irregular peaks in accordance with the Scheme and program of development the unified Power System of Russia for 2016-22.							
2022 year, MW	155680	37930	17096	16831	15151	39266	30126
2. The operational reserve of power when using the power consumption mode in the form of a schedule for December, which lasts all year (Figure 2, right), normative IA $J_{\text{д}j}^{\text{опт.}} = 0.004$							
Power reserve, MW	15610	4894	1078	1619	1629	3365	3025
%	10,02	13,09	6,31	9,62	10,75	8,57	10,04
3. Balance sheet reliability indicators using this condition power consumption mode							
$J_{\text{д}j}^{\text{опт.}}$	0,00836	0,00419	0,00417	0,00417	0,00419	0,00418	0,00419
LOLH	73,26	36,69	36,54	36,54	36,70	36,65	36,66
4. Balance sheet reliability indicators when using other load graphs (12 seasons with one daily load profile)							
$J_{\text{д}j}^{\text{опт.}}$	0,00307	0,00082	0,000818	0,00082	0,000821	0,00082	0,000827
LOLH	26,99	7,27	7,23	7,28	7,26	7,26	7,30
5. The operational power reserve at use of daily schedules of 12 months (fig. 2, at the left) normative IA $J_{\text{д}j}^{\text{опт.}} = 0{,}004$							
Power reserve, MW	11985	3894	678	1204	1254	2565	2400
%	7,69	10,42	3,97	7,15	8,28	6,51	7,97
6. Balance sheet reliability indicators using this power mode representation condition							
$J_{\text{д}j}^{\text{опт.}}$	0,007964	0,00421	0,00416	0,00417	0,00418	0,00417	0,00420
LOLH	63,983	37,14	36,72	36,78	36,86	36,79	37,03
7. Balance sheet reliability indicators when using other load graphs (daily schedule for December, lasting all year)							
$J_{\text{д}j}^{\text{опт.}}$	0,02706	0,0200	0,0199	0,0198	0,200	0,0199	0,0198
LOLH	237	175	174	173	175	174	173
8. The operational power reserve when using the power consumption mode in the form of the schedule of December, which lasts the whole year (figure 2, right), normative IA $J_{\text{д}j}^{\text{опт.}} = 0.004$ for the conditions of accounting for the dependence of scheduled and unscheduled repairs							
Power reserve, MW	14100	4411	970	1441	1502	3031	2744
%	9,05	11,80	5,67	8,56	9,92	7,72	9,11
9. Operational power reserve using daily charts of 12 months (Figure 2, left), normative IA $J_{\text{д}j}^{\text{опт.}} = 0.004$ for the conditions of accounting dependence scheduled and unscheduled repairs							
Power reserve, MW	10500	3486	470	991	1102	2306	2144
%	6,74	9,32	2,75	5,89	7,28	5,87	7,12

The domestic probabilistic indicator of adequacy in the form $J_{\text{д}j}^{\text{опт.}}$ in all j-th territorial zones decreased from the normative value of 0.0042 to 0.00082, foreign *LOLH* from 36 hours to 7.25 h.

It is clear, and this has already been written above, that the results presented are obtained with a conditional representation of the electricity consumption regime. To these conventions it is possible to carry not absolutely correct representation of seasonal reduction of loading, its casual deviations, the account of days off and holidays, the account of capital repairs and many other things. Nevertheless, the experience of calculations shows that these conventions will not have much effect on those given in line 4 of Table 1 results. This again suggests that a change in the representation of the power consumption regime should lead to a change in the normative indicator of adequacy.

The next two lines (8 and 9) characterize the distribution of the operational (compensation) power reserve in the UPS of Russia, obtained by changing the indicators of adequacy of the composition of the generating equipment taken into account (the third moment). In the calculations given in these lines, the composition of the generating equipment was determined based on the conditions for exceeding the generating power corresponding to the composition considered above the load on the component of the operational (compensation) power reserve. Such a composition in real conditions is not so easy to pick up, so certain assumptions were made. In this regard, much can be more correctly implemented when using retrospective information on the conduct of long-term power debits[4].

It should be noted that taking into account the actual composition of the generating equipment in the construction of binominal series of probabilities of emergency power reduction leads to an almost percentage reduction in the operational (compensation) power reserve. This reduction in the power reserve exceeds by more than a quarter the mathematical expectation of the generated generating equipment (power surpluses, planned repairs) from the total available composition corresponding to the installed power of the territorial zones of the UPS of Russia. Naturally, this feature of the presentation of generating equipment should be taken into account when justifying the operational power reserve of the territorial zones of the UPS of Russia.

4 Main findings

- The significant influence of the representation of the annual power consumption regime on the estimation of the indicators of adequacy the territorial zones of the UPS of Russia and their means of providing – power reserv.
- The change in the principle of accounting for the electricity consumption regime from one daily average schedule for December for 250 working days of the year for all 8760 hourly load changes will lead to a change in the normative indicator of adequacy adopted in the 2003 guidelines as an integral probability of failure-free operation $\bar{p} = 0.996$. Studies show that the most acceptable standard values of indicators of adequacy can be taken in Western European countries $LOLH = 3 \div 8$ hours, or in translation to the probabilistic index $\bar{p} = 0.9991 \div 0.9997$.
- In modern conditions, the presence of large surplus power in the UPS of Russia and the transition to scheduled repairs of generating equipment by its condition when justifying the operational (compensatory) component of the normative (full) power reserve, it is necessary to approach more strictly the definition of the actual composition of the generating equipment.

[4] Decree of the Government of the Russian Federation of December 27, 2010, No. 1172 (Edited on 19.01.2018) "On Approval of the Rules for the Wholesale Electricity and Capacity Market and on Amending Certain Acts of the Government of the Russian Federation on the Organization of the Wholesale Electricity and Capacity Market".

References

1. Methodical recommendations on designing the development of power systems. (Approved by Order of the Ministry of Energy of Russia of June 30, 2003, No. 281). - M: Ministry of Energy of the Russian Federation, SO 153-34.20.118-2003.
2. Wolkenau IM, Zeyliger AN, Khabachev LD Economics of the formation of electric power systems. - Moscow: Energoatomizdat. 1981. - 320 s.
3. Volkov G.A. Optimization of reliability of electric power systems. - Moscow: Nauka, 1986. - 117 p.
4. Kovalev GF, Lebedeva LM Reliability of power systems. - Novosibirsk: Science, 2015. - 224 p.
5. Chukreev Yu.Ya. Models for ensuring the reliability of electric power systems. - Syktyvkar: Komi Scientific Center of the UrB RAS, 1995. -176 p.
6. Chukreev Yu.Ya., Chukreev M.Yu. Models for estimating the indicators of balance reliability in the management of the development of electric power systems. - Syktyvkar: Komi Scientific Center of the Ural Branch of the Russian Academy of Sciences, 2014. – 207 p.
7. Billinton R., Allan R.N. Reliability Evaluation of Power Systems. Second Edition. New York and London, Plenum Press, 1996. 509 p.
8. Chukreev Yu.Ya. Provision of balance reliability in relation to the current conditions of long-term planning of the UPS of Russia // Methodological issues of reliability research of large energy systems. Issue.68. - Irkutsk: Research and provision of reliability of energy systems - ISEM SB RAS, 2017. - P. 56-65.

The assessment of probabilities and risk of rare events in electric power industry

Boris Papkov[1]*, *Aleksandr* Kulikov[2] and *Vladimir* Osokin[1]

[1]Nizhny Novgorod State Engineering-Economic University, Knyaginino, Nizhny Novgorod region, Russia
[2]Nizhny Novgorod State Technical University named after R.E. Alexeev, Nizhny Novgorod, Russia

Abstract. Issues related to the assessment of probabilities for significant but rare events, extreme and catastrophic situations in the electric power industry are herein examined and analyzed. Special attention is paid to methods of the quantitative assessment of technological risk and its factors. Certain examples are given.

1 Introduction.

In view of possible disastrous consequences of failures and incidents in the operation of complex technical systems (CTS) the problem of their functioning reliability assessment both at the stage of design, the selection of basic technical solutions and during the operation is one of the priority objectives when studying the problems of the human and environmental safety assurance. The task is significantly complicated for unique CTS because of the lack or limitation of statistical data on the reliability of such systems under actual operating conditions and any unavoidable uncertainties in the assessment of their behavior in case of any rare design and "beyond-design" situations associated with equipment failures, human errors, various external influences caused through the incomplete knowledge about real operating conditions.

The problem of the assurance of human safety and potentially hazardous facilities has been and remains very urgent. It turned to the scientific category because of revealing severe after-effects of rare events arising with a "near-zero" probability.

The safety is in most cases defined as a condition when under threat the level of possible harm to property or personnel is assessed through a risk category with its values not exceeding the permissible (standard) value [1]. The essence of the problem is the quantitative assessment of the degree of danger (risk) and its admissible levels. Danger is herewith considered as an objectively existing possibility of a negative impact on the society, an individual and environment, which may result in social harm, technological, economic or environmental damage. Risk here characterizes any potential danger, the possibility for the implementation of negative events (any worst alternatives) causing harm.

The assessment of the risk of extreme, emergency and catastrophic situations occurring in energy systems consists in the need to assess their probabilities. The main complexity of such an assessment is in the small amount and uncertainty of the relative unreliability of retrospective information. So, it practically does not enable us to obtain the effective, unbiased and consistent statistical estimates of reliability, safety, and survivability indicators. The threat of acts of God, natural calamities and man-made disasters, terrorism, cyber attacks, various kinds of conflicts and any other rare events has turned into a system regularity of the late twentieth century and the early twenty-first century [2]. It is evoked by the modern technological mode of life with new technical, economic, social, political and other artificial systems, which may be more vulnerable because of their complexity to all kinds of accidental and deliberate disturbances and consequently less persistent, less reliable, less safe.

Any extreme (catastrophic) events are associated with rare emissions of a random process, for which no reliable statistics is available. The pulse occurrence time and amplitude are random and indefinite values.

Traditional factors of the possibility of a rare (unpredictable) event are uncertain. The values of a series of dynamic quantities, which determine them, lie on the "tails" of distributions getting within an unlikely (critical) region, which is generally "fuzzy". Therefore, we have to shift to the Fuzzy Sets methods, confining ourselves to the conservative estimate of this area, deliberately expanding it due to the uncertainty which is inevitably associated with more or less justified extrapolation based on "conservative (safe)" expert appraisals.

An increasing risk of the hazardous situation occurrence at electric power facilities is accompanied with the sharp raising cost of compensations for their technical, economic and social consequences with simultaneously reducing safety for operating personnel and the population as a whole.

After the Chernobyl accident (1986) some extraordinary safety precautions were taken all over the world. Risks have significantly and sharply decreased. We hope that such large-scaled accidents will not occur frequent-

* Corresponding author: boris.papkov@gmail.com

ly. But the question is how unoften? The statistical analysis carried out by British and Swiss mathematicians on 216 accidents and emergencies registered for the entire nuclear power engineering history showed that accidents in fact became far less frequent. Notwithstanding, based on the available data [6], it is assumed that Fukushima-like or larger accidents will occur every 60–150 years with probability of $p = 0{,}5$.

2 Elements of Risk in Electric Power Systems.

The reason for off-design and hypothetical accidents is a chain of events or a scenario with a system getting to a dangerous state, the probability thereof is irrelevant, if damage is significant and inadmissible for system users [2, 3]. Moreover, it is incorrect to introduce the concept of "average time till disaster" for rare events. The main task is related to the search for "vulnerability windows" and its solution is performed by methods of the multicriterial evaluation of such generalized factors like "effectiveness", "reliability", "security", "vulnerability", "acceptability", "efficiency", "ergonomics," "competitiveness" and so on. [1].

Danger is characterized not only by the probability (relative frequency) of the crisis (emergency) situation commencement but also by the severity of its consequences. It is shown in earlier publications [2] that the numerical value of risk R in failure of the normal operating mode of equipment in power engineering systems may be assessed as

$$R = pY,$$

where p is the probability of a hazardous event (accident, failure, incident, etc.); Y is the mathematical anticipation of damage to be resulted from this event.

The quantitative risk level is consistent with the intuitive perception thereof and it divides its evaluation procedure in the following two stages: 1) the determination of probabilities for adverse outcomes; 2) the estimation of damages following the said outcomes.

To solve the technotronic security problem, the following two concepts [4] are available: deterministic (of "absolute" safety) and probabilistic (of acceptable risk) ones.

The first concept is based on a normative approach and a logical analysis of the technological process features and sources of possible accidents, the elaboration of a system of mandatory measures (emergency protection, instructions, personnel training). The main drawback of this concept is the lack of the quantitative assessment of safety and sensitivity to specific features of a particular facility as the actual effectiveness of a set of mandatory security measures remains undetermined.

The second concept is based on the statistical analysis of the hazard (risk) level, the results of simulation modeling to identify possible ways for the transition of emergency situations to accidents, the identification of acceptable risk levels. To this end, information is needed on probabilities for the occurrence of accident initiating events and the readiness of protective means, the representativeness thereof is obviously insufficient. When analyzing the risk some difficulties arise related to the initial data completeness and reliability, the adequacy of models of the facility transition to an emergency condition.

The both concepts contain therewith at least two kinds of uncertainties [4]: 1) the uncertainty in assessing the adequacy of security measures taken; 2) the uncertainty associated with the unenumerability of the accident initiating events at power facilities. The said uncertainties are also reflected in recommendations for the critical risk assessment. So, in [5] it is noted that the drift of key safety factors over a critical level is considered disastrous and in [6] that the risk level of the hazardous situation occurrence currently admissible to society in terms of reasonable damage and costs is *acceptable* or *permissible*. The identification and prevention thereof enables you to reduce damage to 10% of an expected one.

3 A Few Examples.

In simulated technical systems (elements and facilities of electric power industry) from time to time emergency situations from time to time occur with rapid nearly instantaneous developing events. When solving the reliability assurance problems an important significance is given to statistical appraisal methods. But the statistical estimates of reliability factors give acceptable results only, if an event is characterized by a mass scale and homogeneity, non-observable in a number of problems.

1). The probability of a design accident on offshore drilling platforms was estimated as one accident per 20 million years ($p = 0.00000005$). However, accidents have already happened at 15 drilling platforms, including the largest one in the Gulf of Mexico in 2010 [5].

2). The estimated probability associated with the possible overflowing of the storage reservoir in the Zeyskaya HPP was $p = 0.001$. Nevertheless, that event occurred in 2007 as a result of heavy rains.

3). By the early 20th century protective structures of the unprecedented 390 cm level were designed in the Netherlands [7]. However, in the course of their construction the height was assumed to be 340 cm; that was by 12 cm higher the maximum recorded for the 25-year monitoring (once per 70 years with the probability of $p = 0.014$). The construction cheapening resulted in the tragedy of 1953. About 2,000 persons perished and huge destructions occurred. Currently hydraulic structures in the Netherlands are built up to the maximum level of 500 cm (once per 10,000 years ($p = 0.0001$)).

4). Moscow, May 2005. The accident occurred at the Chagino-500 kV substation. Its consequences were: 4 million human sufferers and the loss exceeded 3 billion rubles.

5). August 2009 the Sayano-Shushinskaya HPP accident. Its consequences were: 75 persons perished, the loss exceeded 40 billion rubles.

6). It was assumed that the probability of an accident at a nuclear power plant under the normal probability law shall be $p = 1 \cdot 10^{-7}$ per year^{-1}, i.e. one accident per

10 million years. But as time has shown there should be other assessments.

The above examples are just random events with the objective assessment of their probabilities to be impeded because of the relatively high reliability of unique energy object functioning and limited retrospective information.

4 The Elementary Estimation of the Rear Event Probability.

Having identified any practically possible risks, their probability and possible consequences are assessed.

1). Supposing that the designed maximum power of a load centre in the electrical system is $P_m = 500$ MW. It is known that it's excess at $\Delta P = 1$ MW (0.2%) is possible with the probability of $p = 0.01$. In this case, all consumers supplied from this center are disconnected because of the input overload. Obviously, in a number of cases such a probability (possibility) for a similar mode may be neglected. Notwithstanding, when solving the problem of the reliability for power supply systems at explosive or fire hazardous industries, life support systems of the maximum transmission capacity of a power transmission line, the probability of $p = 0.01$ should be considered sufficiently large and to be taken into account when designing the structure of power supply systems. To the point, in 1965 in the USA, the cutting off of the power transmission line with a load of about 2.5% of the capacity of the power system components connected thereby resulted in the bulk failure and a long-term power supply interruption in an area with 30 million inhabitants.

2). In addition to the usually considered numerical characteristic of the position of the random variable (RV) X – the mathematical expectation $M(X)$ – the need often arises to use initial moments of its distribution. The initial moment of the sth order for RV X shall be determined by the mathematical expectation of the sth degree of this RV as

$$a_s\ X\ =\ {}_{i=1}^{n} p_i x_i^s .$$

Let us analyze the consequences of a sudden power failure of different duration for a conditional production resulting in its idle time t_{idle} with the following probability distribution

t_{idle}	0.1	0.5	1.0	8.0
p_i	0.5	0.3	0.15	0.05

Mathematical expectation for RV t_{idle}:

$$M\ t_{idlei}\ =\ {}_{i=1}^{4} t_{idlei} p_i = 0{,}75.$$

Mathematical expectation for squared RV t^2_{idle}:

$$M(t_{idle}^2) =\ {}_{i=1}^{4} t_{idlei}^2 p_i = 3{,}43.$$

Thus, the transition from $M\ (t_{idle})$ to $M\ (t^2_{idle})$ enables to take into account with a lesser degree of risk the influence of a possible value of (t_{idle}), which is large and has a small probability. If RV X had several large and unlikely values, then the transition to the calculation of the initial moments of higher orders $a_3(X)$, $a_4(X)$, etc. would further strengthen their role.

3). Let us consider the case when on the studied electric power facility no extreme event should occur within the supposed time of its operation t_{op}=50 years. So, in accordance with [8] and the Poisson distribution law, the lower confidence limit of failure-free time tn may be determined for the confidence probability α. If $\alpha = 0.95$ is taken, what is typical for the majority of technical solutions, the lower limit for the time of the non-failure operation will be t_n = 16.6 years. Under more stringent conditions with $\alpha = 0.99$ we obtain t_n = 10.8 years.

4). A system (a two-transformer substation) consists of two parallel operating elements. Each of them fails in accordance with the exponential distribution law and the parameter λ. In case of independent failures the probability of failure within time t shall be

$$q\ t\ = (1 - e^{-\lambda t})^2. \qquad (1)$$

In accordance with [9] we assume that independent failures of elements occur with intensity λ_0 and with intensity λ_1 critical failures happen, under the impact thereof each element may become faulty with a probability of $\alpha = 1 - \beta$. With $\lambda_0 + \lambda_1 \alpha = \lambda$ the exponential property is preserved but with $\alpha > 0$ the reliability of the system significantly changes. If $p_i(t)$, $i = 0{,}1$ there is the probability for availability of i faulty elements at the time t, then in [9] a system of two Kolmogorov's differential equations is given, their solution under the initial conditions $p_0(0) = 1$; $p_1\ 0\ = 0$ takes the form of

$$p_0\ t\ = e^{-\ 2\lambda_0 +\ 1-\beta^2\ \lambda_1\ t}$$
$$p_1\ t\ = 2e^{-(\lambda_0 + \alpha\lambda_1)t}[1 - e^{-\ \lambda_0 + \alpha\beta\lambda_1\ t}], \qquad (2)$$

where the probability for failure of the system is determined as:

$$q\ t\ = 1 - p_0\ t\ - p_1(t). \qquad (3)$$

In the limiting case with $\lambda_0 = 0$, $\alpha = 1$, $\lambda_0 = 0$ we obtain: $p_0\ t\ = e^{-\lambda t}$; $p_1\ t\ = 0$, $p_0\ t\ = e^{-\lambda t} p_1\ t\ = 0$ i.e., $q\ t\ = 1 - e^{-\lambda t}$

If failures are independent and $\lambda t = 0{,}01$ acc. to (1) $q(t) \approx 10^{-4}$ In fact, according to (3) and taking into account (2) $q(t) \approx 10^{-2}$.

The results are self-explanatory. Therefore, it is necessary to check the conditions for the possibility of disaster occurrence in case of the disability of the electric power facility under the exposure to certain flow of damaging factors.

5 Discussion.

Ideally, the cumulative impact of multiple unrecordable factors (sudden failures causing the cascade-like development of accidents, harmful emissions, explosions, fires, rainfalls, their intensity, heat and moisture exchange of the atmosphere with the surface, etc.), according to the central limit theorem of the probability theory corresponds to the normal distribution.

For the majority of technical systems, the normal (Gaussian) and exponential random parameter distributions "with short tails" are typical. For example, the distribution of power load maxima of stably working consumers is close to normal and their numerical values do not fit generally within the range of $\pm 3\sigma$ with the probability of $p = 0.0028$. But their values herewith fall beyond the limit of 5σ with the probability of $p \approx 0.000001$.

In [5] it is recommended to use the exponential or power law distribution, which is referred to the class of distributions with "heavy (or long) tails", since the statistics of the major natural and man-made accidents, catastrophes, disasters such as earthquake, flood, snowslides, hurricanes, tornadoes, droughts have a power law of probability distribution. This is due to the fact that all natural and man-made systems apt to major accidents, catastrophes, crises are quite complex, contain a lot of interrelated and mutually dependent elements subjected to the exposure to many random factors with a wide range of parameter values. Their description may not be reduced to a simple sum of a great number of independent terms and the normal distribution law, since the consequences of such events may not be disintegrated into a set of independent sub-processes. It is known that even minor nonlinearities of a dynamic system shall essentially change the "tails" of distributions and, consequently, the estimates of probabilities of catastrophes. Their integral (systemic) research and description is herein required.

Since hardly probable accidents and catastrophes within a limited period of time are quite possible, the neglect of values of random variables falling into the "tail" of such distributions is already inadmissible. To illustrate the foregoing, Fig. 1 shows a typical form of the distribution of the random variable probability densities under the normal, exponential and power distribution laws and it is seen therefrom that the "tail" of the power-series distribution is significantly "heavier", changing very slowly.

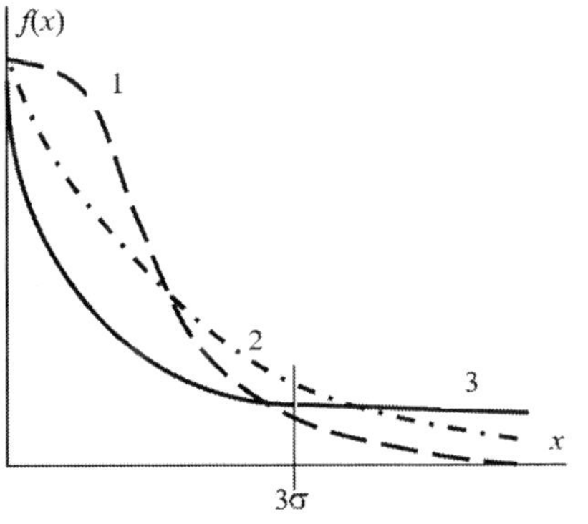

Fig.1. Comparative densities of the normal – 1, exponential – 2, power – 3 laws of the random value distribution

The power law of the probability density distribution has the following form:

$$P\ x\ \sim x^{-\ 1+\alpha},\ \ \alpha \sim 1.$$

The less is α, the more dangerous is the process. For all problems considered $\alpha < 1$.

The difference between the normal and power distributions is of a principal nature. In terms of the safety and risk assessment, the "tail" of the distribution corresponds to the so-called hypothetical accidents and catastrophes; the possibility thereof is practically zero. But as for lack of statistics the "tail" zone is not determined and the reliability of analytical calculations causes "distrust", it is necessary to assume that the actual value of the distribution density functions is unknown in this field.

The availability of the power law of the probability distribution radically changes the perception of the system reliability, possible consequences of extreme events, risk.

The statistics described by power distributions is distinguished by the fact that rare events occurring at the "tail" of the distribution are not such seldom as to be ignored. For the power distribution, the probability of disastrous consequences may be by an order of magnitude more than the probabilities calculated on the basis of the exponential or normal distribution.

6 Comparative Results.

1. Comparative results. To illustrate catastrophic nonlinear effects, in literature data are presented on the results of natural spontaneous phenomena (disasters). In some publications [1, 7, 10, 11] the probabilities for the occurrence of catastrophic events related to extreme floods in accordance with the power and gamma probability distributions were calculated. Their comparison is presented in Table 1.

Table1. Estimated Probabilities of Anomalous Events

Facility	Gamma distribution	Power-series distribution
1	0.00005	0.015
2	0.00036	0.0039
3	0.001	0.059
4	0.0025	0.009
5	0.000036	0.012
6	0.00011	0.098
7	0.0015	0.006
8	0.0055	0.026
9	0.0019	0.0114
10	0.01	0.029

It is noted in [7, 11] that the exponential and gamma distribution densities in the studied small probability region nearly coincide, but the gamma distribution has a greater "flexibility" and therefore it is preferable. The densities of the power and gamma distributions well coincide in the middle part and differ greatly in the region of anomalous events. It explains the difference in their expected repeatability. In other words, the occurred catastrophic floods are not near incredible event but have a fairly high probability of recurrence even during the lifetime of the current generation. Therefore, such distributions should be taken into account, when assessing the reliability and safety of facilities and systems of vital importance.

7 Conclusion.

The proposals presented in this paper open up the prospect for the further study of the extremely important problem of estimating the probabilities of rare events and catastrophes in the electric power industry. Time series characterized by the slow decrease in the number of rare events should be represented by the power distribution. Statistics described by the power distribution states that:

- events falling on the "tail" of the distribution are not so rare as to be ignored;
- catastrophic events are not exceptional but have rather a high probability, which is to be taken into account;
- for the power distribution the probability of catastrophic consequences may exceed the probabilities calculated on the basis of a normal, exponential or gamma distribution by an order of magnitude or more.

Reference

1. E. Kuklev, Determination of Risks for Hazardous Approaches of Marine Vessels in Emergency Situations by Predicting Fuzzy «Vulnerability Windows». Transport of the Russian Federation. №**4** (65) P. 28–31 (2016)
2. B. Papkov, A. Kulikov, The System Theory and System Analysis for Power Engineers. – M.: Publishing Office Yureit, (2016)
3. Reducing Risks of Cascade Faults In Electric Power Systems. – Novosibirsk: Publishing Office SO RAS, (2011)
4. B.Volik, On Technotronic Security Concepts. Automatics and Telemechanics, № **2**. P. 165–170. (1998)
5. I. Prangishvili, System Approach and Management Efficiency Enhancement. – M.: Science, (2005)
6. N. Chura, Technogenic Risk. – M.: KNORUS, (2015)
7. A. Sholomitsky, Theory of Risk. Choice under Uncertainty and Simulation of Risk. – M.: Publ. Office of SU HSE, (2005)
8. Y. Shor, F. Kuzmin, Tables for Reliability Analysis and Control. M.: Publ. Office Soviet Radio, (1968)
9. I. Kovalenko, N. Kuznetsov, Methods of High-Reliable System Design. M.: Radio and Communication, (1988).
10. I.Naydenov, I. Kozhevnikova, Why Do Floods So Frequently Occur, Nature, №**9**, (2003)
11. E. Gumbel, Statistics of Extreme Values. M.: Mir, (1965)

Corporate Governance in Russian electric power industry in terms of consumer requirements to its funding sources

G.I. Sheveleva[1]*

[1]Melentiev Energy Systems Institute, Siberian Branch of the Russian Academy of Sciences, Irkutsk, Russia

Abstract. The paper highlights a strong interest of energy consumers in attracting investment in the development of Russian power generation companies. The importance of corporate governance for enhancing the investment attractiveness of these companies is emphasized. An in-depth evaluation of their current corporate practices was carried out within the framework of the existing ownership structure. The study identified the indicators of corporate governance quality for the benefit of modern investors that are the least observed by the overwhelming majority of power companies. The indicators were obtained on the basis of whether or not the companies satisfy the criteria of the new Russian Corporate Governance Code, and the criteria of the methodologies of Standard & Poor's, Spencer Stuart and Transparency International. The study shows a slight increase in the transparency of the companies in the post-reform period and compares it with the information disclosure by the major corporations of Great Britain, the USA and Europe. The study shows high correlation of the approach and composition of the identified indicators of the corporate governance quality for Russian power generation companies with the 2017 Russian Corporate Governance Index. This Index is based on the international Good Governance Index methodology adapted to the Russian conditions.

1 Introduction

The main financial burden of construction of power plants after the reform of RAO "UES of Russia" was carried by industrial energy consumers under the program for power delivery contracts. Upon completion of the program, they oppose the further use of such a mechanism for financing the development of generating companies. Consumers are interested in attracting investment from these companies [1].

Creation of an environment of trust, transparency and responsibility necessary to stimulate long-term investments is facilitated by quality corporate governance [2]. In the world practice, it proved its importance in enhancing the investment attractiveness of foreign corporations, getting more affordable, inexpensive and stable sources of financing.

The domestic experience of applying the generally accepted standards of advanced corporate practices in the world is relatively small. The basic milestone in their adoption was the Code of Corporate Conduct (2002), which clearly formulated the requirements for the proper conduct of Russian joint-stock companies regarding investors. The past years have seen significant changes, including the changes in the Russian corporate practice. They required the development of a new Russian Corporate Governance Code (hereinafter - the new Code). Corporate governance in the new Code is also defined as the most important measure "... to improve the stability and efficiency of joint-stock companies, increase the flow of investment in all sectors of the Russian economy from both domestic sources and foreign investors" [3].

A slow increase in the quality of corporate governance in Russian power generation companies was noted. In most of these companies, it is formal, declarative, and has a temporary nature of improvement. Furthermore, it does not correspond to the universally recognized standards of the best corporate practices of the states of the Organization for Economic Cooperation and Development: justice, responsibility, transparency and accountability.

2 Indicators of corporate governance quality in Russian electric power industry

The study identified the indicators of the corporate governance quality in the Russian power generation companies according to the formed ownership structure [4-5].

The indicators are based on whether or not these companies fulfill the criteria of the new Russian Code of Corporate Governance, and the criteria of Standard & Poor's, Spencer Stuart, and Transparency International

* Corresponding author: sheveleva@isem.irk.ru

methodologies that evaluate the corporate governance quality, proceeding from the most essential objective parameters for the investor [6-23].

The results of this study made it possible to additionally assess the significance of the membership balance and the performance of Boards of Directors (BODs), to compare the transparency of the investigated companies with the information openness in major corporations in Great Britain, the USA, and Europe.

2.1 Identification of indicators by criteria

2.1.1 Standard & Poor's

The criteria of Standard & Poor's methodology allowed comparing the current transparency indices of the Russian power generation companies with their values calculated by the Standard & Poor's for these companies in the first year after the liquidation of RAO "UES of Russia" (2009). These criteria concerned: ownership structure; rights of shareholders; financial and operational information; composition and procedures of the BOD and management; fees for the management and Directors.

The values of the indices obtained by the criteria of Standard and Poor's indicate a minor improvement in the transparency and, hence, quality of the corporate governance in the majority of the Russian power generation companies over the post-reform period.

As before, the ownership structure is disclosed poorly. Beneficiaries and "other shareholders" of the companies are less disclosed. This is especially characteristic of the companies with the assets belonging to the Russian private holders having foreign offshore companies for this purpose.

The number of such companies and latent "other shareholders" has increased over the post-reform years. The increase in the offshore ownership is especially typical of the PJSC "TGK-2", PJSC "Quadra", PJSC "T Plus", and "Siberian Generating Company" LLC [5].

Low values of disclosing the content of their Code of Corporate Governance and Code of Ethics are mainly caused by the fact that they have not been updated. The Code of Corporate Governance has not been changed since 2006 in 85% of the companies, the Code of Ethics – in 63% of the companies, despite the new 2014 Russian Corporate Governance Code.

The information on BODs is classified. The policy of self-evaluation and external evaluation of the BOD performance is not disclosed. There is no uniform methodology of such an evaluation. It is not present in the new Code, whose recommendations are restricted to a simple enumeration of individual criteria, like «... professional and personal qualities of BOD members, their independence, congruence, and degree of their participation in activity» [3]. Frequently, the self-evaluation comes to simply questioning the BOD members on various organizational issues. As a result, neither shareholders, nor BODs themselves know what to do with the results of such evaluations and abandon them. In 2016, 8% of the investigated companies and 19% of the Russian companies involved the external third party to evaluate the BOD performance. According to Spencer Stuart, this is an obvious growth for the Russian companies compared to 11% and 6% in 2015 and 2014, respectively [7-9]. However, the above estimates are far from the similar indices for major public companies in some other countries. In 2016, these indices were 43.3% for Great Britain, 29% for Italy, and 28% for the Netherlands [9].

There is no detailed information on the fees for each BOD member. As a rule, there are only total data on the fees for the BOD members in the Quarterly Reports of the companies.

In the companies, there is, practically, no relationship between the fees for the BOD members and the company performance, which could bring together the financial interests of BOD and long-term interests of investors. Such a relationship was noted in the PJSC "Unipro", where the annual fees for the BOD members depend on the Earnings Before Interest, Taxes, Depreciation and Amortization (EBITDA). Herewith, the EBITDA corresponds to the values of annual financial statements of the company under the international standards, confirmed by an external auditor. The advantage of this index, compared to other financial indices, is leveling the differences between the evaluated company and its analogs in terms of taxation and consideration of depreciation charges. The annual fees are not paid to the BOD members if the ratio between actual and planned EBITDA indices is under 0, 8. This approach of the PJSC "Unipro" to the evaluation of the BOD performance, in our opinion, is more promising for the benefit of the investor, than the Code recommendations on the fixed annual fees for the BOD members.

According to the criteria of Standard & Poor's, we identified the least disclosed information components in the Russian power generation companies (less than 35%). They are referred to modern indicators of corporate governance quality for these companies:

- Number of the shareholders possessing over 10% of shares each and information about them.
- Data on the beneficiary shareholders.
- Content of the Corporate Governance Code.
- Content of the Code of Ethics.
- Announcement of recommended dividends before the date of record.
- Calendar of the future important events for shareholders.
- Information on some non-auditor services rendered by Auditor.
- Exact conditions of transactions with related parties.
- Audited financial statement according to International Financial Reporting Standards before late April.
- Social reporting (as per The Global Reporting Initiative, GRI).
- Detailed information on non-financial agreement provisions.
- Policy for the evaluation of the BOD performance.
- Detailed information on the fees for the BOD members.
- Detailed information on the fees for the management.
- Information on linkage between the fees for BOD members and the company performance.

2.1.2 New Code, Spencer Stuart, Transparency International

These criteria enabled a more thorough evaluation of the corporate governance quality in the investigated companies. As a result, we obtained additional (to those identified by Standard & Poor's) governance quality indicators. These indicators show whether the company:

❖ Provides shareholders with the information about who nominated candidates to the BOD.

❖ Has no less than 1/3 of an independent member in the BOD.

❖ Discloses the information on the attendance of the meetings of the BOD and its committees by individual directors.

• Has the principles and approaches to the organization of a system for risk management and internal control.

❖ Notifies the BOD of intentions or appointments in the governing boards of other organizations.

❖ Develops and introduces the policy on fees for the BOD members, executive bodies, and other executives, including transparency of mechanisms for determining the fees.

• Has a list of criteria for determining materiality of events.

• Evaluates the effectiveness of the risk management and the internal control system.

❖ Addresses the most important issues at the BOD in-person meetings.

❖ Pays attention whether all the members of the BOD Audit and Remuneration Committees are independent, and at least one of the independent members of the Audit Committee has the experience and knowledge of financial statement preparing, analyzing, and auditing.

❖ Involves independent directors in heading the BOD Committees.

❖ Involves the BOD in addressing the issues of corporate governance practice.

❖ Involves the BOD in addressing the issues of observance of the information policy.

❖ Has the procedures for assessing if the number of the BOD members meets the company's needs, if its individual members, committees and the entire BOD are effective, and for analyzing the needs associated with the BOD members' qualifications, their experience, knowledge and business skills, absence of conflict of interests, etc.

❖ Provides shareholders with the results of assessing all the nominees to be elected to the BOD.

❖ Involves independent directors in the procedure for assessing and approving material corporate events before their implementation.

❖ Has an expanded list of the grounds, by which the BOD members are considered to be interested in the companies' transactions.

• Approves the policy on anti-corruption enforcement.

• Discloses a complete list of the affiliated and associated companies, joint ventures, and other subsidiary structures.

• Provides the data on offshore activity zones of the affiliated and associated companies, joint ventures, and other subsidiary structures, financial expenditures in these zones.

The indicators are listed in decreasing order of the number of companies disclosing such information (under 35%) and, hence, corresponding to these indicators. Three indicators from the Transparency International that are used to assess the transparency of corporate reporting in the Russian companies finish the list. These indicators are especially significant for the investor in the Russian power generation companies with a high share of the offshore ownership. As noted above, the PJSC "T Plus" and the "Siberian Generating Company" LLC refer to such companies. They were studied by the Transparency International and showed zero disclosure of data on the offshore activity zones of their affiliated and associated companies, joint ventures, and other subsidiary structures, as well as on the financial expenditures in these zones. This study indicated a low mean index of the corporate reporting transparency in these companies for all the three indicators of the Transparency International. This index was 4.4 for PJSC "T Plus", and 0.9 (of 10 points for the companies with the highest transparency) for the "Siberian Generating Company" LLC [10].

The overwhelming majority of the above indicators refer to the BODs of the companies (marked with a diamond). They also include the indicators concerning independent directors that are especially significant. Such significance is caused by the high concentration of the ownership in the Russian power generation companies, and the absence of the effective external corporate control from banks and stock market. The studies indicate that the mean index of the independent directors in BODs of these companies was relatively low (under 27% as of 2016). This value is below the mean index in the Russian companies (36.7%) and the lowest among similar indices in the companies of Europe and the USA. The highest percentage of independent directors (according to Spencer Stuart) are noted in the BODs of the companies in the Netherlands, Finland, and Switzerland (84%), and the USA (85%) [9].

The other indicators are also indirectly related to BODs, because, within the delegated authorities, it is the BODs that are responsible for developing, approving, and implementing all the internal documents in the interests of shareholders.

3 Correlation of results with Russian Corporate Governance Index

The identified indicators of corporate governance quality in the Russian power generation companies showed a high correlation with the indicators of 2017 Russian Corporate Governance Index [25].

3.1 Russian Corporate Governance Index

Russian Corporate Governance Index (hereinafter - Index) was compiled twice (in 2016 and 2017) by the Association of Independent Directors in cooperation with the National Research University "Higher School of

Economics", Bank of Russia, Moscow Exchange, and Russian Union of Industrialists and Entrepreneurs [24-25].

The index was developed to understand, what, indeed, good corporate governance for the Russian companies is through the eyes of present-day investors.

The Index was based on the Good Governance Index international methodology that was developed by the British Institute of Directors and Cass Business School (2015), and adapted for the Russian conditions.

There was only one invariable basic principle of this international approach, namely, to analyze the compliance of the companies with the best practices of corporate governance by using:

- Objective signs (indicators) of corporate governance from reports and other open sources concerning the activities of the companies.
- Data on the quality of corporate governance in the companies from the online survey based on their perception by representatives of professional, investment, and expert communities.

The international partners compiling the Good Governance Index discovered which objective requirements of the corporate governance in the world practice should be met by a company for the investor to positively perceive it. In the world practice, these requirements concern audit, risks, and interactions with stakeholders. Estimates of the corporate governance objective signs within these areas highly correlated with the data of the online survey of the stakeholders.

The indicators from the complete final list of the 2017 Index indicators mainly reflected the areas of the corporate governance recommended by the British Institute of Directors: the BOD composition and activity; audit and risk assessment; remuneration; relations with shareholders and stakeholders. To reflect the Russian specificity, the indicators additionally considered changes in the listing rules of Moscow Exchange, adoption of the new Russian Code of Corporate Governance norms, and possibilities to obtain information from open sources.

The 2016 Index demonstrated high significance of the BOD membership balance and effective activity as an objective requirement for corporate governance that affects its positive perception by the investor in the Russian conditions [24]. In the 2017 Index, "... only one factor had a positive and relatively considerable correlation - and that was the Board of Directors." Investors "evaluate more positively the companies, whose BODs approach the best corporate practices, i.e. have independent chairmen, sufficient number of independent directors, who, in turn, head committees, etc." [25].

3.2 Correlation of results

3.2.1 Approach to the identification of indicators

The objective signs of corporate governance in the 2016 Index methodology were mainly related to the new Code [24]. In the 2017 Index, the developers «... managed to depart from excessive attention to the Code of Corporate Governance and - following the recommendations of the British Institute of Directors - to more widely use the corporate behavior indicators promoted by known analytical and news agencies» [25].

Such an approach to improve the list of corporate governance quality indicators completely correlates with the approach of forming the indicators in the research done.

3.2.2 Central role of the Boards of Directors

In our study, conducted for the Russian power generation companies, we obtained a comparable result reflecting the central role of the BOD in improving the quality of their corporate governance.

3.2.3 Importance of independent directors in the Boards of Directors

The conclusion made on the high significance of the presence of independent directors in these companies proved close to the 2017 Index results. More than 30% of the corporate governance quality indicators in that Index concerned independent directors.

3.2.4 The number of the corporate governance quality indicators

The study also indicated a high correlation between the results related to the number of the corporate governance quality indicators. The number of indicators in our study is 35 compared to 34 indicators in the 2017 Index.

3.2.5 Content of indicators

The coincidence in the content of these indicators was, practically, 80%. Non-coincidences concerned, mainly, the presence of additional corporate governance quality indicators (identified by the Transparency International criteria and not considered in the 2017 Index) for the Russian power generation companies.

References

1. Industrial consumers offered their version of updating power capabilities. [Online]. Available: http://www.bigpowernews.ru/print/81977.phtml
2. OECD (2016). Principles of corporate governance G20/OECD, OECD Publishing, Paris. [Online]. Available: http://dx.doi.org/10.1787/9789264252035-ru
3. Corporate Governance Code. Normative acts and operational information of the Central Bank of the Russian Federation, Bulletin of the Bank of Russia, **40** (2014)
4. RBC: 500 largest revenue companies. [Online]. Available: https://www.rbc.ru/rbc500/?utm_source=PR+%D0%A0%D0%91%D0%9A&utm_campaign=876c961

74d-EMAIL_CAMPAIGN_2017_09_21&utm_medium=email&utm_term=0_640f3b0abb-876c96174d-167404361

5. G.I. Sheveleva, ECO **6**, 102 (2017)
6. Investigation of information transparency of Russian companies in the electricity sector 2009: Standard & Poor's Corporate Governance Rating Service. [Online]. Available: http://www.bigpowernews.ru/photos/63/11663_cq85sku0tcccYmQ8VDYSTi6YvsXma40c.pdf
7. Russian Board Index 2015. Spencer Stuart. [Online]. Available: http://www.spencerstuart.com
8. Russian Board Index 2016. Spencer Stuart. [Online]. Available: http://www.spencerstuart.com/research-and-insight/board-indexes
9. Russian Board Index 2017. Spencer Stuart. [Online]. Available: http://www.spencerstuart.com/research-and-insight/board-indexes
10. Transparency of corporate reporting: the report of Transparency International - Russia. [Online]. Available: https://transparency.org.ru/special/trac2018russia/
11. Report on the compliance of PJSC "TGC-1" with the principles and recommendations of the Corporate Governance Code. [Online]. Available: http://www.tgc1.ru/fileadmin/ir/reports/annual/2016/godovoi_otchet_tgk-1_2016_eng.pdf
12. Report on compliance of JSC "TGK-2" with the principles and recommendations of the Corporate Governance Code. [Online]. Available: http://www.tgc-2.ru/upload/iblock/f8d/areport_2016.pdf
13. Report on compliance with the principles and recommendations of the Code of Corporate Governance recommended by the Bank of Russia Letter No. 06-52 / 2463 of 10.04.2014 "On the Code of Corporate Governance". [Online]. Available: http://www.quadra.ru/shareholders/information_for_shareholders/meeting_of_shareholders/gosa2017_AnnualReport.pdf
14. Information on compliance of PJSC "Enel Russia" with the principles and recommendations of the Corporate Governance Code. [Online]. Available: https://www.enelrussia.ru/content/dam/enel-en/documents/en/investors/annual/Annual_report_eng.pdf
15. Report on compliance of PJSC "Unipro" with the principles and recommendations of the Corporate Governance Code. [Online]. Available: http://www.unipro.energy/shareholders/reports/annual_reports/
16. Report on compliance of PJSC "OGK-2" with the principles and recommendations of the Corporate Governance Code. [Online]. Available: http://www.ogk2.ru/eng/si/infodisclosure/year/
17. Report on compliance of PJSC "T Plus" with the principles and recommendations of the Corporate Governance Code. [Online]. Available: http://www.tplusgroup.ru/fileadmin/f/voltgc/ir/disclosure/documents/GO/T_Pljus_PAO_GO_2016.pdf
18. Report on compliance of PJSC "Mosenergo" with the principles and recommendations of the Corporate Governance Code. [Online]. Available: http://www.mosenergo.ru/d/textpage/f9/249/mosenergo-ar-2016-19-05-17.pdf
19. Information on compliance of OJSC "FORTUM" with the Corporate Governance Code. [Online]. Available: https://www.fortum.com/countries/en/analyst/reporting/annual-report/pages/default.aspx
20. Report on compliance of PJSC "TGC-14" with the principles and recommendations of the Corporate Governance Code. [Online]. Available: http://www.tgk-14.com/finances/disclosures/annual-report.php
21. Information on compliance of PJSC "Inter RAO" with the Russian Corporate Governance Code. [Online]. Available:http://www.interrao.ru/upload/docs/Inter_RAO_Annual_Report_2016.pdf
22. Information on compliance of PJSC "RusHydro" with the Russian Corporate Governance Code. [Online]. Available:http://www.rushydro.ru/upload/iblock/686/APPENDICES_ENG_2016.pdf
23. Report on compliance of PJSC "Irkutskenergo" with the principles and recommendations of the Corporate Governance Code. [Online]. Available: http://www.irkutskenergo.ru/gi/24163
24. 2016 Russian Corporate Governance Index. [Online]. Available: http://www.nand.ru/upload/medialibrary/Russian_Corporate_Governance_Index_2016.pdf
25. 2017 Russian Corporate Governance Index. [Online]. Available: http://nand.ru/upload/ux/corporate_governance_index_2017.pdf

Impact of deicing agents on heat network corrosion

Ivan Postnikov[1,*], *Valery* Stennikov[1], *Vitaly* Shamansky[1], *Alexander* Kozlov[1], *Pavel* Sokolov[1]

[1]Melentiev Energy Systems Institute of Siberian Branch of the Russian Academy of Sciences, Laboratory of Heat Supply Systems, 664033 Irkutsk, Russia

Abstract. This research is caused by the need to make a scientific and technical analysis and obtain informed assessments of the consequences of deicing agents application to urban roads for the heat network infrastructure. The main objectives of this research were to assess the impact of the deicing agent on initiation and acceleration of corrosion of the heat network equipment (pipelines, nonmobile and sliding bearings), the effect of this corrosion on reliability of heat supply to consumers, and to determine the economic damage caused by the decline in service life of pipelines. In addition, the study identified additional circumstances of the detrimental effect of deicing agents on the heat network infrastructure associated with destructive changes in the concrete structures of the heat network thermal chambers.

1 Introduction

A significant part of urban heat networks (HN) in countries with cold climatic conditions are laid under roads, including long distance pipelines of large diameter. In this regard, every year the number of damaged pipelines due to the impact of deicing agents (DIA) increases.

Thus, informed assessments of the consequences of DIA application to urban roads for the HN infrastructure are needs. The main objectives of this research were to assess the impact of the DIA on initiation and acceleration of corrosion of the heat network equipment (pipelines, nonmobile and sliding bearings), the effect of this corrosion on reliability of heat supply to consumers, and to determine the economic damage caused by the decline in service life of pipelines.

To perform a physicochemical analysis, we selected fragments of salt deposits, pipelines and structures of thermal chambers affected by DIA related corrosion. The experiments were carried out using specialized and certified devices. All samples were obtained by inside examination of HN thermal chambers and sections of the district heating system (DHS) in Irkutsk. The techno economic analysis of the actual and predicted consequences induced DIA damage to the HN was conducted on the basis of statistical information from incident logs, data on the HN state and other documents, as well as reference data.

2 Physicochemical study on the DIA related corrosion of HN equipment

Sampling was carried out at the HN facilities of DHS of Irkutsk. Fig. 1 presents the photographs from some of the facilities exposed to DIA.

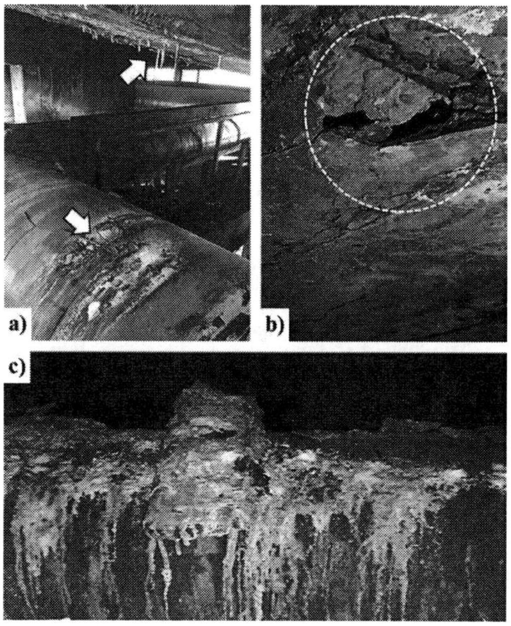

Fig. 1. View of the HN structures and pipelines exposed to DIA: a) concrete structures and pipeline under the Glazkovsky Bridge, Irkutsk (arrows point to mineral stalactites and the pipeline surface under them, that undergoes intensive destruction); b) concrete structure of one of the thermal chambers located under the road; c) the state of the thermal-insulation surface of the pipeline subjected to intensive impact of DIA.

In order to carry out a physicochemical analysis, we selected the fragments of salt deposits, thermal chamber structures (slabs, supporting beams), pipelines, fittings,

*Corresponding author: postnikov@isem.irk.ru

https://doi.org/10.1051/e3sconf/20185802005

thermal-insulation and anticorrosion layer of pipelines damaged by DIA related corrosion.

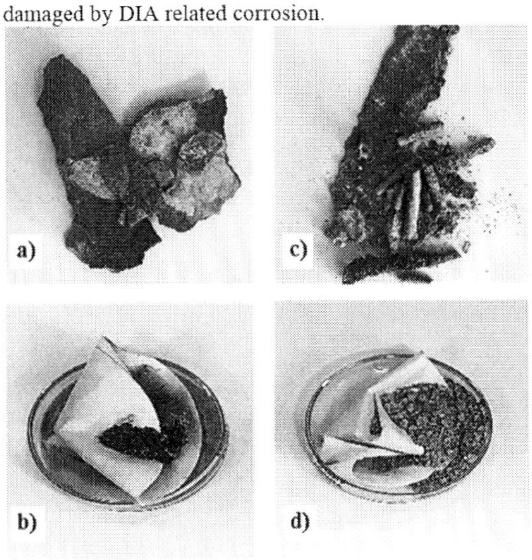

Fig. 2. Samples for physicochemical analysis: a) and b) are the view of a sample of a concrete wall of thermal chamber of HN and undissolved residue obtained from its research; c) and d) are the view of a sample of the destroyed thermal insulation layer of HN and undissolved residue obtained from its investigation.

Fig. 2-a and 2-c demonstrate the samples prepared for laboratory studies (experiments), and Fig. 2-b and 2-d – their undissolved residues after the experiment. The characteristics obtained as a result of the experiment with 4 different samples are presented in Table 1.

Table 1. Results of chemical analysis of solutions of the studied samples

Index		Sample No.			
		1	2	3	4
Content of insoluble part, %		1.0	87.3	0.1	84.7
Content of soluble part, %		99.0	12.7	99.9	15.3
Concentration of solution, %		4.2	0.62	4.8	0.78
pH		7.55	9.13	7.93	10.2
Composition	Ca	613	55	870	58
	Mg	90	0.28	70	12
	Na	17640	143	20417	4657
	K	39	10	120	514
	Cl^-	29419	215	33793	8392
	$SO4^{2-}$	357	185	3193	2013

These characteristics are further used to theoretically analyze various stages of corrosion of the HN pipelines. One of the key factors that determines these processes is the pH index. The values of this index are presented in Table 1. They indicate an alkaline medium for all the samples under study.

Destruction of concrete structures. Chemically, concrete of the most common building brands is an alkaline composition with a pH value of 11–12. When water percolates through concrete, lime is removed because of its solubility (up to 1.3 g/l) and, consequently, the strength of structure decreases, which subsequently leads to gradual destruction of the cement stone. Thus, the reduction in the amount of free lime by 20% leads to a considerable decrease in the mechanical properties of concrete structures [1–3]. At the same time, the branched fine capillary structure of concrete (capillary porosity of 7–10%) makes the water filtration process rather slow.

The main components of DIA used on roads are sodium, calcium and magnesium chlorides. Magnesium chloride is affected by acid hydrolysis to a greater extent: MgCl2 + H2O = MgOHCl + HCl. Getting to the surface of concrete structures, these components neutralize free lime in concrete and dissolve CaCO3. In this case, the expanded pores of concrete are filled with salt solutions, and after water evaporation, they are filled with solid crystals. The crystals are hygroscopic, therefore they absorb water and form crystalline hydrates, for example: MgCl2.2H2O, Na2SO4.10H2O, etc. The pressure of crystallization and hydration arises, which leads to the formation of microcracks inside and on the surface of concrete. This phenomenon is greatly enhanced by cycles of climatic changes in temperature that lead to so-called concrete scaling, i.e. flaking or peeling of concrete surface. This process is clearly seen in Fig. 1-b.

Destruction of thermal insulation layer and anticorrosion coating. After penetration through concrete structures or surface due to washout of Ca(OH)2 from the concrete, the DIA solution acquires alkaline properties. This is also confirmed by the results of a chemical analysis of samples taken from the pipelines and walls of thermal chambers (Table 1).

The thermal insulation of the considered pipelines is made on the basis of polyurethane foam, and the anticorrosion coating is based on the complex polyurethane coating. The heart of the solid polyurethane coating is the so-called prepolymer, which has two types of bonds responsible for the polymeric structure: the amide bond -CO-NH-R'- and the ester bond -CO-O-R-. The latter is unstable because the esters are highly susceptible to hydrolysis reaction, which leads to the breakdown of the bond and formation of an alcohol and an acid: R-CO-O-R'+ H2O <=> R-COOH + R'-OH. It is known [4] that the rate of hydrolysis in alkaline medium compared to neutral medium (pH=7) is increased by $10^{(pH(b)-7)}$ times, where pH(b) corresponds to the contacting solution. In the considered case the pH value of DIA measured after the salt is removed from the fragments of the destroyed thermal insulation, exceeds 10 units (see Table 1). This means that the rate of hydrolysis (i.e. the polymer destruction) rises at least by $10^{10-7} = 1000$ times.

Electrochemical corrosion of metal pipelines under the effect of salt deposits. The main type of corrosion of metal pipelines is pitting corrosion, which is characterized by localized destructions observed in the cases where the corrosion rate in some areas is higher than in others [7–11]. It is also worth noting that in addition to pitting, pipelines are subject to stress corrosion cracking [12]. The effect of pitting processes can be seen on the fragments of the pipelines shown in

Fig. 4. The picture "a" demonstrates pitting damage leading to through holes, while "b" one shows a section with remaining parts of polyurethane thermal insulation.

The rate of electrochemical corrosion [12–16] is proportional to the corrosion current, which, in accordance with Ohm's law, is proportional to the difference in potentials between the cathodic and anodic areas of the sur-face. In general, the surface to be corroded is a branched electrical (electrochemical) circuit in which the corrosion currents connect different anodic and cathodic areas. In this case, currents can flow between any differently charged micro-areas of a non-uniform metal surface. The existence of such heterogeneities is caused both by the nature of metal (heterogeneity of chemical composition) and by the mechanical state of metal (micro-inhomogeneity, dislocations, etc.).

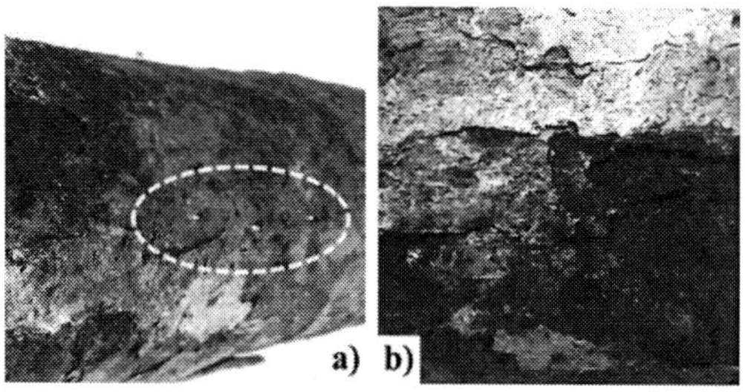

Fig. 3. Fragment of a pipe damaged by corrosion: a) section with pitting corrosion (dotted through holes are visible); b) section with surface corrosion and fragments of destroyed polyurethane thermal insulation.

In the anodic areas of the pipeline metal surface, in the absence of moisture, the iron ions remain on the surface and form the so-called electrical double layer with zero leakage current. However, in the presence of moisture, these ions are hydrated and removed from the structure of the double layer, which initiates their replenishment due to the formation of new ions, and the process of corrosion begins. This process consists of two successive reactions that result in well-known brown rust (ferric hydroxide): 1) Fe2 + 2OH- = Fe(OH)2 and 2) 4Fe(OH)2 + O2 + 2H2O = 4Fe(OH)3. It is known that the presence of chlorides in the solution considerably accelerates corrosion [8, 9] because ferrous ions form highly soluble complexes with chloride ions (FeCl2+, FeCl2+, FeCl3, FeCl4-). These complexes are highly strong and provoke dissolution of the iron hydroxide film that inhibits corrosion. As a result, corrosion of the pipeline surface is increased.

According to the assessment based on all the considered factors and conditions, the corrosion rate of HN pipelines exposed to DIA increases by approximately 7.5 times. As for the chemical corrosion under the effect of air oxygen [17], at this temperature this process is much slower than electrochemical corrosion [12].

3 Techno economic and reliability analysis of DIA impact on HN pipeline

The result of processing the data on the damage in HN of Irkutsk city associated with the use of DIA is shown in Table 2. The damage incidents that occurred in the non-heating season are mainly related to the hydraulic tests and have a long restoration period.

Table 2. DIA induced damage indices in HN of Irkutsk (2017)

Index	Value
Total number of damage incidents in HN	888
Total number of damage incidents associated with the use of DIA	307(35%)
Total repair time of damage related to the use of DIA, h	3897
Average repair time of damage associated with the use of DIA, h	13
Total heat undersupply* due to DIA related damage, Gcal	12176
Specific heat undersupply per 1 incident associated with the use of DIA, Gcal	40
Total cost of repairing damaged areas, thousand rub	162043
Total cost of repairing DIA related damage*, thousand rub	14578(9%)
Average cost of repairing damage due to the use of DIA, thousand rub	47
Specific cost of repairing damage due to the use of DIA, rub/Gcal	1197

*indices calculated using the reference data [18] at a temperature difference of 130/70 °C and the specific friction pressure loss of 15 kgs/(m^{2}m)

As can be seen from the Table 2, 307 damage incidents related to the use of DIA were recorded in HN, which is 35% of the total number. It took 3897 hours to restore the areas damaged by DIA, with an average repair time of 13 hours. The total heat undersupply during the repair of damaged pipelines affected by DIA amounted to 12176 Gcal. Fig. 4 demonstrates the distribution of the pipeline repair duration and heat undersupply by month in 2017.

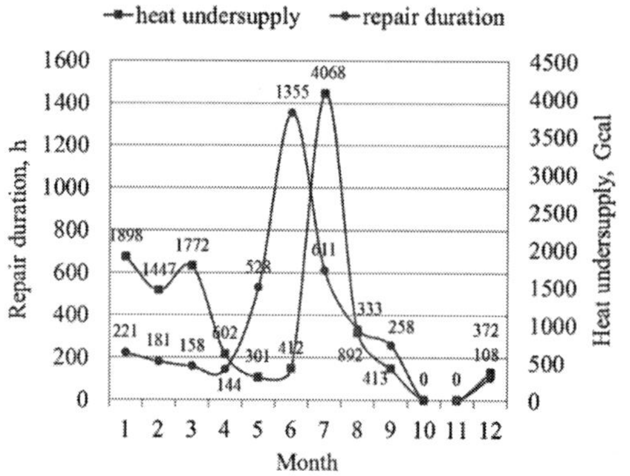

Fig. 4. Distribution of repair duration and heat undersupply due to HN pipeline damage caused by DIA in Irkutsk by month 2017

During this period, more complete reconstruction work is performed on damaged sections (branches) of HN. As seen from Fig. 4, this causes a considerable heat undersupply due to long-term interruptions of hot water supply (months 5–9).

Reliability analysis of HN considering DIA induced damage. Table 3 presents the results of an integrated reliability analysis based on the assessment of the availability factor (AF) for HN of Irkutsk in terms of pipeline damage associated with the use of DIA. In this case, AF is interpreted as an average value of the part of the heating period, during which the indoor air temperature in the customer's premises does not fall below the specified limit value (t_{min}) [19]. The assessment was carried out for three reliability criteria, corresponding to t_{min} = 20 °C, t_{min} = 18 °C and t_{min} = 12 °C. It took into account the slack time related to heat accumulation in buildings and employed nomographic charts of cooling of residential premises [20].

Table 3. Integrated reliability analysis of HN in Irkutsk, considering pipeline damage related to the use of DIA (2017)

Index	Criteria t_{min}		
	20°C	18°C	12°C
Total number of disconnections	127	32	7
Total time of disconnections, h	811	305	89
Average time of repair, h	6.4	6.4	6.4
Availability factor	0.86	0.95	0.98
Required reduction in disconnection time, h	637	131	0
Required reduction in the number of damage incidents	100	21	0

Note: all the indicators are calculated for a heating period of 5784 h; disconnection is interpreted as an interruption of heat supply to consumers in compliance with the accepted criterion; AF is calculated for DHS only with respect to DIA related damage; minimum values for ensuring standard reliability that corresponds to the value of AF 0.97

According to [21], the standard value of AF is 0.97. As seen from Table 3, the required level of reliability under the existing number of DIA induced damage incidents is achieved only if the minimum indoor air temperature of the consumer is at a level of 12 °C. Based on the calculations, the necessary conditions were also established to ensure AF standard value for the specified criteria (Table 3).

Analysis of DIA induced corrosion effect on service life of HN pipelines and their replacement cost overruns. A physicochemical analysis of the increase in DIA induced corrosion intensity in HN (section 2) allows us to confirm the acceleration of the pipeline wear by at least 7.5 times. Taking into account this factor, the data on the length of the Irkutsk HN located under the roads, and the "age" structure of the networks, we made a forecast (up to 2024) of HN pipelines wear and tear (Fig. 5). The starting point of the forecast (2017) is 62 km (14% of the total length of HN in the city) and corresponds to the length of worn-out HN under the roads.

According to the obtained results, without annual replacement of pipelines, the length of worn-out networks located under the roads by 2024 will be 73 km (16% of the total HN length) disregarding DIA impact, and 87 km (19%), considering the effect of DIA. At the rate of network replacement equal to that in 2017 (3.9 km/year) and the absence of DIA, the deterioration of the networks will decrease to 46 km (10%) by 2024, and in the case of DIA use, its level will practically not change (Fig. 5). For the level of wear achieved at the current replacement rate and the impact of DIA, it is necessary to increase the rate of pipeline replacement to 5.8 km.

Based on the indices of the pipeline restoration rates and data on the cost of the pipelines replacement, we compared the expected cost of replacing the worn HN in Irkutsk until 2024 with and without consideration of DIA effect. The level of wear at the end of the considered period was taken equal to 46 km (10%), which corresponds to the current replacement rate of 3.9 km without DIA effect. The results of this assessment are given in Table 4.

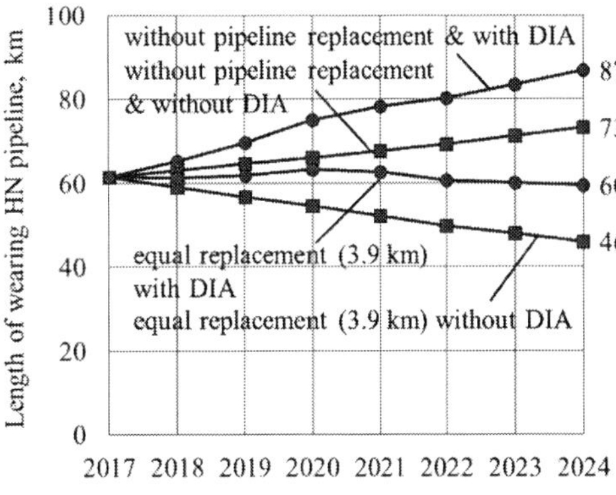

Fig. 5. The forecast dynamics of wear and tear in Irkutsk HN at various annual rates of pipeline replacement with and without using of DIA

The existing level of the pipeline replacement costs disregarding DIA is about 347 million rub/year. At the same time, maintaining the replacement rate of 5.8 km/year that provides the same level of wear at the end of the calculation period (46 km) will require much higher costs that are estimated at about 516 million rub/year. Thus, the cost overruns related to the replacement of the under road pipelines worn-out due to DIA induced corrosion, are quite significant and amount to 49%, which corresponds to 169 million rub/year.

Table 4. Comparative (with and without the consideration of DIA) forecast of the replacement cost for Irkutsk worn-out HN located under roads, with their length reduction to 46 km (10%) by 2024, with the cumulative total, million rub

Index	Year						
	2018	2019	2020	2021	2022	2023	2024
Replacement rate of 3.9 km	347	693	1040	1387	1734	2080	2427
Replacement rate of 5.8 km	516	1031	1547	2062	2578	3094	3609
Overrun	169	338	507	676	845	1013	1182

The cost in Table 4 for each subsequent year is shown in the format of a cumulative total, i.e. includes the sum of costs for the previous years. As a result, the value of the cost for 2024 corresponds to the amount of costs for the entire period in question.

4 Conclusions

The main objectives of research were to comprehensively assess the effect of DIA on corrosion of HN, including a physicochemical analysis of corrosion processes and a techno economic analysis of the consequences of corrosion related damage to HN pipelines. The studies were carried out at HN of Irkutsk city, and the samples taken directly from thermal chambers of HN were used as initial materials for a laboratory analysis.

The physicochemical analysis of the samples revealed the indisputable factors of the destructive effect of DIA on polyurethane anticorrosion and thermal insulation coatings of the pipelines. The calculations demonstrated at least a 1000-fold increase in the rate of hydrolysis, i.e. destruction of polymer insulation, under the operating temperatures of HN pipelines (70–130°C). Moreover, the process of corrosion-related wear of metal pipelines is accelerated by about 7.5 times.

The techno economic analysis of the consequences of DIA use for HN pipelines involved generalized and structured statistical data on the incidents in HN of the city. Based on the calculations using statistical analysis and interpolation methods, we obtained the indices necessary to assess the consequences of pipeline damage caused by DIA in terms of the duration of repairs, heat undersupply, restoration costs, etc. The findings indicate that with the existing number of such damage incidents, the required level of reliability is achieved only for the minimum indoor air temperature at a level of 12 °C. Additional reliability measures will be necessary to meet higher comfort requirements of the consumer in case of pipeline accidents.

The decrease in the service life of the Irkutsk under road HN pipelines damaged by DIA related corrosion has been assessed. The assessment indicates that with the current rates of pipelines replacement, the level of their wear and tear related to DIA application is not reduced. Moreover, the reduction in HN depreciation associated with DIA effect will require a significant increase in funding their reconstruction.

In conclusion, it is worth noting that the conducted studies confirmed the negative effect of DIA on HN of the city. The scale and extent of this effect call for additional in depth investigation of this issue and measures to eliminate the negative corrosive effect of DIA on the heat supply facilities of the city.

The study was carried out at Melentiev Energy Systems Institute SB RAS within the framework of research projects III.17.5.3 №AAAA-A17-117030310450-3 and III.17.1.2 №AAAA-A17-117030310448-0 of the Program for Fundamental Research of SB RAS, and involved the equipment of the Multiple-access Center "High temperature plant" at Melentiev Energy Systems Institute SB RAS.

References

1. Bazhenov Yu.M. Technology of concrete / M.: Publishing house of the ASV, 2002, 500 p.
2. Ramachandran V., Feldman R., Beaudoin J. Concrete Science. Physicochemical concrete study / M.: Stroiizdat, 1986, 279 p.
3. Gorchakov GI, Bazhenov Yu.M. Building Materials / Moscow: Stroiizdat, 1986, 689 p.
4. Lurie Yu.Yu. Handbook of Analytical Chemistry / M .: Khimiya, 1971, 455 p.
7. Ulig G. G., Revi R.U. Corrosion and fight against it / Leningrad: Khimiya, 1989, 456 p.
8. Corrosion, Volume 1: Metal / Environment Reactions / Ed. L.L. Shreir, Elsevier Ltd., 1976, 1232 p.
9. Kruger J., Begum S. Corrosion of Metals: Overview // Reference Module in Materials Science and Materials Engineering, 2016.
10. Chaves Igor A., Melchers Robert E. Pitting corrosion in pipeline steel weld zones // Corrosion Science, 2011, vol. 53 (12), p. 4026–4032.
11. Hunkler F., Frankel G.S., Bohni H. On the mechanism of localized corrosion // Corrosion, 1987, vol. 43, p. 189–191.
12. Vorobieva G.Ya. Corrosion Resistance of Materials in Aggressive Media of Chemical Production / M.: Khimiya, 1975, 816 p.
13. Zhuk N.P. A course in the theory of corrosion and protection of metals / M .: Metallurgy, 1976, 472 p.
14. Maocheng Y.A.N., Jin X.U., Libao Y.U., Tangqing W.U., Cheng S.U.N., Wei K.E. EIS analysis on stress corrosion initiation of pipeline steel under disbonded coating in near-neutral pH simulated soil electrolyte // Corrosion Science, 2016, vol. 110, p. 23–34.
15. Guo H.X.. Lu B.T., Luo J.L. Interaction of mechanical and electrochemical factors in erosion-corrosion of carbon steel // Electrochimica Acta, 2005, vol. 51 (2), p. 315–323.
16. Lu B.T., Luo J.L. Synergism of Electrochemical and Mechanical Factors in Erosion-Corrosion // J. Phys. Chem. B, 2006, vol. 110(9), p. 4217–4231.
17. Kusmierek E., Chrzescijanska E. Atmospheric corrosion of metals in industrial city environment // Data in Brief, 2015, vol. 3, p. 149–154.
18. Design of heat networks. Reference book of designer. Ed. by Nikolaev A.A. / M.: Publishing House for Literature on Construction, 1965, 400 p.
19. Reliability of heating systems / Reference book. Ed. by Sennova E.V. Novosibirsk: Science, 2000, 360 p.
20. Areshkin A.A., Moskalenko A.V., Gorobets N.V. General principles of heat supply backup in heating systems // Novosti Teplosnabzhenia (Heat Supply News), 2011, vol. 1, p. 26–34.
21. Construction Norms and Regulations P 41-02-2003 "Heat networks" / M: Ministry of Regional Development of the Russian Federation, 2012, 75 p.

E3S Web of Conferences **58**, 02006 (2018) https://doi.org/10.1051/e3sconf/20185802006
RSES 2018

Methods for calculation of the marginal exploitation lifespan of power transformers 35 rV and higher based on the state index

Alexandr Nazarychev[1], *Dmitriy* Andreev[1], *Alexey* Tadjibaev[1], *Svetlana* Vysogorets[1] and *Ilia* Sulynenkov[1].

[1]PEIPK, SPb, Russia

Abstract. A methods for determining the marginal exploitation lifespan of a transformer based on a complex diagnostic survey was developed. The integral evaluation of the technical state of the transformers is performed according to the value of the condition index. An example of the marginal exploitation lifespan calculation of the transformer according to the real initial data is given. The application of the methods will allow to proceed to the planning of technical re-equipment and reconstruction of energy facilities taking into account the technical condition of the equipment.

1 Introduction

In the context of the course on digital technologies in the Russian power industry, the development and implementation of smart grid elements, power transformers and autotransformers (hereinafter transformers) 35 kV and higher become one of the most important elements on the electric power systems on which reliability and safety of consumers' power supply depend [1]. In the power grid complex of Russia, a significant part of the power transformer park is approaching the exhaustion of the standard service life, and a considerable part of transformers is currently operated outside of it. This makes necessary to determine the opportunity of extending the standard service life of transformers and to calculate the marginal exploitation lifespans.

The main way to ensure operational reliability and safety of power facilities is to carry out a diagnostic survey of transformers in order to obtain objective and reliable information about their technical condition (TC) based on diagnostic methods. Diagnostics involves the use of various methods and devices for monitoring the parameters of transformers, which would allow a detailed analysis of the presence of defects, their location and determine the integral indicator of technical condition (IITC). In [2 - 7] it is shown that at the present time as the quantitative IITC of transformers, the indicator - state index (SI) is applied. To assess the SI the methods [8,9] is approved by the Decree of the Government of the Russian Federation and by the order of the Ministry of Energy of Russia.

To determine the marginal exploitation lifespan of transformers based on CDS, taking into account the earlier studies [2-7], a Methods was developed whose main purpose is to establish unified, for electric grid companies in Russia, the calculation of the marginal exploitation lifespan of transformers.

In this article, the basic provisions of the Methods are discussed aiming to explain its main principles and to understand the implementation of the inner algorithm for calculating the technical resource and the marginal exploitation lifespan of the transformers. The Methods is based on the methods, algorithms, models, automated systems and other elements of the production assets management system (PAMS) already developed, implemented and approved in the Russian power grid complex.

For transformers, a single text algorithm for calculating the technical state indexes (TSI) is generated, which, on the basis of the whole set of diagnosed parameters, can be done through logical and mathematical formulas:

- calculate the transformer SI that takes into account operating conditions and effects on the vehicle during maintenance and repair activities (MRA) by means of technical parameters determined by CDS results, routine diagnostics, tests, inspections and measurements;
- establish recommended types of necessary actions on the transformer, for its reliable functioning.

Calculations based on TSI are based on the use of reference books of equipment groups, parameters, normative values, defects and other reference books regulated in a single centralized system of regulatory and reference information (RRI) of the electric grid company. The following requirements are imposed on the indicator of transformer SI:

- exponent: The SI should provide an accurate picture of the general condition of the transformer, including in retrospect, and the possibility of continuing the operation of the transformer;
- objectivity: SI should be based on objective, instrumental measurements, calculations, and not on subjective observations and assumptions;
- simplicity: The SI should be understandable and easily interpretable;

* Corresponding author: rector@peipk.spb.ru

https://doi.org/10.1051/e3sconf/20185802006

- compliance with current legislation [8, 9]:

Thus, the SI must reliably reflect the level of the transformer's TC and its change within the established limits, and also have a clear technical sense and unambiguous interpretation in all electric grid companies.

The state of the transformer is marginal when the SI is 0. The marginal state also corresponds to the moment when the transformer's resource is fully exhausted, i.e. the actual generated resource is equal to the normative one. The state of the transformer, at which the SI = 1, corresponds to the state of the new transformer, the operation of which has not yet begun. For this transformer, the actual generated resource is 0.

The actual transformer TC is degraded during exploitation. When carrying out the actions within the existing MRA, the transformer TC is improved. Nevertheless, the overall trend of transformer TC changes over the whole period of its operation has a decreasing character.

In some basic (normative) exploitation conditions, the TC deteriorates as planned (in accordance with the technical documentation of the manufacturing plants for the transformer). Under light operating conditions, the transformer TC is deteriorating less intensively, and in the heavier ones it is more intensive, in comparison with the base conditions.

In [10], a functional relationship was established between the technical resource, the service lifespan, the calendar operating time, the deadline for operation and the level (value) of the transformer TC.

Sudden failures occur under the influence of various sudden factors, for example, such as: natural effects (unaccounted wind loads, "ice rain", ice formation, etc.); vandalism; failures caused by a malfunction of equipment installed near other equipment, etc.

Wear failure is due to an unacceptable decrease in the level of the transformers TC. Wear failures occur, as a rule, on the basis of accumulation and development of defects during operation, i.e. they develop gradually.

The technical resource allows to take into account wear failures due to bringing the actual operating conditions of the transformers to the normative ones by using the factual operating time (factual operated resource) in the calculation expressions instead of the calendar operating time.

To bring the operating conditions into effect, the calculation expressions developed in [10] are used that relate the calendar work, the factual technical resource and the values of the transformer SI. The methodology for calculating the actual exhausted resource is given in [10, 13]. Based on the received value of the actual exhausted resource, the normative residual or factual residual life is calculated. Summarize the factual and residual resource and bringing the unit of measurement of the operating time to the time, the transformer marginal exploitation lifespan is calculated.

Mathematical models for calculating the transformers marginal exploitation lifespan are given in the Methods, taking into account the different completeness and composition of the initial data. Next, consider the general model for calculating the deadlines for operation, which takes into account all possible options for changing the transformer SI. The factual operated resource generally depends on the operating r and the change in the value of the SI according to the function S(r) and is determined in accordance with [10, 13]. In this case, the calculation should be performed in time units of the operating time. The factual spent resource R for working tc (point of control) corresponds to the factual lifespan of transformer Tc during calendar time tc. Normative residual lifespan is determined by the expression:

$$Tres.0=T0-Tc, \quad (1)$$

where T0 – normative lifespan of the transformer.

If the transformer is further operated in the standard exploitation conditions, the transformer marginal exploitation lifespan Tm will be determined as follows:

$$Tm=tc+Tres.0 \quad (2)$$

or:

$$Tm=tc+T0-Tc \quad (3)$$

If the transformer is further operated in conditions other than the normative, the transformer marginal exploitation lifespan will be determined as follows:

$$Tm=tc+Tm.res, \quad (4)$$

where Tm.res – marginal residual lifespan (calendar time) of transformer service, corresponding to normative residual lifespan T0.res, is determined by numeric method from equation:

$$T_{0.res} = \int_{T_c}^{T_{m.res}+T_c} \frac{1-S(t)}{1-S_0(t)} dt \Rightarrow T_{m.res} = \dots \quad (5)$$

The expression for the value of Tm.res from equation (5) determines the time that the transformer will still work in the expected operating conditions before the transition to the limiting state, i.e. when the factual residual resource (the actual lifespan Tres) will reach the normative residual resource (the normative residual lifespan T0.res).

Since for a transformer it is difficult to predict the conditions and modes of its exploitation in the future, it is advisable to assume that after the control point the transformer will operate in the same modes as before the control point. Therefore, its TC and, accordingly, SI will change in a similar way. The determination of the exploitation extension limits for a new transformer Tm under known future exploitation conditions is carried out by numerical methods from equality:

$$T_0 = \int_0^{T_m} \frac{1-S(t)}{1-S_0(t)} dt \Rightarrow T_m = \dots \quad (6)$$

It was shown in [10] that the absence of full-fledged retrospective data on the values of transformer SI can be a significant problem in determining the marginal exploitation lifespan. To be able to apply the Methods during its implementation, validation, and accumulation of the necessary initial data on the SI, it is possible to use the calculation model by applying the linear approximation of the function of changing the SI S(r):

$$\frac{1-S(t)}{1-S_0(t)} = \frac{m}{m_0} = A, \quad (7)$$

where m, m0 – coefficients of linear approximation of the data set intended to obtain the SI change functions for the factual exploitational conditions S(t) and for the basic regulatory operating conditions S0(t) respectively.

The factual operated resource in the general case depends on the operating time r and the change in the value of the SI according to the function S(r) [10]. In this case, the calculation should be performed in time units of the operating mesurement (r = t). The factual operated resource R for working tc (point of control) corresponds to the factual service life of the transformer for a calendar time tc.

Tc=Atc, (8)

Then the normative residual lifespan of the transformer:

Tres.0=T0-Tc, (9)

where T0 – normative lifespan of the stransformer.

или

Tres.0=T0-Atc. (10)

If the transformer is further operated in conditions other than the normative, the transformer marginal exploitation lifespan will be determined as follows:

Tm=tc+Tres.0 (11)

or

Tm= T0-tc(1-A). (12)

If the transformer is further operated in conditions other than the normative, the transformer marginal exploitation lifespan will be determined as follows:

Tm=tc+Tm.res (13)

where Tm.res – marginal residual lifespan (calendar time) of transformer service, corresponding to normative residual lifespan T0.res, is determined as follows:

Tm.res=t0.res / A (14)

Then marginal exploitation lifespan of transformer is determined by expression:

$$T_m = t_c + \frac{T_{0\,res}}{A} \quad (15)$$

The value Tm.res from equation (5) is the time that the transformer will still work in the expected exploitational conditions before the transition to the marginal state, i.e. when the factual residual resource (the factual service life Tres) will reach the normative residual resource (the normative residual service life T0.res).

The determination of the exploitation extension limits for a new transformer Tm under known future exploitation conditions is carried out by expression:

Tm=T0 / A. (16)

In the first approximation, in the absence of data to obtain the coefficient m0, it is advised to use the relation:

$$m_0 = \frac{1}{T_0} \quad (17)$$

The determination of the functional dependence of the SI values on the operating time is performed in accordance with [10]. In this case, the time t is taken as the measuring unit of the operating r.

The recalculation of the marginal service life of the transformer is recommended in the electric grid companies at each significant change in the values of the transformer SI (or when obtaining a new SI value), including calculations before and after each current, medium and major overhaul. Thus, the periodicity of the calculation is recommended to be taken in accordance with the frequency of calculation of the transformer SI. Approximation of the data by the values of the SI with the aim of obtaining the function of changing the SI from the operating time, respectively in this case, must be performed after each update of the SI data (obtaining new values, updating old ones, etc.).

In addition, it is recommended to recalculate the values of the technical resource of the transformer when other data, including normative ones, included in the calculation model of the transformer marginal lifespan are changed.

Since the value of the transformer marginal exploitation lifespan is a predictable value, the final procedure for calculating the marginal exploitation lifespan of transformers in electric grid companies is determined by internal instructions. In the PAMS, the results of the marginal exploitation lifespan evaluation of power transformers are used in strict accordance with the approaches set forth in the normative technical documentation (NTD).

Calculation of the marginal exploitation lifespan of transformers begins with the preparation of the initial data. The basic initial data for the calculation of the marginal exploitation lifespan of the transformer are information on the values of the SI S, S0 of the equipment unit (EU) under consideration. If the functions S(t) and S0(t) are known, then additional preparation of the initial data is not required and one can proceed to the calculation. If one of the functions S(t) or S0(t) is not known, then they should be determined by [10] by approximating the data by the SI values.

Depending on the completeness and quality of the input data, a model should be chosen to calculate the marginal exploitation lifespan of the transformer. In the case of known functions S(t) and S0(t), a general model is used to calculate the transformer marginal exploitation lifespan. At the initial stage of the implementation of the Methods, in the absence of full-fledged historical data on the functions of changing the IS values from the time of the transformer, a model of the calculation of the service life limit with linear approximation of the data by the SI values for each EU should be applied.

After the initial data have been prepared and the model for calculating the marginal exploitation lifespan has been selected, the date (hereinafter the "calculation date") when it to be calculated is to be determined. For the transformer under consideration, the following should be known:

- function S0(t) = S0.T(t), where S0.T(t) is the basic function of the SI change depending on the operating time for the transformer;
- function S(t) = ST(t), where ST(t) is the factual function of the SI change depending on the operating time for the transformer;

Knowing the date of commissioning of the transformer, it is necessary to determine the calendar lifespan of the transformer from the date of commissioning to the date of calculation in years.

t = tc. (18)

The calculation date cannot be earlier than the commissioning date of the transformer. The calculation of the transformer marginal exploitation lifespan is carried out in the following order.

1. The calculation of the factual operated resource (factual service life) of the transformer is performed in accordance with [10].

2. Depending on the model chosen, the value of the normative residual resource (the normative residual service life) of the transformer is determined by expression (1) or (10).

3. If further operation of the transformer is planned in the normative exploitation conditions, the marginal residual lifespan of the transformer corresponds to the normative residual service life, and the total marginal exploitation lifespan is determined by the expression (2) - (3) or (11) - (12) according to selected model.

4. If further operation of the transformer is planned in conditions other than the normative, the marginal residual lifespan of the transformer is determined by numerical methods from equality (5) or expression (14), and the total marginal exploitation lifespan is determined by the expression (4) or (15).

For a new newly installed transformer, the following procedure for calculating the marginal exploitation lifespan should be followed. After preparing the initial data and selecting the model for calculating the marginal exploitation lifespan, it is necessary to determine the date (hereinafter the "calculation date") for which the calculation will be performed. For the transformer under consideration, the following should be known:

- function S0(t) = S0.T(t), where S0.T(t) is the basic function of the SI change depending on the operating time for the transformer;

- function S(t) = ST(t), where ST(t) is the factual function of the SI change depending on the operating time for the transformer;

The marginal exploitation lifespan of the transformer is determined from equation (6) or expression (15), depending on the model chosen.

Let us consider an example of calculating the service life of a transformer TMN-6300/110/10. Figure 1 shows the trend of the change in the SI of the indicated transformer.

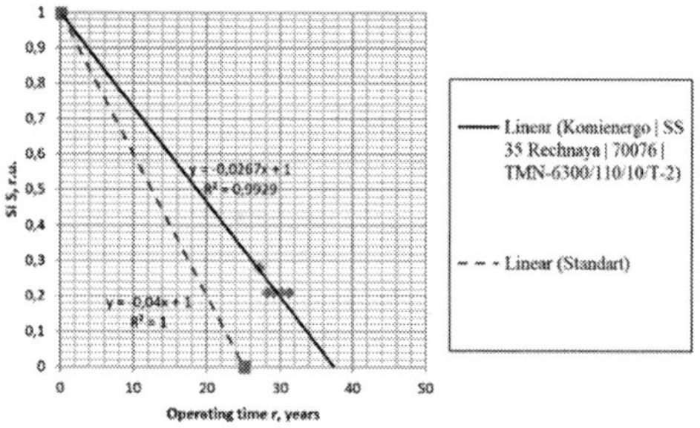

Fig. 1. Trend of change of the SI over time for the transformer TMN-6300/110/10 (the standard service life is 25 years)

The dependence of SI (in r.u.) on time for the transformer under consideration with linear approximation has the form:

S(t)=1-0,0267t (19)

Taking into account expression (7), the coefficient A for this transformer is 0,0267 / 0,04 = 0,67.

The calendar service life of this transformer is tk = 32 years. The actual service life, taking into account expression (8), is determined as follows:

Tc=0,67*32=21,44 years (20)

Then the normative residual life of the transformer, taking into account expression (9), is defined as:

Tres.0=25-21,44=3,56 лет (21)

If the transformer is subsequently operated under standard conditions, the deadline for the operation of the transformer is determined by the expression (11):

Tm=32+3,56=35,56 лет (22)

If the transformer is subsequently operated in conditions other than the normative (in this case, the same as it was used earlier in all 32 years), the deadline for the operation of the transformer will be determined by the expression (15):

$$T_m = t_c + \frac{T_{0\,res}}{A} = 32+3.56/0.67=32+5.13=31.13 \text{ years} \quad (23)$$

2 Conclusion

The article considers the Methods for calculating the marginal exploitation lifespans of power transformers 35 kV and higher based on the use of an integral estimation of the factual state - the state index. Basic principles of the developed Methods and a block diagram of the calculation of the transformers marginal exploitation lifespan are shown. A general calculation model and a model for calculating the marginal exploitation lifespan of transformers based on linear approximation of the values of the state index change function are given. A methodical approach is presented for the use of the Methods in the PAMS. An example is given of calculating the marginal exploitation lifespan of a power transformer, showing the possibility of using the Methods in electric grid companies of the Russian energy industry.

References

1. *The concept of providing reliability in the electric power industry.* / Voropai N.I., Kovalev G.F., Kucherov Yu.N., and others - M.: LLC ID «ENERGY», 2013. - 304 p.

2. A.I. Tadjibaev, A.N. Nazarythev, D.A. Andreev. *The Analysis of Methods For Assessing The Integral Indexes of Electrical Equipment Technical Status* // Научни Известия на НТСМ «Дни на безразрушителния контрол 2016, Созопол – България, Година XXIV, Брой 1(187), Юни 2016. – pp. 405 – 407.

3. Aleksandr Nazarythev, Aleksei Tadjibaev, Dmitry Andreev. *The Methodology of Determining the*

Probability of Electrical Equipment Failures Considering the Technical Status Index // Научни Известия на НТСМ «Дни на безразрушителния контрол 2017, Созопол – България, Година XXV, Брой 1(216), Юни 2017. – pp. 160 – 165.
4. Nazarythev A.N., Novomlinsky E.V., Andreev D.A. *Valuation of the electrical equipment technical condition based on the calculation of integrated indicators* // Methodological issues of reliability research of large energy systems. Issue. 67. Problems of reliability of energy systems / ed. by N.I. Voropay, Yu.Ya. Chukreev. - Syktyvkar: Publishing house of LLC "Komi republican printing house", 2016. - pp. 171 - 179.
5. Berdnikov R.N., Gvozdev D.B., Kuzmin I.A., Nazarythev A.N., Andreev D.A., Tadjibaev A.I. *Methods for valuating the probability of failures of the main electric grid equipment considering its technical condition* // Collection of scientific and technical articles of the employees of the "Rosseti" Group of Companies. - M.: "Electricity. Transmission and distribution ", 2017. - pp. 151 - 163.
6. Nazarythev Aleksandr, Tadjibaev Aleksei. *Inspection and Diagnostics Methods as a Basis to Control the Technical Status of Electrical Equipment* // International Seminar on Learning from Differences in the Energy Sector, Proceedings. Edited by A. Nazarythev and M. Huhtinen. - Saint-Petersburg, Kotka: PEIPK, KYAMK, 2016. -pp. 63 - 72.
7. Alexander Nazarythev, Alexey Tadjibaev and Dmitry Andreev. *Evaluation of electrical equipment service life on the basis of its status index trend* // E3S Web Conf. Volume 25, 2017. Methodological Problems in the Reliability Study of Large Energy Systems (RSES 2017). Bishkek, Kyrgyzstan, September 11-15, 2017 N. Voropai, S. Senderov, A. Michalevich and H. Guliev (Eds.) Article Number 04004. Number of page (s) 7., Section: Influence of quality of electric energy on reliability of power supply. Technical condition of the equipment. Published online 01, December 2017. DOI https://doi.org/10.1051/e3sconf/20172504004
8. Decree of the Government of the Russian Federation of December 19, 2016 No. 1401 "Of the comprehensive determination of indicators of the technical and economic status of power facilities, including indicators of physical wear and energy efficiency of electric grid facilities, and the procedure for monitoring such indicators" (Meeting of Legislation of the Russian Federation Federation, 2016, No. 52, Article 7665).
9. Order of the Ministry of Energy of Russia of July 26, 2017 No. 676 "On approval of the methodology for assessing the technical condition of the main technological equipment and power transmission lines of power plants and electrical networks."
10. Andreev D.A., Nazarythev A.N., Tadjibaev A.I. *Determination of the probability of failure of equipment of electric grid enterprises on the basis of technical condition assessment*, Ed. by A.N. Nazarythev: FSEE "PEIPK". - SPb., 2017. - 194 p.
11. Nazarythev A.N., Andreev D.A. *Methods and mathematical models of a comprehensive assessment of the technical condition of electrical equipment* / Ivan. state. power. un-t. - Ivanovo, 2005. - 224 p.
12. Nazarythev A.N., Andreev D.A. *Methodological basis for determining the marginal exploitation lifespan and the order of technical re-equipment of power facilities* / Ivan. state. un-t. - Ivanovo, 2005. - 168 p.
13. Nazarythev A.N., Tadjibaev A.I., Andreev D.A. *Evaluation of the electrical equipment resource by the trend of the state index change* // Methodological issues of reliability research of large energy systems. Issue. 68. Research and provision of reliability of energy systems / Ed. by N.I. Voropai - ISEM SB RAS, 2017. - pp. 613 - 623.

Management of power transformers reliability by technologies of control on resource characteristics of liquid dielectric

Svetlana Vysogorets[1*], *Alexandr* Nazarychev [1], *Ilia* Gorec [2], *Alexey* Tadjibaev[1].

[1]PEIPK, SPb, Russia
[2]NRU "HSE SPb", SPb, Russia

Abstract. The relationship between the resource of a power transformer and the resource characteristics of a liquid dielectric is shown. Determined the key quality indicators, which allow to control the resource of the liquid dielectric. For the first time, "The method of experimental determination of the liquid dielectric resource and measures for its restoration" was developed.

1 Introduction

In power transformers, insulating oil performs a number of functions: dielectric, cooling agent, protective environment, information (diagnostic) environment. For reliable operation in electrical equipment, transformer oils must have the following exploitational characteristics [1]:

- high electrical strength and low electrical conductivity to ensure reliable insulation;
- Sufficient viscosity-temperature properties to provide cooling of the active part of electrical equipment and (or) its start-up at negative temperatures;
- good chemical stability against aging, i.e. ability for a long time not to change its properties in the process of exploitation, ensuring the duration of the service life;
- chemical resistance, i.e. the ability for a long time not to corrupt the solid insulation.

In the course of long-term exploitation in the power transformer, the transformer oil changes its physico-chemical and exploitational properties ("ages") and its quality indicators reach limit values. The change in the properties of transformer oils occurs under the influence of several factors. The main process that determines the change in the properties of the oil is its oxidation [2, 3, 4].

The complex chemical composition of transformer oils explains the ambiguity of their "behaviour" in exploitation, which ultimately causes difficulties in predicting the early stages of oil aging due to the variety of oxidation products being formed. The diversity of the chemical composition of different groups of transformer oils indicates the need to search for methods for evaluating the quality of exploitational oils, taking into account their individual characteristics, working conditions. At the same time, the need to search for methods for recognizing the early aging of insulating oils is dictated by the need for timely corrective (preventive) measures aimed at improving the efficiency of oil regeneration units – thermosyphon and adsorption filters; prevention of the formation of aggressive aging products that cause the destruction of cellulose insulation; reducing the cost of the repair campaign - reducing the complexity of work in the part of transformer cellulose insulation washing from aging products, excluding the costs of oil change; increasing the environmental friendliness of production - reducing the amount of overused oil that is in exploitation of energy companies.

2 Residual resource of oil and its influence on the resource of electrical equipment

Along with the most common methods for estimating the degree of transformers aging, such as measurement of the degree of polymerization and furan derivatives, such estimation methods as evaluation of molecular weight distribution with thin-layer (penetrating) chromatography, spectroscopy methods (fluoroscopy, ultraviolet, infrared and near infrared) and polarization methods (the method of recoverable voltage and the method of measuring the tangent of the dielectric loss angle of insulation at various frequencies - parameters depending on the condition of the insulation and do not depend on its geometry). At the same time, none of the existing methods for evaluation the degree of aging of the transformer can provide enough reliable (precise) information. Estimation of the residual resource of the transformer is possible only on the basis of a results analysis of a number of diagnostic methods, accounting the characteristics of the oil and the design features of the transformer. Work in this direction is actively carried out by various researchers [5]. According to the data [6], generator transformers with a voltage of 110 kV (with respiration through a silicagel air dryer) and transformers of higher voltage classes (with film protection) have annual damage due to internal short circuit of less than

* Corresponding author: s-151075@yandex.ru

0.5% for 25 years, after 25 years, it increases to 0.75%, and after 35 years to 2.5%, which is including a consequence of decrease in the resource of the transformer grid insulation.

The specified data underline influence of resource characteristics of the transformer isolation on reliability of the equipment. In works [4, 7] the efficiency is shown of resource management of a power transformer through the management of the resource of transformer oil filled into it.

Stability - oil stability against oxidation and sludge formation during exploitation. An indicator of the oil quality, determining its resistance to aging factors, is the "antioxidant stability of the oil."

Residual resource of transformer oil – the total exploitational time of transformer oil from the moment of its quality control to the transition to the limiting state, i.e. a condition in which the oil is incapacitated to perform the specified functions, while the restoration of its operable state is impossible and (or) impractical [4,8]. Accordingly, a method that allows to qualitatively evaluate the oil resource is the method of measuring "stability against oxidation." One of the approaches widely used in electric power industry for evaluating stability against oxidation is the standardized method GOST 981-75 [9]. The stability of oils, measured according to the specified method [9], is estimated by a set of indicators - acid value of oxidized oil (hereinafter referred to as "AVOO") and sediment formed after oxidation (hereinafter "Sediment"). By measuring the antioxidant stability of the exploitational transformer oil by forced oxidation of the oil sample in the laboratory, we estimate how much the condition of the examined oil, at a given time, is resistant to the influence of key factors that determine the course of oxidation processes, which take place in real conditions. Accordingly, the worse the indicators that determine the stability against oxidation, the greater the proportion of the initial resource of oil is lost. Comparative analysis of the change in the parameters of antioxidant stability before and after the exploitational impact on the oil makes it possible to evaluate the effectiveness of the selected / conducted steps to restore its resource.

Transformer oils are a complex multicomponent system obtained by cleaning distillates of petroleum at a certain temperature. In the process of exploitation, the oils undergo profound changes in the chemical and electrophysical parameters, which are characterized by the general concept of "aging" (oxidation). As a result of the oxidation process, an insoluble sediment (sludge) is released from the oil, acids, water and other aging products appear, incl. aggressive ones. The speed of aging in oil is not constant. At the first stage of exploitation, the speed is small - this is the induction period of oxidation. In the future, as the accumulation of aging products happens, the rate increases, then increases drastically.

Aging mechanisms are applicable for all types of petroleum oils, regardless of the raw materials used for their production, the percentage composition of the components, the methods for cleaning petroleum distillates and other factors. However, the speed of aging processes significantly depends on a number of factors, including temperature, oxygen, metal (oxidation catalyst), etc.

Modeling in laboratory conditions of various methods for restoring the properties of transformer oils, followed by a comparative evaluation of their qualitative changes, is the basis for the first developed "Method for the experimental determination of the liquid dielectric resource and measures for its recovery."

The essence of the method is the comparative oxidation of oil in the VTI appliance under the influence of oxygen at an increased temperature in the presence of a catalyst. The initial sample is subjected to oxidation before processing according to established rules, and then followed by comparison with the accepted standardized values and with the results of oxidation of the oil sample after the exposure to the sample, according to a selected scheme. The results of the oxidation of the oil are characterized by AVOO and Sediment. The method provides experimental evaluation of the transformer exploitational oil resource with the subsequent experimental laboratory determination of the materials volume, nomenclature and the order of works for the regeneration of transformer exploitational oil.

3 Experimental evaluation of exploitational transformer oil resource

A qualitative evaluation of the exploitational transformer oil resource is proposed to be conducted by measuring the antioxidant stability according to the GOST 981-75 [9] method using an apparatus for determining the stability of oils against oxidation of APSM-1. The VTI appliance is an integral part of the APSM-1 device, in which oxidation of oils under the influence of oxygen occurs at increased temperatures in the presence of a catalyst. After the process of laboratory forced oxidation of the oil is completed, the measurement of the formed sediment and the acid value of the oxidized oil is conducted, followed by a comprehensive analysis of the results obtained.

In the course of previous research work [4,7], it was established that the threshold values can be taken as the values of AVOO - not more than 0.1 mgKOH/g; Sediment - not more than 0.01% of the mass. If the threshold values are exceeded, it is recognized that the examinated exploitational transformer oil shows signs of a chemically unstable liquid dielectric, indicating a deterioration in the resource. The presence of the AVOO exceeding the accepted standardized value with the absence of Sediment usually indicates a smaller proportion of the resource expended, in comparison with situations, when there is a presence of Sediment (incl. the excess of the allowable value) during the oil oxidation in the laboratory conditions. Accordingly, according to the results of the tests, two categories of the state of the oil resource are determined: the resource is reduced, and the resource is not reduced.

Examination of the tests results for evaluation the residual resource of a liquid dielectric is carried out in a complex manner, taking into account the results of measurements of such quality indicators as acid value, the content of water-soluble acids and alkalis, the tangent of the dielectric loss angle of oil, the content of the antioxidant additive (BHT), the total content of the sludge. These quality parameters are indicative towards aging products formed during oxidation processes in transformer oil [1,3,4,10]. Based on the results of a comprehensive review of the measured oil resource and the quality indicators listed above, an algorithm is determined for conducting an experimental laboratory determination of the materials volume, nomenclature and procedure for performing oil regeneration work.

4 Experimental laboratory determination of the materials volume, nomenclature and procedure for performing regeneration of exploitational transformer oil (mineral oil)

Due to the fact that the chemical composition of transformer oils of different brands is complex and specific, and individual oil conditions exist in each power transformer, the nomenclature of the necessary preventive steps for restoring the oil resource is justifiable to determine by modeling processes in the laboratory (a special laboratory experiment).

To solve the task, the possible and most common combinations of preventive works used in power grid organizations were identified. The impossibility of developing one universal set of measures for the restoration of the oil resource is conditioned by the fact that for a certain degree of oil aging, the effectiveness of its recovery will depend both on the exact set of measures, and on the order of their implementation, the duration of these steps.

Accordingly, the cost of carrying out the work for various sets of measures to restore the resource of the oil will be different. Formulas for calculating costs are listed in Table 1. In this case, the emerging costs (W_{Mi}) will have strict lower and upper limits, which will depend on the proportion of the lost oil resource.

Table 1. Calculation of costs for carrying out preventive works to restore the resource of oil

Work option number	Nomenclature of works	The formula for calculating costs (W_M)
First option	1. Replacement of silicagel in a thermosyphon filter. 2. Oil stabilization by additive	$W_{M1} = (C_s * m_s) + (C_{BHT} * m_{BHT}) + P_1 + (P_2 * m_c)$
Second option	1. Replacement of silicagel in a thermosyphon filter. 2. Oil regeneration by special installation. 3. Oil stabilization by additive	$W_{M2} = (C_s * m_s) + (C_{BHT} * m_{BHT}) + P_1 + (P_3 * m_{o1})$
Third Option	1. Major overhaul of the transformer in the second group of complexity (with the washing of the active part from aging products). 2. Replacement of exploitational oil for fresh one	$W_{M3} = C_{OH} + (m_{o2} * C_o)$.

Notation:

C_s - ruble, the cost of 1 kg of silicagel.

m_s - kg, mass of silicagel filled in a thermosyphon (adsorption) filter;

C_{BHT} - ruble, the cost of 1 kg of BHT.

m_{BHT} - kg, mass of BHT necessary for input into the tank of a power transformer.

P_1 - ruble, cost of works on replacement of silicagel in thermosyphon (adsorption) filter (determined in accordance with base prices [11] - repair of thermosiphon filters of II group of complexity depending on the type of filter).

P_2 – ruble per 1 ton, the cost of processing transformer oil (determined in accordance with the base prices [11] - oil filling in the transformer №01080301).

m_c - tn, mass of the filled concentrated solution.

P_3 - ruble for 1 ton, the cost of processing transformer oil (determined in accordance with the base prices [11] - regeneration with stabilization of transformer oil №01080701).

m_{o1} - tn, mass of the processed oil.

S_{OH} - ruble, the cost of overhaul of a power transformer in Group II of complexity.

m_{o2} - tn, the mass of insulating oil filled into the oil tank of the power transformer.

C_o - ruble, cost of 1 ton of oil.

A key criterion for evaluation the effectiveness of the oil resource restoration [4,10,12] is the change in the antioxidant stability before and after exposure to oil. Effective step to change (restore) the resource characteristics of the oil are all the cases of improvement in two or more times in indicators (AVOO and Sediment), characterizing the antioxidant stabilityn.

5 Conclusion

The relationship between the resource of a power transformer and the resource characteristics of a liquid dielectric is shown. Key quality indicators have been identified that allow monitoring the resource of a liquid dielectric.

A "Method for the experimental determination of the liquid dielectric resource and measures for its recovery" is developed. The method provides experimental determination of the transformer exploitational oil resource with the subsequent experimental laboratory determination of the materials volume, nomenclature and the order of works for the regeneration of exploitational transformer oil.

References

1. *Handbook on the repair and maintenance of power transformers*. - M.: IPKgossluzhby, 2008. – 852 p.
2. Ivanov V.S. *Issues of testing and exploitation of transformer oils* / V.S.Ivanov. - M.: BTI ORGRES, 1962. - 116 p.
3. *Power Transformers: reference book* / Ed. by S.D. Lizunova, A.K. Lohanina - M.: Energoizdat, 2004. – 616 p.
4. Vysogorets S.P. *Development of the new methods and quality valuation algorithm of 35-110kV power transformer exploitational oil. Dissertation for the degree of Candidate of Engineering Sciences* - SPb., 2012 – 260 p.
5. Lutke H., Hohlein J., Kachler A. J. *Transformer ageing research on furanic compounds in insulation oil.* CJGRE, 2002, rep. 15-302
6. Vanin B.V., Lvov U.N. and others. *Issues of increasing the reliability of block transformers.* Electrical stations № 7, 2003.
7.
8. Vysogorets S.P. *Forecasting the residual resource of power transformers oil on the basis of analysis of the liquid dielectric quality* / S.P.Vysogorets // Scientific and Technical Journal «Safety and Reliability of Power Industry» №1(32)2016 – Moscow, 2016, Pp.50-54.
9. GOST 27.002-89 (CMEA Standard 3519-81) *Reliability in engineering. Basic concepts. Terms and definitions* / M: Izdatelstvo Standartov, 1990–32p.
10. GOST 981-75 *Petroleum oils. Method for determination the stability against oxidation* / M: Издательство стандартов, 1992 – 9 p.
11. Brai, I.V. *Regeneration of transformer oil* / I.V.Brai - M.: Publishing House «Chemistry»,1972. -168 p.
12. *Base prices for work on the repair of power equipment, adequate to the conditions of functioning of a competitive market and technical re-equipment.* // Part 6. Base prices for works on repair of transformers and reactors. – M.: 2003. P.54.
13. Aptov, I.S. *Chemical materials in electricity industry* / I.S.Aptov, M.V. Homyakov - M.: «Energia», 1969. – 280 p.

Reliability and service in the power industry

Yulia Zhilkina[1*], and Dmitry Vodennikov[2]

[1] Candidate of economic sciences, Assets management Department, Federal Grid Company of Unified Energy System, Moscow, Russia
[2] Deputy Chairman of the board- Chief Engineer, Federal Grid Company of Unified Energy System, Moscow, Russia

Abstract. Development of scientific and technical progress, introduction of new technologies promotes specialization and isolation of the highly professional service markets in economy. In power industry this market is the most natural object of development of the market relations. Nevertheless, today service structures carry out mainly internal orders on repair of the equipment of the power enterprises in the status of affiliated dependent or indirectly affiliated organizations. The principles of formation and transition to service considering a current state of power industry the main requirements imposed to service activity, her potential and providing balance of interests of market subjects are developed (commercial, reliability and energy efficiency). The main models of the organization of service are proved and recorded. The realized models of service will allow reaching a positive effect.

1 Background

Development of scientific and technical progress, introduction of new technologies promotes specialization and isolation of the highly professional service markets in economy. In power industry this market is the most natural object of development of the market relations. Nevertheless, today service structures carry out mainly internal orders on repair of the equipment of the power enterprises in the status of affiliated dependent or indirectly affiliated organizations.

Despite the considerable volume of researches in the field of the organization of service in a power network complex, a set of aspects in the field remain insufficiently studied, both in foreign, and in domestic science. The main theoretic-methodological problem is absence uniform for subjects of a power market of strategy of service which taking into account national peculiarities of power industry could provide forward development of all branch in general.

2 Reliability and service

The all technical sense of the concept «reliability» [1] is a property of an object to keep steadily in the set limits of value of all parameters characterizing ability of an object to perform the required functions in the set modes and conditions of its application, maintenance, storage and transportation. If to consider power industry as uniform technical system, then reliability should be considered one of technical properties of such system.

Reliability in power industry consists of two components – system reliability and reliability of separate units of equipment. Universal solution of the specified problems is such approach to operation and the maintenance of the equipment at which each unit of equipment is considered as continuous risk factor for business challenges of the enterprise. A main objective of activities for the maintenance of the equipment is the greatest possible reduction of risk in the available conditions and restrictions. At such formulation of the question management of reliability turns from the providing expensive activity into the tool of the solution of the current business challenges – increases have arrived due to more effective use of the equipment, decrease in laid on expenses at preservation of load of the equipment.

Reach compromise between economy on costs of the equipment and desire to avoid accidents, idle times and losses from them is the main administrative problem connected with ensuring reliability.

The situation is aggravated with objective factors as: serious wear of fixed assets, age and technological heterogeneity, decrease in professional level of the operating and repair personnel, decline in quality of the made spare parts and accessories, the operation of the equipment in the limit and ultra-boundary modes of loading which was outdate, not answering to realities regulatory and procedural base of processes of operation, service and repairs. It's possible to solve the problems stated considerably above by modernization of technological base, with transition to service.

2.1 Service advantages

The analysis of theoretical approaches to the organization of service of the electrotechnical equipment in power industry has shown that today in science and practice there is no standard branch definition of service. There are many definitions of the concept "service" which give both lawyers, and economists. The Ministry

* Corresponding author: zhilkina-yv@fsk-ees.ru

of Energy of the Russian Federation on the official site is given definition of service: service as a commercial product is provided by specialized companies who are ready to offer the best ready-made solution to the task of reducing costs, which, with relatively small investments, can give the best result in the shortest possible time.

One of the most essential advantages of service is the flexibility and competent risk management. Transferring a part of the business processes of service company, also the accompanying risks are automatically transferred. At the same time the service company having a number of clients endures economic crises easier that allows keeping quality of goods and services up to standard.

The main strategic advantages in general obtained from service are [2]:

✓ concentration on increase in competitiveness, satisfaction of needs of clients and development of technologies, carrying out optimum reengineering, stable increase in overall performance of the company by continuous improvement of key indicators of work, profitability, including due to release of internal resources, reduction of investments into non-basic funds;

✓ cutting of costs for the business functions transferred to service is reached due to specialization of the service company in narrow area and also to scale effect, that leads to decrease in costs in comparison with a complex of expenses, including tax and social, on creation and support of regular jobs;

✓ improving quality and reliability the functions transferred to service, since when solving similar tasks, service companies accumulate experience and can introduce modern technologies at the customer enterprise, use specialized equipment and highly qualified personnel content whose use which is not rational in the client's environment;

✓ division and transfer of risks due to use performer of the competition forcing mechanisms to react market changes flexibly, to increase efficiency by internal processes optimization.

Positive results to transfer maintenance and repair of company equipment to service organization are:

• transparency and optimization costs of maintenance and repair equipment;
• idle times reduction (increase in productivity) of the equipment;
• existence of guarantee certificates from the service organization;
• quality production improvement (decrease in level of marriage and inadequate quality production cause of unsatisfactory technical condition equipment);
• quality improvement of the rendered services.

Maintenance of reliability characteristics equipment of which has developed a settlement resource up to standard is possible only at the rational organization of his service (capital repairs, modernization and so forth) as his fast replacement by the new equipment isn't feasible for the economic reasons.

Such situation will remain long enough. In general input rates of new power objects will hardly advance growth rates of energy consumption, and in a number of regions become a limiting factor of further economic development today. Thus, service system improvement of electric equipment is one of components of a common problem of increase in reliability of power supply in general. At the same time improvement process in the conditions of sector restructuring breaks up to a number of tasks interconnected among themselves.

Reliability by the following factors is influenced:
1. Loss quality of maintenance and repair;
2. Violation of the rules of operation and technology of repair;
3. Design flaws;
4. Aging, wear;
5. Transition processes;
6. Weather conditions (working conditions);
7. Temperature drops.

The modern equipment allows creating the interactive control system which isn't demanding special dispatching divisions. In this case the automated system is on a continuous communication with the engineer-operator, allowing him to monitor work online and if necessary reporting about probable failures on a pager or the mobile phone. It substantially facilitates service and systems control. At the majority of companies, technical regulations include planned preventive maintenance (hereinafter referred to as "PPM") of complex equipment. Usually it is caused by the fact that repair cost upon accident significantly (sometimes up to 10 times) is more expensive than PPM. At the same time the principle of planned character assumes preventive orientation of a stop and equipment repair [3].

Such service system of the complex equipment consists that by means of constant technical diagnosing the analysis of a condition of knots and the unit in general is made and the forecast of necessary maintenance and repair becomes. At the same time diagnosing by various criteria can be made. It is the simplest to organize control on change of admissible level of one or several parameters. More difficult options include not only admissible size of parameter control, but also forecast of level of reliability knot or the unit in general.

The main complexity maintenance and repair on actual state consists in the organization of collecting and data processing at equipment operation. Despite presence at a considerable part of the systems complex equipment allowing automating all processes not everywhere it becomes, and not any organization is capable such system organize. However, serious producers usually with great attention treat service of the production. It is possible to claim that what the manufacturing firm is better known that has better organized service and the more opportunities for the professional service organization on actual state.

In particular service model cardinally solves different problems of the power company:

• ensuring full readiness of the operation equipment during all term of stay him at the consumer (production aspect);
• consultation of potential consumers before allows services of the power company by them allowing to make him a conscious choice; collection of information how service at "competitors" - a benchmarking is organized; analysis of claims (marketing aspect);

- training personnel of the power company for the most effective and safe operation of the acquired equipment (the training aspect);
- expeditious delivery of spare parts and contents purpose necessary network of warehouses, close contact with manufacturers of the equipment and accessories (logistic aspect);
- methodical ensuring service activity and recommendations regarding improvement of services development; introductions of technological innovations (innovative aspect).

2.2 Model service organization

2.2.1 Service model (example of Federal Grid Company of Unified Energy System)

The basic (autotransformers, transformers, reactors, gas-insulated switchgear, gas-insulated circuit breakers, etc.) and auxiliary (compressors, rechargeable batteries, etc.) the equipment with estimated long (more than 3 years) operation term on objects is transferred to service.

Main objectives of service of the equipment of PJSC "FGC UES": [2]

- increase in reliability of the equipment in the conditions of optimization of costs maintenance and repair (improvement of quality maintenance and repair);
- achievement of economic effect;
- need of extension of the guarantee certificates for the equipment issued by manufacturer at supply of equipment (for the new equipment), and establishments of guarantee certificates of manufacturer for the equipment on which works within service are performed (for the equipment with last warranty period).

In agreeing the model of services for PJSC FGC UES, the following features have been taken into consideration:

• diversity of the equipment installed, most of which no longer has the factory warranty, and a considerable portion has been in operation beyond its specified life;

• highest requirements concerning the emergency response in the event of damage to equipment or devices (speed of emergency response) to ensure stable operation of the UNEG;

• sufficient number of skilled in-house personnel, for many years performing repair and diagnostics of equipment and ready to improve their skills, master new equipment, and perform simple repairs;

• availability of the affiliate specialized in repair and complex diagnostics of equipment – JSC Specialized grid service company Unified national electric grid (JSC Elektrosetservice UNEG), staffed with qualified personnel and modern technological equipment;

• geographical distribution of the UNEG facilities in the territory of 73 constituent entities of the Russian Federation, which complicates the creation of factory service centers that would meet the requirements of emergency preparedness;

• the network of branches of PJSC FGC UES and JSC Elektrosetservice UNEG, covering almost the entire territory of Russia, which is an additional argument in favor of bringing their own staff to perform certain types of work within the service after its training and certification.

Given these peculiarities, PJSC FGC UES organizational document has been approved and will act as a model service organization:

1. Execution of service in the format of bilateral contract between the Customer and the Contractor. The contractor may be the organization is the manufacturer of equipment (devices) or his official representative on the territory of the Russian Federation with authorization for the supply of spare parts and provision of services, with the implementation of certain types of works in the justification and agreement with the Customer by the certified personnel of the Subcontractor. As part of a pilot implementation of the service model PJSC FGC UES in 2010 and 2011. Contracts were signed for execution of works on maintenance with the leading companies-manufacturers of the electrotechnical equipment. Pilot projects on implementation the service models found to be successful.
2. The first experience of outsourced service made by a number of key domestic and foreign manufacturers of electrical equipment allowed to carry out the analysis of this model and its adjustment for the transition to the long-term form of service (5 years).

The development of the concept of service at the facilities of PJSC "FGC UES": [3]

A) Contract for service compressor equipment service
-797 compressors was transferred for service;
-equipment inspections by representatives of the manufacturer are organized;
-an overhaul of the 231 pieces of compressors.

B) Contract for service maintenance of electrical equipment manufactured by ABB
-training of Customer's personnel (122 people) in the service training center with the issuance of certificates;
- creation of emergency reserve circuit breakers;
-15 switches have been overhauled;
-supply of spare parts;
-reducing the time of emergency recovery operations.

3 Conclusions

Introduction of the equipment service model aimed at improving the quality and efficiency of maintenance and repair in terms of cost optimization is one of the most important areas of activity for PJSC FGC UES, where a positive effect has been achieved.

The results obtained in the implementation of the equipment service concept of PJSC FGC UES are:

- extension of warranty issued by the manufacturer;
- ensuring the participation of manufacturers of the main equipment in the maintenance of its normative status throughout the life cycle;
- availability of emergency supplies close to the UNEG facility warehouses, taking into account the recommendations of manufacturers of equipment;
- minimum time for the elimination of faults due to the availability of the necessary resources (qualified personnel and emergency reserve warehouses) and

guaranteed services of the chief engineer of the manufacturer;

- improving qualifications of own maintenance personnel.

In general, the transition to the approved model of service in PJSC FGC UES has shown its transparency and the availability of a mechanism of control over investments and the resulting effect, which is a decisive factor in the development of the industry as a whole.

References

[1] B.A. Raizberg, L.S. Lozovsky, E.B. Starodubtseva "Modern economic dictionary" (2nd ed., rev. M.: INFRA-M, 1999)

[2] Order of JSC FGC UES dd. 19.10.2015 no. 410/349 "On adopting the Regulation on service of equipment of PJSC FGC UES"

[3] Y.V Zhilkina "Organization of service in the electricity industry" (Special issue of the journal ELECTRICITY. Transmission and distribution (No. 2(5) June 2017) for JSC "Rosseti")

Transformation of the model of estimation of balance reliability UPS of the Russia in the account of correlation of the consumption regime

Yuri Chukreyev[1,*]

[1]Komi Scientific Center of the Ural Branch of the Russian Academy of Sciences, ISE and EPN, 167000, Russia

Abstract. The characteristics of the electric power consumption regime are used when determining the indicators of adequacy in the management of the development of the UPS of Russia. The influence of the changed conditions of their representation on the probabilistic indicators of the balance reliability of system-forming links between the territorial zones (UPS) of the computational model of the UPS of Russia is shown. On this basis, we proposed the transformation of the model of estimating the balance reliability indicators, which allows to increase the computational efficiency.

1 Introduction

When planning the development of the industry, the so-called balance of power for the UPS of the country as a whole and its unified EES (UPS) [1, 2] was developed earlier, is being developed today and will be developed in the future. Its form includes two positions (Figure 1): the expenditure part "demand" (maximum load, export and standard power reserve); the incoming "cover" (installed power, unused power for the period of passage of the maximum, power inputs after passing the maximum, underutilization of power).

The issue that concerns both consumers and suppliers of electrical energy is to justify the regulatory power reserve, which is one of the components of the expenditure part of the balance sheet. In the tasks of long-term planning of EES it is called the full (normative) reserve, and conditionally divided into three components: repair, strategic and operational, intended to compensate for unscheduled (emergency) conclusions of the main generating and network equipment for repair. Therefore, in order not to confuse the concept of operational reserve in the current planning and short-term (long-term), this component of the full reserve has recently been called the compensation reserve.

Market relations in the electric power industry exacerbate the issues of justifying the power reserve. With short-term planning (from 3 year to 4 years), the price for capacity during long-term (competitive) power take-off depends on the requirements for a full (normative) reserve in the conditions of market relations.

The difficulty in determining the total capacity reserve is mainly related to finding one of its three components - the operational (compensation) power reserve, which depends on a variety of factors and events, including those randomly determined [3 - 6, etc.]. These include:

- model of the calculation scheme of the UPS;
- disposable capacity structure of generating capacities;
- scheduled repairs of equipment;
- reduction of generating capacity of territorial zones due to emergency damage to power plant units;
- regular and irregular maximum loads of territorial zones and graphs of their changes in the context of the year and day;
- random deviations and irregular load fluctuations;
- reserves of communication capacity in normal and emergency modes between territorial zones.

The article considers the influence of modern possibilities of receiving and processing of probabilistically determined information on power consumption modes and the models of the UPS of Russia calculation schemes used to transform the methodology for estimating indicators adequacy (IA) and its means of support.

2 Representation of a mode of power consumption

Important, having a significant impact on IA and means of their provision in planning the development of EES is the question of presenting the power consumption regime during the settlement period, which usually is a year. The receipt of an hourly perspective schedule of power consumption of the territorial zones of EES for the settlement year at all times caused considerable difficulties.

Studies of balance reliability in our country began at the turn of the 60s of the last century. A generalization of these studies to the concentrated EES is given in [7]. As a yearly load graph, a load schedule was used for the duration in the hour section (8760 hours). As a criterion for making a decision to justify the reserve of power, the

* Corresponding author: chukreev@iespn.komisc.ru

indicator of the integral probability of no-deficit work $p = 1 - J_{д}^{опт.} = 0{,}999$. The index of the integral probability of occurrence of a power shortage $J_{д}^{опт.} = 0{,}001$ had its feasibility study and was determined by the expression $J_{д}^{опт.} = з_{R}^{уд.}/(y_0 \cdot 8760)$ [7]. In this expression, the specific reduced costs for reserve capacity ($з_{R}^{уд.}$) were taken equal to 5 rub. / kW, specific losses from non-supply of electricity to consumers $y_0 = 0.6$ rub. / kW · h.

In the same work, for the first time in the country, taking into account foreign experience, the concepts of the regular maximum of the load and its random deviations due to irregular oscillations and prediction errors were introduced. These concepts are included in almost all approved normative technical documents (NTD) - methodological instructive for the design of the development of power systems (hereinafter MI), used in our country and now. It should be noted that in these NTD introduced the concept of the average daily load schedule of the maximum power consumption of the days of December [2]. The distribution of this schedule is proposed for all 250 working days of the year (Figure 1). In those days, the reasons for justifying such an approach were understandable and understandable to the designers and the scientific community. In what it manifested itself.

In the methodical plan there was a transition from the certain seasonal interval of its decline) to the December maximum and the annual schedule of seasonal load decreases equalized. This was the main reason for the spread of the December average daily load schedule for all working days of the year. It should be noted that the same happened abroad, for example, in the power systems of North America, the balance sheet reliability standard is still applied in the form of *LOLE* = 0.1 days / year, obtained on the basis of analysis of not 8760 hour changes in the load, but only it maximum values for each of 365 days of the year.

At present, the processes of obtaining the initial information on the levels of power consumption and the schedules for their change are significantly different from those that took place in the development of guidelines and MI. Today, it is possible to obtain real-time schedules of load changes in the power systems of the UPS of Russia in any time interval due to information:

- within the retrospective period about the actual capacity at the end of the hour[1]; actual daily average and average monthly temperatures; actual annual consumption of electrical energy;
- within the forecast period about the annual power consumption, set in the annual work of the schemes and programs of long-term development UPS Russia[2].

The presence of such information allows to form the necessary data packet. It includes:

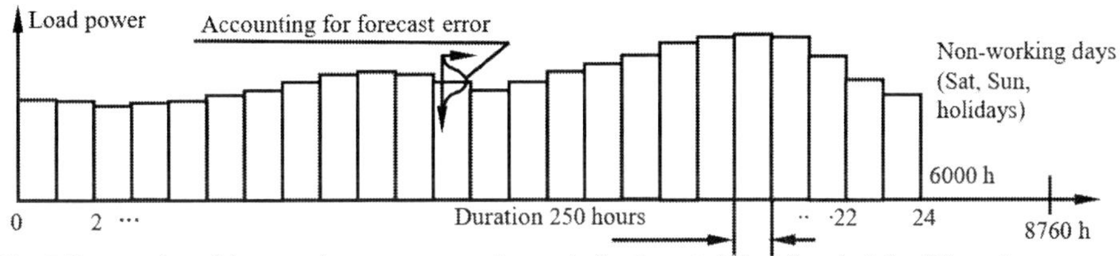

Fig. 1. Presentation of the annual power consumption mode the characteristic daily schedule of December.

consideration of the UPS of the country not by a concentrated association, but by a multi-zone one with the separation as zones of joint EES. This required the development of a new approach to modeling the random states of generating capacity and the load in them with subsequent evaluation for satisfaction of all consumers. Considering 8760 hourly load changes has become, beyond time, an impossible task. The size of the decision criterion for justifying the capacity reserve also changed $p = 1 - J_{д}^{опт.} = 0{,}999$. The indicator of the integral probability of the appearance of a power shortage, by virtue of a change in the principles of determining unit costs for standby capacity (the operating costs began to be taken into account) was adopted by the value $J_{д}^{опт.} = 0{,}004$ [2].

The deficit of generating capacity in the country's UPS that was observed at that time did not allow for capital repairs during peak hours, which, unfortunately, is not observed today. They always fit into seasonal power failures. Thus, the load in them increased (assuming that repairs - this is an addition to the load in a

- the form of forecast daily average load charts;
- random deviations of the load from the mean values;
- the magnitude of the forecasted regular load maxima.

A description of the procedure for obtaining the above information is given in detail in [8]. A few words about the random component of load deviations from mean values. Her, with respect to the problem of balance reliability, at all times associated mainly with fairly well-established probabilistically determined events, caused by two reasons. The first one is caused by the appearance of deviations from the mean values of the daily schedule load (Figure 1), obtained as a result of processing a huge set of information on power consumption modes for a number of working days of the December month of the retrospective period. The second most significant reason is connected with taking into account the influence of the temperature factor on the

[1)] Daily dispatching lists of the operational information complex (OIC) of the main dispatch center (MDC) of SO UES JSC.
[2)] Scheme and program for the development of the UPS of Russia for 2016-2022.

power consumption mode. It is no accident that in the unapproved Ministry of Energy of Russia in guidelines 2011 and in the Order of the Ministry of Energy of Russia[3], the prospective values of the maximum loads are given to the average annual temperatures of the outside air of the territory under consideration.

Statistical processing of power consumption modes for the 6-year period from 2010 to 2015, carried out by the specialists of JSC "SO UPS" [8], allowed, according to the specified in the - within the forecast period about the annual power consumption, set in the annual work of the schemes and programs of long-term development UPS Russia for 2016-2022 annual electricity consumption volumes for the considered territorial zones (UPS):

- regular load maxima;
- random deviations of the load from the mean values;
- average graphs of changes in daily loads;
- correlation dependencies of random load deviations.

The received information on the first two items for 2022 is given in Table 1. In the same table, the differences between regular and irregular load maxima are shown using different techniques. It can be seen that these differences are quite significant.

It should be understood that the methodology for obtaining a regular maximum based on the analysis of retrospective information is not entirely perfect in terms of using retrospective information on power consumption not listed in the middle temperature standards. Studies show that to describe the power consumption mode, you can use the shape of the load curve when using an irregular maximum as a maximum. The last two lines of the table show the values of the random deviations of the load (σ_n), obtained by using different methods. It can be seen that they, obtained by processing retrospective information, are almost 40% higher, given in the 2003 and 2011 MR. The same trend is observed for individual UPS, except for the Middle Volga UPS.

Coefficients of consumption covariance between the territorial zones identified in the form of UPS as a result of the processing of the retrospective information (the last of the items listed above) fluctuate within fairly wide limits. Of the 15 possible coefficients (6 UPS without the UPS of the Far East), more than 10 exceed the threshold of 0.5. This indicates a strong dependence of random deviations in the load at once in several UPS, which largely affects IA and the means of their provision [8].

Table 1. Some load characteristics for the 2022 forecast year

Options	UPS of Russia without UPS of the East	Name of UPS					
		Ural	Middle Volga	South	Nortwest	Centre	Siberia
Load of territorial zones of the forecasted year, MW							
irregular maximum ($P_{нерег}$)	155860	37390	17096	16831	15151	39266	30126
regular maximum (P_{max})	148592	36444	15949	16091	13789	36404	29915
Differences between irregular and regular highs (pu)							
by the method of definition P_{max} [8]	0,0466	0,0253	0,0671	0,0439	0,0900	0,0729	0,0070
in accordance with MU 1981 *	0,0126	0,0161	0,0185	0,0189	0,0187	0,0153	0,0170
Random deviations of the load (σ_n) from the irregular maximum in pu							
according to the above procedure	0,0177	0,0409	0,0327	0,0468	0,0482	0,0389	0,0407
as in the MR-2003 and 2012	0,0127	0,0259	0,0382	0,0385	0,0406	0,0252	0,0288

* the difference between irregular and regular highs $\Delta P_{нерег} = 0{,}01 \cdot P_{max} + 1{,}24 \cdot \sqrt{P_{max}}$

3 Peculiarities of forming the model of the settlement scheme of the UPS of Russia for the tasks of adequacy.

Under the zones of the same reliability (hereinafter the territorial zone is a concentrated system[4]) controlled in the analysis of balance reliability both under the conditions of a vertically integrated system and under the market management principle, a set of sets of territorial zones of EES for which network restrictions do not affect IA should be understood. In the future, this term will also be characterized by the phrase territorial area, i.e. the territory in which network restrictions do not affect the restriction of consumers in power and electricity.

With the long-term planning of the development of the UPS of Russia at all times, a number of various assumptions were adopted that made it possible to simplify the solution of the problem. This applies to the representation of the electricity consumption regime (only in December), the accounting for scheduled repairs and the composition and limitations of generating

[3] Order of the Ministry of Energy of Russia of 07.09.2010, No. 431 (Edited on August 17, 2017) "On approving the Regulations on the procedure for determining the amount of demand for capacity for conducting long-term power take-off ..." (Registered in the Ministry of Justice of Russia on 29.09.2010, No. 18578).

[4] A concentrated system is called a power system, within which there are no restrictions on power transmission in any possible emergency situations caused by unreliability of the generating and network equipment.

equipment, and many other issues. The simplifications introduced during the long-term planning naturally concern also the issues of aggregating the scheme of electrical connections of the main generating and network objects.

It is clear that the introduced simplifications that have a certain influence on the decisions taken to manage the development of EES should have a certain justification. In any case, the model of the settlement scheme of the UPS of Russia for determining IA in order to justify the reserve capacity values should be represented by a graph containing links to nodes (territorial zones) of contiguity. The territorial zone is a kind of association in which network restrictions do not affect the power distribution. Linkages are understood as a set of power transmission lines between territorial zones. The communications of the calculated graph should contain information on maximum permissible overflows (MPO) of capacity (connectivity capacity) in the forward and reverse directions for normal and emergency operation.

In recent years (2010-2015) in the circles of reservation researchers there have been judgments that in the calculations aimed at determining the requirements for the levels of the operational reserve capacity for the short-term period (from 1 to 7 years), the UPS scheme of Russia should be more detailed with the inclusion of up to 100 or more territorial zones. When developing power and power balances for a more distant future, there were no doubts that territorial zones could be represented in a more aggregated form, for example, in the form of united power system (UPS) or free power transfer zones (FPTZ) [6], with possible breakdown of them on sections that limit emergency mutual assistance within them. In any case, the number of territorial zones did not exceed 20.

It is clear that the correct solution of these issues requires the conduct of relevant studies. In the period 2012-2016 such studies in solving a set of issues related to the development of the UPS of Russia were carried out at Joint Stock Company «Scientific and Technical Center of Unified Power System»., which is the basic scientific center of JSC «SO UPS» [9]. Unfortunately, at the present time, the information necessary for the scientific substantiation of these issues is practically closed to researchers dealing with issues of adequacy. In this article, the authors relied on information support received in the course of a number of works with JSC "SO UPS of Russia" in the period 2011-2016.

In 2011-2012. on the instructions of JSC «SO UES» JSC "STC UPS" performed work on the formation of the model of the settlement scheme of the UPS of Russia for solving the problems of ensuring their adequacy [9] It is quite obvious that in order to evaluate the IA of the actual composition of the generating equipment, their accidents, and the projected maximum loads, such schemes are of some interest from the perspective of obtaining IA in the territorial zones and identifying "weak" links in the system. At the same time, solving the task of justifying the means of ensuring adequacy - operational (compensatory) capacity reserves for such schemes is, in our opinion, rather problematic.

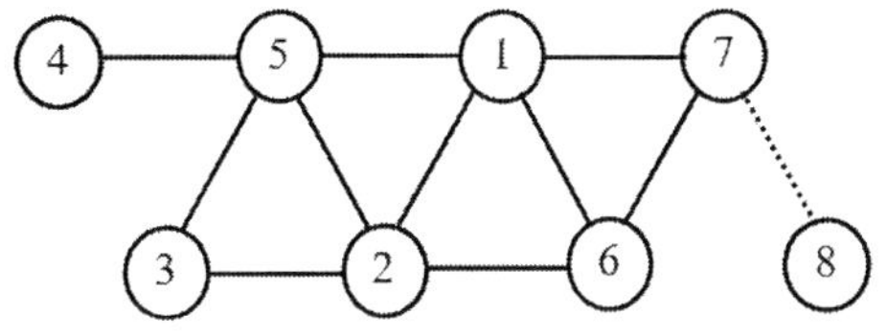

Fig. 2. Model of the settlement scheme of the UPS of Russia:
1 - UPS Urala; 2 – UPS of the Middle Volga;
3 - UPS of the South; 4 - UPS of the North-West;
5 - UPS of the Center; 6 - Kazakhstan; 7 - UPS of Siberia; 8 - UPS East.

On the basis of the developed 56-node model of the design scheme of the UPS of Russia, with the participation of experts from the regimes service of JSC "SO UPS", a design scheme consisting only of UPS was developed (Figure. 2). Information filling on the territorial zones (the composition of the generating equipment, the rate of its failure rate, the form of the change in the daily load schedule, the forecast errors and their correlation dependencies), on the connections (throughput and accidents) was obtained partly work from the schemes and programs of long-term development Russian Unified Power System for 2016-2022, partly as a result of processing of retrospective information on the reporting parameters of the system operation for 2010-2016.

4 Opportunities for transforming the model for justifying the operational reserve of capacity, in relation to the current conditions for the development of the UPS of Russia

Analysis of the results of determining the operational (compensation) capacity reserve, presented in Table 2 and 3 for 2022 (the schemes and programs of long-term development Russian Unified Power System for 2016-2022) allows us to take a slightly different look at the formulation of the task of estimating IA in planning the development of the UPS of Russia in the current conditions for receiving and processing retrospective information on power consumption regimes. Specifically, this relates to the retrospective information revealed as a result of statistical processing, the dependence of the change in the power consumption mode due to random errors in forecasting the load between different territorial zones. The effect of this factor in Table 2 and 3 are termed with and without correlation.

Presented in Table 2 the results of the distribution of the operational capacity reserve for UPS of Russia are of a research nature. The reason is insufficiently worked out and practically not discussed at the conferences the questions on the formation of the initial information on the maximum permissible power flows in the models of the settlement scheme of the UPS of Russia presented by the JSC "SO UPS".

Table 2 - Distribution of the operational (compensation) reserve for the ECO with different presentation of the model of the settlement scheme of the UES of Russia for 2022. (information from the work of CIPR UES of Russia for 2016-2022)

Options		UPS of Russia without UPS of the East	Name of UPS					
			Ural	Middle Volga	South	Nortwest	Centre	Siberia
Irregular maximum loads		155860	37390	17096	16831	15151	39266	30126
Taking into account the correlation of the load								
56 territorial zones, the complete scheme of the UPS of Russia [9]								
Operational reserve,	MW	15950	4914	1112	1682	1713	3393	3136
	%	10,23	13,14	6,50	9,99	11,30	8,64	10,41
7 territorial zones (6 UPS of Russia + Kazakhstan, figure 2)								
Operational reserve,	MW	15610	4894	1078	1619	1629	3365	3025
	%	10,02	13,09	6,31	9,62	10,75	8,57	10,04
without taking into account the bandwidth of communications (UPS of Russia as one node)								
Operational reserve,	MW	14754						
	%	9,47						
Differences in% to original		5,48						
Without taking into account load correlation								
56 territorial zones, the complete scheme of the UPS of Russia [9]								
Operational reserve,	MW	12735	4075	125	1280	1700	2885	2670
	%	8,17	10,90	0,73	7,61	11,22	7,35	8,86
7 territorial zones (6 UPS of Russia + Kazakhstan, figure 2								
Operational reserve,	MW	11915	3894	-25	1189	1609	2793	2455
	%	7,64	10,41	-0,16	7,06	10,62	7,11	8,15
without taking into account the bandwidth of communications (UPS of Russia as one node)								
Operational reserve,	MW	10915						
	%	7,00						
Differences in% to original		8,39						

Table 3. Bandwidth of communications (MW) and ratio multiplicity optimal integral probability of power shortage in the UPS (0.004) to the value of the integral probability of link overloading

		The numbers of ties in the UPS of Russia in the form of UPS (Figure 2)								
		1-2	1-5	1-6	1-7	2-3	2-5	3-5	4-5	6-7
Throughput	in direct	3175	687	3133	840	990	2185	1490	2400	3285
	in the opposite	3870	828	1472	500	485	4315	1620	2665	2890
Without taking into account the correlation		27	22	7,4	7,4	12	120	5,6	39	9999
Taking into account the correlation		88	84	41	41	70	2700	73	103	9999

The results given in Table 2 show that when the levels of the MPO of communications both inside the UPS (here not shown) and between them are received at JSC "SO UPS" (Table 3), the differences in the reservation levels of the territorial zones are extremely insignificant. The explanation for this is quite simple: in power system associations, the MPO (transfer capacity) of connections practically do not limit the transfer of power from one node to another. Moreover, attention is drawn to the fact that when the correlation dependencies between random load deviations in the territorial zones (UPS) are taken into account, the influence of the MPO of the bonds on the IA decreases drastically. In Table 3 shows such relationships, for example, if the integrated probability of a power shortage in the UPS is close to 0.004 for the optimal variant of the development of the UPS of Russia, then for the Ural-Center connection (1-5, in Figure 2) without taking into account the correlation, the probability of overloading its MPO is 0,000119 (multiplicity 0,004/0,00018 = 22, in Table 3). Taking into account the correlation leads to a reduction of more than threefold the probabilities of MPO overloading. For the communication in question, it was 0.000048.

All this leads to the possibility of considering the UPS of Russia in the form of a concentrated power system in which the simulation of random states of generation and especially the load is made at the level of territorial zones, taking into account the dependencies of their random deviations between themselves. The evaluation of the generated state for the purpose of providing the load is made by simply comparing the total for all territorial zones of the model of the calculation scheme of the UPS of Russia, the values of generation and load.

Adopting such an approach dramatically reduces the time required to obtain a preliminary result. Cost reduction occurs due to the exclusion from the consideration in the IA estimation model of the most costly in the temporal aspect of the block of estimating the random state of the system [5, 6]. Then this result can be corrected already taking into account the maximum allowed power flows. From Table 2 that the introduced simplifications in the variant taking into account the correlation lead to a decrease in the compensation power reserve by only 0.75% in relation to the irregular maximum of the load, and if to the reserve capacity, it is only 5.5%.

References

1. Methodical recommendations on designing the development of power systems. (Approved by Order of the Ministry of Energy of Russia of June 30, 2003, No. 281). - M: Ministry of Energy of the Russian Federation, SO 153-34.20.118-2003.
2. I.M. Wolkenau, A.N. Zeyliger, L.D. Khabachev Economics of the formation of electric power systems. - Moscow: Energoatomizdat. (1981) - 320 p.
3. G.A. Volkov . Optimization of reliability of electric power systems. - Moscow: Nauka, (1986) - 117 p.
4. G.F. Kovalev, L.M. Lebedeva Reliability of power systems. - Novosibirsk: Science, (2015) - 224 p.
5. Yu.Ya. Chukreev. Models for ensuring the reliability of electric power systems. - Syktyvkar: Komi Scientific Center of the UrB RAS, (1995) -176 p.
6. Yu.Ya. Chukreev, M.Yu. Chukreev. Models for estimating the indicators of balance reliability in the management of the development of electric power systems. - Syktyvkar: Komi Scientific Center of the Ural Branch of the Russian Academy of Sciences, (2014) - 207 p.
7. I.M. Markovich. Modes of energy systems. Moscow: Energia, (1969) 351 p.
8. Yu.Ya. Chukreev, M.Yu. Chukreev, V.S. Chuprov Characteristics of the electric power consumption regime applied to the problems of balance reliability in the management of the development of electric power systems. Results / Methodological issues of reliability research of large energy systems. Issue. 68. - Irkutsk: ISEM SB RAS, (2017) - P. 328-337.
9. Yu.Ya. Chukreev. Features of the evaluation of the balance-of-life indicators for the long-term development of the UPS of Russia / / Methodological issues of reliability research of large energy systems. Issue.65. - Irkutsk: Reliability of liberalized energy systems - ISEM SB RAS, (2015). - P. 213-222.

E3S Web of Conferences 58, 02010 (2018)
RSES 2018
https://doi.org/10.1051/e3sconf/20185802010

SAIFI reliability calculation for the distribution line

Ivan Bandurin[1,*], *Alexey* Khaimin[1]

[1]Pskov State University, Faculty of Computer Science and Power Engineering, 180000, Russia, Pskov, Lenin Square, 2

Abstract. The article for the first time shows the method of calculating the SAIFI reliability index for the distribution line at the design stage. Formulas and examples of calculation of SAIFI indicators for distribution lines of different configurations are given.

1 Introduction

Government resolution No. 1220 of 31.12.2009 "on the determination of long-term rates of reliability and quality of goods and services supplied" and the relevant order of the Ministry of energy of the Russian Federation of 29 November 2016 №1256 "on approval of Guidelines for the calculation of the level of reliability and quality of goods and services provided for the organization of management of the unified national (all-Russian) electric network and territorial grid organizations" established a mechanism for the relationship between the level of actual reliability and tariffs for electricity transmission grid companies.

The order of the Ministry of energy of the Russian Federation №1256 uses two indicators of the level of reliability of services provided for territorial grid organizations (TGO) – an indicator of the average frequency of termination of electricity transmission to the point of supply of SAIFI and the average duration of the termination of electricity transmission to the point of supply of SAIDI. This situation required the development of methods for calculating the reliability of electricity supply to domestic consumers. In this article the method of calculation of SAIFI reliability index for end users of the distribution line is offered.

When designing a new or reconstructing an operating distribution line, the problem arises of estimating the indices of its reliability. This assessment is necessary primarily for the feasibility study of the construction or reconstruction of the distribution line. In this regard, the purpose of this work is to obtain a method of calculating the SAIFI reliability indicator of the distribution line at the design stage. This confirms the validity of this article.

We will assume that it is advisable to use the same reliability indicators for the distribution line as for the TGO. Reducing the SAIFI value for the distribution line will also reduce the SAIFI value for the entire TGO.

A review of Russian and foreign publications [4-8] showed that the SAIFI reliability index can be determined only by statistical data of operation. Thus, it is possible to obtain an assessment of the reliability of the line before the reconstruction.

2 Methodology for calculating SAIFI's reliability index

The mathematics of reliability:

$$SAIFI = \frac{\sum_{j=1}^{J} N_j}{N_t}, \tag{1}$$

where N_j - the number of points of supply to consumers of network organization services, in respect of which there was a j-th power cut-off within the framework of technological violations, pcs.;

N_t - the maximum number of points of delivery to consumers of services by the network organization for the t-th settlement period of regulation, pcs;

J - the number of stoppages in supply of electricity in respect of that cheque delivery consumer services network in the t-th current control period, pcs.

The 10 kV distribution network consists of mutually redundant lines of tree configuration with cross sections of wires decreasing to the ends of the line, and TP 10/0.4 kV mainly on the branches. 10 kV line redundancy can be full and partial. In case of partial reservation, only a part of consumers will receive high-quality electricity.

In rural networks the most commonly used complete transformer substations 10/0.4 kV outdoor installation (CTS) and simplified gap-free design switchgear of 10 kV with transformers 63-250 kVA. For power critical loads using two CTS, between which is mounted the automatic partitioning or a complete checkpoint transformer substation 10/0,4 kV that is appended to line 10 kV according to the scheme of input-output.

It is not possible to use formula (1) to calculate SAIFI reliability during the design phase. Therefore, using the basic provisions of the probability theory, we convert the formula (1) to a convenient form for calculations.

In the derivation of the formula, we assume that the distribution line has consumers connected to the scheme

* Corresponding author: bandurin_ivan@mail.ru

https://doi.org/10.1051/e3sconf/20185802010

of the 2nd and/or 3rd category of reliability. In most cases, the distribution line has only one circuit breaker mounted at the beginning of the line. And then, if a short-circuit occurs anywhere in the distribution line, the power switch disconnects all consumers at once. However, with the aim of improving the reliability of a line may be set; and additional circuit breakers (reclosers) in the trunk and/or branch.

Divide a distribution line on M elementary parcels. The simplest part of the line is:

1) part of a line that does not contain switches,
2) part of a line that contains only one switch.

The accident of the *i*-th section occurs with the frequency of $\lambda_{\Sigma,i}$ and leads to the disconnection of consumers, the number of which is equal $N_{\Sigma,i}$. The total number of consumers of the distribution line is N_Σ.

Let's make an assumption that simultaneous failure of 2 sites is not possible. This assumption is valid for the Poisson distribution law. From a number of publications [2, 3] it follows that the discrete case value-the number of trips is subject to the Poisson's law of distribution.

Due to the fact that the failures of the sections of the distribution line represent independent events, then on the basis of the form of the total probability and the formula (1) we obtain:

$$SAIFI = \frac{\sum_{i=1}^{M} \lambda_{pl,i} N_{pl,i}}{N_\Sigma}. \tag{2}$$

On the basis of formula (2), we propose the following methods-dick of calculating the indicator SAIFI:

1. Divide a distribution line on M elementary parcels.
2. Find for each *i*-th section of the line the number of failures λ_i and the number of disconnected N_i consumers;
3. Find SAIFI for the distribution line using the formula (2).

Note that the reliability of the power switch is significantly higher than the reliability of the air line. The failure rate of the outdoor unit switch in the stationary state is 0.008 off/year, while the specific rate of sustained air bounce lineages 10 kV, 1 km long and 0.25 off/year [3]. So the indicator of failure rate at the switch turned out to be 30 times lower, in order to simplify the calculations, it is permissible to consider the switch to be ideal and not to take it into account in the method of calculating the SAIFI indicator.

3 Examples of SAIFI reliability calculations

Consider the different distribution line schemes and determine the SAIFI indicator for each of them. The source data for all the schemes of calculation: λ_{L1} = 3 off/year; λ_{L2} = 4 off/year; N_{L1} = 3 PCs; N_{L2} = 10 PCs. Data are taken arbitrarily for convenience and clarity of calculations.

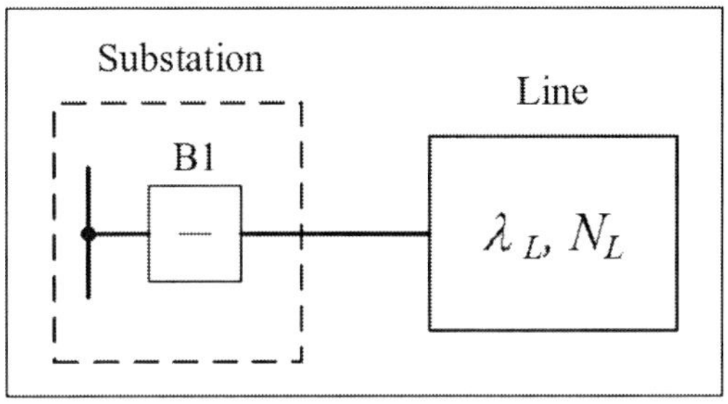

Fig. 1. Block diagram of the simplest distribution line.

Example №1. Let us calculate the SAIFI indicator for the simplest distribution line (Fig. 1). The simplest line is the line that does not contain switching devices.

For the given case $N_j = N_t$. Therefore,

$$SAIFI = \frac{\sum_{j=1}^{J} N_j}{N_t} = \frac{JN_t}{N_t} = J = \lambda_L, \tag{3}$$

where λ_L - he number of failures in the line for the year.

Perform the calculation of the SAIFI index for a distribution line, partitioned tables (Fig. 2).

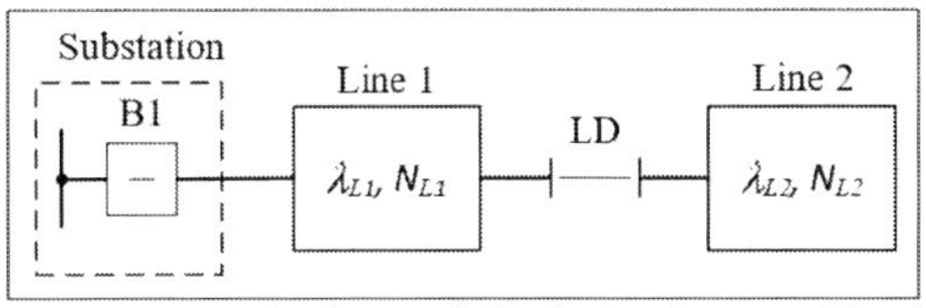

Fig. 2. Structural scheme of distribution lines, partitioned disconnector.

The total number of electricity consumers will be $N_t = N_{L1} + N_{L2}$. In case of a short circuit on the first or second line, all consumers will be disconnected from the power supply $N_{jL1} = N_{jL2} = N_t = N_{L1} + N_{L2}$. Thus, the SAIFI index for a distribution line, a partitioned switch:

$$SAIFI = \frac{\sum_{j=1}^{J} N_j}{N_t} = \frac{\lambda_{L1} N_t + \lambda_{L2} N_t}{N_t} = \tag{4}$$
$$= \lambda_{L1} + \lambda_{L2} = 3 + 4 = 7 \ \frac{\text{outages}}{\text{year}}.$$

Example №2. Find the SAIFI indicator for the switch partitioned distribution line (Fig. 3).

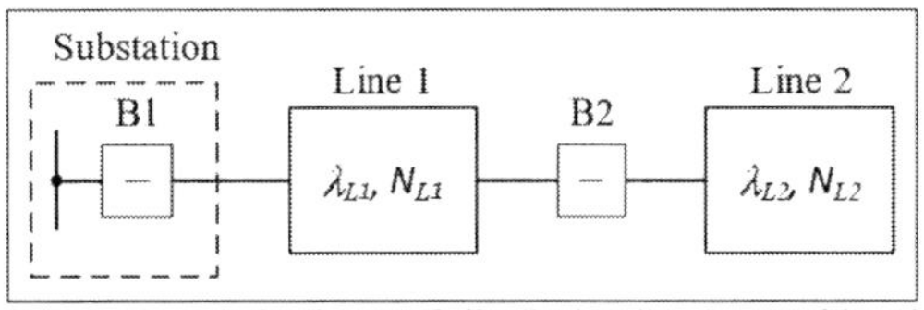

Fig. 3. Structural scheme of distribution lines, a partitioned switch.

The total number of consumers remained the same. In case of an accident on the first line, the number of consumers who will be stopped from supplying electricity $N_{jL1} = N_t = N_{L1} + N_{L2}$. In case of an accident on the second line $N_{jL2} = N_{L2}$. Thus, the SAIFI index for a distribution line, a partitioned switch:

$$SAIFI = \frac{\sum_{j=1}^{J} N_j}{N_t} = \frac{\lambda_{L1} N_t + \lambda_{L2} N_{jL2}}{N_t} = \qquad (5)$$

$$= \lambda_{L1} + \lambda_{L2} \frac{N_{L2}}{N_{L1} + N_{L2}} = 3 + 4\frac{10}{3+10} = 6{,}08 \ \frac{\text{outages}}{\text{year}}.$$

Example №3. Find the SAIFI indicator for the switch-partitioned distribution line, if there is a two-way power supply (Fig. 4).

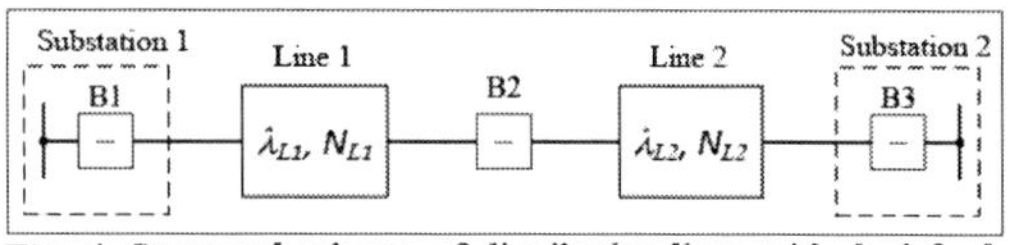

Fig. 4. Structural scheme of distribution lines with dual feed, are interconnected via a switch.

In case of an accident on the first line, the number of consumers who will be stopped from supplying electricity $N_{jL1} = N_{L1}$. In the event of a second line failure $N_{jL2} = N_{L2}$. So SAIFI performance for the switch partitioned distribution line, in the presence of a two-way power supply:

$$SAIFI = \frac{\sum_{j=1}^{J} N_j}{N_t} = \frac{\lambda_{L1} N_t + \lambda_{L2} N_{L2}}{N_t} = \lambda_{L1} \frac{N_{L1}}{N_{L1} + N_{L2}} + \qquad (6)$$

$$+ \lambda_{L2} \frac{N_{L2}}{N_{L1} + N_{L2}} = 3\frac{3}{3+10} + 4\frac{4}{3+10} = 3{,}77 \ \frac{\text{outages}}{\text{year}}.$$

Having considered the examples, you can compare the performance of SAIFI.

Table 1. Comparison of reliability indices SAIFI.

N	The event is to improve the reliability	SAIFI
1	The partitioning of the line using the isolator	7
2	The partitioning of the line using the switch	6,08
3	The partitioning of the line using the switch when the presence of a bilateral power	3,77

The analysis of the results shows the effectiveness of measures to improve the reliability of the distribution line, which is confirmed by a decrease in the SAIFI indicator. That, in turn, can be used for a feasibility study of the reconstruction of the distribution line.

Next, we obtain the formulas for calculating the SAIFI indicator of typical distribution line schemes. First, we obtain a formula for calculating the SAIFI indicator for the distribution line partitioned by switches in the trunk.

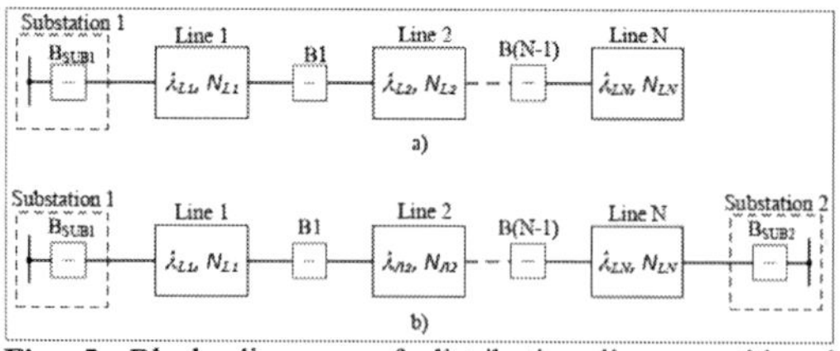

Fig. 5. Block diagrams of distribution lines, partitioned switches in the line a) one-way power supply, b) two-way power supply.

The SAIFI index for the scheme of Fig. 5a:

$$SAIFI = \frac{\sum_{i=1}^{N}\left(\lambda_{Li}\sum_{j=i}^{N} N_{Lj}\right)}{N_\Sigma}. \qquad (7)$$

The SAIFI index for the scheme of Fig. 5b:

$$SAIFI = \frac{\sum_{i=1}^{N} \lambda_{Li} N_{Li}}{N_\Sigma}. \qquad (8)$$

Get the formula for calculating the SAIFI indicator for the distribution line, partitioned switches on the branches (Fig. 6).

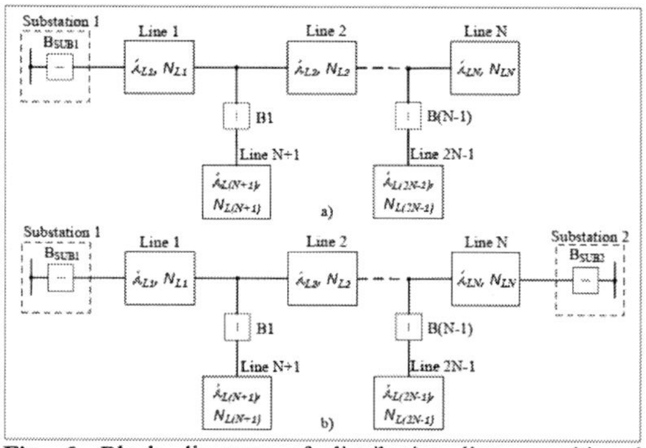

Fig. 6. Block diagrams of distribution lines, partitioned switches on branches a) one-way power supply, b) two-way power supply.

The SAIFI index for the scheme of Fig. 6a and 6b:

$$SAIFI = \sum_{i=1}^{N} \lambda_{Li} + \frac{\sum_{i=N+1}^{2N-1} \lambda_{Li} N_{Li}}{N_\Sigma}. \qquad (9)$$

Get the formula for calculating the SAIFI indicator for the distribution line, partitioned switches in the trunk and branches (Fig. 7).

The SAIFI index for the scheme of Fig. 7a:

$$SAIFI = \frac{\sum_{i=1}^{N}\left(\lambda_{Li}\left(\sum_{j=i}^{N} N_{Lj} + \sum_{j=N+i}^{2N-1} N_{Lj}\right)\right)}{N_\Sigma} + \frac{\sum_{i=N+1}^{2N-1} \lambda_{Li} N_{Li}}{N_\Sigma}. \qquad (10)$$

The SAIFI index for the scheme of Fig. 7b:

$$SAIFI = \frac{\sum_{i=1}^{2N-1} \lambda_{Li} N_{Li}}{N_\Sigma}. \qquad (11)$$

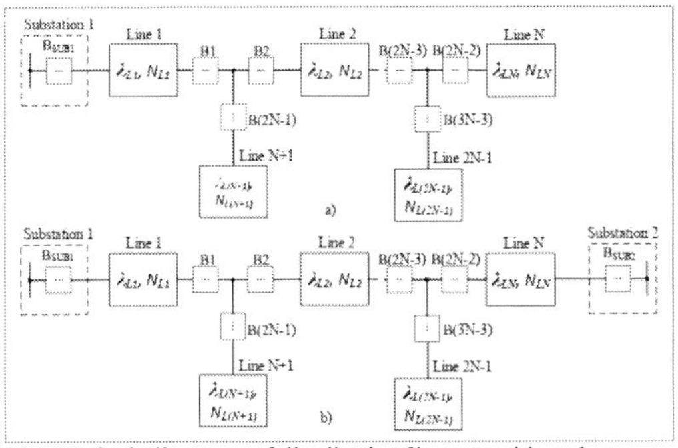

Fig. 7. Block diagrams of distribution lines, partitioned switches in the trunk and on the branches a) one-way power supply, b) two-way power supply.

Example №. 4. Consider the schematic diagram of the distribution line prototype, which was one of the existing 10 kV distribution line of the Pskov region. The final data for the calculation of reliability were taken from the literature [3]. These data reflect the average level of reliability of the elements of rural networks in the European part of the country. As shown by the relevant studies carried out by the author [4] in JSC "firm ORGRES", the indicators of damage to the elements of electrical networks have adequate reliability, because they are not subject to administrative influence.

The results of SAIFI calculation (table 2) for the distribution line are presented below.

Table 2. Calculation of the average number of outages.

№	Equipment	Number	Failure rate, year^{-1}	Average number of failures per year
1	High voltage line with AC wire	20,55	0,25	5,14
2	High voltage line with SIW wire	0,492	0,1	0,05
3	Cable line	0,5	0,075	0,04
4	Linear disconnector	4	0,01	0,04
5	CTS	14	0,07	0,98
Subtotal:				6,24

As follows from table 2, the average number of emergency breaks in power supply to consumers of the distribution line λ_L = 6.24 outages/year.

Thus, according to the formula (4) for the distribution line, we obtain that, as shown by the experimental data for the 10 kV distribution line of the Pskov region, the SAIFI indicator is λ_{Le} = 5.7 outages/year. The theoretical value of the SAIFI indicator differs from the value of the operation data by no more than 10%. This confirms the adequacy of the method of calculation of the SAIFI indicator.

4 Conclusion

In the article, the methodology for calculating the SAIFI reliability indicator of the distribution line for the design phase was developed for the first time. Formulas and examples of calculation of SAIFI indicators for typical distribution lines are given. The developed methodology can be used for the feasibility study of the construction of new distribution lines or the reconstruction of existing lines in order to improve reliability.

References

1. Nepomnyashchy V. A. Reliability of work of elements of electric networks of 1150-35 kV / / Energoekspert. - 2012. - №4. - P. 26-40.
2. Polovko A. M., Gurov S. V. *Fundamentals of reliability theory*. 2nd ed., Rev. and extra – SPb.: BHV-Petersburg, 2006. – P.43-45.
3. Pruss V. L., Tislenko V. V. *Improving the reliability of rural electrical networks*. – L.]. Leningr. otd., 1989 – 208 p.
4. Tereshko O. A. Innovative methods of control of reliability of distribution electric networks / / Methodical questions of research of reliability of big systems of power: Collection of scientific. art / ed. N. I. Voropay, M. A. Korotkevich, A. A. Mikhalevich. - - Minsk: publishing House of BSTU, 2015. - Issue. 66. - C. 206-210..
5. *IEEE Guide for Electric Power Distribution Reliability Indices* - Redline," in IEEE Std 1366-2012 (Revision of IEEE Std 1366-2003) - Redline , vol., no., pp.1-92, May 31 2012
6. M. Papic and R. N. Allan, "Comparison of alternative techniques for the reliability assessment of distribution systems," 1991 Third International Conference on Probabilistic Methods Applied to Electric Power Systems, London, 1991, pp. 174-179
7. R. Allan and R. Billinton, "Power system reliability and its assessment. III. Distribution systems and economic considerations," in Power Engineering Journal, **vol. 7**, no. 4, pp. 185-192, Aug. 1993.
8. P. K. Goswami, S. Chowdhury, S. P. Chowdhury, Y. H. Song and J. K. Das, "Reliability evaluation of distribution system," 2007 42nd International Universities Power Engineering Conference, Brighton, 2007, pp. 158-166

 https://doi.org/10.1051/e3sconf/20185802011

Optimization of the power supply system for connection of the end user to electric grids according to the scheme of the third category of reliability

Ivan Bandurin[1,*], *Alexey* Khaimin[1]

[1]Pskov State University, Faculty of Computer Science and Power Engineering, 180000, Russia, Pskov, Lenin Square, 2

Abstract. The article first developed a mathematical model of optimization of the power supply system for connection of the end user to electric grids according to the scheme of the third category of reliability. The mathematical model allows to choose the optimal type and brand power system equipment.

1 Introduction

At present, the Russian Federation uses a normative approach to ensure reliability. In total there are 3 categories of reliability of electric receivers: I-st, II-th and III-th [4]. By the degree of importance of the category number of consumers increases from the III-th to I-st. Consumers, for whom it is important to ensure uninterrupted power supply, must be connected in accordance with scheme I of the category and, on the contrary, unrelated consumers are connected according to scheme III of the category. Note that the majority of end-users are now connected according to scheme III reliability category.

However, in the article Vorotnitskiy VV [1] for example, it is shown that the reliability of electricity supply to a consumer of category III can be higher than a consumer of category II. It follows that the power supply reliability category can no longer be an expression of that same reliability, so it is better to use the reliability indices. The main indicators of the reliability of power supply to the end user are:

– average number of outages per year;
– average time of restoration of power supply.

Thus, the reliability of electricity supply to consumers in market conditions becomes the subject of contractual relations between the subjects of the energy market and the goods with the corresponding price realized through market services. In this regard, the urgency of the problem of choosing rational schemes of power supply to ensure the contractual reliability of end-users of electricity is growing [1-3].

The article considers a mathematical model for the selection of a rational power supply system for connection of the end user to electric grids according to scheme III of reliability category. The power supply system must provide a certain level of reliability of electricity supply. A review of publications [3, 5-10] showed that in such a direction the problem is solved for the first time.

2 Mathematical model

A schematic diagram of the power supply system for connecting the consumer to electrical networks is shown in Fig. 1. We will compile a mathematical model for choosing a rational system of electricity supply. For this we introduce the initial data, the objective function, constraints objective function.

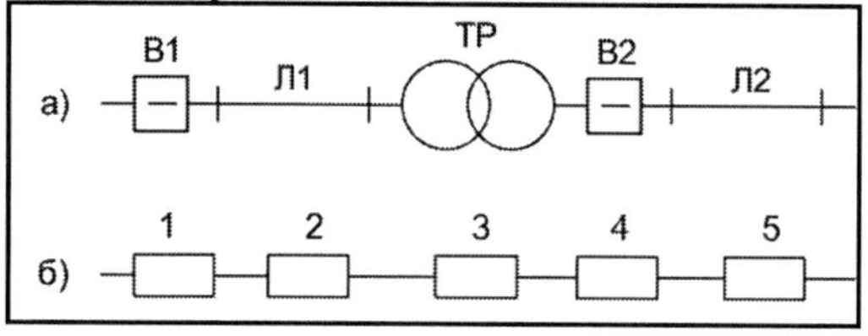

Fig. 1 - Schematic diagram of power supply (a) and the structural diagram of reliability of the consumer of the III reliability category (b), where 1, 4 - circuit-breaker, 2 - high voltage power line section, 3 – power transformer, 5 - low voltage power line section

Initial data (the input variables of the model):

L - distance from the nearest power center to the consumer, km;
P_{max} - maximum power of the consumer, kW;
U_n - consumer nominal voltage, kV;
$cos\varphi$ - power factor of the consumer;
N - average number of outages, outages/year;
T - average time of restoration of power supply, h;
ΔU - voltage loss in the power supply system, %;
ΔP - power loss in the power supply system, %.

2.1 Variables of the mathematical model

1 High and low voltage line type:

* Corresponding author: bandurin_ivan@mail.ru

$$y_{UWli} = \begin{cases} 1, & \text{if the line is made of uninsulated wire,} \\ 0, & \text{otherwise.} \end{cases};$$

$$y_{SIWli} = \begin{cases} 1, & \text{if the line is made of self-supporting insulated wire (SIW),} \\ 0, & \text{otherwise.} \end{cases};$$

$$y_{CLli} = \begin{cases} 1, & \text{if the line is cable,} \\ 0, & \text{otherwise.} \end{cases}.$$

2 Length of high and low voltage line:
L_{UWli} - length of a line of uninsulated wire, km;
L_{SIWli} - length of the line made by SIW, km;
L_{CLli} - length of the line made by the cable, km.
3 Cross-section of high and low voltage line:
F_{UWli} - cross-section of a line of uninsulated wire, mm^2;
F_{SIWli} - cross-section of the line of the SIP, mm^2;
F_{CLli} - cross-section of the line made by cable, mm^2.
4 Type of power transformer:

$$y_{OILtr} = \begin{cases} 1, & \text{if the transformer is oily,} \\ 0, & \text{otherwise.} \end{cases};$$

$$y_{DRYtr} = \begin{cases} 1, & \text{if the transformer is dry,} \\ 0, & \text{otherwise.} \end{cases};$$

$$y_{SF6tr} = \begin{cases} 1, & \text{if gas-insulated transformer,} \\ 0, & \text{otherwise.} \end{cases}.$$

5 Type of high and low voltage circuit-breaker:

$$y_{AIRbi} = \begin{cases} 1, & \text{if the breaker is airy,} \\ 0, & \text{otherwise.} \end{cases};$$

$$y_{VACbi} = \begin{cases} 1, & \text{if the breaker is vacuum,} \\ 0, & \text{otherwise.} \end{cases};$$

$$y_{OILbi} = \begin{cases} 1, & \text{if the breaker is oily,} \\ 0, & \text{otherwise.} \end{cases};$$

$$y_{EMbi} = \begin{cases} 1, & \text{if the breaker is electromagnetic,} \\ 0, & \text{otherwise.} \end{cases}.$$

2.2 The objective function

The optimization criterion is the minimum of the capital costs C and the expenses E of the operation of the power supply system for connection of the consumer to the electrical networks

$$E_n C + E \rightarrow min \tag{1}$$

where $E_n = 0{,}125$ - normative coefficient of efficiency of investments for electric power industry.

Since the term of the construction of most of the projects for connecting consumers to electrical networks is less than one year, it is permissible to use a simpler criterion - the method of reduced costs instead of the discounted cost method.

The damage was not included in the objective function for the following reasons. Economic sanctions against the power grid company arise only in cases when the factual values of the reliability indices of the end-user power supply N_f or T_f exceed the contractual N_c or T_c values [5]. On the other hand, the consumer assesses his risks and chooses the most appropriate level of reliability for him according to the tariff plan. According to the design task, the power supply system must provide reliability indices N and T, which should not exceed the contractual values of N_c and T_c. Therefore, the amount of damage will be close to zero and can be neglected to simplify the mathematical model.

Let us consider in detail the components of the objective optimization function.

Capital costs for the construction of a section of high or low voltage line

$$C_{li}(y_{lij}, F_{lij}, L_{lij}) = \sum_{j=1}^{3} y_{lij} c_{lij}(F_{lij}) \cdot L_{lij}, \tag{2}$$

where c_{lij} - the functions of specific capital costs for the construction of a line with uninsulated wires, SIP or CL per 1 km, cu / km.

The cost of operating a section of high or low voltage line

$$E_{li}(y_{lij}, F_{lij}, L_{lij}, P_{max}) = \sum_{j=1}^{3} y_{lij} e_{lij}(F_{lij}, P_{max}) \cdot L_{lij}, \tag{3}$$

where e_{lij} - the functions of unit operating costs for the construction of a line with uninsulated wires, SIP or CL per 1 km, cu / km.

Capital costs for a high-voltage or low-voltage power circuit breaker

$$C_{bi}(y_{bij}, I_{nij}, I_{scij}) = \sum_{j=1}^{4} y_{bij} C_{bij}(I_{nij}, I_{scij}), \tag{4}$$

where C_{bij} - the functions of capital costs for the installation of air, vacuum, oil or electromagnetic circuit breaker, cu; I_n - rated current of the circuit breaker, A; I_{sc} - rated short-circuit breaking current (SC) of the circuit breaker, kA.

The cost of operating a high-voltage or low-voltage circuit breaker

$$E_{bi}(y_{bij}, I_{nij}, I_{scij}) = \sum_{j=1}^{4} y_{bij} E_{bij}(I_{nij}, I_{scij}), \tag{5}$$

where E_{bij} - the functions of operating costs for of air, vacuum, oil or electromagnetic circuit breaker, cu.

Capital costs for power transformer

$$C_{tr}(y_{trj}, S_{ntrj}) = \sum_{j=1}^{3} y_{trj} C_{trj}(S_{ntrj}), \tag{6}$$

where S_{ntr} - rated power of the transformer, kVA; C_{trj} - functions of capital costs for installation of oil, dry or SF6 transformer, cu.

Costs of operating a power transformer

$$E_{tr}(y_{trj}, S_{ntrj}, P_{max}) = \sum_{j=1}^{3} y_{trj} E_{trj}(S_{ntrj}, P_{max}), \tag{7}$$

where E_{trj} - functions of costs for operation of oil, dry or SF6 transformer, cu.

Thus, the total capital cost for the power supply system

$$C = C_{b1} + C_{l1} + C_{tr} + C_{b2} + C_{l2}. \tag{8}$$

Total costs for the power supply system

$$E = E_{b1} + E_{l1} + E_{tr} + E_{b2} + E_{l2}. \tag{9}$$

2.3 Restrictions of the objective function

1 Restrictions on variables.

1.1 The length of the lines must be positive

$$L_{UWli}, L_{SIWli}, L_{CLli} \geq 0. \tag{10}$$

1.2 Variable types of lines, transformers and switches - binary variables

$$y_{UWli}, y_{SIWli}, y_{CLli} \in 0,1,\quad y_{OILtr}, y_{DRYtr}, y_{SF6tr} \in 0,1,\quad y_{AIRbi}, y_{VACbi}, y_{OILbi}, y_{EMbi} \in 0,1. \tag{11}$$

2 Only one of the line types, transformers or breakers must be used

$$\begin{cases} y_{UWl1} + y_{SIWl1} + y_{CLl1} = 1, \\ y_{UWl2} + y_{SIWl2} + y_{CLl2} = 1, \\ y_{AIRb1} + y_{VACb1} + y_{OILb1} + y_{EMb1} = 1, \\ y_{AIRb2} + y_{VACb2} + y_{OILb2} + y_{EMb2} = 1, \\ y_{OILtr} + y_{DRYtr} + y_{SF6tr} = 1. \end{cases} \tag{12}$$

3 The length of the line must be equal to the specified length

$$L_{UWl1} + L_{SIWl1} + L_{CLl1} + L_{UWl2} + L_{SIWl2} + L_{CLl2} = L. \tag{13}$$

4 The loss of voltage in electric-supply system must not exceed a predetermined value

$$\sum_{i=1}^{2}\sum_{j=1}^{3} y_{lij}\Delta U_{lij}(F_{lij}, L_{lij}, P_{\max}) + \sum_{j=1}^{3} y_{trj}\Delta U_{trj}(S_{ntr}, P_{\max}) \le \Delta U, \tag{14}$$

where ΔU_{ij} - the voltage loss function on the corresponding element (line or transformer),%.

5 According to the maximum allowable current in the line

$$I_{\max lij}(y_{lij}, F_{lij}) \ge \frac{P_{\max}}{\sqrt{3}\cdot U_{nli}\cdot\cos\varphi}, \tag{15}$$

where U_{nli} - the nominal voltage of the high or low voltage line, kV.

6 Restriction on reliability.

6.1 Number of failures. The average number of failures in the power supply system should not exceed the set value

$$\sum_{i=1}^{2}\sum_{j=1}^{3} y_{lij}\omega_{lij}L_{lij} + \sum_{j=1}^{3} y_{trj}\omega_{trj} + \sum_{i=1}^{2}\sum_{j=1}^{4} y_{bij}\omega_{bij} \le N, \tag{16}$$

where ω_{ij} - the failure rate of the corresponding element (line, transformer or switch), failure / year.

6.2 Recovery time. The average recovery time in the power supply system should not exceed the set value

$$\frac{1}{N}\left(\sum_{i=1}^{2}\sum_{j=1}^{3} y_{lij}\omega_{lij}L_{lij}t_{lij} + \sum_{j=1}^{3} y_{trj}\omega_{trj}t_{trj} + \sum_{i=1}^{2}\sum_{j=1}^{4} y_{bij}\omega_{bij}t_{bij}\right) \le T, \tag{17}$$

where t_{ij} - the recovery time of the corresponding element (line, transformer or breaker), hour.

7 Power loss in the power supply system must not exceed the set value

$$\sum_{i=1}^{2}\sum_{j=1}^{3} y_{lij}\Delta P_{lij}(F_{lij}, L_{lij}, P_{\max}) - \sum_{j=1}^{3} y_{trj}\Delta P_{trj}(S_{ntr}, P_{\max}) \le \Delta P, \tag{18}$$

where ΔP_{ij} - the power loss function on the corresponding element (line or transformer), %.

8 For breaking capacity.

8.1 The rated breaking current of the circuit breaker must exceed the maximum current

$$I_{nb}(y_{bij}) \ge \frac{P_{\max}}{\sqrt{3}\cdot U_{ni}\cdot\cos\varphi}. \tag{19}$$

8.2 The rated short-circuit breaking current of the circuit breaker must exceed the short-circuit current

$$I_{SCb}(y_{bij}) \ge I_{P0i}, \tag{20}$$

where I_{SCb} the maximum value of the switched-off short circuit current, kA; I_{P0} the initial value of the periodic component of the short circuit, kA.

8.3 The breaker must disable the aperiodic component of the short-circuit current

$$i_{a\tau j} \le i_{a.norj}, \tag{21}$$

where $i_{a\tau}$ the estimated value of aperiodic current component of the three-phase short-circuit in circuit breaker in time of short-circuit τ, defined by the expression

$$i_{a.\tau} = \sqrt{2} I_{P0} e^{-\frac{\tau}{T_a}}; \tag{22}$$

$i_{a.nor}$ normalized value of the aperiodic component in the switch-off current, determined by the expression

$$i_{a.nor} = \frac{\sqrt{2}\beta I_{SCb.}}{100}, \tag{23}$$

where β content of the aperiodic component in the switch-off current, %.

8.4 According to the enabling capacity, the test shall be carried out according to the condition

$$I_{P0j} \le I_{incj};\ i_{sj} \le i_{incj}, \tag{24}$$

where I_{inc} rated current of the breaker (the current value of the periodic component); i_s shock current of short circuit in the circuit of locates; i_{inc} the highest peak current of inclusion in the catalog.

The shock current is found by the formula

$$i_s = \sqrt{2} k_s I_{P0}, \tag{25}$$

where k_s impact coefficient.

8.5 The electrodynamic resistance of the switch is checked to limit the through current of short-circuit

$$I_{P0j} \le I_{dynj};\ i_{sj} \le i_{dynj}, \tag{26}$$

where i_{dyn} the highest peak (the rated peak withstand current) in the catalog, kA; I_{dyn} the current value of the periodic component of the limit through-fault current, kA.

8.6 For thermal short-circuit resistance, the switch is checked by condition

$$B_K = I_{P0j}^2 \left(t_{discon.bj} + T_{aj}\right) \ge I_{ther.bj}^2 t_{ther.bj}, \tag{27}$$

where $t_{discon.b}$ full-time short-circuit switch c; T_a time constant of the aperiodic component of the fault current, c; $I_{ther.b}$ pre-sensible thermal current circuit breaker, kA; $t_{ther.b}$ duration of the current thermal resistance, s.

3 Getting the solution

The mathematical model allows for the input data to determine:

- Type high and low voltage line can be undertaken, from bare wire, SIW or CL;
- The length of the line of high and low voltage;
- Wire section of each line;
- Type power transformers (oil, dry, SF6);
- Power of the power transformer;
- Type of circuit breaker (air, vacuum, oil and electromagnetic) and brand.

The algorithm for obtaining a solution for the mathematical model for the rational power supply system is as follows:
1. Enter the initial data;
2. Calculate functions that depend on the source data;
3. We present a mathematical model to the standard form;
4. We solve the problem by one of the methods of nonlinear programming;
5. We analyze the obtained solution.

4 Conclusion

The article for the first time developed a mathematical model and optimization algorithm of the power supply system to connect the customer to the power grid according to the scheme of the III category of reliability. The obtained on the basis of mathematical model, rational schemes of power supply can be used by the electric grid organizations for connection of final consumers to electric networks.

References

1. Vorotnitsky V. V. Reliability of power supply as a tool of regulation of relations between suppliers and consumers of energy // Energy and Management -2009. - **№3** - pp. 11-15.
2. Oseichuk, V. A. Reliability of supply in a market economy // news electrical engineering: information publishing / founder of ZAO "news elektrotekhnicks". – SPb., 2011. - **№ 1**. - pp. 20-22.
3. Oseichuk, V. A. Nepomniaschy V. A. Regulation of reliability and quality of electricity supply // Innovatika and expert-2016. - **Issue 1 (16)** - pp.175-185.
4. *Rules of electrical installation* (PUE). 7-e Izd. – M.: NC ENAS, 2004. -222 p.
5. Tereshko O. A. System of economic contract responsibility for the reliability of electricity supply to end consumers//Energoekspert. - Moscow, 2014, **№ 05.** – pp. 12 – 17
6. Fedotova G., Voropay N. Optimization of reliability of power supply of consumers // RT&A. 2007. **№2 (6).** URL: https://cyberleninka.ru/article/n/optimizatsiya-nadezhnosti-elektrosnabzheniya-potrebiteley (дата обращения: 20.07.2018).
7. L. Zemite, J. Gerhards, M. Gorobetz and A. Levchenkov, "Optimization of distribution system reliability," 2016 IEEE 16th International Conference on Environment and Electrical Engineering (EEEIC), Florence, 2016, pp. 1-6.
8. Ching-Tzong Su and Guor-Rurng Lii, "Reliability optimization design of distribution systems via multi-level hierarchical procedures and generalized reduced gradient method," Energy Management and Power Delivery, 1995. Proceedings of EMPD '95., 1995 International Conference on, 1995, pp. 180-185 vol.1.
9. V. Parada, J. A. Ferland, M. Arias and K. Daniels, "Optimization of electrical distribution feeders using simulated annealing," in IEEE Transactions on Power Delivery, **vol. 19**, no. 3, pp. 1135-1141, July 2004.
10. R. C. Lotero and J. Contreras, "Distribution System Planning with Reliability," in IEEE Transactions on Power Delivery, **vol. 26**, no. 4, pp. 2552-2562, Oct. 2011

Improved control algorithm for shunt reactors using fuzzy logic

Etibar Pirverdiyev[1], *Arif Hashimov*[2], *Aytek Babayeva*[2], *Huseynqulu Guliyev*[2]*

[1]Azerenerji OJSC, AZ 1005, A.Alizade str.,10, Azerbaijan
[2]Azerbaijan Scientific–Research and Designed–Prospecting Institute of Power Engineering, AZ1012, Zardabi ave, 94, Azerbaijan

Abstract. The task of provision of the efficiency of fuzzy controller action, regulating the power of shunt reactors, is considered for the stabilization of the voltage values in high-voltage buses, as well as to increase the regime reliability of the power grid. A new algorithm is proposed for controlling the power or inductive resistance of reactor depending on the load resistance, rate of change and voltage on the network buses. The results of the designed experiment based on computer simulation confirm the effectiveness of the proposed control algorithm.

1 Introduction

In extended high-voltage power lines (110-500 kV) and cable systems a shunt reactor (SR) is widely used to compensate for reactive power (charging power). Use of SR allows for solving such problems as unloading of power grid from the charging power in order to reduce losses, smooth and fast control of reactive power, damping of oscillations of active power and voltage on high-voltage buses, in general, the implementation of the tasks of improving the static and dynamic stability of the power system.

SR can be directly connected to the power transmission lines, buses, three-phase transformer (autotransformer) permanently or by means of automatic switch [1,2]. To improve the control of the required reactive power, the reactor must have adjustable parameters.

If the load changes gradually (seasonally, daily or hourly), the use of controlled shunt reactor (CSR) may be one of the best problem solutions. At the same time, in the conditions of short-term and relatively long-term sharp fluctuations of the transmitted power and, accordingly, the load of the high-voltage power grid, the need for effective control of the SR, i.e. the optimal control of the required reactive power is created. In uncertain initial information conditions, the SR control management is subject to a number of difficulties. Under these conditions, the existing SR control systems cannot form a rational control action due to the problems of fluctuations and sharp deviations of voltage on high-voltage buses, as well as the occurrence of overvoltages associated with the implementation of frequent unreasonable switchovers [3,4].

One of the options for the problem is the use of the transformer type SR produced by foreign (for example, ABB) firms with a thyristor control system, which has a core with an air gap filled with oil. As the studies show, when setting the controlled parameters in the conditions of real initial information data, the use of the SR with the control system implementing a particular traditional algorithm does not fully provide the desired results [5, 6].

In this regard, under the above conditions, the creation of the algorithm and control system for SR on the basis of the theory of fuzzy sets and fuzzy logic proposed by the famous Professor L. Zadeh is topical and of scientific interest [6]. The authors of this paper proposed such a system [7]. In this system the primary input parameters are the load resistance Z and its rate of change dZ/dt, and the controlled output parameter is the reactive power of the CSR. Despite the effectiveness of the CSR operation on the basis of the proposed fuzzy logic controller, the intellectualization of the facility is not fully provided, and in some cases there is a problem with the stabilization of the voltage on the high-voltage buses of the power grid. To eliminate these shortcomings, the task of improving and developing of the proposed control algorithm for the fuzzy logic controller (FLC) is assigned. For this purpose, the input and output vectors of linguistic variables are increased by one order, i.e. the current values of bus voltage were added to the input of the FLC, to which the CSR is connected, and to the output – the delay time of the control system action depending on the nature of the load resistance change (Fig.1).

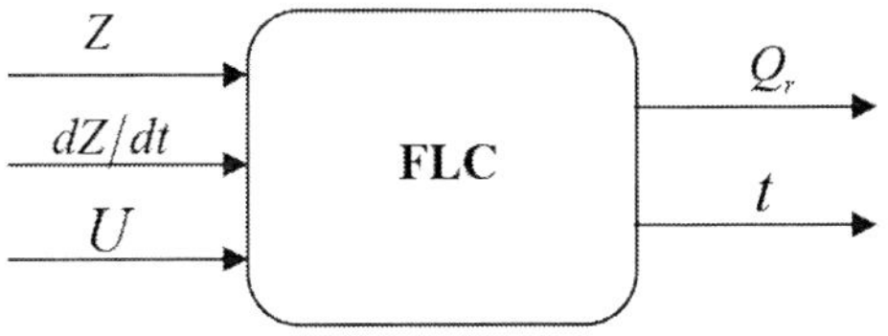

Fig.1. Fuzzy logic controller for CSR

Thus, the synthesized new type of FLC has a control system with three input and two output linguistic parameters, which allows for effective control of the output reactive powers of the CSR installed at the beginning and at the end of power transmission line of the power grid. In this paper, the synthesis and implementation of the FLC with three input and two output parameters for the CSR output power control in the initial fuzzy-probabilistic information environment are considered.

2 Fuzzy control algorithm for CSR

It is known that the dependence of the reactive power Q_r required by the CSR on the line parameters is expressed by the formula [2, 3]:

$$Q_r = P_{nat} \cdot \lambda \cdot \left[1 - \left(\frac{P}{P_{nat}} \right)^2 \right] \cdot L \tag{1}$$

where P – active power transmitted over power line; P_{nat} – natural power of power line; L – length of power line; λ – wave length.

When controlling a reactor with a traditional control system, realizing the control algorithm worked-out based on the formula (1), it is impossible to effectively regulate the reactive power of the CSR due to the above reasons. The proposed FLC realizes the $\tilde{Q}_r = f(Z, dZ/dt)$ algorithm.

As mentioned above, this control algorithm cannot fully provide a voltage stabilization on the buses. For this purpose, one more parameter was added in the input - the bus voltage, and one more parameter in the output - the FLC action delay time, which is set depending on the load resistance change rate. The proposed FLC realizes two laws of fuzzy control:

$$\begin{cases} \tilde{Q}_r = f(U, Z, dZ/dt) \\ t = f(U, Z, dZ/dt) \end{cases} \tag{2}$$

It is known that any FLC consists of fuzzificator, fuzzy logic output mechanism and defuzzificator, and depending on the task, it realizes the appropriate algorithm [3-6].

In this case, the fuzzy control algorithm (2) should be worked-out based on a fuzzy model of linguistic approximation with three input and two output parameters:

$$if\ X_1 = A_1\ and\ X_2 = A_2\ and\ X_3 = A_3\ then\ Y_1 = B_1\ and\ Y_2 = B_2 \tag{3}$$

where X_1, X_2, X_3 – input linguistic variables of the controlled objects; A_1, A_2, A_3 – fuzzy subsets defined in the universal sets E_1, E_2, E_3, respectively, consisting of terms of linguistic variables X_1, X_2, X_3; Y_1, Y_2 – output linguistic variables of the controlled objects; B_1, B_2 – fuzzy subsets defined in the universal sets E_4, E_5, respectively, consisting of terms Y_1, Y_2, where $\forall x_i\ A_i \in E_i,\ i = 1...3$ и $\forall y_j\ B \in E_j,\ j = 1,2$.

For input linguistic variables X_1, X_2, X_3, the membership functions of Gaussian (qaussmf), for output variables – Z - shaped (zmf), S - shaped (smf) and trapezoidal (trapmf) membership functions are adopted [6, 8]:

qaussmf function:

$$\mu_{ki}(x) = \exp\left(\frac{-(x_i - m_{ki})}{2\sigma_{ki}^2} \right), \quad i = \overline{1,n} \quad k = \overline{1,m} \tag{4}$$

trapmf function:

$$\mu_{ki}(x) = \begin{cases} 0, & x_i \le a_{ki} \\ \frac{x_i - a_{ki}}{b_{ki} - a_{ki}}, & a_{ki} \le x_i \le b_{ki} \\ 1, & b_{ki} \le x_i \le c_{ki} \\ \frac{d_{ki} - x_i}{d_{ki} - c_{ki}}, & c_{ki} \le x_i \le d_{ki} \\ 0, & d_{ki} \le x_i \end{cases} \tag{5}$$

zmf and smf functions

$$\mu_{ki}(x) = \begin{cases} 1, & x_i \le a_{ki} \\ nonlinear approximation, & a_{ki} < x_i < b_{ki} \\ 0, & x_i \ge b_{ki} \end{cases} \tag{6}$$

where m – coordinate of the maximum; σ – concentration factor; a, d – fuzzy set carrier; b, c – fuzzy set kernel, $\mu_{A,i}(x): X_i \rightarrow [0,1]$

After determining the fuzzy implication and the type of membership function on the basis of fuzzy approximation between the input and output vectors, the output signals are formed.

3 Determination of parameters of membership function and drawing up of the control algorithm

$$\begin{array}{lll} E_{41}=Max & \text{(maximum)} & \underline{\underline{\Delta}}(Q_p,\mu_{41}(Q_r)) \\ E_{42}=B & \text{(big)} & \underline{\underline{\Delta}}(Q_p,\mu_{42}(Q_r)) \\ E_{43}=M & \text{(mean)} & \underline{\underline{\Delta}}(Q_p,\mu_{43}(Q_r)) \\ E_{44}=S & \text{(small)} & \underline{\underline{\Delta}}(Q_p,\mu_{44}(Q_r)) \\ E_{45}=ZR & \text{(zero)} & \underline{\underline{\Delta}}(Q_p,\mu_{45}(Q_r)) \end{array} \quad (10)$$

Mamdani algorithm was used in the preparation of the fuzzy logic controller (FLC) algorithm. When programming according to the Mamdani algorithm ("reactive power consumption of the reactor - line load" relationship) for the input (voltage, load resistance and its rate of change) and output (delay time and reactive power of the reactor) parameters, the following fuzzy term-subsets are accepted:

- Term-set $T_i(Z)$ for the linguistic variable "Load resistance", where $Z_i \in E_{1i}$ with $i=\overline{1,5}$:

$$\begin{array}{lll} E_{11}=VB & \text{(very big)} & \underline{\underline{\Delta}}(Z,\mu_{11}(Z)) \\ E_{12}=B & \text{(big)} & \underline{\underline{\Delta}}(Z,\mu_{12}(Z)) \\ E_{13}=M & \text{(mean)} & \underline{\underline{\Delta}}(Z,\mu_{13}(Z)) \\ E_{14}=S & \text{(small)} & \underline{\underline{\Delta}}(Z,\mu_{14}(Z)) \\ E_{15}=N & \text{(natural)} & \underline{\underline{\Delta}}(Z,\mu_{15}(Z)) \end{array} \quad (7)$$

For 500 kV transmission line in the natural mode, the wave resistance of 287 Ohms ($P_{nat}=870MVt$) [1,5,6] is accepted, and the values of the linguistic variables Z "Load resistance" are taken in the interval of $Z=(287 \div \infty)\,Om$. The generated random values of variable Z correspond to the power limit of $P=(870 \div 0)MVt$.

- The second input parameter for the FLC is "Dynamics" of the fuzzy linguistic variable, where $T_j(dZ/dt)$ is term-subset, where $(dZ/dt)_j \in E_{2j}$ and $j=\overline{1,5}$.

$$\begin{array}{lll} E_{21}=VB & \text{(very big)} & \underline{\underline{\Delta}}(dZ/dt,\mu_{21}(dZ/dt)) \\ E_{22}=B & \text{(big)} & \underline{\underline{\Delta}}(dZ/dt,\mu_{22}(dZ/dt)) \\ E_{23}=ZR & \text{(zero)} & \underline{\underline{\Delta}}(dZ/dt,\mu_{23}(dZ/dt)) \\ E_{24}=S & \text{(small)} & \underline{\underline{\Delta}}(dZ/dt,\mu_{24}(dZ/dt)) \\ E_{25}=VS & \text{(very small)} & \underline{\underline{\Delta}}(dZ/dt,\mu_{25}(dZ/dt)) \end{array} \quad (8)$$

Values $T_j(dZ/dt)$ of term-subset are accepted in the interval $(-10 \div +10)$.

- The third input parameter for the FLC is "Voltages" of the fuzzy linguistic variable, where $T_k(U)$ is term-subset, where $(U)_k \in E_{3k}$ and $k=\overline{1,5}$.

$$\begin{array}{lll} E_{31}=VB & \text{(very big)} & \underline{\underline{\Delta}}(U,\mu_{31}(U)) \\ E_{32}=B & \text{(big)} & \underline{\underline{\Delta}}(U,\mu_{32}(U)) \\ E_{33}=M & \text{(normal)} & \underline{\underline{\Delta}}(U,\mu_{33}(U)) \\ E_{34}=S & \text{(small)} & \underline{\underline{\Delta}}(U,\mu_{34}(U)) \\ E_{35}=N & \text{(very big)} & \underline{\underline{\Delta}}(U,\mu_{35}(U)) \end{array} \quad (9)$$

- The first output parameter of the FLC is "Reactive power" for the fuzzy linguistic variable, $T_l(Q_r)$ - term-subset, $Q_{rl} \in E_{4l}$ and $l=\overline{1,5}$:
- The second output parameter of the FLC is "Delay time" for the fuzzy linguistic variable, $T_\gamma(t)$ - term-subset, $t_\gamma \in E_{5\gamma}$ and $\gamma=\overline{1,5}$:

$$\begin{array}{lll} E_{51}=VST & \text{(very short)} & \underline{\underline{\Delta}}(t,\mu_{51}(t)) \\ E_{52}=ST & \text{(short)} & \underline{\underline{\Delta}}(t,\mu_{52}(t)) \\ E_{53}=MT & \text{(mean)} & \underline{\underline{\Delta}}(t,\mu_{53}(t)) \\ E_{54}=LT & \text{(long)} & \underline{\underline{\Delta}}(t,\mu_{54}(t)) \\ E_{55}=VLT & \text{(very long)} & \underline{\underline{\Delta}}(t,\mu_{55}(t)) \end{array} \quad (11)$$

The values of the linguistic variable "Reactive power" are accepted in the interval $P=(870 \div 0)MW$ (7)

In the relationships (7)-(11) the $\mu_{1i}(Z)$, $\mu_{2j}(dZ/dt)$, $\mu_{3k}(U)$, $\mu_{4l}(Q_r)$, $\mu_{5\gamma}(t)$– are the membership functions of term-sets of linguistic variables Z, dZ/dt, U, Q_r and t. These variables are respectively defined in the universal set $E_{1i}, E_{2j}, E_{3k}, E_{4l}, E_{5\gamma}$.

The FLC brings into action the reactor power control devices (Q_{r1}, Q_{r2}) depending on the input signals Z, dZ/dt, U and implement 125 fuzzy rules of type (3) based on the Mamdani algorithm.

The definition area of input and output linguistic variables, values of parameters of membership function of separate terms are given in the Table 1.

https://doi.org/10.1051/e3sconf/20185802012

Table 1. Terms of linguistic variables, membership functions and their parameters.

Term-subsets	Membership functions	Parameters
Load resistance , Z		
Very big	gaussmf	[36,24 287]
Big		[29,38 383]
Mean		[50,22 575]
Small		[182,75 1150]
Natural		[425,86 2300]
Load resistance change rate, dZ/dt		
Very big	gaussmf	[2,1 10]
Big		[2,1 5]
Normal		[2,1 0]
Small		[2,1 -5]
Very small		[2,1 -10]
Voltage		
Very big	gaussmf	[3,2 -15]
Big		[3,2 -7,5]
Normal		[3,2 0]
Small		[3,2 7,5]
Very small		[3,2 15]
Reactive power		
Maximum	gaussmf	[7,33 274]
Big		[10,54 257]
Mean		[21,66 206]
Small		[33,27 120]
Zero		[42,03 0]
Delay		
Very short	zmf	[0 0,5]
Short	trapmf	[0 0,05 0,15 0,2]
Mean	trapmf	[0.15 0,2 0,4 0.45]
Long	trapmf	[0.4 0,45 0,85 0.95]
Very long	smf	[0,8 0,9 1 1]

The control matrix-algorithm of reactor power in the format of fuzzy linguistic model "If-then" on the basis of linguistic models (7) - (11) is given in the Table 2. Fuzzy control matrix algorithm consists of 125 rules. The antecedent part of each rule represents three, and the consequent part – two implications.

For defuzzification, i.e. for transition from the phase of fuzzy control actions to a crisp control action, the centroid method [8] is used.

Table 2. Algorithm-matrix of SR fuzzy control

Options	Input parameters			Output parameters	
	Z	dZ/dt	U	Q_r	L
1	N	VS	VS	ZR	VLT
2	N	S	VS	ZR	MT
3	N	ZR	VS	ZR	ST
.	.	.	.	.	.
.	.	.	.	.	.
60	S	VB	S	S	LT
61	M	VS	S	ZR	ST
62	M	S	S	S	ST
.	.	.	.	.	.
.	.	.	.	.	.
123	VB	ZR	B	Max	VST
124	VB	B	B	Max	VST
125	VB	VB	B	Max	VST

4. Simulation results

Calculations, simulating the operation of OS-100/500 U1 type shunt reactors with a capacity of 280 MVA, connected at the ends of power lines with a voltage of 500 kV and length of $L = 300\,km$ are made during the simulation.

The surfaces $Q_r = \tilde{f}(U, dZ/dt), Q_r = \tilde{f}(Z, U)$, $L = \tilde{f}(U, dZ/dt)$ and $L = \tilde{f}(Z, U)$ obtained by using the Fuzzy Logic Toolbox module in the Matlab environment, reflecting the relationships "output parameters of the controller - input ones", which show the controller action when controlling the power consumption of the SR, are shown on the Fig. 2. As can be seen from the surface relationships, at $Z = Z_d$ $(P = P_{nat})\, Q_r \Rightarrow \min$ and at $Z = \infty$ $(P = 0)$ $Q_r \Rightarrow \max$, and at intermediate load values with fuzzy control of the SR power, the Q_r value is in the range of $Q_r = 0 \div \max$.

The SR fuzzy control algorithm has been tested for 500/330 kV intersystem power grid connecting power systems of Azerbaijan and Georgia, the fragment of the electric circuit is shown on Fig.3.

On the basis of simulation modeling of the proposed fuzzy algorithm for SR control, the calculated experiments for different load modes of the electrical network are carried out. The obtained results for the voltage at the Samukh node of 500 kV Samukh-Gardabani PTL are shown on the Fig.4. As can be seen from the figure, at the fuzzy SR control depending on the overcurrent along the lines, the voltage on the 500 kV buses is within the prescribed norm.

The curves of the change of power losses during the control depending on the transmitted power in the line and the CSR power are shown on the Fig.5. As can be seen from the figures, at the fuzzy control of the reactive power flow, the active power losses in the considered network are reduced in the range from 0.26 to 0.178, i.e. to 31.5%.

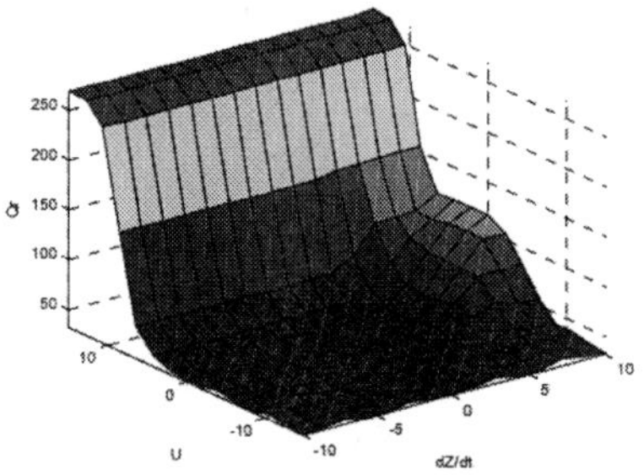

a)

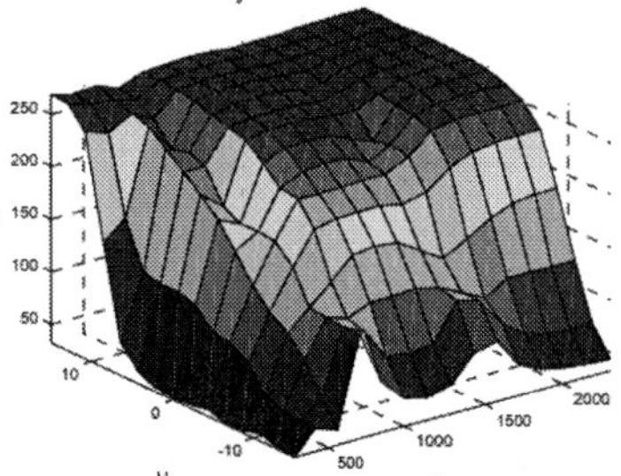

b)

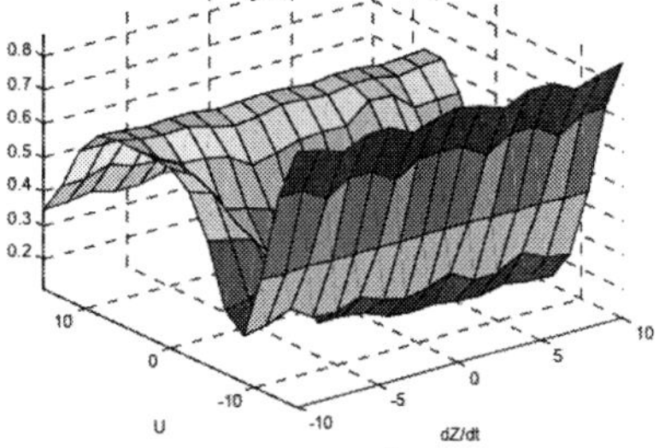

c)

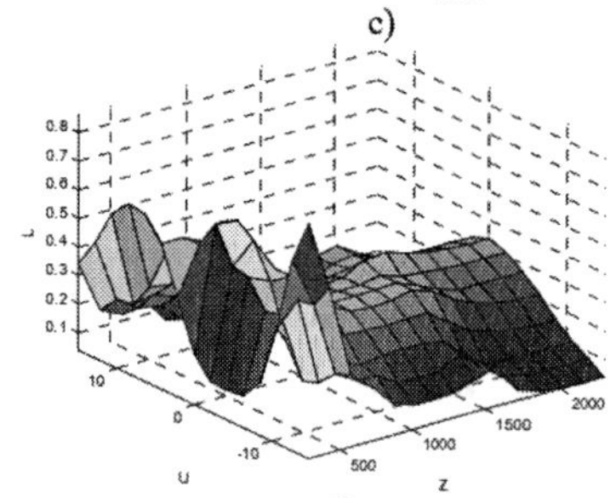

d)

Fig.2. Control surfaces $a - Q_r = \tilde{f}(U,\ dZ/dt)$; $b - Q_r = \tilde{f}(U,\ Z)$; $c - L = \tilde{f}(U,\ dZ/dt)$; $d - L = \tilde{f}(U,\ Z)$

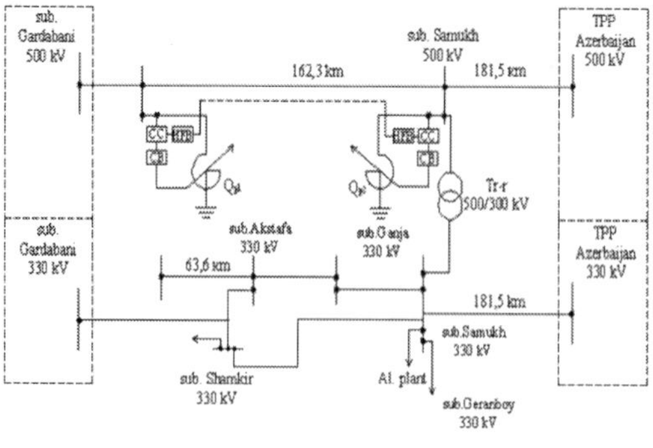

Fig.3. Fragment of the electrical circuit of 500/330 kV network of JSC "Azerenerji" JSC's system

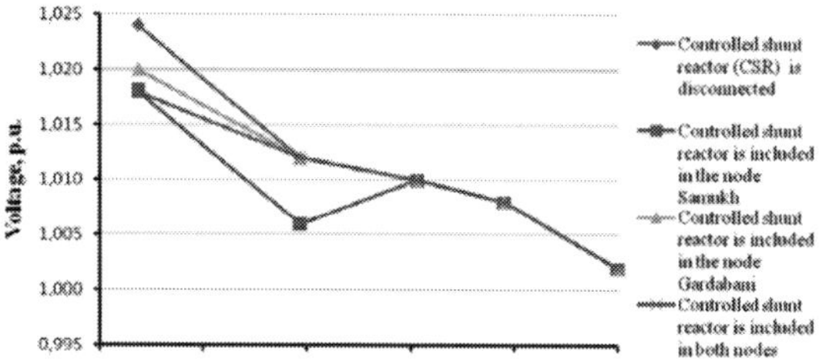

Fig.4. Curves of change of voltage for 500 kV Samukh - Gardabani PTL

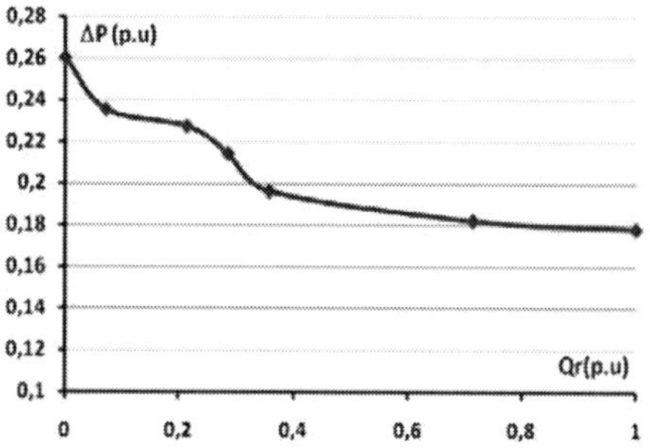

a)

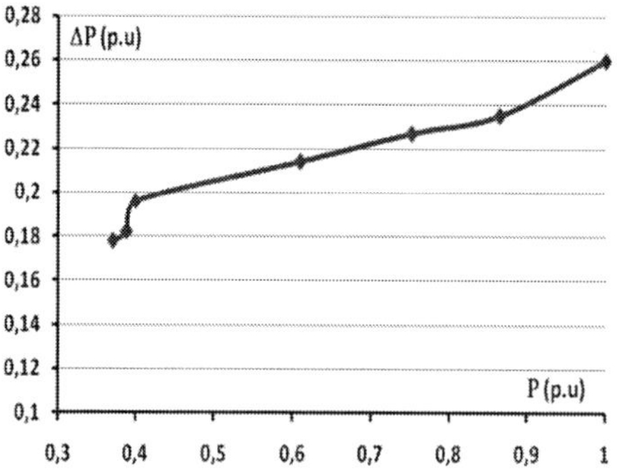

b)

Fig.5. Change in power losses depending on the power of CSR (a) and the power flow (b)

https://doi.org/10.1051/e3sconf/20185802012

5 Conclusion

1. It is proposed a simple and easily implemented algorithm for the control of power of shunt reactors installed at the ends of power line of the high-voltage power grid on the basis of fuzzy sets and fuzzy logic depending on the values of load and voltage resistance, for damping the voltage fluctuations on the buses of the high-voltage power grid.
2. The results of simulation modeling for the intersystem real electric network of power grids, which implements the proposed control algorithm, confirm its effectiveness. At that, the voltages in the nodes of the high-voltage grid are within the specified permissible limits, and the power losses from the power flow are significantly reduced.

References

1. V. I. Kochkin, O. P. Nechaev *Application of static reactive power compensators in electric networks of power grids and enterprises*. M., Publishing house of SC ENAS, 248 (2002)
2. A. Dolgopolov *Controlled shunt reactors*. M., 112 (2012)
3. P. Bulykin, V.I. Kochkin, L.P. Kubarev, L. L. Fedosov *Controlled shunting reactor of new generation*. Energoekspert, №3, 46-51 (2011)
4. A.V. Romodin, D. Yu. Leyzgold *Approach to the development of models of elements of flexible power lines on the basis of volt-ampere characteristics*. Fundamental research, №8, 137-142 (2015)
5. A.M. Hashimov, H.B. Guliyev, A.R. Babayeva *IJTPE Journal International Journal on Technical and physical problems of engineering*, Issue 30, Vol. 9, No. 1, 18-22 (2017)
6. L.A. Zadeh *Fuzzy Sets*. Inform. and control, vol. 8, 338-353 (1965)
7. E.S. Pirverdiyev, A. M. Hashimov, H.B. Guliyev, A. R. Babayeva *Development of models of fuzzy logic for the control of shunt reactors. Methodical questions of research of reliability of large power systems*. Collection of scientific papers. Issue. 68. Research and provision for the reliability of energy systems. Bishkek, 242-251 (2017)
8. A.V. Leonkov *Fuzzy modeling in MATLAB and fuzzyTECH environment*. BHV Petersburg, St. Petersburg, 736 (2003)

Resilience Assessment of the State Estimation Software under Cyber Attacks

*Nikolai*Voropai[1], *Irina*Kolosok[1],and *Elena*Korkina[1]*

[1]MelentievEnergy Systems Institute, Irkutsk, Russia

Abstract. Taking into account Smart Grid creation, cyber security appears as critical problem. In spite of wide set of technical tools and organisational decisions, which use for protection of electrical objects against cyber attacks, it is impossible to prevent them completely. Therefore, the ability of Smart Grid to resist cyber attacks is important and necessary property. Thus, the cyber resilience is ability of complicate technical-information system to keep operable state under contingencies including cyber attacks. This paper deals with general problem of electric power system resilience and its cyber resilience, determination methods to calculate quantitative measure of Smart Grid resilience. The state estimation software resilience under cyber attacks is discussed.

1 Some terminology considerations

Present-day electric power systems (EPSs) are most complicated technical man-made objects, when using innovative technologies to generate, transmit, deliver and store electric power. They permanently develop affected by many objective factors. During EPS development, their properties vary, there is a necessity to study new properties of transforming EPSs and new problems in operation of these systems, and, accordingly, a necessity to use new tools to maintain EPS normal operation.

Recently, in English publications, a new term, resilience, reflecting EPS property transformation has been actively discussed [1-6 etc.]. The most comprehensive definition of the term "resilience" is given in [2], where it is defined as "as the ability of the system or systems-of-systems (SoS) to withstand a change or a disruptive event by reducing the initial negative impacts (absorptive capability), by adapting itself to them (adaptive capability) and by recovering from them (restorative capability)." It is important to note that this definition refers to any systems irrespective of their nature. In [4], the discussed property is also represented for systems of any nature, and it is addressed as complex, including ecological, organizational, and system components. Herewith, the latter features as an ability to minimize the value and duration of deviations from target metrics of a system.

Studies [1, 3, 5, 6] address this property with reference to EPSs, and in [3, 5], the focus is on external extreme perturbations (for example, hurricanes, etc.). Over the recent years, researchers actively investigate the problem of cyber attacks as potential external disturbances affecting the information-communication subsystem of present-day cyber-physical EPSs [7-9, etc.]. At the same time, in the review [3] - besides external extreme disturbances, and in [1, 6] – exclusively, it is a problem of resilience in relation to the cascade system collapses. Herewith, in [1], to corroborate the trend of growing importance of the addressed property, there is information on the increase in scales of after effects of the cascade system failures for users in time based on the statistics about the USA EPS over 1991-2005 This legitimacy of growth of scales of after effects of system collapses features any developing EPS.

The Russian term corresponding to the notion resilience is survivability. Applicable to EPS, survivability is a property of a system to withstand disturbances without allowing their cascade development with a mass interruption of supply to users, and to recover the system initial state, or the one close to it [10]. From this definition, one can see that, in the survivability property, shown are the absorptive and adaptive abilities of a system noted in [2], as well as its ability to recover.

In the English literature [e.g., 11] the term vulnerability is used instead of survivability. At that, vulnerability is related to dynamic mode of reliability, to dynamic security. The differences between the terms survivability and vulnerability are obvious: survivability assumes a certain "activity" of a system when resisting perturbations due to a rationally organized structure, expedient operation modes, and efficient control. Vulnerability reflects as if a "passive" response of a system to perturbations. Taking this into account, vulnerability is a complementary (opposite) property of a system as compared with survivability [10, 12].

It is expedient to note some nuances related to the PES survivability concept. Those nuances are reflected in Figure 1, where 1 is the system operational capacity normal level φ, 2 is the limiting state (for details, see

* Corresponding author: korkina@isem.irk.ru

[10, 12, 13]), 7 is the restoration stage. According to the figure, there are two cases possible. In the first case, an emergency process starts with an ordinary perturbation, and then, due to failures of control units and staff mistakes, there occurs a cascade development of emergency 3. At each stage of such a development, the emergency control system attempts to interrupt the emergency cascade development 6. Upon reaching the limiting state 2, the emergency cascade development becomes non-reversible 4, an avalanche process develops rapidly, the emergency control schemes either have no time to response, or have already exhausted their possibilities by this stage. It is this case that the resilience property is addressed in [1, 6]

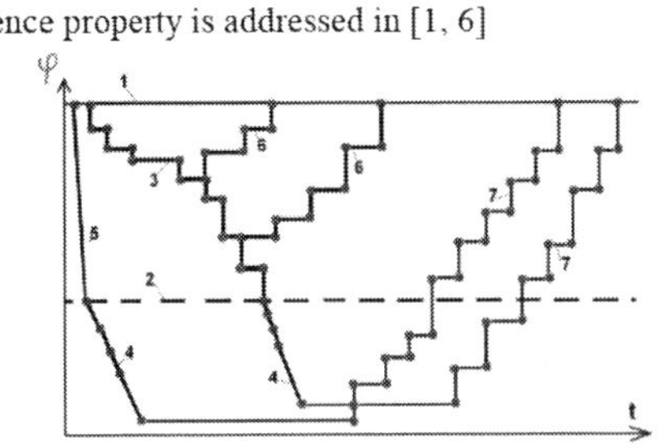

Fig. 1. Illustration for a EPS behavior in terms of survivability.

The other case applicable to the survivability property is related to an extreme non-ordinary initial disturbance 5 (hurricanes, tornadoes, etc. refer to such disturbances [3, 5], as well as cyber attacks (CAs) [7-9]), after which the system appears below the limiting state, and there occurs an avalanche uncontrollable development of an emergency process 4.

2 Smart Grid cyber resilience

Creating Smart Grids (SG) provides:

- using new technologies widely and gages to measure mode parameters (digital measuring transformers and converters, devices to measure vector values -phasor measurement units (PMU), and so forth),
- creating wide area measurement systems (WAMS), protection systems (WAPS), and control systems (WACS) based on the high-precision synchronized phasor measurements.
- present-day means of communication that principally change the processes of data acquiring, processing, and transmitting,
- substation digitization and automation,
- changing properties of the observational object (power system) due to a grid saturation with active elements (FACTS, HVDC),
- using technologies of a distributed computing,
- high-performance hardware.

Introducing obvious advantages of using such technologies and hardware, a Smart Grid also has its defects: higher vulnerability of the entire Smart Grid and its local infrastructures to cyber attacks, due to which introducing new technologies should be accompanied by allowance for requests of the Smart Grid information protection from adverse external effects. Therefore, there emerge problems of providing Smart Grid cyber security, along with traditional problems of raising the efficiency of production, transformation, transmission, and power distribution, Smart Grid reliability, security, and survivability. Solving the Smart Grid cyber security problems requires supplementing and expanding the "reliability" property with new concepts, such as "vulnerability," "survivability," "cyber resilience," etc., which is necessary for revealing and assessing mechanisms for cyber attack effect on electric power industry objects.

In the Russian sources [9, 14], the term durability is used and its interpretation is provided: 1) ability of Smart Grid and its elements to withstand external effects and to operate in the normal mode under cyber attacks; 2) ability to maintain normal operation (in standard modes or those close to the latter) during and after disruptive events (DEs), including cyber attacks.

In [15], the author, based on the analysis of situation in cyber security over the last 2-3 years, notes that cyber resilience is the property of an information system enabling it to exist under continuous or constant attacks. The term survivability prompts us that, to solve the problems set before the system, it is not necessary to attempt to protect everything. It is sufficient to focus on the critical components, providing their redundancy, duplication, staff qualification, and other problems that, applicable to information systems, may be taken from the reliability theory.

Apparently, the Smart Grid survivability principles may and should be used when studying the Smart Grid cyber resilience problem. On the other hand, it is obvious that providing cyber resilience entirely and immediately is difficult; therefore, it is necessary to solve this problem systemically, starting the phase of developing preventive measures and ending with the restoration phase after a cyber attack.

3 Main regularities of stable behaviour of the system

The concept of resilience has been developed and investigated in various fields, but there has been no uniform standard definition so far. Like shown in Section 1, under the effect of external disturbances, a system should have abilities to absorb effects, adapt to the latter, and restore. These abilities may be regarded as three essential properties of a system resilience: enhancing any of them will enhance the system resilience.

Figure 2 borrowed from [16] shows the main regularities of the system steady and unstable behaviour under effects of damage factors. Apparently, the system behaviour character shown on Figure 2 is similar, in many respects, to the EPS behaviour illustrating the survivability property (Figure 1).

Depending on the EPS ability to adapt, self-organize, and restore, possible are different versions of final restoration:1) by the mode not different from pre-emergency state (robust behaviour), 2) incomplete recovery with the mode relaxed parameters (flexible behaviour), 3) transition to the non-operable state (destructive behaviour), 4) with the mode parameters surpassing those of the pre-emergency mode (recovery with adaptation).

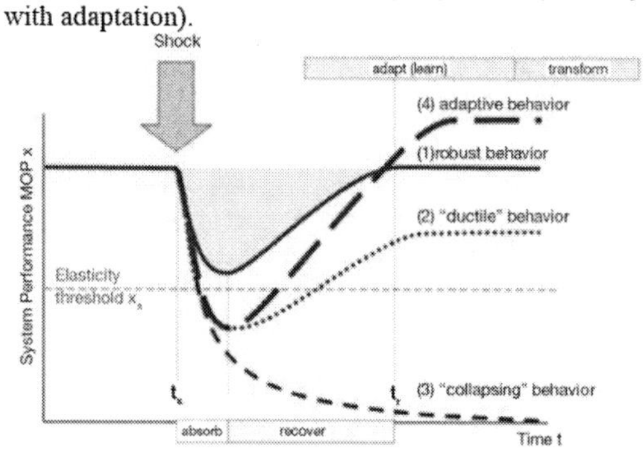

Fig. 2. Main regularities of the system stable and unstable behaviours.

4 Quantitative measure of resilience

Assessing the EPS resilience helps find the best strategy to restore the system operation and minimize originating losses. Over the last decade, various methods to quantify the assessment of resilience to external disturbances have been offered.

In the reliability theory, developed were some approaches to assess the reliability quantity indicators. The main of the latter is – failure-free operation - is the property of an object to keep operable state for a certain time.

To assess reliability, important is the duration under operational loads, i.e., the time factor. To assess the resilience to external disturbances, important is the effect level, at which there can occur a failure, i.e. the kind of disturbance and its intensity.

Resilience to external disturbances has a non-linear nature of a threshold kind: the action of the disruptive mechanism manifests itself only after the disturbance level exceeds a threshold value, the limiting condition of the system shown in Figures 1 by line 2 or the elasticity threshold Xs in Figure 2. An increase in the main characteristic of the system resilience - the time to reach the limiting condition - enables to decrease the risk of developing critical (emergent) situation and to provide security.

In foreign studies, proposed was the General resilience metric (GR) to determine quantity resilience indicator with a possibility to take into account the criteria of individual properties.

To illustrate this parameter, we use Figure 3 from [2] that shows the main phases and transitions when analysing the system resilience to external disturbances. The Y-axis represents the measure of performance (MOP).

The first phase is the initial steady phase (t<td), at which the system performance assumes its target value.

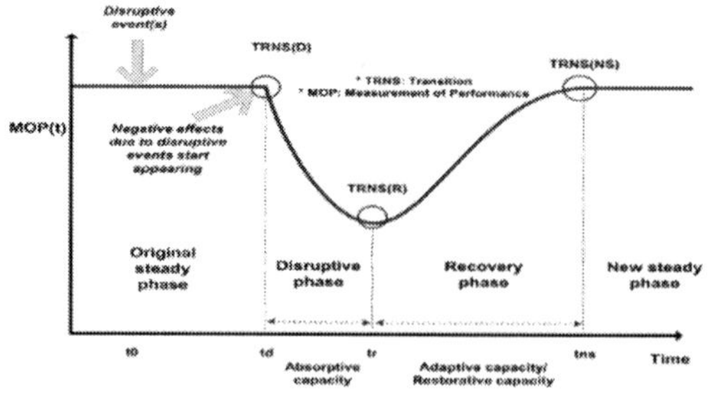

Fig. 3. System phases and transitions at a DE effect [2].

The second phase is the destructive (disruptive) phase ($t_d \leq t < t_r$), at which the system performance starts dropping, until it reaches the lowest level at time tr. During this phase, one can assess the system absorptive capability. Robustness (R) is the measure that enables to assess this ability being the MOP minimal value.

The third phase is the restoration phase ($t_r \leq t < t_{ns}$), at which the system performance starts to increase to a new steady level. During this phase, one can assess the system adaptive and restorative capabilities by introducing the corresponding measures: $RAPI_{DP}$ (rapidity) and TAPL (time averaged performance loss). The reached new steady level may be equal to the previous one, but it can be lower or higher than that level. To take this into account, developed was a quantitative measure - Recovery Ability (RA).

Considering these measures, a formula was obtained to calculate the integrated metric for the system resilience to external disturbances (GR):

$$GR = f\,(R, RAPI_{DP}, RAPI_{RP}, TAPL, RA) =$$

$$= R\left(\frac{RAPI_{DP}}{RAPI_{RP}}\right)(TAPL)^{-1}\ RA$$

where: R is the robustness (MOP minimal value of),
$RAPI_{DP}$ is the disruptive phase rapidity,
$RAPI_{RP}$ is the restoration phase rapidity,
TAPL is the time averaged performance loss,
RA is the recovery ability.

The GR metric is dimensionless and is an extremely useful measure to compare resilience of engineering systems. For example, it may be used to compare the resilience of different systems to the same DE. For more resilient system, a value of this metric will be higher. This metric may be also used to compare the resilience of the same system to different DEs. A higher GR indicates that the system is more resilient to certain DEs. Besides, the GR metric may be used to compare various versions and develop more resilient system to a certain DE as per various mitigation and protection strategies.

Yet, to calculate this metric, one should have statistics on the number of CAs over a certain time interval, and on the duration of individual phases of effect and restoration. Therefore, in a number of studies, for example, in [17], proposed were some simplified versions to measure a similar metric. We term it the cyber resilience index (CRI). Figure 4 shows two

https://doi.org/10.1051/e3sconf/20185802013

versions of the figure from [17], similar to Figure 2, but in a simpler view.

From this figure, one can see that the CRI value (R in the figure) is inversely proportional to the A triangle area restricted by the system normal performance line, by the line of the performance drop at the destructive phase, and by the line of the performance growth at the restoration phase. Also, from the figure, one can see that, in the b-case, the system CRI is higher, than that in the a-case.

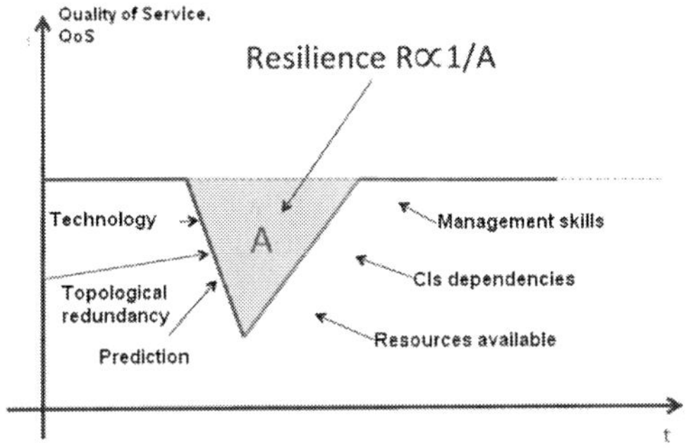

a)

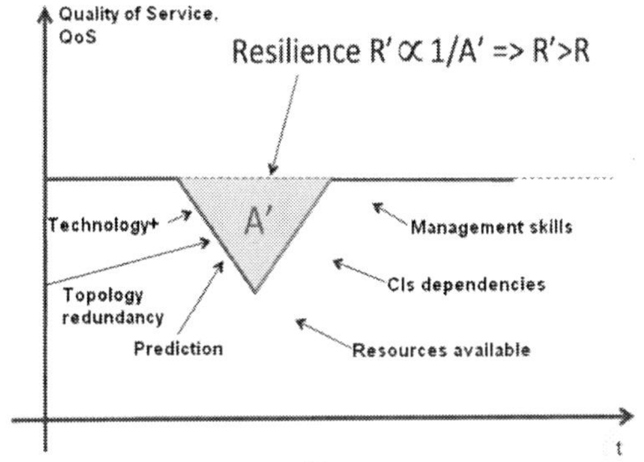

b)

Fig. 4. Illustration for the cyber resilience index (CRI).

5 Case study

As a case study, we assessed the cyber resilience for the OTSENKA (State Estimation) software developed at the Melentiev Energy Systems Institute of the Siberian Branch of the Russian Academy of Sciences [18].

The primary problems solved at the EPS state estimation:

- forming the current scheme from the SCADA;
- checking the observability;
- bad data detection (BDD) in SCADA-measurements;
- filtering random SCADA errors (obtaining SCADA estimates and calculating non-measured parameters).

In a reliable state estimation software, all these functions should be doubled. Thus, in the OTSENKA software, checking the observability of the scheme is executed twice: based on the topological method until solving the state estimation problem (estimation calculation) and during the estimation calculation at the Jacobian matrix triangular resolution. BDD is done a priori, i.e. prior to solving the state estimation problem through the Test Equation method [18], or during calculating the estimates by non-quadric (robust) criteria, or after the state estimation by estimate residuals. Estimating may be performed through state estimation linear algorithms, if the scheme is observable through PMU-measurement, non-linear algorithms through the weighed least square method from the SCADA and PMU data, or by using robust criteria to suppress erroneous measurements.

Figure 5 illustrates the state estimation software resilience under cyber attacks.

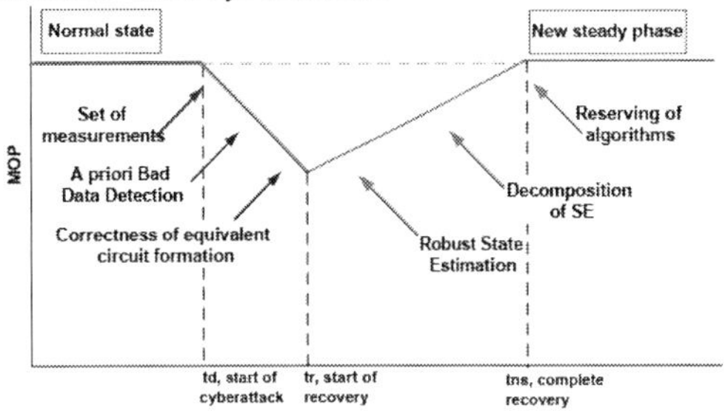

Fig. 5. EPS state estimation software cyber resilience.

The curve length and inclination within the $[t_d; t_r]$ interval depend on the composition of the processed measurements, on the aprioristic BDD efficiency, on the Jacobian matrix conditionality, and on other factors. The curve length and inclination within the interval $[t_r; t_{ns}]$ interval (restoration phase) depend on the possibility to duplicate the applied algorithms.

Figures 6-8 show the analyses of probable state estimation software attacks.

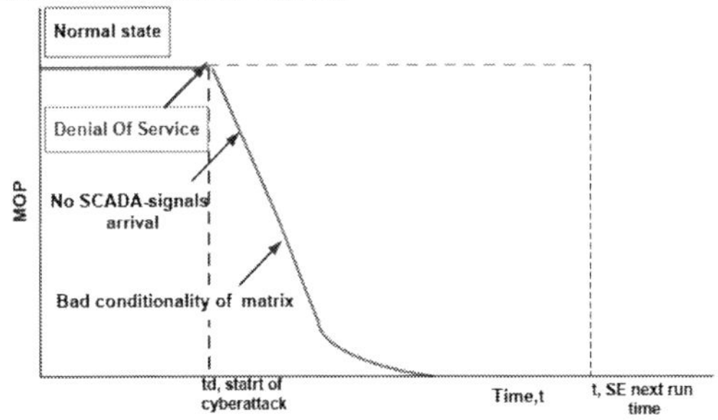

Fig. 6. Impossibility to recover SE software due to a denial of service (DoS) attack.

With no SCADA signals due to a DoS attack, the scheme is impossible to correctly design (Figure 6). The generated matrix thereof has a bad conditionality, the iterative process diverges. There emerges an issue to develop special alarms for such cases.

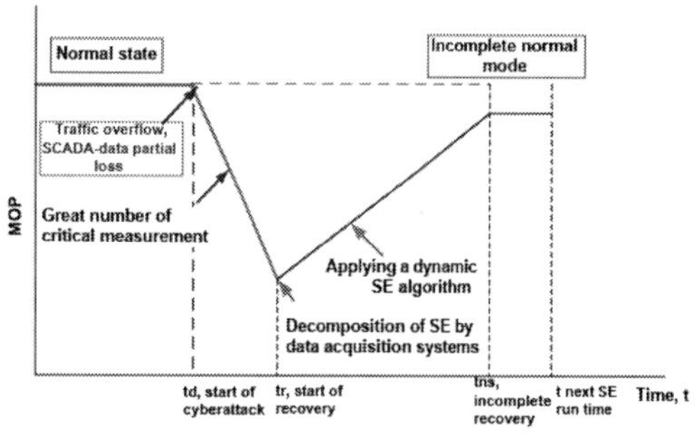

Fig. 7. State estimation software restoration under a man-in-the-middle (MITM) attack.

Figure 7 shows the consequences of a MITM attack: the attacker is inside the network and by repeated sending of the same packages overwhelms the traffic. Therefore, the transmission of other necessary packages is blocked. A scheme part becomes unobservable. The state estimation software performance persists due to the decomposition algorithm from SCADA and WAMS measurements [19] and due to using the estimations of the unobservable scheme part the previous cycle.

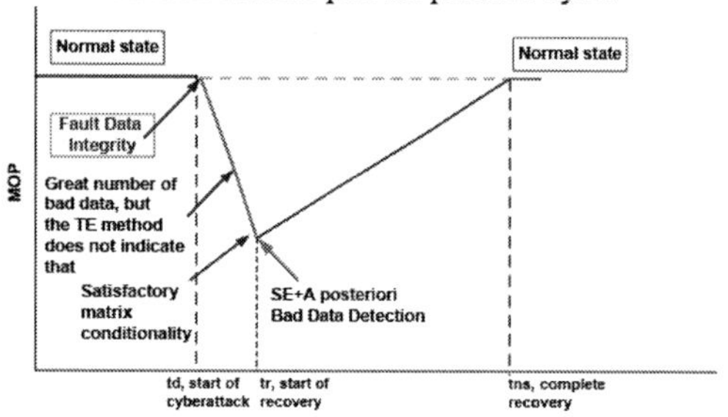

Fig. 8. State estimation software recovery under a fault data integrity (FDI) attack.

Figure 8 shows the consequences of an FDI attack: the attacker injects false data that cannot be detected through the Test Equation method. The state estimation software performance resilience persists due to a posteriori BDD.

6 Conclusions

Resilience is the ability of a system to withstand the effect of disruptive events (cyber attacks), to reduce initial negative effects, to adapt to the latter, and to restore after them.

This concept supplements and expands such EPS (Smart Grid) properties, as "reliability," "security," "flexibility," "vulnerability," etc.

Resilience has a non-linear nature of a threshold kind. To assess a degree of a PES resilience, one should develop quantitative metrics.

Proposed was a general resilience metric (GR) to quantify resilience. The GR takes into account the absorptive, adaptive, and restoration capabilities of a system.

It is important to study the Smart Grid resilience problems systemically, i.e., by considering all its capabilities and possible cyber attacks, in order to solve the problem completely and not to admit negative effects of cyber attacks on the Smart Grid vulnerability.

This study is supported by the Siberian Branch of the Russian Academy of Sciences (Project III.17.4.2) of the Federal Program of Scientific Research (№ AAAA-A17-117030310438-1).

References

1. Massoud A. Challenges in reliability, security, efficiency, and resilience of energy infrastructure: Toward smart self-healing electric power grid // IEEE PES General Meeting, Pittsburg, USA, July 20-24, 2008, 5 p.
2. Cen Nan, Sansavini G., Kroeger W. Building an integrated metric for quantifying the resilience of interdependent infrastructure systems // 9th Intern. Conf. on Critical Information Infrastructure Security, Limassol, Cyprus, October 13-15, 2014, 12 p.
3. Yezhou Wang, Chen Chen, Jianhui Wang, Baldick R. Research on resilience of power systems under natural disasters - A review // IEEE Trans. Power Syst., 2016, Vol. 31, No. 2, pp. 1604-1612.
4. Zhonglin Wang, Nistor M.S., Pickl S.W. Analysis of the definitions of resilience // 20th IFAC World Congress, Toulouse, France, July 9-14, 2017, pp. 11136-11144.
5. Panteli M., Mancarella P., Trakas D.N., Kyriakides E., Hadziargiriou N.D. Metrics and quantification of operational and infrastructure resilience in power systems // IEEE Trans. Power Syst., 2017, Vol. 32, No. 6, pp. 4732-4741.
6. Kezunovic M., Overbye T.J. Off the beaten path: Resiliency and associated risk // IEEE Power and Energy Magazine, 2018, Vol. 16, No. 2, pp. 26-35.
7. Mehrdad S., Mousavian S., Madraki G., Dvorkin Yu. Cyber-physical resilience of electrical power systems against malicious attacks: A review // Current Sustainable / Renewable Energy Reports, <https://doi.org/10.1007/s40518-018-0094-8>
8. Kolosok I.N., KorkinaE.S., and GurinaL.A. Analyzing security of state estimation results from PMU under WAMS cyberattacks // Methodical issues of studying reliability of the bulk energy systems. Issue 66. Minsk: BNTU, 2015, P. 66 - 75, in Russian.
9. Papkov B.V., Kulikov A.L., and OsokinV.L. Cyber threats and cyber attacks in electric power industry.Nizhnii Novgorod: NRU RAE&PS, 2017, P. 80, in Russian.
10. Reliability of energy systems (Collection of recommended terms). Moscow: Energiya (Energy) Publ., 2007, P. 152, in Russian.
11. Fouad A.A., Zhou Qin, Vittal V. System vulnerability as a concept to access power system dynamic security // IEEE Trans. on PS, Vol. 9, No. 2, pp.1009 - 1015.
12. Voropai N.I. EPS Survivability: methodical bases and methods of research // Izvestiya (Bulletin) of the

USSR AS. Energetics and Transport, 1991, No. 6, Pp. 52 - 59, in Russian.
13. Antonov G. N, Cherkesov G.N., Krivorutskiy L.D., et al. Methods and models for studying survivability of energy systems. Novosibirsk: Nauka (Science) Publishers, 1990, P. 285, in Russian.
14. Bakulin V.M, MalkovS.Yu., Goncharov V.V., and Kovalyov V.I. Managing the provision of complex engineering system durability. Moscow: FIZMATLIT (Physical-Mathematical Literature) Publishers, 2006, P. 304, in Russian.
15. Lukatskiy A. Cyber resilience, cyber survivability, cyber reliability, cyber continuum. https://www.securitylab.ru/blog/personal/Business_without_danger/342751.php, in Russian.
16. Heinimann H.R. Future resilient systems. Singapore-Zurich: ETH Risk Center, 2014, 600 p.
17. Tofani A., Alessandroni S., D'Agostino G., Di Pietro A., Onori G., Pollino M. and Rosato V. Operational Resilience Metrics for a Complex Electrical Networks // Intern. Conf. on Critical Information Infrastructure Security CRITIS 2017, Lucca – Italy.
18. Gamm A.Z., and Kolosok I.N. Detecting bad data of SCADA-measurements in power systems. Novosibirsk: Nauka (Science) Publishers, 2000, P. 152, in Russian.
19. Kolosok I.N., Korkina E.S. Decomposition of power system state estimation problem as a method to tackle cyberattacks // The1st IEEE Intern. Conf. ICPS-2018, Saint-Petersburg, 15-18 May 2018, SF-004928

Estimating the effect of equipment reliability indices, schedules, and regular overhaul scopes on reliability and efficiency of combined heat and power plants

Elena Stepanova [1,*], *Alexey* Maksimov [1]

[1] Melentiev Energy Systems Institute of Siberian Branch of Russian Academy of Sciences, Russia

Abstract. We have developed a technique and a programming-computing suite (PCS) to estimate the effect of equipment reliability indices, schedules, and regular overhaul scopes on reliability and efficiency of combined heat and power plants (CHPPs). We describe the approach to predict heat and electric loads for the investigated CHPP operation period, taking into account the features of the power cogeneration. We performed optimization studies of two operation periods (different in overhaul resources) for an industrial-heating CHPP.

1 Introduction

The combined heat-and-power plants (CHPPs) maintained in the Russian Federation (RF) are often the only heat sources for their users. Currently, observed is a high percent of their equipment wear, which increases hazard of emergencies, reduces the energy and economic efficiency of their operation [1, 2]. Regular overhaul increases the duration and reliability of equipment operation at power plants. Besides, one should take into account the cogeneration sector transition to market relations in many RF regions, where CHPPs are difficult to compete with hydro- and condensation power plants (CPPs) in benefitting, when participating at the Wholesale Electricity Market (WEM) [3-5].

2 Problem statement

Reducing overhaul intervals and increasing overhaul operation scopes enable to decrease the number of equipment emergency failures and undersupply of heat and electricity for users, but increases the down time and equipment regular overhaul expenditures that are an essential part of operating costs. Reducing regular overhaul scopes and increasing the overhaul intervals reduces the equipment down time, but increases the number of possible failures that, considering climate features of many RF regions, may lead to serious consequences. Therefore, to solve this problem, it is relevant to determine optimal overhaul interval durations and regular overhaul scopes for CHPP turbo- and boiler units, at which all requirements for the power supply reliability and operation features of each power installation are met. Performing such studies may be implemented through mathematical simulation and optimization [6-9].

Solving this problem required developing a technique to estimate the effect of equipment reliability indices, schedules, and overhaul scopes on the CHPP operation, and a programming and computing suite (PCS) implementing that technique. Figure 1 presents the PCS structure comprising four main blocks.

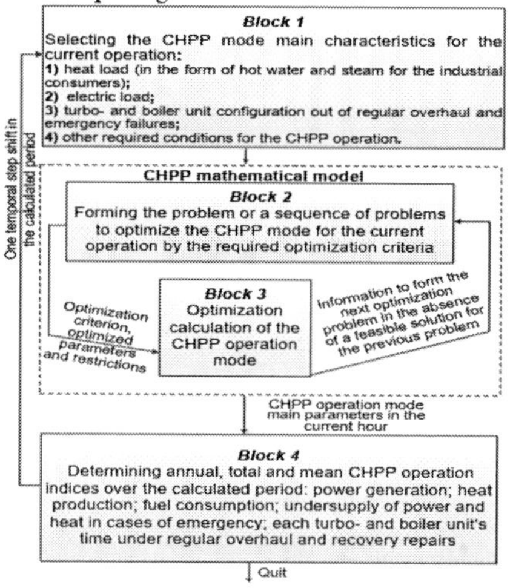

Figure 1. PCS structure to estimate the effect of the equipment reliability indices, schedules, and overhaul scopes on CHPP operation reliability and efficiency.

In our study, we accepted the following general provisions. Each turbo- and boiler unit may be in the following conditions: operation, redundancy, regular overhaul, and emergency repair. Regular overhauls are held in certain number of operation hours with certain equipment idle hours. Simultaneous different regular overhauls [10] are admitted. The turbo- or boiler unit emergency shutdown, with a subsequent regenerative

* Corresponding author: step@isem.irk.ru

repair, is addressed as a random event. After an emergency failure, a post-damage repair follows for a given renewal time. To simulate emergency failures, we use random-number generators. To more precisely estimate the selection of the overhaul schedule, it is desirable to address the CHPP operation period in 10 and more years. Over the addressed time interval, the CHPP operation mode main conditions are formed depending on necessity of detailing the problem solution for each hour, three, six, or twelve hours.

3 Block 1

PCS Block 1 forms the CHPP mode main characteristics for the operation at the current instant. Figure 2 presents the PCS Block 1 structure.

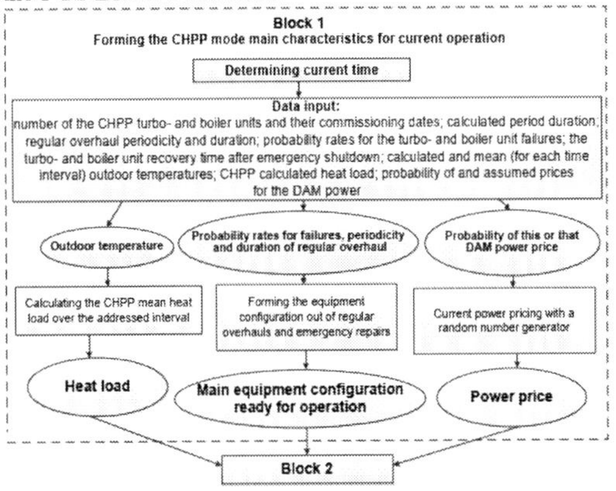

Figure 2. Structure of the PCS Block 1 to estimate the effect of equipment reliability indices, schedules, and overhaul scopes on CHPP reliability and efficiency.

The set parameters for Block 1 are: the number of CHPP turbo- and boiler units, and their commissioning years; calculated period duration; periodicity and durations of regular overhauls; turbo- and boiler units' failure probabilities determined by the parameters of failure flows and recovery time for turbo- and boiler units after an emergency shut-down (statistical data); calculated and mean air temperature for each time interval, the CHPP calculated heat load; the day-ahead-market (DAM) power estimated prices and probabilities of their appearance. The computed parameters are: the CHPP heat load (in the form of hot water), the price of the electric power and the composition of the capital equipment that is not under regular overhaul and emergency failures within the current calculated interval.

Due to the features of the CHPP combined power generation, we use the following approach to determine the heat and power loads at a power plant. Because the heat load of the hot-water supply (HWS) and of the CHPP heating depends directly on the air temperature, we address a year monthly, a month is split into three time intervals with the mean air temperature characteristic of each interval. The mean air temperature is obtained from meteorological data for the certain region in the last 10 years and more. Further, based on the air temperature, user's calculated load, and the HWS share, we determine the HWS and the heating mean loads for the calculated time interval [11]. The heat energy supply in the form of steam for industrial consumers is accepted equal to the contract heat load in steam.

To determine the CHPP electric load, we use the approach explicitly described in [12, 13]. Its essence is as follows. For the WEM participants, the power scopes not covered by regulated contracts are sold at non-regulated prices within free contracts, at the DAM, and the balancing market. I.e., there is an electricity price uncertainty that causes an uncertainty of the CHPP electric power. The CHPP operation within electric load maxima is sufficiently difficult due to a poor maneuvering of major steam-electric generating units, especially for the plants with an inadequate number of peak hot-water boilers. At the same time, the electric power generated at a CHPP in the condensation mode is expensive; therefore, it is profitable to sell it within its price maxima periods.

In our study, we address a version of CHPP possible participation at the DAM. The CHPP electric output at the DAM (N^{CHPP}) depends on the DAM price ($Ц^{DAM}$). If $Ц^{DAM}$ is not less, than the price, profitable for CHPP in the condensation mode ($Ц_{con}^{CHPP}$), the CHPP will bear the greatest possible electric load. If $Ц^{DAM}$ is less than $Ц_{con}^{CHPP}$, but not less than the price, at which the CHPP operating in the heat-extraction mode ($Ц_{\max h}^{CHPP}$) agrees to sell its electric power, the plant will bear the greatest electric load (possible for the heat-extraction mode) equal to $N_{\max h}^{CHPP}$. Otherwise, the CHPP load will be minimally possible ($N_{\min h}^{CHPP}$). The specified assignment rules N^{CHPP} are determined through the following system of logical conditions.

$$N^{CHPP} = \begin{cases} N_{con}^{CHPP}, & Ц_{con}^{CHPP} \le Ц^{DAM} \\ N_{\max h}^{CHPP}, & Ц_{con}^{CHPP} > Ц^{DAM} \ge Ц_{\max h}^{CHPP} \\ N_{\min h}^{CHPP}, & Ц_{\max h}^{CHPP} > Ц^{DAM} \end{cases} \qquad (1)$$

Based on the statistical data analysis or predicted studies of supply point groups (SPGs), to which the addressed CHPP relates to, one may determine the probability of implementing of various values of $Ц^{DAM}$. As a result, we obtain probabilities P_1, P_2 … P_n, corresponding to various DAM values $Ц_1^{DAM}$, $Ц_2^{DAM}$ …, $Ц_n^{DAM}$, where n is the greatest possible DAM price. Herewith, the conditions $0 \le P_i \le 1$, $i = 1..n$, $\sum_{i=1}^{n} P_i = 1$ should be met.

4 Blocks 2 and 3

Like one can see from Figure 1, after the formation of the CHPP main operation conditions, there occurs a transition to optimization problems. Optimization calculations are performed through a CHPP mathematical model created by the Computerized Programming System (CPS PCS) developed earlier at the Melentiev Energy Systems Institute of the Siberian Branch of the Russian Academy of Sciences [7, 14]. The CHPP model comprises mathematical models for CHPP individual equipment elements. The models describe in detail the processes occurring in the equipment, and are based on physical laws and existing methods of calculating heat-and-power installations (HPIs) (energy and material balances of energy carriers, processes of expansion and compression of working substances, standard method for thermal calculation of a boiler unit, etc.).

The first problem to solve is the maximizing profit problem, when optimizing the main operating configuration created in Block 1 (the configuration out of regular overhauls and emergency failures). The technique of optimizing the equipment configuration is described in detail in [15, 16]. In the CHPP mathematical model under creation, we introduce auxiliary, material variables, whose variation from 0 to 1 (for one boiler or turbounit) enables to estimate the necessity to include/exclude this or that generating equipment in/out the CHPP operation.

Mathematical statement of the problem to gain additional profit from selling electric power at the DAM (when optimizing the equipment configuration) has the following view

$$\max_{x,d,N^{CHPP}} \Pr , \tag{2}$$

under the conditions

$$y = H(x,Q,N^{CHPP}),\ G(x,y,Q,N^{CHPP}) \geq 0 , \tag{3}$$

$$B^{CHPP} = \varphi(x,y,Q,N^{CHPP}), \tag{4}$$

$$N_{\min h}^{CHPP} \leq N^{CHPP} \leq N_{con}^{CHPP} ,\ \underline{x} \leq x \leq \overline{x} , \tag{5}$$

$$0 \leq d \leq d^{\max} , \tag{6}$$

where y is the vector of dependent optimized parameters (direct steam discharge for turbines, steam discharges from non-controllable turbine withdrawals, electric powers of turbines, pressure in non-controllable turbine withdrawals, etc.); x is the vector of independent optimized parameters (steam discharges in turbine condensers and steam discharges from regulated turbine withdrawals, pressure ahead of turbine regulating diaphragms, etc.); H is the p-dimensional vector-function of restrictions-equalities; it includes the equations describing technological links among the scheme elements, energy and material balances of the CHPP technological scheme elements, etc.; Q is the vector of the power plant's heat load values; B^{CHPP} is the CHPP boiler-unit total fuel consumption; $\underline{x}$ and $\overline{x}$ are the vectors setting the intervals for the variation in the optimized parameters of the installation; d is the determined operating equipment configuration.

If the first problem has a feasible solution, then the second problem of the profit maximization is solved at already fixed equipment configuration obtained as a result of solving the first problem. Fixation occurs as follows. The equipment configuration obtained by rounding the auxiliary variables to the proximal entire quantities may be considered optimal, if, as a result of solving the first problem, the $\left|d_i - d_{N_i}\right| \leq \varepsilon$ condition is met for the obtained values of the operating aggregate auxiliary variables. Here, d_{N_i} is an entire quantity meeting the $0 \leq d_{N_i} \leq d^{\max}$ condition, $d^{\max}$ is the maximal given configuration of the operating equipment. In the optimization calculations through the given methodical approach, the value ε was accepted equal to 0,15 [16]. At the numerical value of the auxiliary variable (upon rounding) of this or that aggregate equal to 0, there occur logical conditions for its exclusion from calculating the CHPP operation mode. The mathematical statement of the second problem will correspond to (1-5), but instead of Condition (6), the next condition is considered:

$$d^t \in D , \tag{7}$$

where d^t is the fixed value for the vector of discrete variables d; D is the assemblage of optimally "suspicious" versions of the equipment configuration (it is formed taking into account the equipment in the operable state).

As a result of solving the second problem, we determine the CHPP operation mode that provides the consumers' needs in heat and power and gains the greatest possible profit at such a mode from selling power. In case of the DAM low price, we obtain the mode with the complete power supply to the consumers, but without CHPP participation at the DAM.

In cases of emergency failures of this or that equipment, or when a regular overhaul coincides with an emergency failure, solving the first problem may show that the CHPP, in such situations, is not able to provide the consumers with necessary supplies (heat load in the form of hot water and steam, and electric power generated by heat consumption). Then, we switch to solving a sequence of problems to minimize the energy undersupply to consumers. It is necessary to clarify in advance, what loads of the CHPP external consumers are subject to reducing, first of all. The CHPP electric and heat loads are ranked by "importance" degrees. Further, one solves (in the ascending order of "importance") a sequence of problems to minimize undersupplies of electric and heat loads of a power plant. In greater detail, these problems are described in [17].

Upon calculating the CHPP operation mode at a current instant, the following main indices are passed from Block 3 into Block 4: CHPP electric and heat loads, electric output for the CHPP internal needs, operating equipment configuration, fuel consumption, undersupplies of electric and heat energy, operation, reserve, regular overhaul, and recovery repair durations for each turbo and boiler unit. Next, by the PCS time meter, there occurs a time shift (1 hr, 3- or 6-hr, etc.), and the calculations continue until the entire calculated period is studied.

5 Blocks 4

In Block 4, the CHPP operation indices - annual, total, and mean over the entire calculated period - are calculated: power generation, heat production, fuel consumption, power and heat undersupplies in emergency cases, operation, reserve, regular overhaul, and recovery repair durations for each turbo and boiler unit.

$$Э_{ann}^{CHPP} = \sum_{i=1}^{n} N_i^{CHPP} T_i \,,\; Э_{c.p.}^{CHPP} = \sum_{k=1}^{K} Э_k^{CHPP} \,, \quad (8)$$

$$Q_{ann}^{CHPP} = \sum_{i=1}^{n} Q_i^{CHPP} T_i \,,\; Q_{c.p.}^{CHPP} = \sum_{k=1}^{K} Q_k^{CHPP} \,, \quad (9)$$

$$B_{ann}^{CHPP} = \sum_{i=1}^{n} B_i^{CHPP} T_i \,,\; B_{c.p.}^{CHPP} = \sum_{k=1}^{K} B_k^{CHPP} \,, \quad (10)$$

$$\Delta Э_{ann}^{CHPP} = \sum_{i=1}^{n} (N_i^{r} - N_i^{ac}) T_i \,, \Delta Э_{c.p.}^{CHPP} = \sum_{k=1}^{K} \Delta Э_k^{CHPP} \,, \quad (11)$$

$$\Delta Q_{ann}^{CHPP} = \sum_{i-1}^{n} (Q_i^{r} - Q_i^{ac}) T_i \,,\; \Delta Q_{c.p.}^{CHPP} = \sum_{k=1}^{K} \Delta Q_k^{CHPP} \,, \quad (12)$$

where, $Э_{ann}^{CHPP}$ is the CHPP annual power generation; $Э_{c.p.}^{CHPP}$ is the CHPP power generation for the calculated period K comprising k years; N_i^{CHPP} is the CHPP power output in the i-mode; T_i is the time interval (1 hr in this paper); Q_{ann}^{CHPP} is the CHPP annual heat production; $Q_{c.p.}^{CHPP}$ is the CHPP heat production over the calculated period K; Q_i^{CHPP} is the CHPP heat production in the i-mode; B_{ann}^{CHPP} is the CHPP annual fuel consumption; $B_{c.p.}^{CHPP}$ is the CHPP fuel consumption over the calculated period K; B_i^{CHPP} is the CHPP fuel consumption in the i-mode; $\Delta Э_{ann}^{CHPP}$ is the annual power undersupply to consumers; $\Delta Э_{c.p.}^{CHPP}$ is the power undersupply to consumers over the calculated period K; N_i^{r} is the required minimally-possible power output generated at the heat consumption by the CHPP operating turbounits in the i-mode; N_i^{ac} is the actual maximally-possible power output in the i-mode; ΔQ_{ann}^{CHPP} is the annual heat undersupply to consumers; $\Delta Q_{c.p.}^{CHPP}$ is the heat undersupply to consumers over the calculated period K; Q_i^{r} is the required heat supply to consumers in the i-mode; Q_i^{ac} is the actual heat supply to consumers in the i-mode; n is the total number of modes per year.

To estimate the effect of various schedule versions, of the regular overhaul scopes, and of the equipment reliability indices on the CHPP operation efficiency in general, one may use the obtained mean-annual values for the entire calculated period, because all the annual indices depend on random values (possible emergency failures and probabilities of this or that DAM price).

$$\overline{Э}_{c.p.}^{CHPP} = Э_{c.p.}^{CHPP} \big/ K \,,\; \overline{Q}_{c.p.}^{CHPP} = Q_{c.p.}^{CHPP} \big/ K \,, \quad (13)$$

$$\overline{B}_{c.p.}^{CHPP} = B_{c.p.}^{CHPP} \big/ K \,, \quad (14)$$

$$\overline{\Delta Э}_{c.p.}^{CHPP} = \Delta Э_{c.p.}^{CHPP} \big/ K \,, \overline{\Delta Q}_{c.p.}^{CHPP} = \Delta Q_{c.p.}^{CHPP} \big/ K \,, \quad (15)$$

where $\overline{Э}_{c.p.}^{CHPP}$ is the arithmetic mean value for the CHPP power generation over the calculated period K comprising k years; $\overline{Q}_{c.p.}^{CHPP}$ is the arithmetic mean value for the CHPP heat production over the calculated period K; $\overline{B}_{c.p.}^{CHPP}$ is the arithmetic mean value for the CHPP fuel consumption over the calculated period K; $\overline{\Delta Э}_{c.p.}^{CHPP}$ is the arithmetic mean value for the power undersupply to consumers over the calculated period K; $\overline{\Delta Q}_{c.p.}^{CHPP}$ is the arithmetic mean value for the heat undersupply to consumers over the calculated period K.

We suggest using the set energy supply assurance coefficient for each kind of energy produced by the CHPP as the CHPP complex reliability indices: $k_c = 1 - \Delta Э \big/ Э$, where $\Delta Э, Э$ are the actual undersupply and the set supply of the energy for a certain period, respectively.

6 Example

The technique to estimate the effect of the CHPP reliability parameters, schedules, and regular overhaul scopes on the reliability and efficiency of its operation was tested by an example of an industrial-heating CHPP. Figure 3 shows its simplified flowsheet.

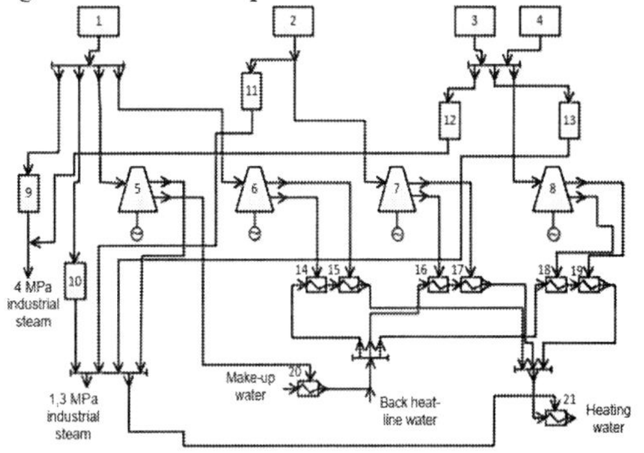

Figure 3. CHPP simplified flowsheet. Designations: 1 - a group of BKZ-420-140 steam boilers (4 pcs); 2 - a group of BKZ-500-140 steam boilers (2 pcs); 3 - a BKZ-500-140 steam boiler; 4 - a BKZ-820-140 steam boiler; 5 - a group of PT-60-130/13 steam turbines (2 pcs); 6, 7 - T-175/210-130 steam turbines; 8 - a T-185/220-130 steam turbine; 9-13 - pressure-reducing desuperheating stations; 14-19 - own network water heaterline heaters for the T-175/210-130 and T-185/220-130 turbines, respectively; 20 - a group of make-up water heaters; 21 - a group of plant main network water heaters..

In our study, we addressed two versions of the CHPP operation period. The versions differed in the overhaul period. Table 1 provides the input data. Table 2 presents the power prospective DAM prices and their appearance probabilities. These data were obtained based on statistical sampling for the characteristic period [12]. The price of the fuel used was 1400 rbl./t.s.f.

Table 1. Calculation input data.

Parameters	Version 1	Version 2
Calculated period/step, yrs/hrs	15/6	15/6
CHPP installed electric power, MW	655	655
Calculated heat load, Gcal/hr	1147	1147
Probability of the boiler unit emergency failure, net	0,0289	0,0289
Boiler unit recovery time after emergency failure, hrs	38	38
Probability of the turbounit emergency failure, net	0,0197	0,0197
Turbounit recovery time after emergency failure, hrs	68	68
General overhaul: - overhaul period, hrs - overhaul duration, CDs	 33000 34000	 46 46
Mean overhaul: - overhaul period, hrs - overhaul duration, CDs	 9000 9500	 24 24
1st class current overhaul: - overhaul period, hrs - overhaul duration, CDs	 2900 2920	 10 10
2nd class current overhaul: - overhaul period, hrs - overhaul duration, CDs	 1900 2000	 5 5

Table 2. Prospective electric power DAM prices and their appearance probabilities.

Items	Values									
Price, rbl./M W hr	0	100	200	300	400	500	600	700	800	900
Proba-bility	0,006	0,011	0,024	0,049	0,164	0,318	0,274	0,120	0,031	0,003

As a result of the optimization calculations for the CHPP operation two versions, we obtained the CHPP operation annual indices, and, based on them, the mean-annual values for a fifteen-year period. Table 3 provides the main mean values for the CHPP operation over a 15-yr period.

Table 3. Main mean-annual CHPP operation indices for 15-yr period.

Parameters	Version 1	Version 2
Power generation, kiloMW hr	2070,9	2091,6
Power generation by heat-extraction cycle, kiloMW hr	1848,3	1866,8
Power generation by condensation cycle, kiloMW hr	222,6	224,8
Power supply, kiloMW hr	1776,8	1794,6
Heat supply, kiloGkal	4025,3	4065,6
Standard fuel consumption, kilotons of s.f.	986,9	996,8
Turbounits' operation hours, hr.	22775	23003
Turbounits' hours in reserve, hr.	15714	15871
Turbounits' hours under overhaul, hr.: - regular; - emergency.	 5312 4662 650	 5157 4500 657
Boiler units' hours in operation, hr.	29751	30049
Boiler units' hours in reserve, hr.	31167	31479
Boiler units' hours under overhaul, hr.: - regular; - emergency.	 9163 7609 1554	 8915 7345 1570
Power undersupply, kiloMW hr	0,149	0,151
Heat undersupply, kiloGcal	37,7	38,1
Power set supply coefficient	0,999	0,999
Heat set supply coefficient	0,990	0,990
CHPP possible additional profit from DAM power selling, mln. rbl.	166,9	168,9

Like one can see from Table 3, Version 2, different from Version 1 in longer overhaul periods for turbo and boiler units, has a little higher production and supply of the power and heat, than that of Version 1. Also, Version 2 features lesser total time for the equipment regular overhaul (4500 hrs), but shows a little longer period for the equipment emergency repair (657 hrs). The set energy supply assurance coefficients for both versions are within 0,990-0,999, which evidences high reliability of supply to consumers. A possible mean extra profit from the DAM sale of power is 2 mln. rbl. (1,1 %) higher in Version 2, than that of Version 1 over the entire period. Under real conditions, the heat and power undersupply damage should be estimated for particular consumers, and it will depend both on economic factors and on the hazard of negative consequences for each consumer. To approximately estimate the damage from emergency breaks in the power supply within a public grid with different consumers, one may use averaged data on reparations to consumers with the value of about 70 rbl/kW hr. [18]. The values for a possible damage from the CHPP power undersupply will be 10,3 mln. rbl and 10,4 mln. rbl. for Versions 1 and 2, respectively. The difference is under 1%. At the Irkutsk Oblast mean heat rate of 985 rbl./Gcal [19], the heat undersupply for Versions 1 and 2 (in monetary terms) will be roughly 37,1 mln. rbl. and 37,5 mln. rbl., respectively. The difference is 1%. Considering that the presented approximate economic estimate from the energy undersupply shows inessential difference between the versions, and that the equipment in Version 2 has a shorter regular overhaul downtime (and, accordingly, lower overhaul costs), one may conclude that Version 2 is more preferable for the given CHPP.

7 Conclusion

We developed a technique to estimate the effect of equipment reliability indices, schedules, and overhaul scopes on the CHPP operation reliability and efficiency. By the created programming and computing suite, conducted were optimization investigations into two versions (different in overhaul periods) of an industrial-heating CHPP operation period.

The study was done within Research Project III.17.1.1. from the basic research programs of the Siberian Branch of the Russian Academy of Sciences, Reg. No. AAAA-A17-117030310433-6

References

1. Pshenichnikov S, and I. Sumskoy. Combustion Plant Energy Equipment Wear. "Russia's United Power Systems" RC ASO rating [e-resource] - 2005. - 12. - Access mode: <http://www.e-m.ru/er/2005-12/22870/?phrase_id=333686>, in Russian.
2. Barinov V.A, V.A. Saveliev, M.G. Sukharev, et al. Reliability of Liberalized Energy Systems // Novosibirsk: Nauka (Science), 2004. - P. 333, in Russian.
3. Semenov V.G. District Heating Cogeneration under present-day market conditions // Energetics Reliability and Safety. 2012. No. 18. Pp. 67-73, in Russian.
4. Sukhareva E.V., and E.M. Lisin. Analyzing problems of decrease in the efficiency of CHPP industrial-economic activities under power market conditions //In: Social Responsibility of Business. International Theoretical and Practical Conference: proceedings. Ed.: Yu.A.Anisimova. 2014. Pp. 268-277, in Russian.
5. Rezinskikh V.F., and A.G. Tumanovskiy. Improving Reliability and Efficiency of CP operating equipment // New in Russian Electric Power Industry. 2011. No. 4. Pp. 5-13, in Russian.
6. Popyrin L.S. Mathematical modeling and optimizing heat-and-power facilities. - Moscow: Energiya (Energy), 1978. - P. 416, in Russian.
7. Kler A.M., N.P. Dekanova, E.A. Tyurina, et al. Thermal Power Systems: Optimizing Research. - Novosibirsk: Nauka (Science), 2005. - P. 236, in Russian.
8. Shchinnikov P.A, G.V. Nozdrenko, O.V. Borush, et al. Optimizing the CHPP generating unit operation modes // Izvestiya RAN. Energetika. (Proceedings of the Russian Academy of Sciences. Power Engineering). - 2014. - No. 3. - Pp. 54-60.
9. Stojiljkovic, M., et al.: Mathematical modeling and optimization of tri-Generation systems with reciprocating engines. THERMAL SCIENCE: Year 2010, Vol. 14. - No. 2. - P. 541-553.
10. SO 34.04.181-2003 "Rules to organize maintenance service and repair of the equipment, buildings, and structures at power plants and grids" Russia's Unified Energy System, in Russian.
11. Sokolov E.Ya. District Heating Cogeneration and Heating Systems. - Moscow: Energoatomizdat (Energy and Nuclear Power Publishers), 1982. - P. 360, in Russian.
12. Kler A.M., A.S. Maksimov, A.V. Chalbyshev, and E.L. Stepanova. Optimizing the CHPP operation modes to maximize profit under the conditions of the electric power balancing market // Izvestiya RAN. Energetika. (Proceedings of the Russian Academy of Sciences. Power Engineering). 2014. No. 2. With. Pp. 71-80, in Russian.
13. Kler A.M., A.S. Maksimov, E.L. Stepanova, and A.V. Chalbyshev. Problems of internal optimization of the CHPP operation regimes at the Day-Ahead Market // Energetik (Power Engineer). 2015. No. 8. Pp. 20-23, in Russian.
14. Kler A.M., N.P. Dekanova, S.K. Skripkin, et al. Mathematical modeling and optimizing in the problems of operational control at combined heat-and-power plants. - Novosibirsk: Nauka (Science). Sib. Enterprise of RAS, 1997. - P. 120, in Russian.
15. Kler A.M., N.P. Dekanova, and E.L. Stepanova. Optimizing the CHPP main equipment configuration of major power sources // Izvestiya RAN. Energetika. (Proceedings of the Russian Academy of Sciences. Power Engineering), 2004. - No. 6. - Pp. 43 - 52, in Russian.
16. Kler A.M., and E.L. Stepanova. Optimizing the CHPP main equipment configuration in normal and emergency operation modes // Metodicheskiye voprosy issledovaniya nadyozhnosti bol'shikh system energetiki (Methodical issues of studying reliability of the power engineering major systems). Issue 64 Reliability of power engineering systems: achievements, problems, prospects // Ed.-in-chf. N.I. Voropay - Melentiev Energy Systems Institute of the Siberian Branch of the Russian Academy of Sciences, 2014. - Pp. 346 - 357, in Russian.
17. Kler A.M., E.L. Stepanova, and A.S. Maximov. Solving the sequence of problems to minimize undersupplies of heat and power for post-emergency conditions of the CHPP operation // Metodicheskiye voprosy issledovaniya nadyozhnosti bol'shikh sistem energetiki (Methodical issues of studying reliability of the power engineering major systems). Issue 65 Reliability of liberalized power systems / Ed.-in-chf. N.I. Voropay - Melentiev Energy Systems Institute of the Siberian Branch of the Russian Academy of Sciences, 2015. - Pp. 415 - 424, in Russian.
18. Reference book on designing power grids / Ed. by D.L.Faybisovich. - 4th Issue, revised and supplemented. - Modcow: ENAS Publishers, 2012. - Pp. 376, in Russian.
19. Irkutskenergosbyt LLC Official Website [e-resource] < https://sbyt.irkutskenergo.ru >.

Online references will be linked to their original source, only if possible. To enable this linking extra care should be taken when preparing reference lists.

https://doi.org/10.1051/e3sconf/20185802015

An algorithm to calculate feasible operating conditions of electrical network, given overhead conductor temperature and sag constraints

Yekaterina Popova [1,*] *Oleg* Voitov and *Lyudmila* Semenova [1]

[1] Energy Systems Institute of Siberian Branch of the Russian Academy of Sciences, Department of Electric Power Systems , 664033 Lermontov str., 130, Irkutsk, Russia

Abstract. The paper presents an algorithm to calculate feasible steady-state conditions of an electrical network. A distinctive feature of the calculation is the consideration of constraints on the temperature of bare conductor and the clearance of a controlled span of overhead line. This consideration is based on forming and solving an equation of conductor heat balance to calculate the conductor temperature depending on the current flowing through the conductor and environmental factors, including the actual values of air pressure and air temperature. The paper presents the results of numerical calculations performed for individual overhead lines and for a test 16-node scheme of electrical network. The results illustrate the algorithm performance and the importance of considering the above constraints.

1 Introduction

The bulk of present-day algorithms and software designed to calculate steady-state and feasible operating conditions of electrical networks do not consider the actual temperature conditions of overhead lines. Such parameters of conductors as active resistance and conductor sag (or clearance) are frequently determined in these conditions considering only the air temperature, and ignoring the effect of many other environmental factors and current load of conductors [1].

The overwhelming majority of domestic and foreign software developed to calculate the feasible load flow (except for the one proposed in this paper) does not take into account mechanical parameters of conductors: the sag and length of the conductor (or clearance). This leads to an incorrect estimation of a capability of transmitting additional power along individual conductors and in the spans of the electrical network.

Meanwhile, at the up-to-date level of hardware and software development, mathematical modeling of an electrical network suggests and allows increasing the accuracy of mathematical description of both the parameters of the electrical network components and the means for control of their operation. The issues of modeling overhead lines whose parameters are complex functions of environment state parameters and electrical network operating parameters are very important. As is known, the design of overhead lines involves consideration of a wide spectrum of factors affecting the line parameters, which enables the safety conditions for people and machinery to be met, and the reliability of the overhead line operation to be provided. However, eventually, both the parameters of components and the design conditions vary. This leads to the necessity of monitoring the mechanical parameters of the overhead lines (sag, clearance) [2,3] that are based on both their real measurements and numeric calculations performed during the intervals between planned measurements.

In this paper, we propose calculating the feasible power flow, considering both electrical (nodal voltage, currents) and geometric (sag, clearance) parameters of lines. This can be done by considering and solving the equations of the conductor heat balance and state, that make it possible to determine the conductor temperature, to specify the value of thermal current rating and to check if the rated sag and clearance values are met. The sag and clearance should be calculated for particular controlled spans, because considering all the spans leads to an excessive increase in the dimensionality of the power flow problem. For example, for a 500-kV overhead line 300 km long with a 500-m span, the number of spans will be about 600. For each overhead line, we suggest considering only the so-called controlled spans, for which it is sensible to monitor the state of clearances, for example, in the sites, where the overhead lines of different voltage levels intersect, or have very long spans, etc. To take into account the controlled spans in the calculation of power flow, it is necessary to additionally specify sag heights and physical-mechanical parameters of conductors. It is also necessary to introduce additional nodes that will split the initial line into sections, whose total length is equal to the line length.

The paper is arranged as follows. Section 1 addresses a general algorithm to calculate power flow in the electrical network, considering environmental factors (air temperature and pressure, wind velocity and

*Corresponding author: elen@isem.irk.ru

direction, solar radiation) that make it possible to determine the temperature of overhead bare conductor. Section 2 briefly describes an algorithm to calculate the conductor temperature, considering environmental factors [4,5]. Section 3 presents an algorithm to calculate the feasible power flow, considering inequality constraints on the load flow variables [6]. Section 4 discusses an algorithm to calculate the current carrying capacity in lines, considering the sag and clearance constraints in a controlled span [7]. Section 5 provides an analysis of the numerical calculation results obtained with the SDO-7 software for individual overhead lines and for a case study [8].

2 An algorithm to calculate feasible power flow in electrical network

The algorithm contains the following stages.

A. Calculate feasible power flow of the electrical network, temperature, sag and clearance in the controlled spans. If the values for the controlled parameters are feasible, the algorithm terminates. Otherwise, we go to step B.

B. Calculate the thermal current ratings at which the conductor temperature corresponds to a standard value and the constraint on the clearance in each controlled span is met. Return to step A and repeat calculations, until the specified accuracy of determining the thermal current ratings is reached.

Let us address each stage of the algorithm in more detail.

3 Calculation of overhead conductor temperature

Based on algebraic transformations of the heat-balance equation in a quadratic form, given in the standard documentation [9], we form a sixth degree polynomial for each $i-j$ line [5]:

$$f(T_{con}, U_i, U_j, z_{ij}, d_{con}, T_{air}, Q_r, Q_c, Q_{rad}, v) = 0. \quad (1)$$

After (1) is solved, the Newton method is used to find the conductor temperature T_{con} for the actual values of: nodal voltages U_i, U_j, magnitude of line complex impedance z_{ij}, conductor diameter d_{con}, air temperature T_{air}, solar radiation power Q_r, convective Q_c and radiation Q_{rad} heat transfer coefficients, and wind velocity v.

The convective heat transfer coefficient of the "traditional" heat-balance equation (1) is determined for normal environmental factors: an air pressure of 760 mm Hg, and an air temperature of 0°C. For other values of the air temperature and pressure, the convective heat transfer coefficient is determined by the expressions [5] obtained by transforming the initial criteria equations of convective heat transfer [9] that characterize the conditions for free convection (at $v \leq 1.2$ km/s)

$$Q_{c1} = 0.39 \cdot d_{con}^{0.75} \cdot (T_{con} - T_{air})^{0.3} \cdot P_{air}^{0.5} / T_{air}^{0.75}. \quad (2)$$

and forced convection (at $v \leq 1.2$ km/s)

$$Q_{c2} = 0.661 \cdot \sqrt{d_{con} \cdot v \cdot P_{air} / T_{air}}. \quad (3)$$

The specified expressions (2), (3) in their explicit form contain the actual environmental factors to be taken into account.

4 Calculation of feasible power flow

We state and solve the non-linear programming problem of the form [6]

$$W(Z, D) = 0, \quad (4)$$

$$Z_{min} \leq Z \leq Z_{max}, \quad (5)$$

$$F(Z) \leq F_{max}, \quad (6)$$

where W is a vector-function of nodal active and reactive power imbalances; Z is the state variable vector including complex numbers of nodal voltages and transformation ratio; D is the vector of nodal loads; F is the vector of functional variables including active and reactive power, power flows, and currents along the lines; $Z_{min}, Z_{max}, F_{max}$ are the vectors of minimum and maximum values of variables.

When solving the matrix equation (4) by the Newton-Raphson method for each controlled $i-j$ line, we consider the relationship between the longitudinal resistances r_{ij} and the magnitude of current in the line [4]. For this purpose, at the point of solution to polynomial (1), which is an implicit function $z_{ij}(U_i, U_j)$, we find derivatives $\partial r_{ij} / \partial U_i$ and $\partial r_{ij} / \partial U_j$ used to create the elements of the Jacobian matrix row of the form $\partial w_k / \partial r_{ij} \cdot \partial r_{ij} / \partial U_{i,j}$, where k is the row number. Solution to problem (4)-(6) contains, in particular, the values of conductor temperature for each controlled line that correspond to the design conditions, i.e. the nodal loads and environmental factors.

5 Consideration of constraints on the clearance of the controlled span and conductor temperature

After the conductor temperature T_{con} is calculated for each controlled span, we build the conductor state equation [10] of the form

$$\sigma_n - \frac{E \cdot \gamma_n^2 \cdot l_{con}^3}{24 \cdot \sigma_n^2 \cdot l} = \sigma_m - \frac{E \cdot \gamma_m^2 \cdot l_{con}^3}{24 \cdot \sigma_m^2 \cdot l} - \alpha \cdot E \cdot (T_{con\_n} - T_{con\_m}), \quad (7)$$

where α is the coefficient of linear thermal expansion for the conductor material, C^{-1}; E is the conductor elasticity modulus, Pa, whose values are assumed according to the data of reference books [11]; l_{con}, l are the conductor length and the controlled span length, m. In equation (7), the values of parameters with the subscript n correspond to the current design conditions, and those with the subscript m correspond to the case, where the conductor is under the maximum permissible tension $\sigma_m = \sigma_p$, and only its own weight $\gamma_m = \gamma_1$ affects it. The conductor temperature T_{con_m} is assumed equal to the lowest air temperature of -40°C, and the assumed wind velocity is $v \leq 1.2$ m/s.

By solving equation (7), we find the tension in the conductor material σ_n that is used [10] to calculate the length of a big equivalent span l_{eq} and the conductor sag f_{ij} in the controlled span $i - j$ (Fig.1).

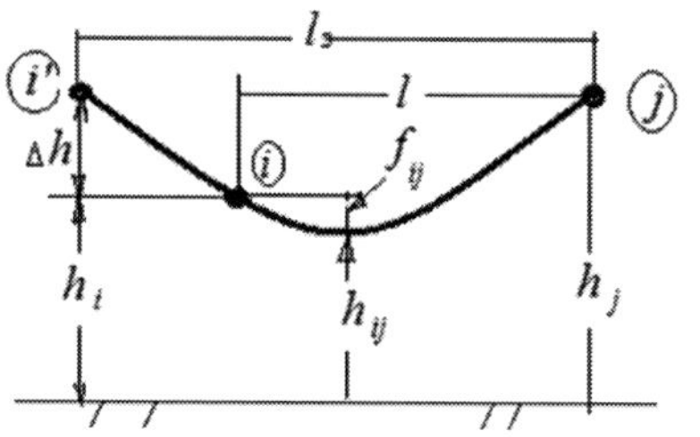

Fig. 1. A sag in the overhead line controlled span

$$\Delta h = |h_i - h_j|, \tag{8}$$

$$l_{eq} = l + \frac{2 \cdot \sigma_n \cdot \Delta h}{\gamma_n \cdot l}, \tag{9}$$

$$f_{eq} = \frac{\gamma_n \cdot l_{eq}^2}{8 \cdot \sigma_n} + \frac{\gamma_n^3 \cdot l_{eq}^4}{384 \cdot \sigma_n^3}, \tag{10}$$

$$f_{ij} = f_{eq} - \Delta h, \tag{11}$$

where h_i, h_j are the heights of the conductor suspension in the controlled span of the overhead line; f_{eq} is a sag of the big equivalent span.

The clearance in the controlled span is calculated by the following expression

$$h_{ij,min} = min(h_i, h_j) - f_{ij}. \tag{12}$$

If the temperature constraints

$$T_{con\_ij} \leq T_{con\_max} \tag{13}$$

and/or the clearance constraints

$$h_{ij} \geq h_{ij,min}, \tag{14}$$

are not met, where T_{con_max} is the maximum allowable conductor temperature assumed equal to 70°C according to [11]; $h_{ij,min}$ is the minimum clearance in the controlled span, we calculate the thermal current ratings of the line, at which constraints (13)-(14) are met.

For this purpose, when condition (13) is not met, we form the equation of conductor heat balance of the form

$$f(T_{con\_max}, I_{ij\_max}, z_{ij}, d_{con}, T_{air}, Q_r, Q_c, Q_{rad}, v) = 0, \tag{15}$$

and by solving (15) we find the value of the thermal current ratings I_{ij_max} for the known temperature value T_{con_max}.

When condition (14) is not met, we use expression (16) to determine the maximum permissible sag f_{ij_p} in the span, at which the clearance condition is met:

$$f_{ij\_p} = min(h_i, h_j) - h_{ij\_min}. \tag{16}$$

Further, from expression (10) for the maximum permissible sag value f_{ij_p}, we find the maximum permissible tension σ_{n_p}, and from the equation of state (7), considering σ_{n_p}, we determine the conductor temperature, at which condition (14) is met. Then, we solve the conductor heat-balance equation (15) to determine the thermal current rating.

6 Algorithm testing results

The proposed algorithm to calculate feasible power flow was tested both for individual overhead lines with conductors from various manufacturers and for the test 16-node scheme.

A. We investigated the effect of the actual air parameters (air pressure and temperature, both considered in the convective heat transfer coefficient in the conductor heat-balance equation (1)), on the overhead conductor parameters (their temperature and resistance per unit length) for the conductor AC-120/27 (Table 1).

The calculations allowed for the wind velocity $v < 1.2$ m/s at its direction perpendicular to the conductor. The radiation heat transfer coefficient was assumed equal to $Q_{rad} = 0$. The values of conductor current varied in a range of $I = \lfloor 100 \div I_{ij_max} \rfloor$A, where I_{ij_max}=375 A is the value of thermal current rating for the AC-120/27 conductor.

We considered the air parameters: $T_{air\,i} = 20^0$C, $P_{air\,i} = (710; 640)$ mm Hg, characteristic of Irkutsk (Irkutsk Region) and of Orlik Settlement (Okinsky District, Republic of Buryatia [12]).

The following notations were used in Table 1: $\Delta T = \left|T_{con}^{n\text{-}norm} - T_{con}^{norm}\right|$ and $\delta r = (r_{0(T_{con})}^{n\text{-}norm} - r_{0(T_{con})}^{norm}) \cdot 100\% \Big/ r_{0(T_{con})}^{n\text{-}norm}$ are the absolute and relative errors in determining the conductor temperature and resistance per unit length. The errors occur due to considering normalized (instead of actual) air parameters in the heat-balance equation; T_{con}^{norm}, $r_{0(T_{con})}^{norm}$ are the conductor temperature and resistance per unit length at normalized parameters P_{air}, T_{air}; $T_{con}^{n\ norm}$, $r_{0(T_{con})}^{n\text{-}norm}$ are the same conductor parameters at the current air parameters $P_{air\ i}$, $T_{air\ i}$.

Table 1. Relationship between the temperature and resistance per unit length of the AC-120/27 conductor and the air parameters in the heat-balance equation at $v < 1.2$ m/s.

I, A	T_{con}, °C	$r_{0(T_{con})}$, Ohm/km	T_{con}, °C	$r_{0(T_{con})}$, Ohm/km	T_{con}, °C	$r_{0(T_{con})}$, Ohm/km	ΔT, °C	δr, %	ΔT, °C	δr, %
	$P_{air} = 760$ mm Hg, $T_{air} = 0^0 C$		$P_{air\ i} = 710$ mm Hg, $T_{air\ i} = 20^0 C$		$P_{air\ i} = 640$ mm Hg, $T_{air\ i} = 20^0 C$		$P_{air\ i} = 710$ mm Hg, $T_{air\ i} = 20^0 C$		$P_{air\ i} = 640$ mm Hg, $T_{air\ i} = 20^0 C$	
100	26,63	0,2586	27,10	0,2591	27,39	0,2593	0,47	0,19	0,76	0,27
200	40,02	0,2721	41,50	0,2736	42,43	0,2745	1,48	0,55	2,41	0,87
300	59,41	0,2916	62,48	0,2947	64,44	0,2967	3,07	1,1	5,03	1,72
375	78,32	0,3107	83,11	0,3155	86,17	0,3186	4,79	1,5	**7,85**	**2,48**

As is evidenced by an analysis of the calculations (similar to the above) for the AC-120/27 conductor parameters at different combinations of $P_{air i}$, $T_{air i}$ in the heat-balance equation, the maximum improvement in the conductor parameters is achieved at: $P_{air\ i} = 640$ mm Hg, $T_{air\ i} = 40^0 C$, and is $\delta r = 3,35\%$ at $v < 1.2$ m/s.

B. We investigated the effect of considering the constraints on conductor sag (or the clearance) in the controlled span of the overhead line (14) on the calculation results of feasible power flow on the example of a test electrical network.

Table 2 presents the parameters for the investigated scheme, where we used the following notations: N_{nod}, N_{OHL}, N are number of nodes, overhead lines and controlled spans; l_{OHL} are the length of overhead lines; $P_{gen,sum}$, $P_{ld,sum}$, are the total active power of generation and load of the electrical network scheme; U_{nom} are the rated voltage levels. The wind velocity in power flow calculation was assumed to equal $v < 1.2$ m/s, the wind direction was perpendicular to the conductor.

Table 2. Parameters of the test scheme

N_{nod}	N_{OHL}	N	l_{OHL}, km	l, m	$P_{gen,sum}$, MW	$P_{ld,sum}$, MW	U_{nom}, kB
16	16	2	50-200	300-500	2164.3	2100	500-20

In the power flow calculations, additionally, the air parameters in the convective heat transfer coefficient of the heat balance equations were varied (2).

Table 3 demonstrates the influence of air temperature and pressure on the thermal current ratings calculated for the AC-240/32 conductor of the overhead line in an electrical network.

Table 3. Dependence of the AC-240/32 conductor thermal current ratings on the air parameters in the heat-balance equation at $v < 1.2$ m/s, and $T_{con} = 60^0 C$

P_{air}, mm Hg	I_{ij_max}, A			
	$T_{air} = -20^0$ C	$T_{air} = 0^0$ C	$T_{air} = 20^0$ C	$T_{air} = 40^0$ C
660	856,82	708,03	547,95	359,79
700	867,19	716,19	553,85	363,27
760	882	727,85	562,29	368,25

As seen from the Table, the values of conductor thermal current rating differs greatly depending on the air parameter variation in the convective heat transfer coefficient.

Table 4 shows the comparison of the thermal current rating values for three overhead lines equipped with AC-240/32, AC-300/64, and AC-600/72 conductors. These values were obtained from the calculation of the feasible power flow based on the above-mentioned method and from the standard documentation [9].

Table 4. Comparison of the conductor thermal current ratings calculated using the 1987 method and the method proposed in the paper (SDO) at P_{air}=760 mm Hg, T_{air}=0⁰C.

T_{con}, °C	I_{ij_max}, A		δI, %	I_{ij_max}, A		δI, %	I_{ij_max}, A		δI, %
	1987 method	SDO method		1987 method	SDO method		1987 method	SDO method	
	AC-240/32			AC-300/64			AC-600/72		
20	390	389,2	**0,2**	450	451,7	**0,4**	715	713,1	**0,3**
40	580	580,9	**0,2**	670	674	**0,6**	1065	1063	**0,2**
60	730	727,8	**0,3**	845	844,5	**0,1**	1340	1332	**0,6**
80	845	849,7	**0,6**	985	985,9	**0,1**	1565	1555,2	**0,6**

Table 4 demonstrates that at comparable calculation conditions (rated air parameters P_{air} = 760 mm Hg, $T_{air} = 0^0$ C that are included in coefficient (2)), the results of calculations differ by less than 1%, which is within the round-off error of calculation results.

Table 5 presents the feasible power flow calculated considering the constraints on temperature (13) and clearance (14) in two controlled spans of the overhead line.

Table 5. Feasible power flow calculated considering constraint on clearance in the spans at P_{air} = 760 mm Hg.

Span nodes	Conductor type	l_{eq}, m	T_{con_max}, ^0C	I_{ij_max}, A	h_i, m	h_j, m	$h_{ij,min}$, m	h_{ij}, m	f_{ij_p}, m	Notes
2-5	AC-400	470	70	825	15	7	3	3	12	$I_{ij\_max} = 0$
10-20	AC-240	602	70	605	15	8	7	7	8	$I_{ij\_max} = 0$
2-5	AC-400	300	45,3	523,1	14	14	7	7	7	
10-20	AC-240	300	63,9	558,6	14	14	7	7	7	

As seen from two upper lines in Table 5, when the specified mechanical clearances in spans l_{eq}, h_i, h_j, h_{ij}, do not correspond to a physical sense (i.e. to the recommended span length and suspension height at standard supports for a given voltage class) [11], the calculated thermal current rating is equal to zero. This is because the calculated conductor sag is much larger than that allowable one under the set clearances, and the calculated allowable conductor temperature is much lower than the air temperature. In such a situation, the feasibility of the operating conditions can be provided by refusing to consider the constraints on mechanical clearances, and instead, specifying the rated values of temperature T_{con_max} and the conductor current I_{ij_max} in spans.

With specified clearances in the same spans that correspond to a physical sense (Table 5, bottom lines), the calculated allowable conductor temperature is lower than the rated value $T_{con\_max} = 70^0 C$. In this case, the transfer capability of the controlled spans should be reduced by 37% and 8%, to provide feasible operating conditions and minimum clearance $h_{ij,min}$, respectively.

7 Conclusion

1. Our study indicates that correction of the convective heat transfer coefficient in the equation of conductor heat-balance by considering the current values of air pressure and temperature rather than the rated ones makes it possible to expand the list of considered environmental factors, and to increase the accuracy of modeling overhead line and operation of the entire electrical network.
2. It is important to improve the algorithm to calculate the feasible power flow of the electrical network by considering an additional inequality constraint imposed on the minimum clearance (or allowable sag) within the

controlled span because it enables the calculation results to be considerably corrected when making decisions about additional current load of the lines.

3. The developed algorithm demonstrated its performance and an agreement between the calculation results for the thermal current rating of conductors and its values from standard documentation in the comparable cases.

References

1. E.V. Popova, Improvement in the algorithms to calculate power flow in electric power systems considering the current temperature conditions of overhead lines: PhD dissertation (Irkutsk, P. 172, 2017).
2. B.I. Mekhanoshin, V.A. Shkaptsov, Elektro **6**, 33 (2006).
3. A.V. Samarin, D.B. Rygalin, A.A. Shklyayev, Natural and Technical Sciences **1,2** (2012).
4. O.N. Voitov, E.V. Popova, Electricity **9**, 24 (2010).
5. E.V. Popova, Power Energy Through the Eyes of Youth: Proceedings of 3rd International Scientific-Technical Conf. Collected papers, Vol. 1, 351 (Ekaterinburg: UrFU, 2012).
6. N.A. Murashko, Yu.A. Okhorzin, L.A. Krumm, et al., *Analysis and Control of Steady States of Power Systems* (Nauka, Novosibirsk, 1987).
7. K.P. Kryukov, B.P. Novgorodtsev, *Design and mechanical calculation of power transmission lines* (Energiya, Leningrad, 1979).
8. Compliance Certificate No. ROSS RU.SP20.N00008 "The SDO software to analyze and optimize steady states, and assess electric power losses in electrical networks" - Certification System of GOST R Gosstandart of Russia. Valid through: 31 Jan 2009 - 31 Jan 2012.
9. *A technique to calculate thermal current ratings for operating transmission lines* (Soyuztekhenergo, Moscow, 1987).
10. E.F. Makarov, *Reference book on 0.4-36 кV electric networks,* Vol. 1 (Papyrus Pro, Moscow, 1999,).
11. http://www.gismeteo.ru <http://www.gismeteo.ru> (accessed 15 Apr 2018).

Minimizing the number of remotely controlled switches when the planning the reconfiguration of a primary distribution network

Irina Golub[1,*], *Oleg* Voitov[1], *Evgeny* Boloev[1] and *Lyudmila* Semenova[1]

[1] Energy Systems Institute of Siberian Branch of the Russian Academy of Sciences, Department of Electric Power System , 664033 Lermontov str., 130, Irkutsk, Russia

Abstract. The work shows the possibility to select the permanent configuration of a primary distribution network at which power losses during a week with maximum load and during a week with minimum load the power losses would be close to losses at optimum hourly reconfiguration. Data on the mix of loop branches obtained during the reconfiguration algorithm operation is used for selecting the tie-switches that allow both the regime restoration after emergency opening of a sectionalized switch, and acceleration of this process, especially if the tie-switch is remotely-controlled. The IEEE 33-node test systems are used to demonstrate the proposed algorithms.

1 Introduction

Reconfiguration of a primary distribution network is the most recognized method of optimizing its topology. The main purpose of reconfiguration is reliable power supply of consumers, which in case of emergency opening of a normally closed sectionalizing switch in the primary distribution network is performed by closing a normally opened tie-switch to ensure operation of a weakly closed primary distribution network as an opened one. Other criteria of optimum reconfiguration may include minimization of power losses and voltage losses, minimization of the network maintenance costs and increase of its transfer capacity.

Reconfiguration as a method for power loss reduction in the city distribution network was first proposed in [1]. Since then a large number of reconfiguration algorithms has been developed that, according to [2], can be divided into four groups according to combination of heuristic and optimization methods, heuristic or meta heuristic methods only, and mathematical algorithms.

In the previous studies [2] on reconfiguration for power loss reduction all the switches were deemed to be remotely controlled, which does not correspond to reality as switches used in conventional primary distribution networks are switched over manually for reconfiguration.

It does not allow frequent switchings and configuration of a large number of distribution networks remains constant during the season. Switchings for losses minimization in this case can be implemented, for example, for the regime of maximum or minimum loads.

With transition to active distribution networks the manual switches shall be replaced by automatic or remotely controlled ones, which would allow more frequent reconfiguration in the hourly, weekly, daily, monthly or seasonal time intervals. Remotely controlled switches are important for power supply restoration after a permanent fault [3]. Occurrence of transient processes and over voltage during switching is a negative factor limiting the frequency of reconfigurations and having a negative impact on the network reliability [4].

The main obstacle that does not allow the use of only remotely controlled switches for reconfiguration is high cost, a limited number of allowable switchings and, hence, high cost of an individual switching. According to [5], cost of a remotely controlled switch is as high as $4,000. Its service life being 40 years, the maximum number of switchings is 10,000 [6], i.e., 250 switchings a year, or 5 switchings a week. Unjustifiably high costs of replacing all the switches by switches controlled from a dispatching center by a distribution network operator make the problem of determining the minimum number of switches used for the network reconfiguration important [7].

According to [8], for hourly reconfiguration of a distribution network that ensures minimum daily power losses, availability of renewable power generation has a greater impact on power losses reduction than reconfiguration, but with introduction of renewable power generation the number of switchings during reconfiguration increases considerably. It also shows the possibility of selecting the configuration of a primary distribution network at which power losses during the day would be close to daily losses at optimum hourly reconfiguration. In the present paper this idea extends to selection of the permanent configuration at which during a week with maximum load and during a week with minimum load the power losses would be close to losses at optimum hourly reconfiguration.

A reconfiguration algorithm offered by the authors [9] is based on the methods of determining of a maximum spanning tree [10] on the network graph and a mix of branches of independent loops by their chords

* Corresponding author: golub@isem.irk.ru

https://doi.org/10.1051/e3sconf/20185802016

[11] with the corresponding normally opened tie switches. A reconfiguration algorithm is included into a fast-acting program for loadflow calculation SDO-7 [12].

Data on the mix of loop branches obtained during the reconfiguration algorithm operation is used for selecting the tie-switches that allow both the regime restoration after emergency opening of a sectionalized switch, and acceleration of this process, especially if the tie-switch is remotely-controlled.

The article is structured in the following way. Section 1 is devoted to study of the impact of optimum hourly reconfiguration on power losses in the test distribution network with and without renewable generation sources. In Section 2 these losses, as reference ones, are compared to losses at selection of a permanent configuration of a distribution network. Section 3 solves the problem of regime recovery after permanent fault at any sectionalizing switch at a permanent configuration of a distribution network. The effort is made to solve the problem of assessing the feasibility and acceptability of the post-emergency regime.

2 Power losses during reconfiguration

To demonstrate the options of a permanent configuration of a distribution network and to analyze factors affecting power losses in it, a 33-node circuit is used Fig. 1 [13] that includes 32 sectionalizing and 5 tie-switches. Data on the rated loads of this diagram were taken from [13].

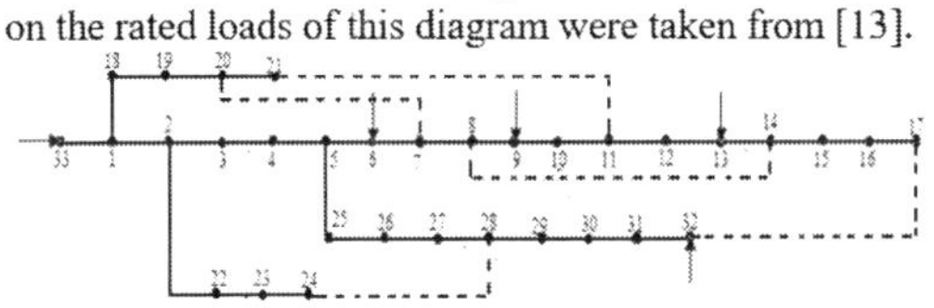

Fig. 1. A diagram of a 33-node distribution network; dashed lines are branches with tie-switches

Fig. 2 shows profiles of hourly curves with maximum Fig. 2 (*a*) and minimum Fig. 2 (*b*) weekly loads that are built using standard graphs of daily change in the industrial, commercial and municipal loads that were proposed in [2]. Daily graphs of hourly change in the active generating power of renewable generation sources in Nodes 6, 9,13, and 32 Fig. 3 that were taken from [2] were deemed to be constant for every weekday.

Fig. 4 presents curves of losses in the distribution network under the lack or availability of renewable generation and reconfiguration in it within weeks with regimes of maximum and minimum loads that confirm notable impact of renewable generation on the reduction of power losses.

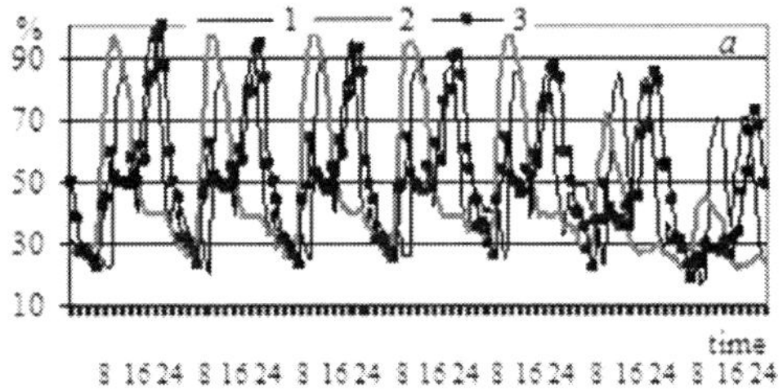

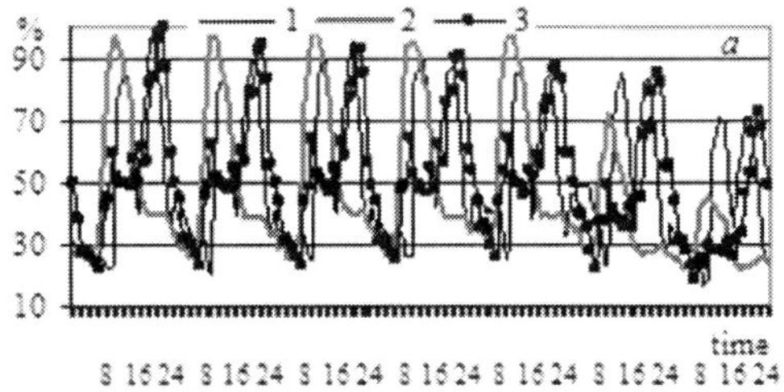

Fig. 2. Curves for maximum (*a*) and minimum (*b*) weekly industrial (1), commercial (2), and municipal (3) loads

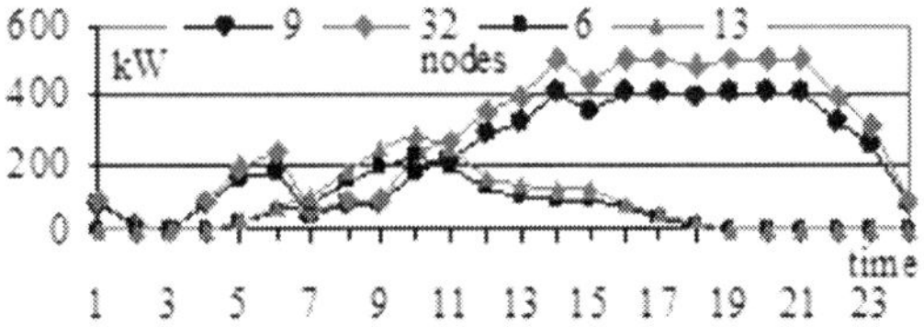

Fig. 3. Change in the active power of renewable generation sources in the distribution network nodes within a day

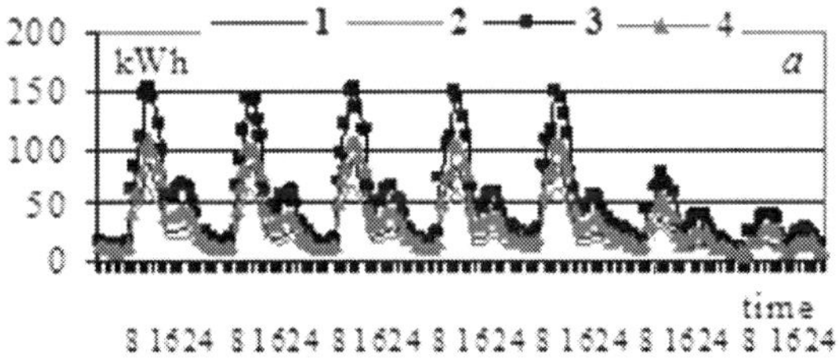

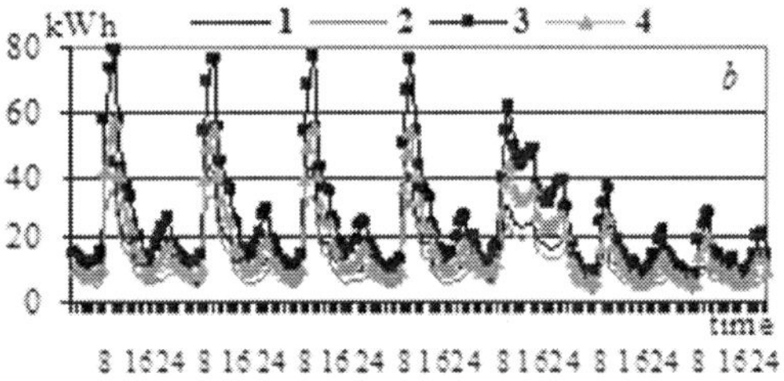

Fig. 4. Change in the power losses in the regime of maximum (*a*) and minimum (*b*) loads in the primary distribution network within a week: With renewable generation sources under the lack (1) and availability (2) of reconfiguration; without renewable generation sources under the lack (3) and availability (4) of reconfiguration

3 Selection of permanent configuration of a distribution network with minimum losses

3.1 *An option with availability of renewable generation sources in the distribution network with maximum loads.*

The number of switchings within a week in the studied network with renewable generation at optimum hourly reconfigurations equals 530, or 76 switchings a day, on the average. Compare: the number of switchings in the network without renewable generation for the same period is 132. Let us study the chances of selecting such permanent configuration of a distribution network that would ensure loses close to those at optimum hourly reconfiguration during a week.

To determine the permanent configuration for each day of the week with maximum loads, the following calculations have been performed. Configuration ensuring the minimum power losses was determined using the given hourly loads and generations Fig. 4. Then daily power losses were calculated for optimum configurations of each hour subject to their invariability within the day.

Fig. 5 demonstrates daily losses for seven days of the week starting from Monday, at permanent configuration of a distribution network that coincides with every hourly configuration obtained for it.

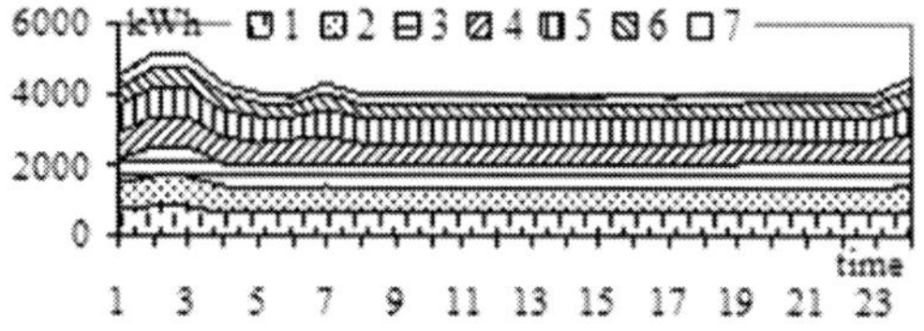

Fig. 5. Values of daily looses for each day of the week at permanent configuration of a distribution network that coincides with optimum hourly configuration obtained for it

For example, minimum losses for the first day of the week correspond to the 15th hour, and network configuration is defined by tie-switches (1: 5-6, 9-10, 13-14, 17-32, 24-28). Configuration with minimum losses of the second day with tie-switches (2: 6-7, 7-8, 13-14, 17-32, 27-28) coincides with configurations of the 4th, 6th and 7th days. Configuration with minimum losses of the third day is determined by tie-switches (3: 6-7, 9-10, 13-14, 24-28, 29-30), and configuration of the 5th day is determined by tie-switches (4: 5-6, 10-11, 13-14, 17-32, 24-28).

Fig. 6 shows changes in the power losses within a week for each out of four selected configurations; Fig. 7 presents aggregate values of weekly losses compared to losses under the lack of reconfiguration and losses at optimum daily reconfiguration.

Comparison of losses Fig. 7 distinguishes configuration of the first day of the week. Subject to its permanency the losses within a week are 257 kWh higher than losses at optimum hourly reconfiguration requiring 530 switchings.

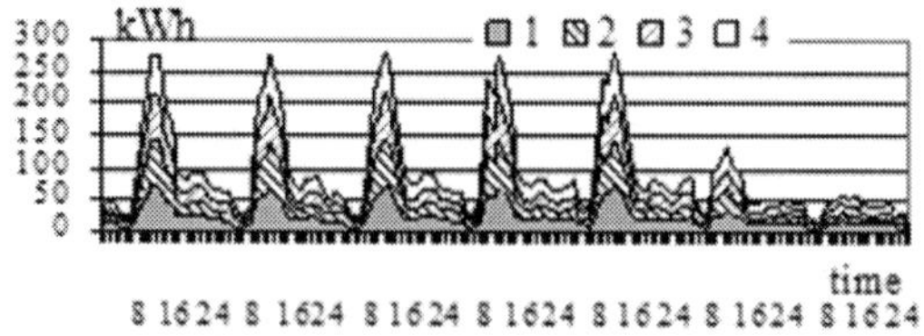

Fig. 6. Change in the power losses within a week for each out of four selected configurations

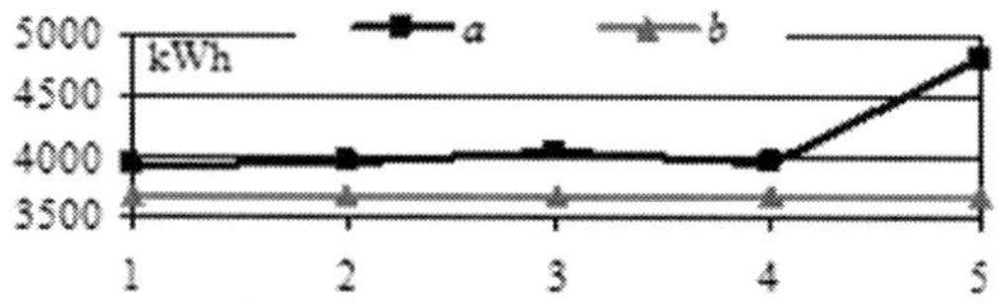

Fig. 7. Comparison of aggregate weekly power losses in the distribution network: *a* – for four options with permanent configuration; 1-4 and 5 – under the lack of reconfiguration; *b* – at optimum hourly reconfiguration

3.2 *An option without renewable generation sources in the distribution network with maximum loads.*

For the option of a distribution network without renewable generation it was found out that on week days it corresponds to configuration with tie-switches (1: 6-7, 8-9, 13-14, 24-28, 31-32), and on weekends the configuration (2: 6-7, 8-9, 13-14, 27-28, 31-32) differed from weekdays configuration by one switch.

Comparison of losses at permanent configurations with losses at optimum weekly configuration Fig. 8 has shown that losses at the first configuration exceed optimum ones by 64 kWh, at second one – by 82 kWh, whereas losses in the network without reconfiguration exceed losses at optimum reconfiguration by 2646 kWh.

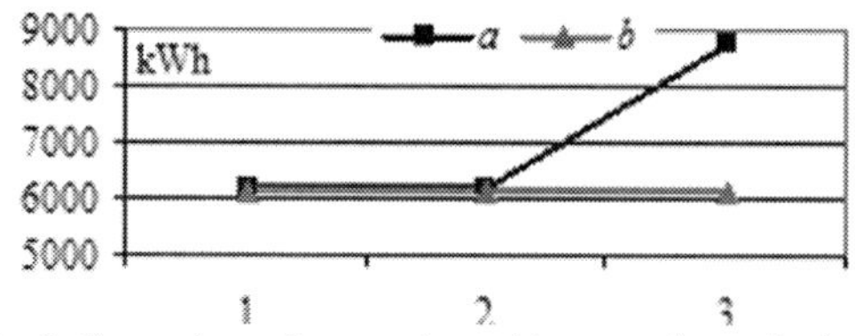

Fig. 8. Comparison of aggregate weekly power losses in the distribution network: *a* – for two options with permanent configuration; 1-2 and 3 – under the lack of reconfiguration; *b* – at optimum hourly reconfiguration

3.3 *An option with the availability of renewable generation sources in the distribution network with minimum loads.*

For the regime of minimum load at optimum hourly reconfiguration of the network within a week the reduction of losses as against losses under the lack of reconfiguration accounted for 612 kWh at 572 switchings.

Three configurations with minimum daily losses were singled out for the considered option. The first configuration is common for the first, second, sixth and seventh days and is determined by tie-switches (1: 6-7, 7-20, 8-14, 17-32, 27-28). The second configuration with switches (2: 6-7, 7-8, 12-13, 17-32, 27-28) corresponds to the third and fourth days. The third configuration corresponds to the fifth day and is determined by tie-switches (3: 5-6, 7-8, 13-14, 17-32, 27-28).

Comparison of aggregate weekly power losses for three options with permanent configuration allowed identification of the second configuration whose losses differ from losses at optimum reconfiguration by 170 kWh Fig. 9.

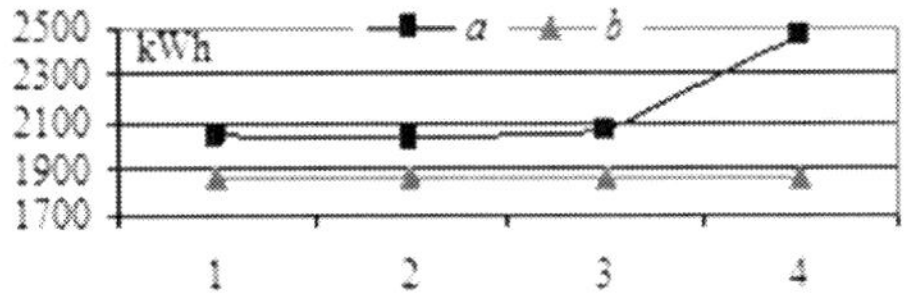

Fig. 9. Comparison of aggregate weekly power losses in the distribution network: *a* – for three options with permanent configuration; 1-3 and 4 – under the lack of reconfiguration; *b* – at optimum hourly reconfiguration

3.4 *An option without renewable generation sources in the distribution network with minimum loads.*

Calculations were made for the regime of minimum load without renewable generation Fig. 10.

Analysis have shown that the first out of two selected configurations corresponding to weekdays include tie-switches (6-7, 8-9, 13-14,24-28,31-32), and the second configuration for weekends differs from it by one tie-switch 27-28 only that shall replace tie-switch 24-28. Thus, configurations with minimum weekly power losses for regimes with maximum and minimum loads coincided.

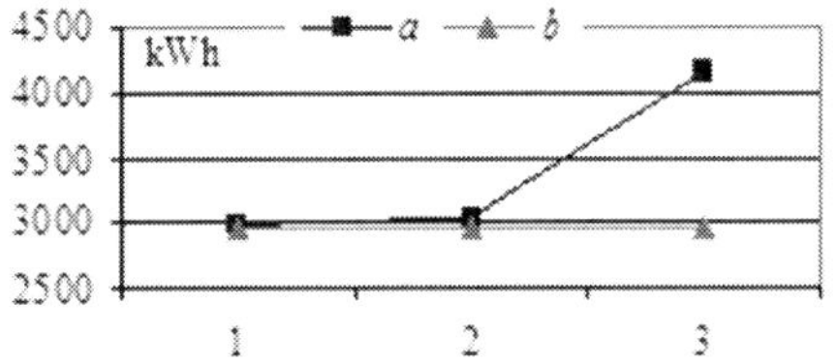

Fig. 10. Comparison of aggregate weekly power losses in the distribution network: *a* – for two options with permanent configuration; 1-2 and 3 – under the lack of reconfiguration; *b* – at optimum hourly reconfiguration

Comparison of aggregate weekly power losses have shown that losses at the first configuration differ from losses in the network with optimum reconfiguration by 27 kWh, at the second configuration – by 63 kWh, whereas losses in the network without reconfiguration exceed losses in the network with reconfiguration by 1197 kWh.

3.5 *Selection of permanent configuration of a distribution network with minimum losses*

Results of selection of a permanent configuration of a distribution network with renewable generation for the regime with maximum and minimum loads are given in Table 1 that also includes results of additional study of weekly losses in the regime of maximum loads for three configurations of the regime of minimum loads, and in the regime of minimum loads for four configurations of the regime of minimum loads.

Table 1. Power losses during the week with maximum loads and the week with minimum loads, aggregate power losses for a two-week period.

№	Mix of tie-switches	Power losses	Power losses	Aggregate power losses
	For regime with maximum load	Maximum load	Minimum load	
1	1: 5-6, 9-10, 13-14, 17-32, 24-28	**3935.454**	2120.017	6055.471
2	2: 6-7, 7-8, 13-14,17-32, 27-28	3948.498	**2028.642**	**5977.14**
3	3: 6-7, 9-10, 13-14, 24-28, 29-30	4056.291	2685.134	6741.425
4	4: 5-6, 10-11, 13-14, 17-32, 24-28	3946.542	2122.655	6069.197
	For regime with minimum load	Maximum load	Minimum load	
5	1: 6-7, 7-20, 8-14, 17-32, 27-28	4077.551	2038.154	6115.705
6	2: 6-7, 7-8, 12-13, 17-32, 27-28	3973.783	**2032.745**	6006.528
7	3: 5-6, 7-8, 13-14, 17-32, 27-28	**3917.894**	2056.139	**5974.033**

Comparison of losses for two weeks with maximum and minimum loads shows that minimum losses correspond to the third configuration of the regime with minimum loads. Minimum weekly power losses for regimes with minimum and maximum loads that equal 6182.563 kWh and 2120.201 kWh, respectively, in the network without renewable generation sources are observed at the same permanent configuration with tie-switches 6-7, 8-9, 13-14, 24-28, 31-32.

4 Selection of tie-switches for the regime restoration at emergency opening of sectionalized switches network

For final selection of permanent configuration of a distribution network, make sure that at emergency opening of any sectionalized switch a short circuit of the corresponding tie-switch would lead neither to occurrence of unacceptable voltage deviations in the distribution network, nor to rated temperature-rise currents. In case of limitations violation the controls that ensure permissible post-emergency regime shall be selected.

A reconfiguration algorithm used in this work is based on building the spanning tree on the network graph and on determination of the mix of branches of independent loops by their chords, and ideally suits determination of tie-switches used for power supply recovery in case of emergency opening of a sectionalized switch.

An algorithm for selection of tie switches options includes the following steps:

1. Branches with the given tie switches specify the mix of the network graph chords, the remaining branches being the branches of a spanning tree.
2. For determining the branches of each independent loop by its chord, a sub-matrix N_t of a block of trees of the second matrix of incidences shall be built [10],

$$N_t = -M_{ch}^{\mathrm{T}}\left(M_t^{\mathrm{T}}\right)^{-1} \qquad (1)$$

the number of rows in which would be equal to the number of chords, and the number of columns would be equal to the number of branches in the spanning tree, where M_t and M_{ch} are sub-matrices of the first incidence matrix $M=(M_t\, M_{ch})$ that correspond to the tree branches and chords. Formula (1) forms the base for a topology algorithm [11] for building both the sub-matrix $\left(M_t^{\mathrm{T}}\right)^{-1}$, and the second incidence matrix $N=(N_t N_{ch})$.

The same branch of the spanning tree can be within several loops; in this case after the emergency opening of a sectionalized switch the regime may be recovered by closing any related general loop of tie-switches. Minimum hourly power losses and minimum voltage deviations may be the criteria for selecting the best option.

Let us analyze the process of network reconfiguration with renewable generation sources whose configuration corresponds to the first mix of tie-switches for the regime of maximum loads Fig. 11 at emergency opening of any sectionalized switch. The studied configuration corresponds to optimum configuration of the 15-th hour of the first day of the week with maximum loads.

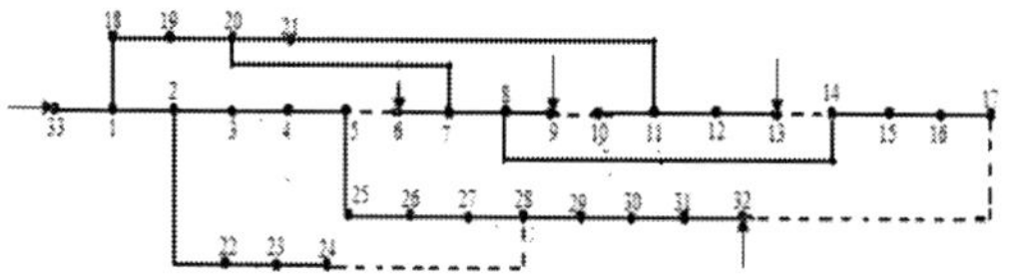

Fig. 11. Configuration of a distribution network corresponding to the first mix of tie-switches for the regime of maximum loads

Table 2. Options of tie switches that can be used for regime recovery at emergency opening of sectionalized switches in the distribution network with renewable generation sources

Sectionalized switches	Options of tie switches 1	2	3	4	Sectionalized switches	Options of tie switches 1	2
1-2	5-6	17-32			14-15	17-32	
1-18	5-6	17-32			15-16	17-32	
2-3	5-6	17-32	24-28		16-17	17-32	
2-22	24-28				18-19	5-6	17-32
3-4	5-6	17-32	24-28		19-20	5-6	17-32
4-5	5-6	17-32	24-28		20-21	9-10	13-14
5-25	17-32	24-28			22-23	24-28	
6-7	5-6				23-24	24-28	
7-8	9-10	13-14	17-32		25-26	17-32	24-28
7-20	5-6	9-10	13-14	17-32	26-27	17-32	24-28
8-9	9-10				27-28	17-32	24-28
8-14	13-14	17-32			28-29	17-32	
10-11	9-10				29-30	17-32	
11-12	13-14				30-31	17-32	
11-21	9-10	13-14			31-32	17-32	
12-13	13-14						

Possible switchings are given in Table 2. For example, at emergency opening of a sectionalized switch 7-20 within four independent loops: 1: **5-6**, 5-4, 4-3, 3-2, 2-1, 1-18, 18-19, 19-20, 20-7, 7-6; 2: **9-10**: 9-8, 8-7, 7-20, 20-21, 21-11, 11-10; 3: **13-14**,:13-12, 12-11, 11-21, 21-20, 20-7, 7-8, 8-14; 4: **17-32**: 32-31, 31-30, 30-29, 29-28, 28-27, 27-26, 25-5, 5-4, 4-3, 3-2, 2-1, 1-18, 18-19, 19-20, 20-7, 7-8, 8-14, 14-15, 15-16, 16-17; power supply can be restored by closing the tie-switch corresponding to the chord (in bold) of any out of four independent loops. Should tie-switches be considered as candidates for substitution by remotely-controlled ones, then from Table 2 it follows that four remotely-controlled switches instead of five ones are sufficient for power supply restoration. Actually, in the emergency situations when switch 24-28 is offered as an alternative for power supply restoration, tie-switch 17-32 can be used instead of it.

Analysis of voltages and currents in the week with maximum loads has shown that the emergency opening of a sectionalizing switch 1-2 constitutes the heaviest case as after power supply recovery by closing the tie-switch 5-6, the voltage drops occurred that exceeded permissible ones

$$(U_{rated} - U)\cdot 100\% / U_{rated} \le 10\% .$$

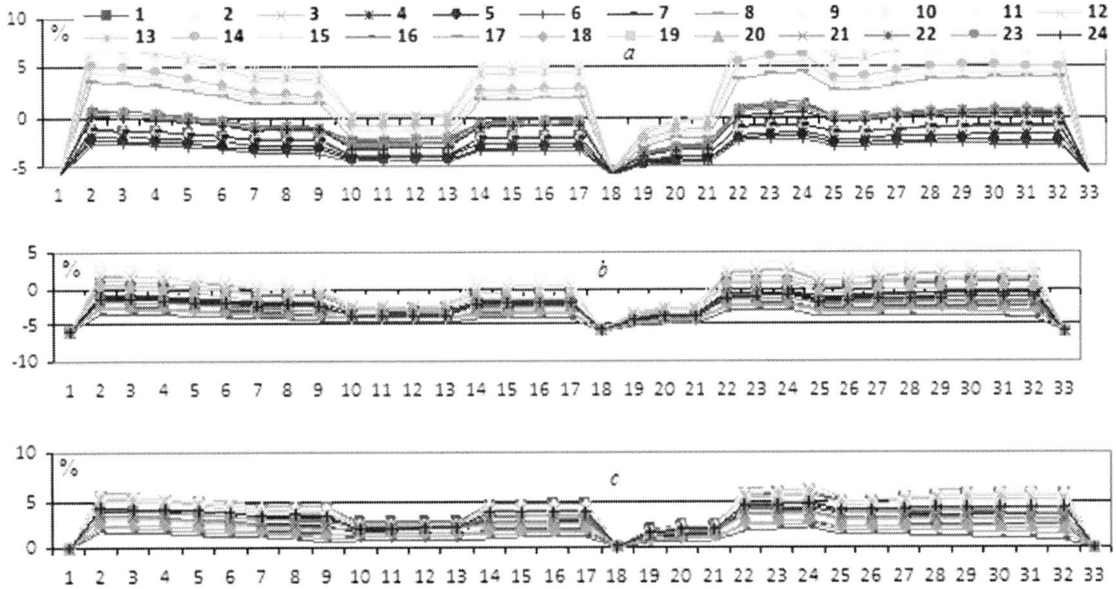

Fig. 12. Deviation of voltages in the nodes of the distribution network Fig. 11 after opening the tie 1-2 and closing the tie 5-6 under 6% voltage increase in the power supply node for the first (*a*) and sixth (*b*) days of the week at rated voltage in the power supply node for the seventh day of the week (*c*)

As a possible control to ensure this requirement was chosen voltage increase in power supply node 33 by 6% in relation to the rated voltage, Fig.12.

5 Summary

The proposed approach to selection of permanent configuration of a distribution network for reducing the power losses in it, and power supply restoration in case of emergencies is one of possible solutions that can be implemented at the intimal stage of transition from a traditional passive to an active distribution network. Comparison of such economic indicators as costs and power losses, switch costs, consumers' profit reduction due to power supply interruption, and assessment of the network reliability before and after reconfiguration shall be the most important criteria of the reconfiguration efficiency. Algorithms of optimum reconfiguration and determination of switchings for power supply restoration in case of emergency have low labor intensity and can be introduced into programs for computing the power flows in the distribution network.

The work is done in the framework of the project III.17.4.2. program of fundamental research SB RAS, registration number AAAA-A17-117030310438-1.

References

1. A. Merlin, H. Back Search, PSCC 5, 1 (1975)
2. M.R. Dorostkar-Ghamsari, M. Fotuhi-Firuzabad, M. Lehtonen, A. Safdarian, IEEE Trans. on Power Systems 31, 1879 (2016)
3. S. Lei, J. Wang, Y. IEEE Trans. on Power Systems 33, 3129 (2018)
4. V. Spitsa, X. Ran, R. Salcedo J.F. Martinez, R.E. Uosef, F. León, D. Czarkowski, Z. Zabar, IEEE Trans. Smart Grid 3, 887 (2012)
5. S. A. Yin, C. N. IEEE Trans. Power Syst. 24, 652 (2009)
6. Switchgear Solutions, Schneider Electric Inc.
7. S. Lei, Y. Hou, F. Qiu, J.Yan, IEEE Trans. on Sustainable Energy 9, 420 (2018)
8. I.I. Golub, O.N. Voitov, E.V. Boloev, L.V. Semenova, Izv. RAS Energetika 1, 25 (2018)
9. I.I. Golub, O.N. Voitov, E.V. Boloev, L.V. On Methodical problems of reliability research of large energy systems 67, 141 (2016)
10. E. Mainika, Algorithms for optimization on networks and graphs (Mir, Moskow, 1981)
11. S.K. Gursky, Algorithmization of condition control problems for bulk power systems (Nauka i tekhnika, Minsk, 1977)

https://doi.org/10.1051/e3sconf/20185802016

12. O.N. Voitov, System studies in energy. Retrospective of scientific schools at SEI-ISEM, 324 (Nauka, Novosibirsk, 2010)
13. M.E Baran, F.F. Wu, IEEE Trans. Power Del. 4, 1401 (1989)

Configuration management of electricity distribution grids 6-35 kV according to the criteria of reliability

Vladimir Vukolov[1,*], *Boris* Papkov[2], *Evgeny* Pnev[2] and *Anton* Kolesnikov[3]

[1]Nizhny Novgorod State Technical University n.a. R.E. Alekseev, 603950, Nizhny Novgorod, Russian Federation
[2]Nizhny Novgorod State University of Engineering and Economics, 606340, Knyaginino, Russian Federation
[3]Branch office ESP-NN-SESH, Electroshchit – TM Samara, GK, CJSC, 603000, Nizhny Novgorod, Russian Federation

Abstract. Questions of a choice of criteria for definition of places of opening of electric networks 6-35 kV are considered in this paper. Recommendations are given on the configuration management of distribution electrical networks based on the hierarchy analysis method.

1 Introduction

In most Distribution Grid Organizations, the task of optimizing the configuration of the distribution grid (6/10-35 kV) is solved only if there are damages to its individual elements participating in power transmission normal mode. Post-accident configuration management, as a rule, is carried out by the criterion of a minimum number of switching operations to eliminate damage and restore power supply to consumers (installation and coordination of relay protection devices is simplified). Automation of the normal mode, reacting to a change in the flow distribution, in grids of 6-35 kV is practically not represented.

The current legislation in the field of tariff formation (loss standard and transport tariff for Distribution Grid Organizations do not depend on grid mode but are determined based on the unit output per unit length of the electric line) practically excludes the account of electric power losses as the main factor in determining the disconnection points in the normal mode of the distribution grid. At best, in practice, the approach is to maximize the reliability of grid due to the use in the normal mode of electrical components with the best reliability. At the same time, for various reliability indices, the optimal open-loop configurations of the 6 - 35 kV electric grid are not the same in the general case.

The actual task is to develop a universal approach to determining the most efficient distribution grid configuration for Distribution Grid Organizations and end consumers, considering the following factors:

- compliance of the parameters of the mode with grid solutions (voltage in the nodes of connection of consumers and long-tolerated power line currents);
- value of the load losses of electric power in the normal steady-state grid mode;
- reliability of the main power transmission elements.

Its implementation will lead to both a reduction in electricity losses during power transmission, and an increase in the reliability of electricity supply to individual consumers and the electric grid. As a mathematical tool of the approach being developed, it is proposed to use the hierarchy analysis method proposed by T. Saati [1].

2 Application of the hierarchy analysis method to assess the reliability of open-loop configurations of a 6-35 kv distribution grid

The hierarchy analysis method allows obtaining a multi-criteria evaluation of the ways of power transmission from the power centers to the end users of the 6-35 kV grid based on pairwise comparisons of both possible configurations and single reliability indicators. Its application allows to include in the analysis all available indicators and characteristics of the elements of the power supply system.

In the problem of assessing the reliability of open-loop configurations of an electrical grid, it is advisable to use this method to determine the optimal path to each consumer individually and the overall configuration of the entire power supply system. To do this, it is necessary to determine the list of criteria by which to evaluate the properties of the object. When comparing the reliability of power transmission lines for such criteria, the following are accepted:

- time of failure-free work (t_{fw}, year);
- parameter of the flow of restoration (ω_r, 1 / year);
- the inverse of the recovery time (recovery rate - $\mu = 1 / t_r$, 1 / hour).

The choice of such reliability indicators is since their influence on each other is insignificant and the maximum values of these indicators are used in further calculations.

The practical implementation of this approach begins with the use of the algorithm for constructing all the

* Corresponding author: vvucolov@mail.ru

trees in the graph of the grid area [2] to determine the possible open configurations of a distribution grid of arbitrary type. Additionally, a complete list of transmission paths from each of the power centers to all load nodes can be obtained.

After this, the transition to the alternate calculation of the selected reliability indicators of all possible paths of electricity transmission to each consumer (load node) of grid under investigation is carried out. For simplicity, all branches of the graph of grid area are set by blocks (for example, electric line element is considered together with the switching equipment at both ends). This approach allows you to better navigate the elements of the simulated grid, determine the reliability of the blocks, and, to a large extent, reduce the number of objects in the future calculation.

For the automation of calculations, an algorithm for transforming an arbitrary configuration grid scheme has been developed, which compares two elements and determines their form of connection among themselves. Then the circuit is converted into one equivalent element and, depending on the type of connection, the new end element is assigned the values of the beginning and the end. This procedure runs through the entire grid and repeats until the last element remains [3].

The hierarchy of choosing the optimal transmission of electricity to a consumer with eight possible options for electricity supply is shown in Fig. 1.

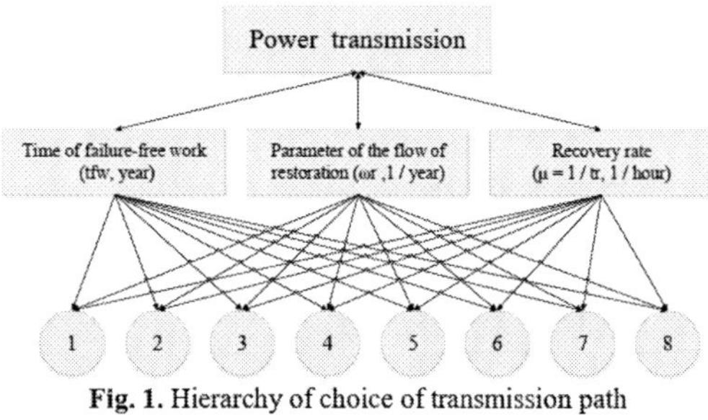

Fig. 1. Hierarchy of choice of transmission path

To select the optimal power transmission, it is necessary to compare the elements of the third level in pairs by the force of their influence on the element of the second level, i.e. put numbers that reflect a pairwise comparison of possible transmission paths in the matrix and find the eigenvector (EV) with the largest value.

When comparing the paths of transmission of electricity to the consumer by one criterion, for example, the time of failure-free work, the matrix is filled with the relations between the values of this reliability index. The next step is to calculate the priority vector for the given matrix. From a mathematical point of view, the EV after normalization will be the vector of priorities. Its value can be obtained by extracting the root of the n-th degree from the product of n elements (B_i) of each row by the expression:

$$\mathrm{EV} - \sqrt[n]{B_1 \cdot B_2 \cdot \ldots \cdot B_n} \qquad (1)$$

The numbers obtained are normalized. The sum of the values of the eigenvector is divided by the sum of the corresponding column of the matrix. By adding the products of the obtained normalized values to the sum of each column of the matrix, the main eigenvalue λ_{max} is determined. The closer λ_{max} to n, the more consistent the result.

The deviation from consistency is expressed by the consistency index (CI), which is defined as:

$$\mathrm{CI} = \frac{\lambda_{\mathrm{max}} - n}{n - 1} \qquad (2)$$

To assess the consistency of the resulting matrix and, correspondingly, to check the adequacy of the comparisons of the variants given in [1], the consistency relation (CR) is used, which is defined as the ratio of the CI to the average random index (RI) of the matrix of the same order. CR $\leq$ 0,1 for the solution of the tasks is considered acceptable [1].

In the presence of statistical or reference information on single indicators of the reliability of grid elements, a complete consistency of the matrix of comparisons of the transmission paths of electricity (CR= 0) should be observed. This is due to the fact that in each comparison the same reliability indicators (time of failure-free work, recovery flow parameter or intensity of recovery) participated.

Since the chosen criteria characterize different reliability properties, in calculating its resulting estimate, it becomes necessary to consider the "weight" of each criterion and the path of electricity transfer. The "weight" of the criterion is determined based on expert assessments by the personnel of the Distribution Grid Organizations electrical service. To obtain a matched matrix, weighting coefficients based on expert estimates are assigned in the range from 1 to 9 in accordance with the scale of relative importance of the criteria [1].

The last step is to calculate the overall reliability estimate for all possible ways of transmitting electric power N for each consumer of grid area in question. For this purpose, the sum of the products of the "weight" of the estimate according to the accepted criterion k_i (i is the number of selected reliability indicators) is found on the normalized value of this criterion. For the selected reliability indicators, the estimation of the j-transmission path to the consumer is determined by the expression:

$$N_j = k_1 \cdot t_{\mathrm{fw.norm}j} + k_2 \cdot \omega_{\mathrm{Bnorm}j} + k_3 \cdot \mu_{\mathrm{norm}j} \qquad (3)$$

where are the normalized estimates of the corresponding reliability index of the j-transmission path to the consumer in question.

The calculation algorithm is implemented for each consumer (load node) of the distribution grid under study.

3 Determination of the places of disconnection in distribution electric grids

Determination of the optimal transmission path to each load node does not guarantee the construction of an optimal configuration for the entire grid area. This is due to the fact that the choice of a certain path for the first consumer, in general, significantly reduces the possible number of configurations for the next in order. It is necessary to specify the priority hierarchy of the load nodes themselves, which are chosen as:

- reliability category of load receptors;
- total installed capacity of consumers (S, MVA).

Based on the results of the ranking of load nodes by their importance, two options are possible for choosing the optimal open distribution grid configuration for reliability:

1) Begin the construction by choosing the optimal configuration from the most significant consumer to the next in order, until all consumers are connected to the power centers;

2) Calculate the reliability estimates of all possible grid configurations considering the priority of load nodes and choose among them the configuration with the maximum value of the obtained index.

Additionally, the configurations obtained by these methods are checked for the possibility of the existence of a stable transmission mode at the maximum and minimum grid loads. The second option allows you to find the configurations that most satisfy the requirements of optimality, but at the same time is more labor-intensive and voluminous from the point of view of the calculations. The total number of options for the solutions under consideration will also be the largest. At the same time, this variant of searching optimal locations for the distribution grid allows obtaining the best configurations not only by the criterion of ensuring the maximum reliability of electricity supply to consumers, but also taking into account the energy losses in the steady-state grid mode.

In general, the sequence for determining the locations of the spread in the distribution grid 6-35 kV is as follows:

1) Based on the algorithm for constructing all the trees in grid area graph [2], a complete list of open grid configurations is defined, as well as possible ways of power transmission for each load node.

2) Calculation of steady-state grid conditions for all configurations (at minimum, maximum and medium loads) is performed, configurations that do not provide conditions for the existence of a long-tolerated regime are excluded from further consideration.

3) Load losses for the medium load regime are calculated and their normalized values are determined.

4) Reliability rating (R_Σ) is carried out for each configuration [5]:

$$R_\Sigma = \sum_{k=1}^{n} R_k \cdot A_k \qquad (4)$$

where R_k is the estimates of the reliability of the transmission path to the k-customer, A_k is the estimate of the priority of the k-consumer, n is the total number of consumers in grid area.

Summation of the numerical values of the reliability estimates of the transmission paths to all consumers in grid area, considering their hierarchy, is mathematically acceptable [4], since for all configurations the reliability indicators of individual elements and node loads are unchanged.

5) The obtained reliability estimates for all considered grid configurations are normalized.

6) The optimal configuration of the distribution grid is calculated considering the weight coefficients of the reliability and loss estimates determined individually for each distribution grid organizations.

4 Implementation of the algorithm for choosing the open configuration of the 6-35 kV power grid region

Approbation of the developed algorithm for locating faults was carried out using the example of a 35 kV electric grid section of the Nizhny Novgorod energy system (Fig. 2).

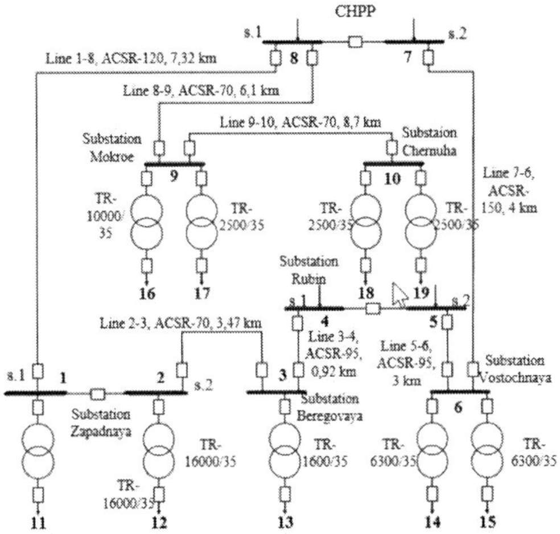

Fig. 2 Single-line scheme of a 35 kV grid section

Under the nodes of grid area is meant the distribution device of high-voltage substations (35 kV) and low (6-10 kV) voltages. Distribution device low-voltage are nodes for connecting the load. The load on the substation side of substations is absent. Under the branches, three types of electrical equipment are understood: transformers and overhead transmission lines in conjunction with the Switch-gear equipment, as well as sectional switches. The power centers of grid section under consideration are the Novogorkovskaya CHPP and the Rubin substation. Data on the average loads of nodes on the annual interval, used in calculating the load losses, are given in Table. 1. Initial data on indicators of reliability of electrical equipment and transmission lines were adopted based on materials of JSC "Firm ORGRES". According to the analysis results, 14 possible grid configurations were obtained, only 12 of them meet the requirements (Table 2)

Table 1. Grid node load

$\underline{S}$**11**=4,318+*j*2,149 MVA	$\underline{S}$**12**=2,569+*j*1,158 MVA	$\underline{S}$**13**=0,485+*j*0,175 MVA
$\underline{S}$**14**=2,363+*j*1,951 MVA	$\underline{S}$**15**=2,35+*j*0,796 MVA	$\underline{S}$**16**=1,218+*j*0,406 MVA
$\underline{S}$**17**=1,197+*j*0,577 MVA	$\underline{S}$**18**=0,945+*j*0,456 MVA	$\underline{S}$**19**=0,397+*j*0,219 MVA

Table 2. Characteristics of possible open grid configurations

№ configuration	Disabled branches	Load losses, MW	Characteristic of the transmission mode
1	7-8, 4-5, 1-2, 5-6	0,141	Meets the requirements
2	7-8, 4-5, 1-2, 7-6	0,143	Meets the requirements
3	7-8, 4-5, 2-3, 7-6	0,181	Meets the requirements, heavily loaded line 1-8 (0,180 kA)
4	7-8, 4-5, 2-3, 5-6	0,174	Meets the requirements, heavily loaded line 1-8 (0,180 kA)
5	7-8, 4-5, 3-4, 5-6	0,192	Meets the requirements, heavily loaded line 1-8 (0,195 kA)
6	7-8, 4-5, 3-4, 7-6	0,193	Meets the requirements, heavily loaded line 1-8 (0,195 kA)
7	7-8, 4-5, 1-8, 5-6	0,182	Meets the requirements, heavily loaded line 3-4 (0,194 kA) и 2-3 (0,183 kA). It is not desirable to work continuously in the maximum load mode (Load line 2-3 is close to the limit)
8	7-8, 4-5, 1-8, 7-6	0,184	Meets the requirements, heavily loaded line 3-4 (0,194 kA) и 2-3 (0,183 kA). It is not desirable to work continuously in the maximum load mode (Load line 2-3 is close to the limit)
9	7-8, 1-8, loss power center 4 и 5	0,320	Meets the requirements.
10	7-8, 4-5, loss power center 4 и 5	0,194	Meets the requirements
11	7-8, 2-3, loss power center 4 и 5	0,184	Meets the requirements
12	7-8, 1-2, loss power center 4 и 5	0,173	Meets the requirements, heavily loaded line 7-6 (0,139 kA),
13	7-8, 4-5, loss power center 7 и 8	0,320	Doesn't match requirements, heavily loaded line 3-4 (0,299 kA) и 2-3 (0,285 kA, overloaded in the maximum mode)
14	7-8, loss power center 4, 5 и 8	0,73	Doesn't match the requirements. voltage decreases below acceptable, overloaded lines 7-6, 3-2

For each consumer of grid region in question (nodes 11-19), single reliability indicators of all possible options for power transmission from the power center were calculated. To obtain the resulting reliability estimates, a pairwise comparison of single indicators was performed (Table 3).

Table 3. "Weighing characteristics" of single reliability indicators

	t_{fw}, year	ω_r 1/y	$1/t_r$, 1/h	EV	Norm. value
t_{fw}, year	1,000	9,000	7,000	3,97905	0,792757363
ω_r 1/year	0,111	1,000	0,500	0,38157	0,076021412
$1/tr$, 1/h	0,143	2,000	1,000	0,65863	0,131221225
Summ	1,254	12	8,5	5,01926	1
λmax	3,0217				
CI	0,0109				
CR	0,0187				

The relative importance of the criteria is chosen by the Distribution Grid Organization's staff, which ensures the operation of distribution networks. The results obtained are acceptable in terms of CR size and are the same for all load nodes of the network area.

Estimates of reliability indicators of possible options for electricity transmission to each consumer. node 11 (Zapadnaya substation, n.1) are given in Table. 4.

Similar estimates were obtained for the remaining load nodes of the network under consideration. Criteria for assessing the priority of consumers are determined from the condition that the total capacity of the node is 3 times less important component than the maximum category of reliability of consumers in the node. This ratio is chosen based on expert assessments of Distribution Grid Organization's personnel, for which the priority is minimal damage to consumers from disruption of electricity supply. Calculation of the priority of load nodes of the network region is presented in Table 5.

Table 4. Evaluation of the distribution of reliability of transmission paths by electric power distribution for node 11

№ s/p	Path to nodes from CP	t_{fw}, year	ω_r 1/y	$1/t_r$, 1/h	Crite-rion	Weight	Rating
1	8-1-11	0,15	0,14	0,16	t_{fw}, year	0,7927	0,1544
2	8-7-6-5-4-3-2-1-11	0,09	0,11	0,09	ω_r 1/y	0,0760	0,0884
3	7-8-1-11	0,15	0,14	0,17	$1/t_r$, 1/h	0,1312	0,1534
4	7-6-5-4-3-2-1-11	0,08	0,11	0,09			0,0810
5	4-3-2-1-11	0,180	0,044	0,126			0,1626
6	4-5-6-7-8-1-11	0,075	0,206	0,120			0,0911
7	5-4-3-2-1-11	0,203	0,045	0,137			0,1822
8	5-6-7-8-1-11	0,072	0,205	0,106			0,0865

Table 5. Comparison of the criteria for assessing the priority of consumers

№ Load node	Category	S, p.u.	Criterion	Weight	Rating
11	0,273	0,271	Category	0,75	0,272220
12	0,273	0,158	*S*, MVA	0,25	0,244084
13	0,030	0,029			0,029961
14	0,152	0,172			0,156632
15	0,152	0,139			0,148449
16	0,030	0,072			0,040741
17	0,030	0,075			0,041371
18	0,030	0,059			0,037449
19	0,030	0,025			0,029088

The obtained results show that the simplest variant of determining the places of disconnection in the considered network is to ensure the maximum reliability of power transmission to the key consumer of the network (node 11), with the subsequent selection of connections to the remaining consumers that are not connected power center.

By this principle, the most rational 7th or 8th network configurations are selected (Table 2). For a more detailed search for optimal locations, a comparative evaluation of the reliability of all possible configurations taking into account the priority of consumers is calculated (expression 4) (Table 6)

Table 6. Evaluation of reliability of all possible network area configurations

Configuration	Node									R_Σ
	11	12	13	14	15	16	17	18	19	
1	0,086	0,106	0,136	0,078	0,078	0,083	0,083	0,083	0,083	0,089543
2	0,086	0,106	0,136	0,094	0,094	0,083	0,083	0,083	0,083	0,094471
3	0,086	0,080	0,136	0,094	0,094	0,083	0,083	0,083	0,083	0,088031
4	0,086	0,080	0,136	0,078	0,078	0,083	0,083	0,083	0,083	0,083103
5	0,086	0,080	0,031	0,078	0,078	0,083	0,083	0,083	0,083	0,079939
6	0,086	0,080	0,031	0,094	0,094	0,083	0,083	0,083	0,083	0,084867
7	0,091	0,106	0,136	0,094	0,094	0,083	0,083	0,083	0,083	**0,095703**
8	0,091	0,106	0,136	0,078	0,078	0,083	0,083	0,083	0,083	0,090775
9	0,045	0,049	0,030	0,078	0,078	0,083	0,083	0,083	0,083	0,061278
10	0,086	0,080	0,031	0,078	0,078	0,083	0,083	0,083	0,083	0,079939
11	0,086	0,080	0,031	0,078	0,078	0,083	0,083	0,083	0,083	0,079939
12	0,086	0,049	0,030	0,078	0,078	0,083	0,083	0,083	0,083	0,072412

According to the results obtained, according to the criterion of ensuring maximum reliability of consumers in the network area in question, configuration 7 is preferable. The final choice of the optimal solution, according to the authors, should be taken considering the load losses in the normal transmission mode.

The priority of losses and reliability is determined by Distribution Grid Organization's personnel, depending on the likely failure to supply electricity (in the example considered, 0.4 to 0.6).

The results of the calculation are given in Table 7.

Table 7. Complex multi-criteria evaluation of reliability and efficiency of all possible network area configurations

Configuration	Reliability rating	Load losses, MW	The reciprocal of losses	Normalized losses	Resulting configuration rating
1	0,089543	**0,141**	7,092	0,107	0,096597
2	0,094471	0,143	6,993	0,106	**0,098955**
3	0,088031	0,181	5,525	0,083	0,086216
4	0,083103	0,174	5,747	0,087	0,084603
5	0,079939	0,192	5,208	0,079	0,079448
6	0,084867	0,193	5,181	0,078	0,082241
7	**0,095703**	0,182	5,495	0,083	0,090636
8	0,090775	0,184	5,435	0,082	0,087318
9	0,061278	0,32	3,125	0,047	0,055657
10	0,079939	0,194	5,155	0,078	0,079123
11	0,079939	0,184	5,435	0,082	0,080817
12	0,072412	0,173	5,780	0,087	0,078389

Configuration No. 2 is the most optimal for long-term power transmission mode. In addition, the obtained estimates allow to determine the preferred solutions for controlling switching equipment in post-emergency and recover mode of the network.

5 Conclusions

The developed control algorithm for the configuration of 6-35 kV distribution networks based on Saati hierarchy analysis method makes. It possible to take multi-criteria solutions for the selection of optimal opening locations. It is recommended to choose the optimal network configurations when jointly considering the losses of electricity in the steady mode of the network and complex estimates of ensuring the reliability of electricity supply to individual consumers.

The work was supported by the Ministry of Education and Science of the Russian Federation, (Agreement No. 14.577.21.0244 on granting a subsidy of 26.09.2017, the unique identifier of the project RFMEFI57717X0244).

References

1. T. L. Saaty *Decision making with the analytic hierarchy process* (Int. J. Services Sciences, Vol. 1, No. 1, 2008)
2. V.U. Vucolov, A.L. Kulikov, B.V. Papkov, *Opredelenie mest razmykaniya raspredelitelnyh jelektricheskih setej v zadache povysheniya nadezhnosti elektrosnabzheniya* (Syktyvkar, OOO «Komi respublikanskaya tipografiya», 2016).
3. V.A. Zorin, E.V. Projdakova., M.A. Fedotkin *Vvedenie v obshhie cepi Markova. Uchebno-metodicheskoe posobie (*Nizhnij Novgorod, NNSTU, 2013)
4. B.V. Papkov, A.L. Kulikov, *Teorija sistemy i sistemnyj analiz dlja jelektrojenergetikov* (Jurajt,2016)
5. Ju.B. Guk, *Teorija nadezhnosti v jelektrojenergetike* (Leningrad, Energoatomizdat, 1990)

 https://doi.org/10.1051/e3sconf/20185802018

Methods of analysis of the power system security

Alexey Osak[1,*], *Daniil* Panasetsky [1], and *Elena* Buzina[1]

[1]ESI SB RAS, 664033 Lermontov str., 130, Irkutsk, Russia

Abstract. Increasing the intelligent level of the PS control systems caused by the implementation of Smart technologies changes the structure and the properties of PS and increases the importance of system reliability analysis. System reliability analysis includes two components – the balance analysis and the regime analysis. On the one hand, there are a large number of studies that assess the reliability of the power system examining various aspects and methods of solving this problem. In practice, the security analysis is limited by the calculations of power flows, static and dynamic stability for a number of forecast periods for the normal and repair circuits considering the most severe disturbances. The existing approach allows defining the requirements and adjusting emergency control systems, but does not allow evaluating and comparing solutions for power grid constructions. The authors propose a new method for power system reliability evaluation, which is suitable for planning development and operation of power systems. The method includes a general description of the algorithm which allows to compare various development scenarios, as well as to assess the reliability level of their implementation.

1 Introduction

The rise of intelligent level of the power system (PS) control due to the implementation of Smart technologies and conception of intelligent power system (IPS) of active-adaptive grid (AAG) changes the structure and characteristics of PS [1-2], simultaneously increasing the importance of assessment power system security, as aspect of system reliability. System reliability (PS reliability) includes balance and regime components.

In Russia, in accordance with [3], power system security is understood as a property of the system to maintain the prescribed modes of operation when conditions change, element failures and sudden perturbations occur. In other countries, the term security has a similar meaning, which is understood as the ability of the system to withstand sudden disturbances [4-6]. There is distinguish between static and dynamic power system regime security. In assessing static regime security, dynamic transient processes in PS are not taken into account, the model of the system is limited to the equations of steady-state. Dynamic regime security is related to the dynamics of PS behavior, when a nonlinear model of transient processes in PS or its linearized version can be considered.

There is a great number of studies on assessment of power system reliability (in particular [7-8]) dealing with various aspects and methods for solution of this problem. In practice, the security analysis is limited by calculations of static and dynamic stability for a number of forecast periods for normal and repair circuits considering the most severe disturbances [9]. Such an approach allows us to define requirements and adjust emergency control (EC). However, it cannot assess and compare solutions for power grid constructions.

The authors propose a new method to determine application of various aspects of security at different stages of development planning and during operation power system. A technique and an algorithm are proposed that allow to determine the magnitude of possible under-supply of electricity for various emergency situations, taking into account the emergency control and the expected actions of the dispatcher. This approach allows us to determine the mathematical expectation of annual damage from all accidents in PS, resulting in a shortage of electricity.

The authors propose the new method of steady state calculation of PS taking into account the discrete and interval characteristics of the parameters of the regime. Using this method it is possible to perform the analysis of system reliability assessing the control of PS and finding different ways of automatic or operational control allowing PS to move from pre-emergency to acceptable post-emergency state in the event of emergency disturbances.

2 Methodology

2.1. Reliability criteria for power system expansion planning

The most important criterion for decision-making in operation and in planning of power system development is system reliability and PS viability. Moreover, both balance and regime aspects of system reliability should

* Corresponding author: osakalexey@mail.ru

https://doi.org/10.1051/e3sconf/20185802018

be provided [5-8]. Planning is carried out at various levels with different time interval, therefore, various aspects of system reliability are considered for different types of planning. When choosing the sufficiency of measures of power grid construction in the development process of power systems, the algorithm given in Fig. 1 is used.

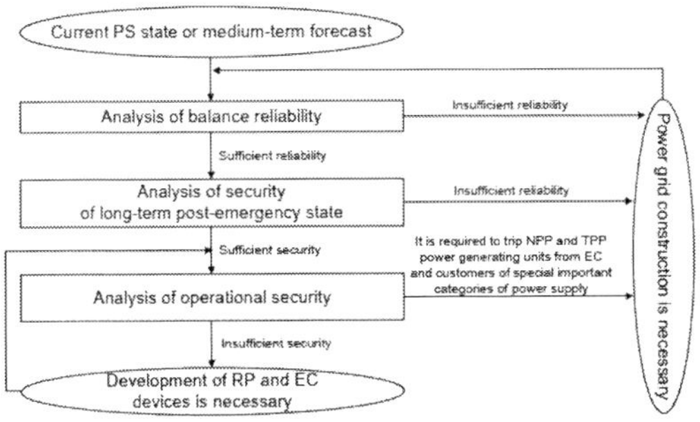

Fig. 1. Algorithm of feasibility studies for measures of power grid construction.

We distinguish two components in assessment of PS security:

Assessment of PS security in the aspect in dynamics process of its transition to post-emergency state after emergency disturbances: evaluation of disturbance consequences (including static and dynamic stability analysis), operation of relay protection (RP), emergency control (EC), automatic voltage regulators (AVR), automatic frequency and power control (AFPC), sources of reactive power and on-load tap-changers (OLTC). Changes of generator power are recorded with primary regulation within the primary reserve.

Assessment PS security in the aspect of its recovery process after the potential emergencies and transition from post-emergency to normal state: changes of generators power are recorded with primary, secondary and operational regulation within primary and secondary reserves.

2.2. Methods and assessment criteria of PS security in the aspect of its recovery process after the potential emergencies and transition from post-emergency to normal state

At the moment of emergency disturbance by RP and EC operation, partial load shedding (power consumers) is possible as well as output of regime parameters (current, voltage and frequency) beyond permissible values. After RP and EC response, automatic regulators and regime automatics (RA) together with the operational-dispatcher personnel mobilize hot reserve at power plants. Regime parameters are inserted into permissible limits and power consumers disconnected during the emergency are connected. According to the rules [10, 11], it takes 20 minutes to carry out these measures. However, if this time is not enough, the forced regime of power system is introduced and other 20 minutes (totally 40 minutes from the beginning of the emergency) are spent for normalization of the post-emergency regime.

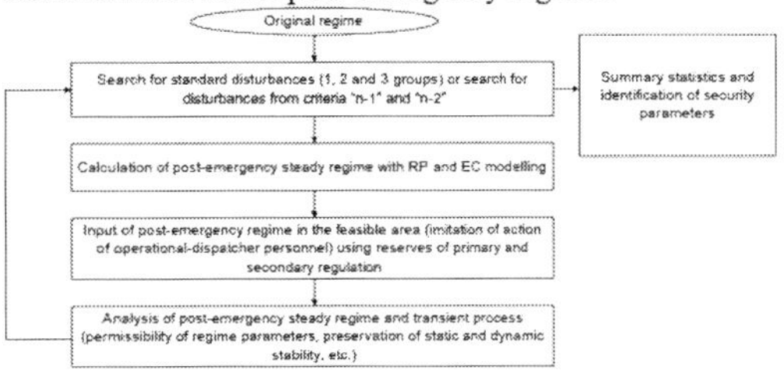

Fig. 2. Algorithm of security analysis during long-term post-emergency state of PS

The authors propose to use as one of the PS security criteria the ability of the PS to recover to normal state after an emergency disturbance for 20 (40) minutes, taking into account the operation of the systems of RS, PA, RA and automatic regulators, and further implementation of operational activities by operational dispatch personnel. It is necessary to check if the electricity supply to consumers of the 2nd and higher categories of reliability comply with the requirements for reliability. If it is provided for particular pre-emergency scheme-regime conditions and particular emergency disturbance, it is considered as a sufficient PS security, i.e. it does not require additional power grid construction for improved reliability. If after 20 (40) minutes the restrictions for consumers remain, the volume and probability undersupply of electrical energy are determined, taking into account the subsequent mobilization of all types of reserves.

The authors propose [12] the algorithm for determining the probabilistic characteristics of electrical energy undersupply by iterating over all emergency disturbances and the integration of the probabilistic volume of undersupply (Fig. 2):

1. Original calculating model is developed. Pre-emergency values of loads and generations in the nodes and state of network elements are determined.
2. Available ranges are specified for regulation of variable regime parameters of static var compensator (SVC), on-load tap-changers (OLTC) of transformers and autotransformers (if automatic or operational tap-change operation is allowed), booster transformers (BT), fixed series compensation (FSC), thyristor-controlled series capacitor (TCSC), as well as generators with active and reactive power. These ranges can be as follows:

• For SVC:
 - variable ranges of shunt of capacitor banks (*Bmin*, *Bmax*);
 - variable ranges of reactive power (*Qmin*, *Qmax*);
 - type of regulation – smooth or discrete (a number of steps or values of changing parameter for each step of control);

• For longitudinal FSC (TCSC) and controlled longitudinal reactors:
 - variable ranges of longitudinal resistance (*Xmin*, *Xmax*);

- type of regulation – smooth or discrete (a number of steps or values of changing parameter for each step of control);

• For OLTC of power and phase-variable transformers (autotransformers):

- variable ranges of transformation coefficient (*Ktr.min, Ktr.max*);

- a number or a step of changing of transformation coefficient;

• For generators:

- permissible values of reactive power (*Qmin, Qmax* or in the form of permissible power);

- permissible values of active power (*Pmin, Pmax* – technological minimum and maximum);

- permissible values of changing of active power for 20- and 40-minute range (*dPmin.20, dPmax.20, dPmin.40, dPmax.40*).

3. Continuous permissible ranges are specified for regime limitations in nodes and branches of the calculating model:

• for voltage (Umin, Umax) and current (*Imin, Imax*).

4. Limitations are specified for capacity in the controlled sections:

• or in the form of values of maximum available load flow / emergency allowable load flow for concrete conditions of the grid elements (a list of switched on or tripped grid elements);

• or in the form of assignment of vector (trajectory) of weighting for calculation of maximum load flow (MLF) and emergency allowable load flow (EALF) of the concrete controlled section.

5. Controllers of EC device effect are specified:

• or with possible maximal volumes of decrease of load and generation;

• or with EC algorithms stating concrete controlling effects for each stage;

• load category is specified, which can be limited.

6. Possible controlling effects are specified for operational-dispatcher personnel (except specified in Item 2) in the following way:

• enumeration of possible maximal volumes of load shedding and generation of nodes with the help of

• operational-dispatcher personnel;

• load category which can be limited.

7. Repair schemes are specified (for a concrete calculation interval: winter/summer etc.):

• or tripping of one grid element (under repair);

• or enumeration of changes of the grid topology (tripping of the equipment under repair, switching on/tripping of the section breakers, etc.), and switching on/tripping of the branches of the calculating model);

• or enumeration of the network topology changes and changes of generation composition (switched on/tripped

• of generators and values of active generation power);

• in some cases, if the calculating model does not switch on distribution systems, besides changes of the grid topology and generation, the load redistribution between the nodes is specified (thus, changes of the distribution grid topology is considered including those of the grids of internal power supply of consumers with several supply centers).

8. Enumeration of emergency disturbances caused by long-term grid tripping is specified. Every disturbance is specified by:

• enumeration of tripped grid elements (branches of the calculating model, e.g. TL with tapping is modelled with several nodes and branches);

• probability of concrete emergency disturbance;

• expected time of recovery.

9. Search for emergency disturbances (Item 8) is specified for the original (normal) scheme and for all specified control schemes (Item 7) by turn:

• pre-emergency regime is assessed and changes are overlaid in the control schemes (Item 7, tripping/switching on of branches and changes of generation and load);

• emergency disturbances are searched by turn (Fig. 2) overlaying tripping specified for a concrete disturbance (Item 8);

• EC operation is modelled during estimation of the steady regime (Item 5, in case of its actuation for a concrete disturbance);

• if the iterative process is non-convergence during estimation of the steady regime, it is necessary to model EC (Item 5, using optimization method and method of directed or complete search) to reach post-emergency steady regime;

• if the obtained steady regime has load limitations (load shedding by EC, Item 5) and/or regime parameters go beyond permissible values (Item 3) or load flow along the controlled sections exceeds EALF (Item 4, specified or determined by the weighting for achieving post-emergency regime), it is necessary to introduce the regime in the allowable area (Items 3 and 4), including the realization of measures available for operational-dispatched personnel (Items 2 and 6) and if possible to recover the load tripped from EC (Item 5);

• if post-emergency regime together with realization of input measures meets all limitations (Items 3 and 4) and if the load tripped from EC but during the input it was possible to recover it, such disturbance does not reduce security;

• if post-emergency regime together with realization of input measures does not meet all limitations (Items 3 and 4), it is necessary to fulfil additional adjustment of control able to be applied by operational-dispatcher personnel (Item 6);

• if post-emergency regime together with realization of input measures meets all limitations (Items 3 and 4) but has non-regenerable load (*dPlim*), it is necessary to calculate electrical energy undersupply (*dWlim*). Duration of electrical energy undersupply is calculated from the expected time of recovery of the tripped emergency element (*Trep*), which can be decreased by time of generation mobilization (*Tmob*), if all permissible values of changes of active power for 40-minute range or together with mobilization of cold reserve (limitations on technological maxima and minima) allow completely or partially recovery of pre-emergency level of load.

10. Integral parameters of electrical power undersupply and damage together with probability are calculated from results of search for all disturbances for all

calculating models. Moreover, category of consumers included in the limitation list is taken into consideration.

11. If several options of the development of power grid are compared, parameters specified in Item 10 are calculated for each option.

12. The duration of electrical power undersupply ($T_{lim.emerg..\ No.\ xxx}$) for a concrete emergency situation is determined from the minimum of mean repair time of the damaged device ($T_{rep.\ emerg.\ No.\ xxx}$) or mean time of mobilization of the generating power ($T_{mob.\ emerg.\ No.\ xxx}$):

$$T_{lim.\ emerg.\ No.\ xxxx} = min(T_{rep.\ emerg.\ No.\ xxx},\ T_{mob.\ emerg.\ No.\ xxx})$$

Therefore, the value of electrical power undersupply ($dW_{lim.\ emerg.\ No.\ xxx}$) for a concrete emergency situation is determined from the value of power limitation ($dP_{lim.\ emerg.\ No.\ xxx}$) and duration of power undersupply ($T_{lim.\ emerg.\ No.\ xxx}$):

$$dW_{lim.\ emerg.\ No.\ xxx} = dP_{lim.\ emerg.\ No.\ xxx} \cdot T_{lim.\ emerg.\ No.\ xxx}$$

The expectation value of the annual monetary loss (ML) from the concrete situation ($ML_{ann.\ emerg.\ No.\ xxx}$) depends on failure flow for a year for concrete damaged equipment during this emergency ($w_{emerg.\ No.\ xxx}$):

$$ML_{ann.\ emerg.\ No.\ xxx} = w_{emerg.\ No.\ xxx} \cdot dW_{lim.\ emerg.\ No.\ xxx} \cdot Y_{0\ RF}$$

$$ML_{ann.\ emerg.\ No.\ xxx} = w_{emerg.\ No.\ xxx} \cdot dW_{lim.\ emerg.\ No.\ xxx} \cdot Y_{0.\ Reg}$$

Where $Y_{0\ RF}$ and $Y_{0\ Reg}$ is the energy intensity of the gross domestic product (GDP) for the country as a whole, and the energy intensity of the gross regional product (GRP) for specific subjects of the Federation. For large or particularly important industrial enterprises, the amount of damage is determined individually, but not below the energy intensity.

The expectation value of the annual loss from all emergencies in the power system associated with the electrical power undersupply will be determined from the arithmetical sum of expectation values of annual losses from concrete emergencies ($ML_{ann.\ emerg.\ No.\ xxx}$):

$$ML_{ann.emerg} = \sum_{i=1}^{n} ML_{ann.emerg.\ №i}$$

Besides the loss of the expected electrical power undersupply, it is necessary to take into account a probable loss from unforeseen tripping of consumers or generation. It is suggested to consider this loss only in those cases and volumes which are stipulated in the agreements (agreements of power supply or technological connection). If these aspects are not stipulated in the agreement, the loss is not considered. The maximum allowed time of electrical power undersupply for concrete consumers should correspond to agreement conditions and category of consumers (according to the decrees of the Russian Federation Government).

2.3. PS steady state calculation with taking into account the discrete and interval characteristics of the regime parameters

The solution of the problem of paragraph 9 is proposed to perform by PS steady state calculation calculating taking into account the discrete and interval characteristics of the regime parameters [13-14].

In general, the balanced steady state parameters as well as their functions should be as close as possible to values $V \rightarrow$ F€, these values are defined as discrete characteristics of the regimes parameters. They can be measurements, planned values of regimes parameters, forecast values, etc. Elements of $v_i \in V$ are characterized by two parameters $\{€_i, d€_i\}$. In the case of measurement, $d€_i$ means the dispersion of $€_i$, in other cases, it may be the permissible deviations of v_i from $€_i$, specified as a percentage.

In addition, the G regimes parameters should be in some intervals of their $G \rightarrow [\underline{G}, \overline{G}]$ values, if possible. Accordingly, each interval variable g_i is characterized by four parameters $\{\underline{g}_j, \underline{\sigma}_j, \overline{g}_j, \overline{\sigma}_j\}$, where $\underline{g}_j$, $\overline{g}_j$ – upper and lower interval limits; $\underline{\sigma}_j$, $\overline{\sigma}_j$ – boundary dispersions. From $\underline{\sigma}_j$, $\overline{\sigma}_j = 0$ it follows that $\underline{g}_j \le g_j \le \overline{g}_j$. The control range of controlled items (see Item 2), security restrictions (see Item 3-4), the volume control actions (see Item 5-6) are specified using these interval limits.

The current load and generation values, as well as various expected (recommended, optimal) values of voltages, generations and loads of nodes, currents and flows through the branches can be set either by discrete parameters of the V, or by interval parameters of G with the values of dispersions (relative deviations) selected by taking into account the ranking of the importance of the recommended constraints or the severity of the control actions. By setting the current load and the active power of the generation by the discrete parameters of V, in the simulation of the emergency perturbation first of all will be found the controlling effects, not related to the change in the load and the generation of active power. And only if it is impossible to bring the regimes into the permissible area, the resource controlling effects will be used to reduce the load and change the generation.

Because of the need to rank the role of variables y_i in the functioning of the system, it is necessary to take them into account repeatedly in the G with different boundary characteristics and percentage rates $(\underline{\sigma}_{ij}, \overline{\sigma}_{ij})$. Accordingly, the wider the range, the lower the percentage rates. The possibility of simultaneous assignments to the same parameter regime of restrictions in the form of discrete values of the expected values of V and several interval ranges G with different values of variance (percentage variance) makes it quite flexible to influence the results of the calculation of the variation of the controlling effects. Using this method, it is possible to perform the analysis of PS security, assessing the control of PS and finding different ways of automatic or operational control allowing PS to move from pre-emergency to acceptable post-emergency mode in the event of emergency disturbances.

To characterize the proximity measure of $V \to V\text{€}$, $G \to \left[\underline{G}, \bar{G}\right]$, the weighted least squares function is commonly used:

$$f = \sum_i^n a_i^2 + \sum_i^m b_i^2 \tag{1}$$

where $a_i = \dfrac{v_i - \text{€}_i}{\delta\text{€}_i} \cdot k_{vi}$;

$$b_i = \frac{g_i - g_i^*}{\sigma_i^*} \cdot k_{gi} \cdot k_i ;$$

The coefficients k_{vi}, k_{gi} take into account the regulating effect of interval variables on the variables v_i и g_i. If there is a regulating effect of some interval variable g_j on another variable, for example, v_i, it exists if the variable g_j does not reach one of the boundaries. Therefore $k_{v_i g_i} = 1 - k_j$.

In the accepted approach, $0 \le d\left(g_i\right) \le 500$ depends on the distance of g_i from the interval boundaries, on the value of f and on the course of the computational process.

$$g_i \le \underline{g}_i \to \left\{k_i = 1;\ g_i^* = \underline{g}_i;\ \sigma_i^* = \underline{\sigma}_i\right\};$$

$$g_i \ge \bar{g}_i \to \left\{k_i = 1;\ g_i^* = \bar{g}_i;\ \sigma_i^* = \bar{\sigma}_i\right\};$$

$$\underline{g}_i < g_i < \bar{g}_i \to \left\{\begin{array}{l} g^*_i = 1;\ g_i^* = g_i;\ \sigma_i^* = \underline{\sigma}_i \cdot \alpha + \bar{\sigma} \cdot (1 - \alpha)\ ; \\ 0 < \alpha < 1; \\ k_i = \dfrac{1}{1 + d\left(g_i\right)} \end{array}\right\};$$

The steady state of PS is usually described by the equations of balance of currents or powers in the grid nodes and in vector form:

$$W\left(X, Y\right) = 0 \tag{2}$$

here, the vector Y is a vector of independent parameters of the regime, which at the stage of solving the formula (2) take specific values. X is the vector of dependent parameters on the vector Y. This implicit dependency is indicated by:

$$X = X\left(Y\right).$$

The solution of the problem is reduced to finding such a vector Y, at which:

$$\min f(Y)$$

and

$$w(X, Y) = 0.$$

We'll solve the problem iteratively. Denote $Y_{k+1} = Y_k + Z$. Linearize f at the point Y_k:

$$a_i\left(Y_{k+1}\right) = a_i\left(Y_k\right) + Z\frac{\partial a_i}{\partial y_k};$$

$$b_i\left(Y_{k+1}\right) = b_i\left(Y_k\right) + Z\frac{\partial b_i}{\partial y_k}.$$

Substituting these expressions in (1), we obtain a quadratic form $f\left(Z\right)$, which has a minimum at $\dfrac{\partial f}{\partial z} = 0$:

$$\frac{\partial f}{\partial z} = A \cdot Z + B = 0 \tag{3}$$

The matrix A is usually non-degenerate, so the system of equations (3) has a solution. The nondegeneracy of the matrix A follows from the fact that all elements of the Y vector belong to V or G, and the elements of Y are independent by definition. Therefore, the rank of A is equal to the rank of Y. The system of equations (3) always has a solution, even if $V = \varnothing$. It is for this reason b_i accounted for on a whole range g_i.

In the next step, we are looking for a solution to the system:

$$W(X_k + dX_k, Y_k + Z \cdot t), t = 0 \tag{4}$$

The solution of the system of equations (4) exists if the following criterion is satisfied:

suppose $\Delta W = W(X_k, Y_{k+1})$,

$$\Delta X = -\Delta W \frac{\partial W\left(X_k, Y_{k+1}\right)^{-1}}{\partial X_k}$$

$$k = 2\sqrt{\left\|\Delta X^{\mathrm{T}} \frac{\partial^2 W}{\partial X^2} \Delta X\right\| / \left\|\Delta W\right\|}\ .$$

For k<1, the solution of the system (4) usually exists. This criterion works well if we are close to the boundary of the limit on the existence of the solution. A more reliable criterion $k < 2/3$. Therefore, if $k > 2/3$ choose $t = \dfrac{2}{3}/k$ and return to the beginning of balancing. It should be noted that the balancing procedure takes 2-3% of the total time of the task.

Given the balancing $\tilde{Z} = Z \cdot t$ define $f\left(Y_k + \tilde{Z} \cdot t\right)$ as a quadratic polynomial of t:

$$f\left(Y_k + \tilde{Z} \cdot t\right) = a_0 + a_1 \cdot t + a_2 \cdot t^2.$$

This polynomial has a minimum at $t = -\dfrac{a_1}{2a_2}$.

If $t < 1$ get a new $\tilde{\tilde{Z}} = \tilde{Z} \cdot t$ and return to the step of balancing steady state.

If $t > 1$, then take $t = 1$ and go to the next step of the solution of the problem.

The calculation process is terminated when the specified accuracy of minimization is achieved:

$$f(Y_k) - f(Y_{k+1}) < \xi .$$

3 Conclusion

This paper presents methods for assessment of system reliability in operation and during planning of power system development. The method proposed contains general description of the step-by-step algorithm with the help of which it is possible to compare options and scenarios of power system development and to assess the level of reliability control for validation of decision making.

The most significant assessment of security is the evaluation of losses during long power limitation of consumers (over 20 and 40 minutes). The algorithm proposed allows the calculation of probable losses for comparative analysis of different options of PS development.

In comparison with the previously published paper [12], in this paper it was possible to adapt the known method of PS steady state calculation [14] by taking into account discrete and interval characteristics of the parameters of the regimes for the analysis of the PS security. Also the wording of the aspects of PS security is refined.

The work was carried out within the project III.17.4.2 (No. AAAA-A17-117030310438-1) of the fundamental research program of the Siberian Branch of the Russian Academy of Sciences.

References

1. N.I. Voropai, D.N. Efimov et al., "Smart Technologies in Emergency Control of Russia's Unified Energy System", Smart Grid, IEEE Transactions on. 2013. Vol. 4, no. 3. P. 1732-1740.
2. N.I. Voropai, A.B. Osak, "Power systems of the future", Energy Policy. No. 5. 2014. P. 60-63. (in Russian)
3. Eds. N.I. Voropai, "Reliability of energy systems (Collection of recommended terms)", Moscow: "Energia", 2007. 192 p. (in Russian)
4. "Reliability concepts in bulk electric power systems", New York: North American Electric Reliability Council, 1985. 96 p.
5. N. Balu, T. Bertram, A, Bose et al., "On-line power system security analysis", Proceedings of the IEEE. 1992, Vol. 80, No. 2. P. 262-280.
6. R.J. Marceau, J. Endrenyi, R. Allan et al., "Power system security assessment: A position paper", Electra. 1997, No. 175. P. 49-77.
7. Ed. N.A. Manov, "Methods and models for studying power system reliability", Syktyvkar: Komi SC UrB RAS. 2010. 292 p. (in Russian)
8. Eds. N.I. Voropai, G.F. Kovalev, "Concept of reliability control in electric power engineering", M.: LLC PH "Energy". 2013. 304 p. (in Russian)
9. "Guidelines for stability of power systems", approved by Order No. 277 of 30.06.2016 of Russian Ministry of Energy. (in Russian)
10. "STO 59012820.29.240.007-2008. Rules for prevention of development and disturbance liquidation of the normal state of electrical annex of power systems". Standard JSC "SO UPS". (in Russian)
11. "STO 59012820.27.010.002-2016. Rules of transition to operation in forced regime in controlled sections of the dispatcher center JSC SO UPS", Standard JSC "SO UPS". (in Russian)
12. A. Osak, D. Panasetsky, E. Buzina, Analysis of the security during power system expansion planning // E3S Web Conf. Volume 25. 2017. Article Number 03004. DOI: 10.1051/e3sconf/20172503004
13. A.B. Osak, A.I. Shalaginov, D.A. Panasetsky, E.Ya. Buzina, "Method for monitoring of PS state and Assessment of regime reliability from criterion of PS control", Methodic problems of studies of reliability of large power systems / Eds. N.I. Voropai, Yu.Ya. Chukreev. – Syktyvkar: LLC "Komi Republican Printing House", 2016, p. 337-346. (in Russian)
14. A.B. Osak, E.Ya. Buzina, "Method and Software for Identifying the Parameters of a Power District". Bulletin of the South Ural State University. Ser. Power Engineering, 2018, vol. 18, no. 1, pp. 53–60. (in Russian)

Adequacy analysis of electric power systems with wind and solar power stations

Dmitriy Karamov[1], *Sergey* Perzhabinsky[1]*

[1]Melentiev Energy Systems Institute of SB RAS, Irkutsk, Russia

Abstract. We developed a new method of adequacy analysis of electric power systems with wind and solar power stations. There are storage batteries in the electric power system. Various types of storage batteries can be used in electric power systems. They are electrochemical, hydroelectric, heat or air storages. The modelling of wind speed and solar radiation is based on software «Local analysis of environmental parameters and solar radiation». The original combination of modern models of meteorological data processing is used in the software. For adequacy analysis of electric power system, we use nonsingle estimation of electricity sacrifice in random hour. Simulation of random values is carried out by the Monte Carlo method.

1 Introduction

Researches of adequacy of electric power systems are presented, for example, [1-5]. The method of adequacy analysis based on estimation of random states of electric power system was suggested in [4]. Simulation of random states in the method is carried out by the Monte Carlo method.

Modern electric power systems contain generators on renewable energy sources. Generation on wind and solar stations has random and oscillated character. Using of storage batteries in the such electric power system allows to stabilize of electricity supply. Appling of traditional method [4] for adequacy analysis of the electric power system is not possible Modification of the method under modern conditions supposes high quality of meteorological data processing, modelling of charging and discharging processes, simulating of random values of solar radiation and wind speed.

Scientific researches of adequacy analysis of electric power systems with wind power generators and different storage batteries are presented in articles [9-19]. However, we concentrate our attention on a research of R. Billinton [13-15] and Y-F. Li [9-11, 16]. Integration of wind power stations and storage batteries in the system along with detailed describing of adequacy analysis is considered in these articles. We also give high estimation of research of Ungjin Oh [12]. Ungjin Oh fully described of charge and discharge processes of storage batteries under solving of the problem of adequacy analysis of electric power system with wind power plants and energy storages.

In spite of high level of presented articles, it is necessary to research of impact of a quality of meteorological data on a system reliability level. In particular, we need to analyze of possibility of an employing of a real meteorological data of FM 12 Synop и METAR to estimate of adequacy of electric power system with power generators on renewable sources and energy storages. In world practice they recommend to use Rayleigh and Weibull distributions for wind speed simulations [20-23].

The aim of our article consists in the development of a new method for adequacy analysis of electric power systems with renewable sources and storage batteries. Various types of storage batteries can be used in electric power systems. They are electrochemical, hydroelectric, heat or air storages. A storage hydroelectric power station is most used as energy storage in large electric power systems. We talk about important things in modelling of work of wind and solar power stations in the first part of our article. The second part of our article contains a description of a modification of the model of electricity shortage estimation. Quadratic losses of electricity under transmission are took into account in the model. Theoretical research of the model [24] speaks about the efficiency of applying of the model under adequacy analysis of electric power systems. We represented results of an experimental research of test scheme of electric power system with wind, solar power generators and storage batteries in the third part of the article. As an object of research we used a real long-term meteorological data from the Russkiy island.

2 Analysis of a long-term meteorological data

One of the most important problem in the modelling of renewable energy sources is a processing of long-term meteorological data. The quality of analyzing of long-term meteorological data impacts on estimation of electricity generation and main parameters of the regime of a wind or solar power station.

* Corresponding author: sergperj@gmail.com

We are going to simulate of wind speed and solar radiation on a base of the earlier developed method for a reproduction of the environment state in the considered location [25]. The given method was represented by software «Local analysis of environmental parameters and solar radiation». Long-term meteorological data of FM 12 Synop and METAR are used in the software.

2.1. The modelling of work of a wind power station

For the computing of generated power on a wind station we used the follow classical formula [22, 23]

$$P_w = \frac{s_w c_w \rho v^3}{2}. \quad (1)$$

where s_w is square of swept surface, m^2, c_w is a power coefficient, p.u., ρ is air density, kg/m^3, v is a wind speed, m/s. The aria of swept surface is computed by formula $s_w = \pi r^2$ for a wind power generator with horizontal axis of rotation where r is radius of wind wheel.

Air density with humidity is found by the follow rule

$$\rho = \frac{\rho_d R_v + \rho_v R_d}{R_d R_v t}. \quad (2)$$

here, ρ_d is partial pressure of dry air, Pa, R_d is a gas constant for dry air (287,058 J/kg·K), t is a temperature of outside air, K, ρ_v is a pressure of vapour, Pa, R_v is a pour constant (461,495 J/kg·K).

Revaluation of a wind speed to a height of wind wheel tower is realized by

$$v_2 = v_1 (h_2 / h_1)^\beta \quad (3)$$

where v_1 is a measured wind speed on a height of 10 m (h_1), v_2 is correct wind speed on a working height h_2, β is a roughness coefficient of an underlying surface.

We analyzed of a wind power plant with horizontal axis of rotation «Enercon E-82 E2» in our research. «Enercon E-82 E2» is often used in littoral and offshore regions of European Union. The height of such wind power plant is equal 80 m and radius of the wind wheel is 40 m.

2.2. The modelling of work of a solar power station

Generated power (W) of a solar panel depends on solar radiation rate and square of working surface. We'll be able to compute generated power by the next formula

$$P_s = s_s c_s I_t. \quad (4)$$

where s_s is square of surface of the solar panel, m^2, c_s is a power coefficient of the solar panel, p.u., I_t is a total solar radiation, W/m^2.

Solar radiation incoming to Earth is inconstant and depends on many parameters. Geographic coordinates, time zone, amount of days of a considered period, the time of the sun's appearance and others are unvarying parameters. Composition of clouds, albedo of Earth's surface, atmospheric mass, thickness of ozone screen, spectral distribution of shine, existence of vapour and aerosol in clouds are variable parameters.

First of all, we find a time of rise, zenith and sunset for every day of a considered period on a selected territory. After that we compute of direct (I_b), diffuse (I_d) and total (I_t) solar radiation. A detailed description of a procedure of solar radiation computing is presented in [25].

Total solar radiation (W/m^2) is defined by the rule

$$I_t = (I_b + I_d) C_r. \quad (5)$$

Here C_r is a coefficient of decreasing of total solar radiation, $C_r = 1 - 0,75(N/8)^{3,4}$.

A value of cloudiness N depends on grade of clouds covering a sky in some moment of time [25].

2.3. The model of electricity shortage estimation

We consider a scheme of electric power system with n nodes and some set of system links. Let N is a given number of modeled hours of electric power system work. An every hour of electric power system work is characterized by a set of random values such as generated electricity $\bar{x}_i^k$ in a node i, electricity demand $\bar{y}_i^k$ in a node i and line capacity $\bar{z}_{ij}^k$ between nodes i and j, $i = 1, \ldots, n$, $j = 1, \ldots, n$, $i \neq j$, $k = 1, \ldots, N$.

It is set minimum $\overline{\Delta x_i}^{\min}$ and maximum $\overline{\Delta x_i}^{\max}$ capacity of a storage battery which located in node i. For k-th hour electricity reserve $\underline{\Delta x}_i^k$ and residual capacity $\overline{\Delta x_i}^k$ of a storage battery are computed on a base of previous computations, $i = 1, \ldots, n$, $k = 1, \ldots, N$.

Let x_i is electricity generated in a node i, Δx_i is change of accumulated electricity in storage battery in a node i, y_i is electricity consumption in a node i, z_{ij} is electricity transmitted from node i to node j, $i = 1, \ldots, n$, $j = 1, \ldots, n$, $i \neq j$.

The problem of electricity shortage estimation in k-th hour, $k = 1, \ldots, N$, is written as:

$$G \sum_{i=1}^{n} y_i + \sum_{i=1}^{n} \Delta x_i \to \max, \quad (6)$$

$$x_i - y_i + \sum_{j=1}^{n} (1 - a_{ji} z_{ji}) z_{ji} - \sum_{j=1}^{n} z_{ij} - \Delta x_i = 0, \quad (7)$$

$$i = 1, \ldots, n, \; i \neq j, \quad (8)$$

$$0 \le y_i \le \bar{y}_i^k, \; i = 1, \ldots, n, \quad (9)$$

$$0 \le x_i \le \bar{x}_i^k, \; i = 1, \ldots, n, \quad (10)$$

 https://doi.org/10.1051/e3sconf/20185802019

$$-\underline{\Delta x}_i^k \le \Delta x_i \le \overline{\Delta x}_i^k,\ i=1,\ldots,n, \quad (11)$$

$$0 \le z_{ij} \le \Xi_{ij}^k,\ i=1,\ldots,n,\ j=1,\ldots,n,\ i \ne j. \quad (12)$$

Here a_{ij} is a nonnegative coefficient of electricity losses under transmission from node i to node j, $i \ne j$, G is a high positive number. The weight coefficient G is necessary for mainstreaming of minimization of electricity shortage. Variables Δx_i can be negative. It means that electricity consumption is bigger than electricity generated in a current hour.

That's way electricity shortage is covered by means of consuming electricity from a storage battery. Let Δx_i^{k-1} is an optimal value of a variable Δx_i in the problem (1) – (6) for $(k-1)-$th hour. Values $\underline{\Delta x}_i^k$ and $\overline{\Delta x}_i^k$, $i=1,\ldots,n$, $k=2,\ldots,N$, are computed recurrently $\underline{\Delta x}_i^k = \underline{\Delta x}_i^{k-1} + \Delta x_i^{k-1}$, $\overline{\Delta x}_i^k = \overline{\Delta x}_i^{\max} - \underline{\Delta x}_i^k$.

We suppose that a random value ζ_i is equally distributed on interval $[\overline{\Delta x}_i^{\min}, \overline{\Delta x}_i^{\max}]$, $i=1,\ldots,n$. Any meaning of ζ can be used as $\underline{\Delta x}_i^0$, $i=1,\ldots,n$.

The described process of charging and consuming of electricity from storage battery is Markov process [15]. We can consider an optimal solution of the problem (6) – (12) as equally probable for every hour. The problem (6) – (12) has a solution because its feasible set is not empty (the point $y_i=0$, $x_i=0$, $\Delta x_i=0$, $z_{ij}=0$, $i=1,\ldots,n$, $j=1,\ldots,n$, $i \ne j$, is feasible) and bounded according to (8) – (12), objective function is linear. Uniqueness of optimal values of y_i, $i=1,\ldots,n$, is followed from theorems [10].

Simulation of random values is realized on their distributions. Reliability indices and procedures of their computing are also discussed in [8].

3 Experimental research

Experimental research of developed method was played out on a test example [8]. The scheme of electric power system consists of 5 nodes и 5 system links (a fig.1).

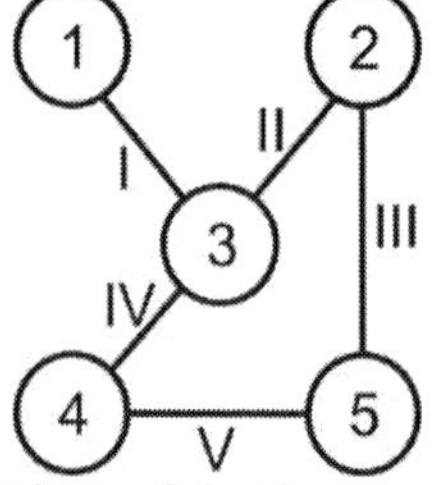

Fig.1. Scheme of electric power system

You'll be able to study of technical characteristics of the electric power system in a table 1. Nodes in the system are linked by electric transmission lines with voltage of 220 kV and a line capacity of 135 MW with an integral coefficient of an accident 0,011 p.u. by 100 km and a coefficient of electricity losses 0,0001661 p.u. Wind, solar power stations and storage batteries are located in the node №1 of the electric power system.

Table 1. Technical characteristics of the electric power system

A node №	Total power in a node, MW	Max; of load, MW	A link №	Quantity of power lines in a link	Length, km
1	600	700	I	2	200
2	1000	700	II	2	100
3	1400	1300	I & II	1	300
4	1000	1000	IV	2	200
5	1200	1100	V	2	300
Total	5200	4900	-	-	-

We have analyzed of meteorological data from the Russkiy island for modelling of wind speed and solar radiation. There are 1892160 observations.

Processing of long-term meteorological data showed that a wind direction is northerly for autumn and winter seasons in the research location. While a wind direction is southerly for spring and summer terms. We used of pressure of air, temperature, humidity for computation of air density. Air density has high meanings for autumn and winter seasons from behind of low temperatures, high atmosphere pressure and low humidity. Opposite meanings of air density are observed for spring and summer seasons. Monthly average wind speed of more 5 m/s on a height 10 m. is observed for autumn and winter seasons.

We presented of monthly average wind speed on a height 80 m. on a fig.2. You'll be able to see from a fig.2 that wind power plant would be worked in nominal regime from September through April.

Results of adequacy analysis of the electric power system with wind power station of 20 MW were obtained in [8]. Air density and storage batteries were not used in computation process of generated electricity. The probability of work without electricity shortage has meaning 0,994920 p.u. in this case. The mathematical expectation of electricity shortage is 0,196116 MWh.

We presented of results of computations of reliability indexes for the system with wind power plants and storage batteries in a table 2. A wind speed and volume of generated electricity were computed by described above formulas.

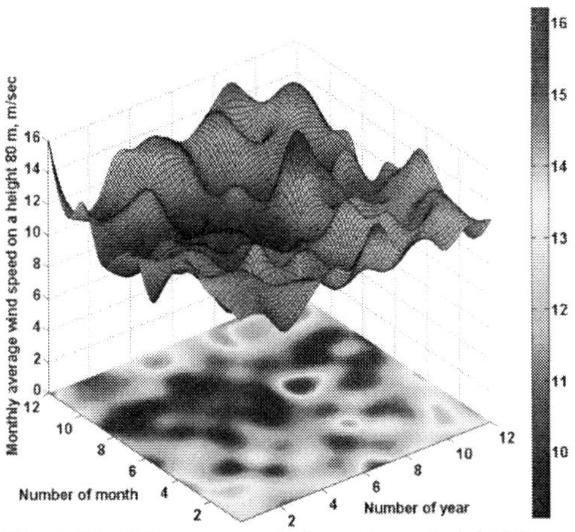

Fig. 2. Monthly average wind speed on a height 80 m., the Russkiy island.

Table 2. Reliability indexes of the electric power system with wind power plants and storage batteries

Structure of electric power system		The probability of work without electricity shortage, p.u.	The mathem; expectation of elec-ty shortage, MWh
Wind power plants, MW	Storage batteries, MWh		
20	-	0,995890	0,135934
	2	0,996120	0,147552
	4	0,996220	0,169415
	6	0,996630	0,123294

Complete accounting of environmental data allowed to obtain more valid reliability indexes. Integration of storage batteries also increased reliability level of the electric power system. Herewith we can place 20 MW of wind power plants and 2 MWh of storage batteries in the electric power system for obtaining of requested reliability level (0,996 for probability of work without electricity shortage). We presented of changing of solar radiation of the Russkiy island for twelve's year observations on a fig.3. You can see that solar radiation is most intensive from April through November. Results of adequacy estimation of electric power system with solar power plants and storage batteries are presented in a table 3.

Table 3. Reliability indexes of the electric power system with solar power plants and storage batteries

Structure of electric power system		The probability of work without electricity shortage, p.u.	The mathem; expectation of elec-ty shortage, MWh
Solar power plants, MW	Storage batteries, MWh		
10	-	0,995002	0,169478
	2	0,995501	0,131489
	10	0,996321	0,106480
	15	0,997403	0,095589

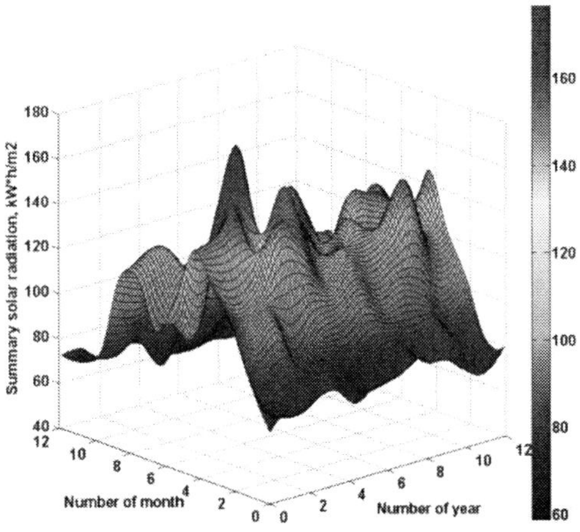

Fig. 3. Changing of solar radiation of the Russkiy island for twelve's year observations

Results of experimental research confirm a fact that for using of solar power plants in electric power system we need bigger storage batteries.

4 Conclusion

The new model of adequacy analysis of electric power system with wind and solar power plants and storage batteries is presented in the article. We use the software «Local analysis of environmental parameters and solar radiation» for detailed modelling of environmental conditions on the considered territory. Complete accounting of environmental data increases of efficiency of modelling of power plants based on renewable sources. Results of our experimental research confirm of correctness of the above statement. Accounting of air density under modelling of wind power plants increases validity of reliability indexes. Integration of storage batteries into electric power systems with wind and solar power plants also increases reliability level and stabilizes of regime indexes. Herewith storage batteries are momentary power reserve. Increase of capacity of storage batteries leads to decrease the mathematical expectation of electricity shortage. Modelling of charging and discharging processes in a storage battery is presented in a view as Markov process. This fact allows to increase computational efficiency of the method.

References

1. G.F. Kovalev, L.M. Lebedeva, Reliability of power systems (Novosibirsk, Nauka, 2015, 224 p.)

2. R. Billinton, R.N. Allan, Reliability Evaluation of Power Systems, Plenum Publishing (New York, 1996)

3. Yu. YA. Chukreev, Model of reliability of electric power systems (Syktyvkar, Nauka, 1995, 176 p.)

4. Y.N. Rudenko, M.B. Chelcov Reliability and redundancy in power systems (Novosibirsk, Nauka, 1974, 263 p.)

5. G.F. Kovalev, L.M. Lebedeva, Complex models of optimizing the design state regimes in assessing the reliability of electric power systems (Irkutsk, ESI SB RAS, 2000, 73 p.)

6. D.S. Krupenev, S.M. Perzhabinskiy, Algorithm for optimizing the balance reliability of electric power systems, J. Proceedings of the Russian Academy of Sciences. Power Engineering, 2, (2014)

7. D.S. Krupenev, S.M. Perzhabinskiy, Applying average dual estimations in algorithm for optimal reliability of electic power system, J. Managing large systems, 54, (2015)

8. D.S. Krupenev, S.M. Perzhabinskiy, Assessment of the reliability of electric power systems with wind power plants, J. Proceedings of the Russian Academy of Sciences. Power Engineering, 2, (2017)

9. Y. Li., E. Zio. Uncertainty analysis of the adequacy assessment model of a distributed generation system // Renewable Energy. – 2012. – V. 41. – P. 235-244.

10. R. Rocchetta., Y.F. Li., E. Zio. Risk assessment and risk-cost optimization of distributed power generation systems considering extreme weather conditions // Reliability Engineering & System Safety. – 2015. – V. 136. – P. 47-61.

11. Y. Li., E. Zio. A multi-state model for the reliability assessment of a distributed generation system via universal generating function // Reliability Engineering & System Safety. – 2012. – V. 106. – P. 28-36.

12. U. Oh, J. Choi, H-h. Kim. Development of Reliability Contribution Function of Power System including Wind Turbine Generators combined with Battery Energy Storage System // The Transactions of The Korean In-stitute of Electrical Engineers. – 2016. – V. 65. – P. 371-381.

13. W. Wangdee, R. Billinton. Reliability assessment of bulk electric systems containing large wind farms // International Journal of Electrical Power & Energy Systems. – 2007. – V. 29. – P. 759-766.

14. K. Xie., R. Bilinton. Energy and reliability benefits of wind energy conversion systems // Renewable Energy. – 2011. – V. 366. – P. 1983-1988.

15. R Billinton, H. Chen, R. Ghajar. Time-series models for reliability evaluation of power systems including wind energy // Reliability evaluation of power systems. – 1996. – V. 36. – P. 1253-1261.

16. R. Ak, Y-F. Li, V. Vitelli, E. Zio. Adequacy assessment of a wind-integrated system using neural network-based interval predictions of wind power generation and load // International Journal of Electrical Power & Energy Systems. – 2018. – V. 98. – P. 213-226.

17. N. Shi, Y, Luo. Energy Storage System Sizing Based on a Reliability Assessment of Power Systems Integrated with Wind Power // Sustainability. – 2017. – V. 9. – P. 395-415.

18. M. N. Scheu, D. Matha, M. Muskulus. Validation of a Markov-based Weather Model for Simulation of O&M for Offshore Wind Farms // The Twenty-second International Offshore and Polar Engineering Conference. – 2012. – P. 1-6.

19. M. N. Scheu, A. Kolios, T. Fischer, F. Brennan. Influence of statistical uncertainty of component reliability estimations on offshore wind farm availability // Reliability Engineering & System Safety. – 2017. – V. 168. – P. 28-39.

20. A. Ilinca, E. McCarthy, J.-L. Chaumel, J.-L. Retiveau. Wind potential assessment of Quebec Province // Renewable Energy. – 2003. – V. 28. – P. 1881-1897.

21. W. Jojola-Talburt, J. Reimann, A. John. Native American Wind Resource Atlas. Office of Indian Energy and Economic Development. Division of Energy and Mineral Development. 2010 Edition. URL: https://www.bia.gov/sites/bia.gov/files/assets/as-ia/ieed/ieed/pdf/idc013229.pdf

22. Nikolaev V.G, Ganaga S.V, Kudryashov Yu.I, National cadaster of Russia wind energy resources and methodology for their determination (Moscow, Atmograph, 2008. 584 p)

23. Nikolaev V.G. Resource and feasibility study of large-scale development of wind power in Russia (Moscow, Atmograph, 2011. 501 p)

24. V.I. Zorkaltsev, S.M. Perzhabinskiy, Model of shortage optimization in electric power system, J. Managing large systems, 30.1, (2010)

25. Karamov D.N. Mathematical modelling of an autonomous power supply system using renewable energy sources. Bulletin of Irkutsk State Technical University, 2015, vol. 104, no. 9, pp. 133–140. In Rus.

26. Karamov D.N. Mathematical modeling of solar radiation based on open access long-term meteorological observation data. Bulletin of the Tomsk Polytechnic University. Geo Assets Engineering, 2017, vol. 328, no. 6. 28–37. In Rus.

27. S.M. Perzhabinsky., D.N. Karamov. Method for analysing the adequacy of electric power systems with wind power plants and energy storages. E3S Web of Conferences, 2017, vol. 25, no. 1. 1–5.

28. Wong L.T., Chow W.K. Solar radiation model // Applied Energy. – 2001. – V. 69. – P. 191–224.

29. Kasten F., Czeplak G. Solar and terrestrial radiation dependent on the amount and type of cloud // Solar Energy. – 1980. – V. 24. – P. 177–189.

Model of adequacy optimization of electric power systems under market conditions

Natalia Aizenberg[1], *Sergey* Perzhabinsky[1]*

Melentiev Energy Systems Institute of SB RAS, Irkutsk, Russia

Abstract. We present a new model of adequacy optimization of electric power systems under market conditions in the article. Optimization is realized by a criteria of maximum of social welfare. Social welfare includes profits of generating companies according penalties of unreliable electricity supply, consumer surplus, costs for development and servicing of electricity grids. Adequacy analysis of variants of development of electric power system is based on multiple estimation of electricity shortage in a random hour of system work. We analyze system work in each of a random hour in two stages. For the first stage we define equilibrium electricity demand in each system node and equilibrium price of electricity according to Cournot model. For the next stage failures of power generating equipment and transmission lines are simulated. We also estimate of electricity shortage in a current hour on the second stage. Reliability indexes and profits of generating companies are formed after a whole cycle of computations. Values of a profit are depended on reliability of electricity supply. Simulation of random values is carried out by the Monte Carlo method.

1 Introduction

Problems of development of electric power systems were traditionally solved on a base of minimization of reduced costs [1, 2]. We should make a note of follow reliability minimization problems such as: a problem of attaining of a given values of reliability indexes with minimal costs [1]; a problem of minimization of reduced costs and the mathematical expectation of financial damage of electricity shortage [2]; a problem of maximization of reliability index under a given level of costs [2]. For solving these problems, they use information about aggregate electricity demand, a structure of generating equipment and configuration of a network of electric power system.

Modern electric power systems are liberalized in many countries. Liberalized energy markets have different features [3] which we must take into account under planning of development of electric power systems. [4]. A type of market interaction is a reason of a choice of behavior criteria of electricity market agents. They are: profit maximization of generating and network companies; maximization of social welfare; maximization of consumer utilities. A result of interaction of agents is market equilibrium with a price and a quantity of electricity supply and demand. This equilibrium is not perfect competitive. It is often oligopolistic. Herewith networks of electric power systems are a monopoly segment of electricity market.

Let's consider review of some adequacy optimization methods for power electric systems in market conditions. A mechanism of forming of payment for generating power was researched in [5]. The payment is a motivation for investment in development of electric power system under conditions of competitive electricity market. The given approach is based on using of reliability index LOLP (Loss of Load Probability) [6]. Cournot oligopoly model [7] is applied as an instrument of adequacy optimization in [8]. According to the model a profit of generating company depends on choosing of a volume of power reserve. But random nature of generation, consumption and transmission capacity of electricity lines are not taken into account in these models. Authors of articles [8, 9] consider scenarios of development of electric power systems and estimate of adequacy level for each scenario.

An aim of our research is to develop of a model of adequacy optimization of electric power system based on imitation of market interaction for each hour of system work and simulation of random values such as electricity demand and failures of electric equipment. We offer sequential search of the best of development variant of electric power system from a set of given variants. Search of the best variants is realized by a criteria of maximum of social welfare. Social welfare includes profits of generating companies according penalties of unreliable electricity supply, consumer surplus, costs for development and servicing of electricity grids. Adequacy analysis of variants of development of electric power system is based on multiple estimation of electricity shortage in a random hour of system work. We analyze system work in each of a random hour in two stages. For the first stage we define equilibrium electricity demand in each system node and equilibrium price of electricity according to

*Corresponding author: sergperj@gmail.com

Cournot model. For the next stage failures of power generating equipment and transmission lines are simulated. We also estimate of electricity shortage in a current hour on the second stage. Reliability indexes and profits of generating companies are formed after a whole cycle of computations.

2 Analysis and optimization of adequacy of electric power system

According to [10] adequacy analysis of electric power system is realized in the next order. First of all, set of random hours of system work is formed. After that electricity shortage is estimated for all simulated hours of system work. Finally, reliability indexes are computed.

Taking account of market conditions under adequacy analysis we have to modificate the given approach. Let's consider that a simulated hour of electric power system work characterizes by set of random functions of electricity demand in system nodes. We assume a low elasticity of electricity demand. It is vindicable because we discuss a long-term period of developing of electric power system. Consequently, we can present a demand function as a linear.

Generating companies maximize their profit with orientation on function of total electricity demand and power capacity of their plants. We suggest to use Cournot model for simulation of described interaction on the electricity market for every hour [7, 11]. A price and quantity of electricity are found as a result of searching of equilibrium on electricity market.

Further we simulate failures of generating and network equipment which determine quantity of generated electricity and capacity of network lines. On this stage it is necessary to estimate of electricity shortage for every hour taking into account equilibrium values of electricity demand, generated electricity in system nodes and real capacity of transmission power lines. Difference between equilibrium and computed values of electricity demand is electricity shortage for the considered hour.

After all computations we define follow reliability indexes such as the probability of electricity shortage, the mathematical expectation of electricity deficit and the mathematical expectation of electricity price. The reliability fee is established on a base on these reliability indexes. We also compute profits of generating companies taking into account possible penalties of low reliability electricity supply.

Optimization of adequacy of electric power system is realized on a base of enumeration and adequacy analyzing of variants of development of electric power system. A variant of system development with highest level of social welfare is optimal. Social welfare is sum of profits of generating companies, consumer surpluses minus network development costs. We suggest that government is social planner which takes himself of network exploitation costs and costs of building of electric power lines.

3 Price setting in the model

Let's consider a scheme of electric power system with n nodes and some set of system links. According to method of adequacy analysis of electric power system it is necessary multiple forming of computed hours of system work. Let N is a given number of modelled hours.

Inverse function of electricity demand is given for k-th hour, $k = 1, \dots, N$, in node i, $i = 1, \dots, n$,

$$p_i^k = \beta_i^k - \alpha_i y_i^k .$$

Here y_i^k is quantity of consumed electricity in a node i for k-th hour, kWh, p_i^k is electricity price in i-th node for one kWh, α_i is a positive coefficient of i-th demand function, β_i^k is meaning of random value β_i with given distribution law, for example, truncated normal law for interval $[\underline{\beta}_i ; \overline{\beta}_i]$, $i = 1, \dots, n$, $k = 1, \dots, N$.

Let's suggest that equilibrium price is equal for all system nodes $p^k = p(y^k)$ where y^k is total amount of consumed electricity for k-th hour

$$y^k = \sum_{i=1}^{n} y_i^k , \; k = 1, \dots, N . \tag{1}$$

We assume that S is amount of generating companies with generating capacities located in different nodes. All generating capacities are selected on competitive market of capacity. Formed price in model includes payment for provided capacity and consumed electricity. Generating company s produces x_{si}^k quantity of electricity in i-th node and x_s^k total amount of electricity along all nodes

$$x_s^k = \sum_{i=1}^{n} x_{si}^k , \; s = 1, \dots, S , \; k = 1, \dots, N .$$

Installed power $\bar{x}_{si}$ is given for every node i. Cost function for s-th generating company depends on volume of generated power $C_s(x_s^k)$, $s = 1, \dots, S$. Profit function of generating company s, $s = 1, \dots, S$, is written as

$$\pi_s(x_s^k, y^k) = p(y^k) x_s^k - C_s(x_s^k) . \tag{2}$$

For searching of equilibrium of Cournot-Nash it is necessary to maximize function (2) for every hour k, $k = 1, \dots, N$, subject to

$$\sum_{i=1}^{n} y_i^k = \sum_{s=1}^{S} \sum_{i=1}^{n} x_{si}^k , \tag{3}$$

$$0 \le x_{si}^k \le \bar{x}_{si}, \; i = 1, \dots, n. \tag{4}$$

Standard method for solving the problem (2) – (4) is Cournot tatonnement method [12]. The main idea of the method consists in sequential maximization of profit of every generating company with fixed values of generated electricity by others companies. Computational process converges if demand functions are linear and cost functions are quadratic [13].

Let's denote by means of $\tilde{x}_{si}^k$, $\tilde{y}_i^k$ equilibrium in the model (2) – (4) in a k-th hour, $s = 1, \dots, S$,

$i=1,\ldots,n$, $k=1,\ldots,N$. From here according to (1) it is defined $\tilde{y}^k$ then price $p(\tilde{y}^k)$ and profits of generating companies $\pi_s(\tilde{x}_s^k, \tilde{y}^k)$ are computed, $s=1,\ldots,S$, $k=1,\ldots,N$.

It is important to note that we don't take into account network constraints in problem (2) – (4). These constraints essentially increase complexity of computation process. Electric power system is considered as a combined node on this stage of computations. Network constraints are taken into account on the second stage of developed approach under searching of electricity shortage in electric power system.

4 The model of electricity shortage estimation

After finding equilibrium in the model (2) – (4) it is necessary to compute optimal distribution of electricity consumption in system nodes for k-th hour, $k=1,\ldots,N$. Herewith we have to simulate meanings of random values of maximal generated electricity $\bar{x}_i^k$ and random values of transmission capacities of power lines $\bar{z}_{ij}^k$ between nodes i and j, $i=1,\ldots,n$, $j=1,\ldots,n$, $i\neq j$.

For estimation of electricity shortage in computed hours we use the follow problem [11]. Generated electricity x_i in node i, kWh, consumed electricity y_i in node i, kWh, transmitted electricity z_{ij} from node i to node j, kWh, are variables in the problem, $i=1,\ldots,n$, $j=1,\ldots,n$. Problem of estimation of electricity shortage in k-th hour, $k=1,\ldots,N$, is written as

$$\sum_{i=1}^{n} y_i \to \max, \tag{5}$$

$$x_i - y_i + \sum_{j=1}^{n}(1-\gamma_{ji} z_{ji}) z_{ji} - \sum_{j=1}^{n} z_{ij} = 0,\ i=1,\ldots,n, \tag{6}$$

$$0 \le y_i \le \tilde{y}_i^k,\ i=1,\ldots,n, \tag{7}$$

$$0 \le x_i \le \bar{x}_i^k,\ i=1,\ldots,n, \tag{8}$$

$$0 \le z_{ij} \le \bar{z}_{ij}^k,\ i=1,\ldots,n,\ j=1,\ldots,n,\ i\neq j. \tag{9}$$

Here γ_{ij} are positive coefficients of electricity losses under transmission from node i to node j, $i\neq j$.

As a rule, the adequacy analysis of electric power system is realized for year. Every hour of work of electric power system is modelled. The failures of generators and power lines are used as random parameters. Time of repair of equipment and fluctuations of load in the year are taken into account under simulating. The rules of simulations of random values such as available capacity of generator, transmission capacities of power lines or load value are discussed in [14].

Let set of $\hat{x}_i^k$, $\hat{y}_i^k$, $\hat{z}_{ij}^k$ is optimal solution of the problem (5) – (9), $k=1,\ldots,N$, $i=1,\ldots,n$, $j=1,\ldots,n$, $i\neq j$. Value of electricity shortage in node i, $i=1,\ldots,n$, according to optimal solution of the problem (5) – (9) is defined by the formula

$$w_i^k = \tilde{y}_i^k - \hat{y}_i^k,\ k=1,\ldots,N.$$

An hour with number k, $k=1,\ldots,N$, is deficit if value

$$w^k = \sum_{i=1}^{n} w_i^k$$

is different from zero. Let's numerate all deficit hours and denote their total amount by means of H. Then the probability of non-shortage work of electric power system is computed by the next formula

$$P = 1 - \frac{H}{N}.$$

The mathematical expectation of electricity shortage in nodes of electric power system is computed by the next rule

$$MW_i = \sum_{j=1}^{H} \frac{w_i^j}{N},\ i=1,\ldots,n.$$

That's way the mathematical expectation of electricity shortage in electric power systems is calculated by

$$MW = \sum_{i=1}^{n} MW_i.$$

The mathematical expectation of price is computed in the same way

$$MP = \frac{1}{N}\sum_{k=1}^{N} p(\tilde{y}^k).$$

The mathematical expectation of maximal price in every system node i, $i=1,\ldots,n$, is found by $M\beta_i = \frac{1}{N}\sum_{k=1}^{N}\beta_i^k$.

5 Modelling of behavior of economic agents taking into account reliability level

As a result of the reliability estimation of every variant of equipment setting in electric power system we find a set of meanings of reliability indices. Willingness of consumers to pay for electricity depends on meanings of reliability indices. Let R is a some reliability index which characterizes fault rate, their duration and depth. In the general case, we can use n-dimensional vector instead of R. Components of the vector are meanings of some reliability index obtained for system nodes. We assume that values of the index R can not be below a certain level $\underline{R}$ and can not be above a level $\overline{R}$. Simulation and analysis of N computed hours for every scenario of development of electric power system allow to estimate volume of satisfied demand $\dot{y}_i$, volume of produced electricity $\dot{x}_i$ in every node and volume of electricity flows $\dot{z}_{ij}$ in the system

$$\dot{y}_i = \frac{1}{N}\sum_{k=1}^{N}\hat{y}_i^k, \ \dot{x}_i = \frac{1}{N}\sum_{k=1}^{N}\hat{x}_i^k, \ \dot{z}_{ij} = \frac{1}{N}\sum_{k=1}^{N}\hat{z}_{ij}^k,$$

$k = 1, \dots, N, \quad i = 1, \dots, n, \quad j = 1, \dots, n, \quad i \neq j$.

Using these indexes, we evaluate of behavior of agents interacting in the electricity market.

Payment for reliability is price which consumer agrees to pay. The payment can be presented in the form of an increasing function $g(R)$. View of function $g(R)$ depends on choice of the reliability index.

5.1 Modelling of electricity demand

The reliability payment is defined either for every type of consumer or every system node. Computation of reliability indexes R_i, $i = 1, \dots, n$, is necessary for every system node in the last case. Price of high reliability electricity supply in node i is presented as function

$$g(R_i) = (R_i)^m MP. \quad (10)$$

Here MP is the mathematical expectation of price of electricity supply without electricity shortages, m is parameter defining of importance of reliability factor for consumer.

5.2 Modelling of generating company behavior

Generating company with number s, $s = 1, \dots, S$, obtains profit depending of reliability indexes R_i, $i = 1, \dots, n$,

$$\pi_s(R) = \sum_{i=1}^{n} g(R_i)\dot{x}_{si} - \sum_{i=1}^{n} C_s(R_i, \dot{x}_{si}). \quad (11)$$

Here R is n-dimensional vector with R_i components, $\dot{x}_{si}$ is the mathematical expectation of generated electricity of s-th generating company in node i, $C_s(R_i, \dot{x}_{si})$ are operating and investment costs of generating company s needed for providing a certain level of reliability R_i and some volume of generated electricity $\dot{x}_{si}$ in a node i, $i = 1,...,n$, $s = 1,...,S$.

Generating company s, $s = 1,...,S$, solves problem of maximization of function (11) subject to constraints

$$\underline{R}_i \leq R_i, \ i = 1,...,n, \quad (12)$$

or

$$R_i \leq \overline{R}_i, \ i = 1,...,n, \quad (13)$$

depending on view of reliability index R_i.

5.3 Modelling of network company behavior

Network component seriously impacts on electricity price. Costs of building of new power lines are given for every variant of development of electric power system. We take into account these costs under computation of social welfare. This fact is explained of usual market structure where network companies are a natural monopoly segment regulated by government. Profit of a such company can be considered zero or normal. Electricity price can be more expensive due to network additive in condition of free market pricing.

5.4 Choice of the best development variant for electric power system

Development variant of electric power system is confessed the best if social welfare is maximal for this one. Therefore, problem of adequacy optimization is written as

$$SW = \sum_{s=1}^{S}\pi_s(R) + CS(\dot{y}, R) - NC \to \max_{R}. \quad (14)$$

Here NC is cost of network development, $CS(\dot{y}, R)$ is consumer surplus which depends on reliability level of electricity supply and satisfied electricity demand in nodes. Consumer surplus is defined as difference between utility of electricity using and the mathematical expectation of electricity price

$$CS(R) = \sum_{i=1}^{n}\int_{0}^{\dot{y}_i}(M\beta_i - \alpha l_i)dl_i - \sum_{i=1}^{n}\dot{y}_i \cdot g(R_i).$$

The problem (14) is solved by enumeration of all possible development variants.

6 Conclusion

We suggest a new approach to adequacy optimization of electric power systems under market conditions supposing of competition of different agents of electricity market. This particularity determines impropriety of implementation of traditional methods for solving the problem.

Adequacy analysis of development variants of electric power system is based on method with using Monte Carlo method for simulation of random values. Analysis of adequacy is realized in two stages. On the first stage of computational process equilibrium values of electricity demand and electricity price are determined according to Cournot model and with taking into account random nature of demand functions. Failures of generating and network equipment are simulated on the second stage. Electricity shortage is also estimated for a current hour on this step. We compute reliability indexes, profits of generating companies taking into account payments for high reliability electricity supply at the end of computational process.

We apply the model with quadratic losses of electricity for estimation of electricity shortage in a computed hour. The given model guaranties of uniqueness of distribution of electricity shortage by system nodes and single-valuedness of reliability indexes.

Further it is necessary to concretize network company's role, realize representative computing experiment and compare different models of consumer behavior.

This work is partially supported Russian Foundation for Basic Research, 16-06-00071a.

References

1. D. Krupenev, S. Perzhabinsky Algorithm of adequacy optimization of electric power system. *Izvestia Rossiskoi Academii Nauk. Energetica*, **2**: 96-106 (2014).
2. D.S. Krupenev, S.M. Perzhabinsky A Reliability Optimization Algorithm with Average Dual Estimates for Electric Power Systems. *Automation and Remote Control*, **78**(12): 2241-2247 (2017)
3. H. Chen, Generating System Reliability Optimization: Thesis of Doctor of Philosophy, University of Saskatchewan (2000)
4. N. Aizenberg, Application of Supply Function Equilibrium Model to Describe the Interaction of Generation Companies in the Electricity Market, *Lecture Notes in Computer Science*, **9869**: 469-479 (2016)
5. M. Assili, M. Hossein, D.B. Javidi, R. Ghazi,Ghazi An improved mechanism for capacity payment based on system dynamics modeling for investment planning in competitive electricity environment, *Energy Policy*, **36**(10): 3703-3713 (2008)
6. R. Billinton, R.N. Allan, Reliability Evaluation of Power Systems, *Plenum Publishing* (New York), 540 (1996)
7. S.V. Podkovalnikov, C.A. Semenov, O.V. Khamisov Development of generating capacities under various structural organization of electric power markets. *Izvestia Rossiskoi Academii Nauk. Energetica*, **4**, 3-14 (2014)
8. A.S. Chuang, F. Wu, P. Varaiya, A game-theoretic model for generation expansion planning: problem formulation and numerical comparisons, *IEEE Trans. on Power Systems*, **16** (4): 885-891 (2001)
9. A. Motamedi, H. Zareipour, M. O. Buygi, W. D. Rosehart, A transmission planning framework considering future generation expansions in electricity markets, *IEEE Trans. on Power Systems*, **25**(4): 1987-1995 (2010)
10. V.I. Zorkaltsev, L. M. Lebedeva, S. M. Perzhabinsky, Model for estimating power shortage in electric power systems with quadratic losses of power in transmission lines, *Numerical Analysis and Applications*, **3**(3): 231-240 (2010)
11. A.A. Vasin, V.V. Morozov Game theory and models of mathematical economics. M.: MAX Press (2005)
12. E. Moulin Theory of games with examples from mathematical economics. Moscow: Mir (1985)
13. R. Baldick Electricity market equilibrium models: The effect of parametrization. *IEEE Transactions on Power Systems*, **17**(4):1170-6 (2002)
14. D.S. Krupenev, S.M. Perzhabinskiy, Assessment of the reliability of electric power systems with wind power plants, J. *Izvestia Rossiskoi Academii Nauk. Energetica*, **2**, 39-47 (2017)

Formation of the Z-form of equations of steady-state modes of energy systems' complex electric networks

Dauren Akhmetbayev*[1], *Arman* Akhmetbayev[2], *Assemgul* Zhantlessova[1]

[1]Power Supply Department, S.Seifullin Kazakh Agrotechnical University, Kazakhstan
[2] Direction of Information Systems of JSC Kaztelecom, Kazakhstan

Abstract. This paper describes the state of the methodological problem of calculating steady-state modes of energy systems' complex electric networks. It also describes the topological method of forming the Z-form of equations of steady-state modes of complex electric networks. The analytical dependence of the node impedance matrix with the matrix of the nodal currents distribution coefficients is established. The matrix of infeed coefficients is determined during the initial data preparation. An analytical approach for determining infeed coefficients topological essence is considered. A simplified method for calculating the driving current distribution coefficients is proposed based on all possible graph trees of a complex electric network. An algorithm for forming infeed coefficients matrix in the environment is developed. A technique for obtaining real solutions of the steady-state mode equations is developed. Steady-state modes direct formation significantly reduces the amount of work performed, increases the visibility of the calculation algorithms performance, and ensures fast and reliable iteration convergence. Increases the level of automation and efficiency of the calculations performed.

1 Introduction

Methodological problems of analysing the electric power system modes are associated with the need to perform a large amount of work on the steady-state modes calculation [1, 2]. In most cases, steady-state modes equations are written in various forms and are nonlinear, which can be solved only by iterative methods [3-5]. Hence, there are many widely used programs for calculating steady-state modes [6-8]. When selecting the most effective of iterative methods, special attention is paid to ensuring and convergence rate of the iterative process.

The conducted studies [9-11] have shown that the difficulties of obtaining real solutions in the steady-state modes calculation can be considerably overcome if we start with inverted form of nodal equations. Complexities in the formation of the inverted form of nodal equations have been overcome to a certain extent by using hybrid matrixes that allow them to combine their advantages by representing them as products of sparsely populated matrixes using various matrix factorization methods.

In this paper, the methodology for the formation of the nodal voltages inverted form is developed based on the network topology using the graphs theory.

2 Rearrangement of the state equation

It is sufficient to have all the infeed coefficients and driving currents coefficients to determine the current distributions in the circuit [12]:

$$\dot{I}=\underline{C}Ч\dot{J}\,, \qquad (1)$$

where $\dot{J}$ - driving currents column matrix; $\underline{C}$- infeed coefficients rectangular matrix.

The values of the currents in the circuit branches remain unchanged if we multiply and divide the right-hand side of equation from the left (1) into the branches resistance matrix, and write in the following form:

$$\dot{I}=\underline{Z}_b^{-1}\underline{Z}_b\underline{C}\dot{J}\,. \qquad (2)$$

On the other hand, the inverted form of the nodal voltages equations has the following form [12]:

$$\dot{U}_\Delta=\underline{Z}\dot{J}\,, \qquad (3)$$

which, taking into account the first Kirchhoff law, can be written in the form:

$$\dot{U}_\Delta=\underline{Z}M\dot{I}=\underline{Z}M\underline{Z}_b^{-1}\underline{Z}_b\underline{C}\dot{J}=\underline{C}^t\underline{Z}_b\underline{C}\dot{J} \qquad (4)$$

The resulting equation (4) is a solution of the equation of nodal voltages and allows us to write the equality in the following form:

$$\underline{Z}=\underline{C}^t\underline{Z}_b\underline{C}=\underline{Y}^{-1}\,. \qquad (5)$$

Thus, the problem of forming an inverted form of nodal equations can be reduced to determining the infeed coefficients matrix.

* Corresponding author: axmetbaev46@mail.ru

3 Infeed coefficients matrix and network topology

The coefficient of the j-node driving current distribution in the i-branch is the ratio defined by the complex number [13]:

$$\underline{C}_{ij}=\frac{\dot{I}_i}{\dot{J}_j} \tag{6}$$

The topology of electric networks makes it possible to establish an analytic relationship between the infeed coefficients and the structure of the straight-line graph [15]. A significant place in the network topology is occupied by such subgraphs as trees and 2-trees. From the values of trees and 2-trees of the graph, numerators and denominators of the driving infeed coefficients of power grids are formed [electric network]. The complexity of searching and singling out the 2-trees of the graph unnecessarily hampered the topological method development in the electric power industry. A new approach for determining the numerators of the infeed coefficients in the function of the network parameters was proposed in [17-19] without searching for and determining the value of the 2-trees of the complex network. Based on all possible trees, specific trees of the oriented graph are formed. The elements of the infeed coefficient matrix are calculated based on the topological expression [17]:

$$\underline{C}_{ij}=\frac{\sum_{k=1}^{M} \underline{F}_{kj}^{i}}{\sum_{k=1}^{N} \underline{F}_{k}} \tag{7}$$

where $\sum_{k=1}^{N} \underline{F}_k$ - the total value of all possible graph trees; $\sum_{k=1}^{M} \underline{F}_{kj}^{i}$ - algebraic sum of the values of the i-branch graph specific trees.

For all infeed coefficients, the denominator is determined once by the total value of the possible graph trees. An algorithm and a program for calculating the infeed coefficients in the Matlab environment have been developed.

4 Formation of Z-form of nodal voltages

Expression (4) allows directly associating the $\dot{\underline{J}}$ driving currents matrix with $\dot{U}_{\Delta}$ voltages matrix at the nodes with respect to the neutral. After replacing the o.c.p. in branches and transverse branches with the corresponding driving currents, the matrix of voltages at the nodes with respect to the neutral is given by [20]:

$$\dot{U}=U_0+\underline{C}^t\underline{Z}_b\underline{C}\overset{*}{U}_d^{-1}\overset{*}{S} \tag{8}$$

where $\underline{C}$ - infeed coefficients complex matrix rectangular; $\underline{Z}_b$ - branches resistance diagonal matrix; $\overset{*}{U}_d$ - nodal conjugate voltages diagonal matrix; $\overset{*}{S}$ - matrix of the column of nodal loads and generators conjugate powers

5 Steady-state mode calculations

The nodal equation (8), after the representation of complex quantities $\dot{S}=P+jQ$; $\underline{C}^t\underline{Z}_b\underline{C}=\underline{Z}$, can be recorded for k-node in the form of [22]:

$$\dot{U}_k=U_0+\sum_{j=1}^{n} \underline{Z}_{kj}\overset{*}{U}_j^{-1}\, P_j-jQ_j \,, \tag{9}$$

where $\dot{U}_k=U_k'+jU_k''$.

After separation into real and imaginary parts, equation (1) has the form of:

$$U_k'=U_0+\sum_{j=1}^{n} Z_{kj}U_j^{-1}\, P_j\cos(\delta_k+\psi_{kj})+Q_k\sin(\delta_k+\psi_{kj}) \tag{10}$$

$$U''_k=\sum_{j=1}^{n} Z_{kj}U_j^{-1}\, P_j\cos(\delta_k+\psi_{kj})-Q_k\sin(\delta_k+\psi_{kj}) \,, \tag{11}$$

where $U_k=\sqrt{(U_k')^2+(U_k'')^2}$ - k -node voltage module; $\delta_k=\mathrm{arctg}\frac{U_k''}{U_k'}$ - k -node voltage phase; $Z_{kj}=\sqrt{(\mathrm{Re}\sum_{j=1}^{m} \underline{C}_{kj}{}^t\underline{Z}_j\underline{C}_{jk})^2+(\mathrm{Im}\sum_{j=1}^{m} \underline{C}_{kj}{}^t\underline{Z}_j\underline{C}_{jk})^2}$ - module of mutual nodal resistance;

$$\psi_{kj}=\mathrm{arctg}\frac{\mathrm{Im}\sum_{j=1}^{m} \underline{C}_{kj}{}^t\underline{Z}_j\underline{C}_{jk}}{\mathrm{Re}\sum_{j=1}^{m} \underline{C}_{kj}{}^t\underline{Z}_j\underline{C}_{jk}}$$

- phase of complex mutual nodal resistance.

The real and imaginary parts of the nodal voltages are determined by iterative methods, which are accompanied by the definitions of the modules and voltage phases of the nodes. Assigning any generator unit with voltage and active power leads to a partial change in the algorithms for calculating the required variables. Suppose that the parameters of the n-th generating unit are assigned with the active power and voltage. Then the required parameters of the mode of the considered node are the phase of the nodal voltage δ_n and the reactive power Q_n. The value of the reactive power of the n-th node is determined based on [18] and the following expression is valid [22] for the k- iteration:

$$Q_n^k=\frac{U_n^2\cos\delta_n^{k-1}-U_0U_n-Z_{nn}P_n\cos(\delta_n^{k-1}+\psi_{nn})-\sum_{j=1}^{n-1} Z_{nj}U_nU_j^{-1(k-1)}\, P_j\cos(\delta_j^{k-1}+\psi_{nj})+Q_j\sin(\delta_j^{k-1}+\psi_{nj})}{Z_{nn}\sin(\delta_n^{k-1}+\psi_{nn})} \tag{12}$$

The imaginary part of the considered node voltage, respectively, is:

$$U_n \sin\delta_n^k = \sum_{j=1}^{n-1} Z_{nj} U_j^{-1(k-1)} \cdot$$

$$\cdot\left(P_j \sin(\delta_j^{k-1}+\psi_{nj})-Q_j\cos(\delta_j^{k-1}+\psi_{nj})\right)+ \quad (13)$$

$$+Z_{nn} U_n^{-1}\ P_n \sin(\delta_n^{k-1}+\psi_{nn})-Q_n^k \cos(\delta_n^{k-1}+\psi_{nn})$$

Hence, the value of the nodal voltage phase for the k-iteration has the form of:

$$\delta_n^k=\arcsin\frac{U_n^{''(k)}}{U_n}, \quad (14)$$

The essence of the iteration consists in correcting the nodal currents. The line's charging capacities, the corona loss, as well as power transformers no-load losses are taken into account by the corresponding currents, depending on the voltage applied to them, which are refined during the iteration.

Algorithms for the steady-state mode formation are realized in the MATLAB environment. All possible graph trees are defined based on the theory of structural numbers, and specific trees based on algorithm specially developed for this purpose. Block diagram of the algorithms implementation is shown in Figure 1.

In order to provide clarity of the above material, consider a typical example of a 110 kV network (Fig. 2), studied for training purposes in [23], where two generator nodes "0" and "1" operate on a common load in a ring scheme with the following parameters: Z_1=10+j20 Ohm; Z_2=10+j25 Ohm; Z_3=15+j30 Ohm provided that the basic node voltage is U_0= 115 kV and the corresponding power of S_1=28,8675+j17,3205 MVA, S_2=-46,1880-j23,0940 MVA.

The total values of the possible graph trees (Fig. 2b) and the specific trees for the first element of the infeed matrix are determined [17, 19]:

$$\Sigma \underline{F}=\underline{Y}_1\underline{Y}_2+\underline{Y}_1\underline{Y}_3+\underline{Y}_2\underline{Y}_3;\ \underline{F}_{11}=-(\underline{Y}_1\underline{Y}_2+\underline{Y}_1\underline{Y}_3).$$

The values of infeed coefficient matrix calculated by the program are presented in the form:

$$\underline{C}=\begin{Vmatrix} -0.7299-0.0073i & -0.4051+0.0109i \\ 0.2701-0.0073i & -0.4051+0.0109i \\ -0.2701+0.0073i & -0.5949-0.0109i \end{Vmatrix},$$

on which the following generalized network parameters, necessary for calculations are determined:

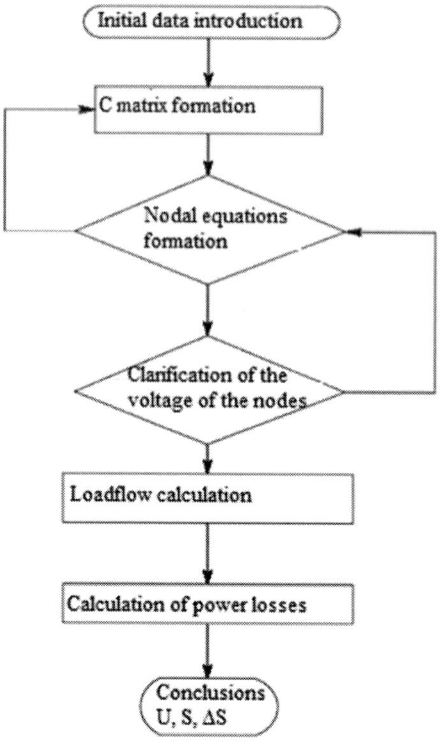

Fig. 1. Steady-state mode formation and calculation block diagram

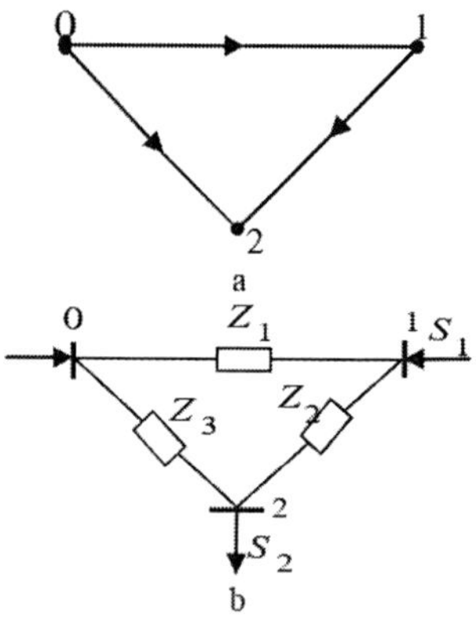

Fig.2 a) Network design scheme; b) Oriented network graph

$$Z_{11} = \sum_{j=1}^{3} \underline{C}^t_{1j}\underline{Z}_j\underline{C}_{j1} = \underline{C}^t_{11}\underline{Z}_1\underline{C}_{11} + \underline{C}^t_{12}\underline{Z}_2\underline{C}_{21} +$$

$$+\underline{C}^t_{13}\underline{Z}_3\underline{C}_{31} = 16,3223; \varphi_{11} = 64,0079;$$

$$Z_{12}=\underline{Z}_{21}=\sum_{j=1}^{3} \underline{C}^t_{1j}\underline{Z}_j\underline{C}_{j2}=\underline{C}^t_{11}\underline{Z}_1\underline{C}_{12}+\underline{C}^t_{12}\underline{Z}_2\underline{C}_{22}+$$

$$+\underline{C}^t_{13}\underline{Z}_3\underline{C}_{32}=9,0618; \varphi_{12}=61,9;$$

$$Z_{22} = \sum_{j=1}^{3} \underline{C}^t_{j2}\underline{Z}_j\underline{C}_{j2} = \underline{C}^t_{21}\underline{Z}_1\underline{C}_{12} + \underline{C}^t_{22}\underline{Z}_2\underline{C}_{22} +$$

$$+\underline{C}^t_{23}\underline{Z}_3\underline{C}_{32} = 20,0292; \varphi_{22} = 64,49.$$

Based on equations (10), (11) and (12), (13), calculations for a given network were made twice, when the first node is assigned with active and reactive powers, and when the same node is assigned with voltage and active power. The calculations are performed by a simple iteration method with accuracy and the results of which are given in Tables 1, 2.

Table 1. Results of the calculation of the network mode in Fig. 6a, when the nodes are assigned with active and reactive powers

Iteration index	0	1	2	3	4	5
U_1', kV	-	115,7126	115,4016	115,4105	115,4096	115,4096
U_1'', kV	-	0,2703	0,2847	0,2777	0,2783	0,2782
U_1, kV	110	115,7129	115,4020	115,4108	115,4100	115,4100
δ_1, grade	0	0,1338	0,1414	0,1379	0,1382	0,1381
U_2', kV	-	109,8411	109,6159	109,6140	109,6123	109,6123
U_2'', kV	-	-4,4186	-4,1153	-4,1479	-4,1456	-4,1459
U_2, kV	110	109,9300	109,6931	109,6924	109,6907	109,6907
δ_2, grade	0	-2,3036	-2,1500	-2,16071	-2,1660	-2,1661

Calculations results show that the developed algorithm of calculation, at assigning the node with voltage and active power provides convergence of the established mode at the fifth step of iteration process, at accuracy of $10^{-5}U_i$. It is known from [23] that the Newton method ensures the convergence of the steady-state mode at the accuracy of $10^{-3}U_i$, at the fifth step of the iteration, whereas the simple iteration on the proposed algorithm, the steady-state mode of the same electric network, also converges at the fifth iteration, but at calculation with accuracy of $10^{-5}U_i$.

Table 2. The results of the calculation of the network mode in Fig. 6a, with the generator node assigned with voltage U_1 = 115,4100 kV and the active power P_1 = 28,8675 MW.

Iteration index	0	1	2	3	4	5
Q_1, MVAр	-	16,4542	17,2586	17,2480	17,2804	17,2805
U_1'', kV	-	0,1903	0,2848	0,2745	0,2749	0,2749
δ_1, grade	0	0,0945	0,1414	0,1363	0,1365	0,1365
U_2', kV	-	109,8151	109,6101	109,6416	109,6405	109,6406
U_2'', kV	-	-4,3641	-4,0988	-4,1269	-4,1249	-4,1251
U_2, kV	110	109,9018	109,7167	109,7192	109,7181	109,7181
δ_2, grade	0	-2,2758	-2,1410	-2,1556	-2,1546	-2,1547

This reveals the advantage of the proposed method. To estimate the accuracy, the results of the fifth iteration of the steady-state mode from [23] is presented: U_1=115,4150 kV, δ_1=0,1352 grade., U_2=109,7210 kV., δ_2=-2,1553 grade, which, as can be seen from Table 3, differ within the rounding error. Therefore, the proposed method can be used both in scientific studies and in the performance of operational calculations of steady-state modes of electric power systems.

6 Conclusions

1. A topological algorithm for the search and calculation of specific trees for a directed graph of a complex electric network was developed.
2. A topological method for the formation of the complex network steady-state mode in the MathLab environment was developed.
3. Calculations of modes of complex networks do not experience any difficulties when changing the form of the task of generating units.
4. The proposed method increases the degree of automation of the ADMS tasks.

References

1. Venikov V.A. Modeling of large systems of electric power industry. Electricity, No. 5, 64-67 (1978)
2. Venikov VA, Golovitsyn BI, Liseev MS, Unarokov AA Calculation of the EPS regime based on telemetry data based on the regularization method // Izvestiya / Energy and transport., No. 2, 39-49 (1976)
3. Barinov VA, Sovalov SA Regimes of power systems: Methods of analysis and control, Moscow: Energoatomizdat, 440 (1990)
4. Stroyev VA, Rokotyan IS Methods of mathematical optimization in power supply problems, M.: MEI, 72 (1993)

5. Khachatryan V.S. Method and algorithm for calculating the steady-state regimes of large power systems. Izvestiya / Energy and transport., No. 4, 45-57. (1973)
6. Geraskin O.T. Diakoptics and sparsity of tasks and calculations of large electric energy systems. News of higher educational institutions of the USSR. Power engineering. No.1. 7-12 (1982).
7. Idelchik V.I. Calculations and optimization of modes of electric grid of power systems, Moscow. Energia, 190 (1977)
8. Gursky S.K. Algorithmization of problems of control of modes of complex systems in the electric power industry. Minsk, Science and Technology, 367 (1977)
9 . A computer in energy problems. TIER, Moscow, Mir. T.62, 239 (1974)
10. Smirnov K.A. On the convergence of iterative solutions of the steady-state regime of the electric power system for given power of nodes // Izvestiya / Energy and transport., No. 3, 120-129 (1975)
11. Manusov BZ, Lykin AV, Sidorkin Yu.M. Algorithms of the Newton-Raphson method for solving nodal equations in inverted form // Izvestiya / Energy and transport., No. 9, 3-7 (1974)
12. H.F. Fazylov, T.X. Nasyrov. Established regimes of electric power systems and their optimization. Lightning. 370 (1999)
13. NA Melnikov. A mathematical method for analyzing electrical circuits. 232 (1972)
14. Melnikov N.A. The application of the distribution coefficients in the calculation of complex circuits for the replacement of electrical circuits. Electricity, No. 2, 9-13 (1961)
15. Geraskin O.T. Application of the theory of signal graphs to the calculation of the current distribution coefficients in electrical networks. Izvestiya / Energy and transport., No. 5, 33-39 (1965)
16. Geraskin O.T. Methods for the determination of 2 - graph trees in the topology of electrical networks. Izvestiya / Energy and transport., No. 5, 106-112 (1967)
17. Dauren S. Akhmetbayev, Daurenbek A. Aubakir, Yermek Zh. Sarsikeyev, Bakhtybek A. Bainiyazov, Mikhail A. Surkov, Vitaliy I. Rozhkov, Gulbahit N. Ansabekova, Assel S. Yerbolova, Azamat T. Suleimenov, Miramgul S. Tokasheva. Development Of Topological Method For Calculating Current Distribution Coefficients In Complex Power Networks, Results in Physics, 7, 1644-1649 (2017)
18. Akhmetbayev D. Method of calculating the steady-state modes of electrical networks based on the current distribution coefficients. Electricity, No. 11, 23-27, (2010)
19. Dauren Akhmetbayev, Arman Akhmetbayev, Azamat Suleymenov, Michal Kolcun. Modeling the Set Mode of Complex Power Grid, Based on Infeed Coefficients. Proceedings of the 9th International Scientific Symposium, StaráLesná, SlovakRepublic, ELEKTROENERGETIKA, 12.-14, 9 (2017)
20. Akhmetbayev D., Akhmetbayev A., Aidarova A. Determination of the rational transformations distribution networks distribution. E3S Web of conferences 25.04003 (2018).
21. Arman Akhmetbayev, Dauren Akhmetbayev, Serik Zhumazhanov, Bauyrzhan Zhakishev. New Modeling of Steady-State Modes of Complex Electrical Grids of Power Systems MATEC Web of Conferences 155, 01043 (2018).
22. Akhmetbayev D., Akhmetbaev A., Berdygozhin A.S. Topological method for the formation of an inverted form of the node equations of electric energy systems. Electricity, No. 5, (2018)
23. Yu.N. Astakhov, V.A. Venikov, V.V. Ezhov, N.I. Zelenokhat, E.N. Zuev, V.Idelchik. I.V. Litkens, S.V. Nadezhdin, E.V. Putyatin, Yu.P. Ryzhov, V.A. Stroev. Electric power systems in examples and illustrations. Edited by V.A. Venikova, Energo atomizdat, 504 (1983)

Calculations of the complex network's steady-state modes by brining it to equivalent open

Dauren Akhmetbayev[1], *Assemgul* Zhantlessova[*1] , *Arman* Akhmetbayev[2]

[1]Power Supply Department, S.Seifullin Kazakh Agrotechnical University, Kazakhstan
[2]Direction of Information Systems of JSC Kaztelecom, Kazakhstan

Abstract. The paper considers the development of the idea of diakoptics as applied to the calculation of the steady-state modes of energy system's complex electrical networks. The well-known goal of diakoptics is to obtain the equations of state for the dedicated part of the system, the study of which is much simpler than the study of the initial system and can be achieved by improving its steady state equations. Technique for dividing a complex-closed system into a set of uncomplicated subsystems was developed based on the inverse form of nodal equations. Analytic expressions for the intersection circuits obtained based on identical equality of voltage at the nodes of the system division into subsystems are proposed. Using the example of 110kV network calculation, the technique for determining the matrixes of generalized parameters of the dedicated subsystems, the sizes of which depend on the number of their broken link is shown. Analytical determination of the equality condition of the voltage of subsystems intersections nodes, allowed analyzing a complex closed network by bringing it to an equivalent open.

1 Introduction

The essence of the method of diakoptics is to solve tasks with a large number of variables in parts, by which properties the basic characteristics of the original system get restored.

Each subsystem, determined using the unmating method is analysed and calculated separately, then a solution for the entire system is determined based on tensors of the partial solutions joints [1].

As a result of the initial system unmating, an additional system- circuits of intersections is formed. All the parameters arising because of separation are collected in this system. Through the circuit of intersections, a link is established between old and new parameters using the coupling equations. Moreover, each subsystem, which research is much simpler than the research of the original system, is analysed separately, then the particular solutions are joined step by step until a solution of the original system is obtained [2, 3].

In some works [4, 5], in the process of iterations, in addition to the matrixes of generalized parameters for individual subsystems, there are also matrixes the order of which determined by the number of boundary nodes of the entire set of subsystems can be very high. Standard methods of forming matrixes of generalized parameters that cannot be determined directly by the scheme are practically unacceptable because of the excessively large volume of computations [6]. Applying the properties of matrix math, for example, dividing matrixes into blocks can significantly reduce the amount of the computational process. In this case, the inversion of a single matrix of large order is replaced by the inversion of several matrixes of lower order, which leads to a significant reduction in the number of computational operations. However, the recalculation of the matrixes of the generalized parameters of the subsystems, corresponding to the restoration of broken links remains complicated. These shortcomings were overcome by the development of the method of diakoptics by a number of authors [7, 8].

For the first time, it was suggested that the schemes be cut by nodes, and not by branches, preserving all branches of the scheme. At the same time subsystems were allocated in such a way that they were connected in series, forming a highway or radial circuit connected to the slack node.

This approach made it possible to form subsystems of approximately similar sizes, by increasing the necessary number of sub graphs of the initial system graph. At each step of the computational process, equivalent matrixes of generalized system parameters are used. They have a low order, corresponding to the sizes of individual subsystems, and the matrix is calculated only once at the stage of formation of conditionally constant information. One of the main advantages of the above approach is the provision of a favourable convergence condition for the computational process, since the links between the subsystems are not violated at any stage of this process [9-11].

The approach developed below is fundamentally different in that the network nodal voltages determined, as a result of individual calculation of individual subsystems, will be as they would be at the boundaries

* Corresponding author: acbizh@mail.ru

of broken links without dividing into parts of the original system. Such an approach can be implemented on the basis of the inverse form of the nodal voltage equations.

2 Inverse form of the nodal voltage equations

By transforming the known equations of state of arbitrary complexity electrical networks, we obtain a matrix equation for the nodal voltage drops with respect to the matrix of driving currents distribution coefficients, which has the following form [11-13].

$$\dot{U}_D = \underline{C}^t \underline{Z}_b \underline{C}\dot{J}, \qquad (1)$$

where $\underline{C}$ - rectangular matrix of a complex of current distribution coefficients; $\underline{Z}_b$ - diagonal matrix of the complex of scheme branches resistances;

$\dot{J}$ - column matrix of the complex of driving currents; t - matrix transposition sign.

The matrix of current distribution coefficients is determined based on all trees of the directed graph of a complex electrical network, and is used as initial data for calculating its steady-state modes [14-16].

3 Conditions for the equality of the subsystems intersections voltage nodes

Complex-closed circuit can be reduced to an open form by cutting the nodes. The choice of the node-cutting place can be arbitrary, the main thing is that all contours are destroyed and the resulting open circuits remain connected to the basic node. The advantage of the proposed approach is that to obtain open networks it is necessary to break considerably fewer links, since cutting one node can replace the cutting of several branches incident to this node.

The conditions for the equivalence of the response of the original complex scheme and the open sub schemes at the nodes of their intersections can be established based on the multiport network shown in Fig. 1.

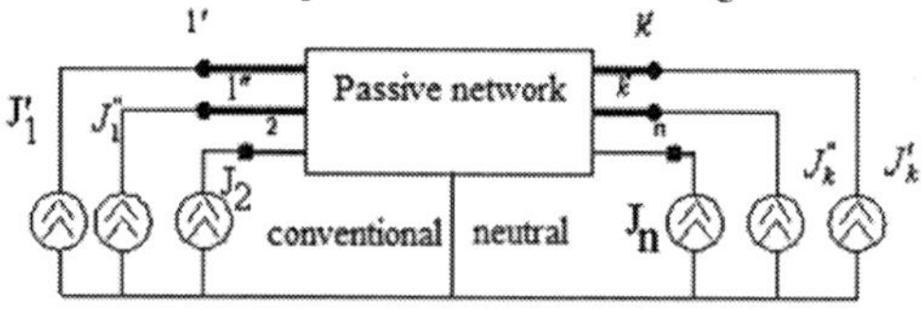

Fig. 1. Multiport network.

Suppose that, as a result of cutting the first k-nodes, a branched open circuit is obtained, where each cut point is replaced by two fictious-points provided with additional indexes (' and "). Thus, the original complex system is divided into two subsystems consisting of an open circuits' scheme.

The relationship between the initial and new disturbances at the intersections nodes of the subsystems is established based on the first Kirchhoff law, which will be written in the matrix format in the form:

$$\dot{J}' + \dot{J}'' = \dot{J}_p. \qquad (2)$$

New circuit responses at the subsystems intersections nodes are determined based on position principle, in the form of a sum of subsystems responses caused by new disturbances applied to the subsystems intersections nodes and disturbances of the subsystem applied to uncut nodes, which are calculated on the basis of matrix equations:

$$\dot{U}'_\Delta = \left\|\underline{C}_p{}^t \underline{Z}_b \underline{C}_p\right\|' \cdot \dot{J}' + \left\|\underline{C}_p{}^t \underline{Z}_b \underline{C}_n\right\|' \cdot \dot{J}_n, \qquad (3)$$

$$\dot{U}''_\Delta = \left\|\underline{C}_p{}^t \underline{Z}_b \underline{C}_p\right\|'' \dot{J}'' + \left\|\underline{C}_p{}^t \underline{Z}_b \underline{C}_n\right\|'' \dot{J}_n, \qquad (4)$$

where $\dot{J}', \dot{J}''$ - column matrixes of fictitious currents applied to subsystems intersections nodes;

$\dot{J}'_n, \dot{J}''_n$ - column matrixes of driving currents, uncut nodes of subsystems; $\underline{C}'_p, \underline{C}''_p$ - matrixes of distribution coefficients of fictitious nodal currents of subsystems; $\underline{C}'_n, \underline{C}''_n$ - matrixes of current distribution coefficients of the subsystems uncut nodes; $\underline{Z}'_b, \underline{Z}''_b$ - diagonal matrix of the branches of the corresponding subsystems; $\left\|\underline{C}_p{}^t \underline{Z}_b \underline{C}_p\right\|', \left\|\underline{C}_p{}^t \underline{Z}_b \underline{C}_p\right\|''$ - matrixes of the system functions of resistance with respect to the subsystems intersections nodes; $\left\|\underline{C}_p{}^t \underline{Z}_b \underline{C}_n\right\|', \left\|\underline{C}_p{}^t \underline{Z}_b \underline{C}_n\right\|''$ - matrixes of system resistance functions with respect to the subsystems uncut nodes.

Response of the circuit of open subsystems intersections will be equivalent to the response of the initial system circuit at the equality of voltage vectors at the subsystems intersections nodes obtained as a result of traversing different branches:

$$\dot{U}'_\Delta = \dot{U}''_\Delta, \qquad (5)$$

equating the right-hand parts (3), (4):

$$\left\|\underline{C}_p{}^t \underline{Z}_b \underline{C}_H\right\|'' \dot{J}''_n - \left\|\underline{C}_p{}^t \underline{Z}_b \underline{C}_n\right\|' \dot{J}'_n = $$
$$= \left\|\underline{C}_p{}^t \underline{Z}_b \underline{C}_p\right\|' \dot{J}' - \left\|\underline{C}_p{}^t \underline{Z}_b \underline{C}_p\right\|'' \dot{J}''$$

or

$$\left\|\underline{C}_p{}^t \underline{Z}_b \underline{C}_p\right\|' \dot{J}' - \left\|\underline{C}_p{}^t \underline{Z}_b \underline{C}_p\right\|'' \dot{J}'' = \dot{U}^{''-'}_{\Delta n}, \qquad (6)$$

where $\dot{U}^{''-'}_{\Delta n} = \dot{U}''_{\Delta n} - \dot{U}'_{\Delta n}$ - algebraic difference of nodal voltages drops caused by currents of uncut nodes of subsystems; $\dot{U}'_{\Delta n} = \left\|\underline{C}_p{}^t \underline{Z}_b \underline{C}_H\right\|' \cdot \dot{J}'_n$ - nodal voltage drops caused by currents of uncut nodes of the subsystem with the index ('); $\dot{U}''_{\Delta n} = \left\|\underline{C}_p{}^t \underline{Z}_b \underline{C}_n\right\|'' \dot{J}''_n$ - the same, the subsystem with the index (").

 https://doi.org/10.1051/e3sconf/20185802022

Determination of the equivalence conditions for the responses of the initial system and the systems obtained as a result of cutting the nodes reduces to determining new disturbances that satisfy the equalities of the responses of the subsystems relative to the nodes of their intersections. For this purpose, we make up a system of matrix equations (2) and (6):

$$\left.\begin{array}{l} \dot{J}'+\dot{J}''=\dot{J}_p \\ \left\|C_p{}^t Z_b C_p\right\|' \dot{J}' - \left\|C_p{}^t Z_b C_p\right\|'' \dot{J}'' = \dot{U}_{\Delta n}^{''-'} \end{array}\right\}, \qquad (7)$$

where the required parameters are new distributions, $\dot{J}',\dot{J}''$ applied to the subsystems intersections nodes.
The system (7) can be written in the form of a matrix equation:

$$\left\|\begin{array}{cc} \dot{E} & \dot{E} \\ (C_p{}^t Z_b C_p)' & -(C_p{}^t Z_b C_p)'' \end{array}\right\| \cdot \left\|\begin{array}{c} \dot{J}' \\ \dot{J}'' \end{array}\right\| = \left\|\begin{array}{c} \dot{J}_p \\ \dot{U}_{\Delta n}^{''-'} \end{array}\right\|, \qquad (8)$$

where $\dot{E}$ is the unity diagonal matrix;
$\left\|\begin{array}{cc} \dot{E} & \dot{E} \\ (C_p{}^t Z_b C_p)' & -(C_p{}^t Z_b C_p)'' \end{array}\right\|$ - a square, nonsingular matrix of the size that is equal to the number of fictitious nodes of intersections circuit.
Then the solution (8) is written as:

$$\left\|\begin{array}{c} \dot{J}' \\ \dot{J}'' \end{array}\right\| = \left\|\begin{array}{cc} \dot{E} & \dot{E} \\ (C_p{}^t Z_b C_p)' & -(C_p{}^t Z_b C_p)'' \end{array}\right\|^{-1} \cdot \left\|\begin{array}{c} \dot{J}_p \\ \dot{U}_{\Delta n}^{''-'} \end{array}\right\|. \qquad (9)$$

If we use the properties of partitioning the matrixes into blocks, then the solution of system (6) can be substantially simplified by eliminating $\underline{J}'$ variables. After simple transformations, the system (7) reduces to the form:

$$\underline{Z}\dot{J}'' = \dot{U}_{\Delta\Sigma}, \qquad (10)$$

where
$\underline{Z} = -(\left\|C_p{}^t Z_b C_p\right\|'' + \left\|C_p{}^t Z_b C_p\right\|')$ - square, nonsingular matrix of the size that is equal to the number of nodes of the original system intersections circuit;
$\dot{U}_{\Delta\Sigma} = (\dot{U}_{\Delta n}^{''-'} - \left\|C_p{}^t Z_b C_p\right\|' \dot{J}_p)$ - column matrix of fictitious nodal voltage drops.
The solution of the matrix equation (10)

$$\dot{J}'' = \underline{Z}^{-1}\dot{U}_{\Delta\Sigma}, \qquad (11)$$

is simpler relatively to $\underline{J}''$ than the solution of the matrix equation (9).

To solve (11), we need to calculate the inverse matrix $\underline{Z}^{-1}$, the size of which is equal to the number of cut nodes, which requires less labor than determining the inverse matrix in expression (9).
The values of the remaining desirable variables are determined by the formula:

$$\dot{J}' = \dot{J} - \dot{J}'', \qquad (12)$$

After determining the fictitious currents of all the subsystems intersections nodes, we can investigate the behavior of the initial system based on the modes of the dedicated subsystems.

In the general case, the source scheme can be divided into n subsystems. Then, to determine fictitious currents at the nodes of their intersections, the following expression is true:

$$\left\|\begin{array}{c} \dot{J}' \\ \dot{J}'' \\ \dot{J}''' \\ \dots \\ \dot{J}^{n} \end{array}\right\| = \left\|\begin{array}{ccccc} \dot{E} & \dot{E} & \dot{E} & \dots & \dot{E} \\ (C_p^t Z_b C_p)' & -(C_p^t Z_b C_p)'' & 0 & \dots & 0 \\ (C_p^t Z_b C_p)' & 0 & -(C_p^t Z_b C_p)''' & \dots & 0 \\ - & - & - & - & - \\ (C_p^t Z_b C_p)' & 0 & 0 & \dots & -(C_p^t Z_b C_p)^{n} \end{array}\right\|^{-1} \cdot \left\|\begin{array}{c} \dot{J}_p \\ \dot{U}_{\Delta n}^{''-'} \\ \dots \\ \dots \\ \dot{U}_{\Delta n}^{n-1} \end{array}\right\|, \qquad (13)$$

where
$\dot{U}_{\Delta n}^{n-1} = \dot{U}_{\Delta n}^{n} - \dot{U}'_{\Delta n} = \left\|C_p^t Z_b C_n\right\|^{n} \cdot \dot{J}_n^{n} - \left\|C_p^t Z_b C_n\right\|' \cdot \dot{J}'_n$ - algebraic difference of nodal voltage drops caused by the currents of uncut nodes of the corresponding subsystems.

4 Calculation of a complex network by bringing it to equivalent open

As an example, we will consider the design diagram of the 110 kV electric network shown in the figure 2. с задающими токами: $\dot{J}_1$=0.16+0,05i; $\dot{J}_2$=0.27+0,05i; $\dot{J}_3$=6+4,2i.

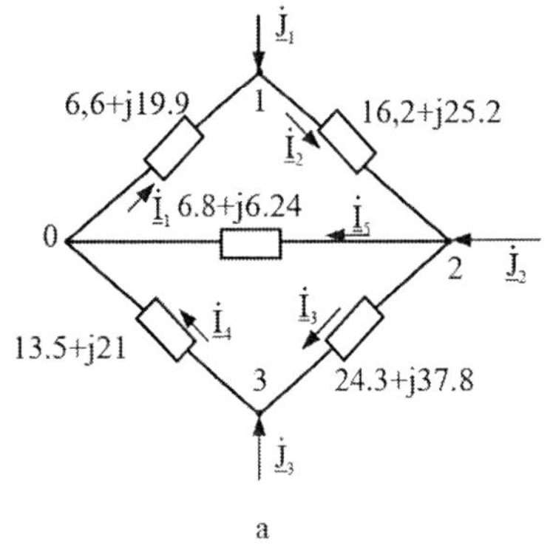

a

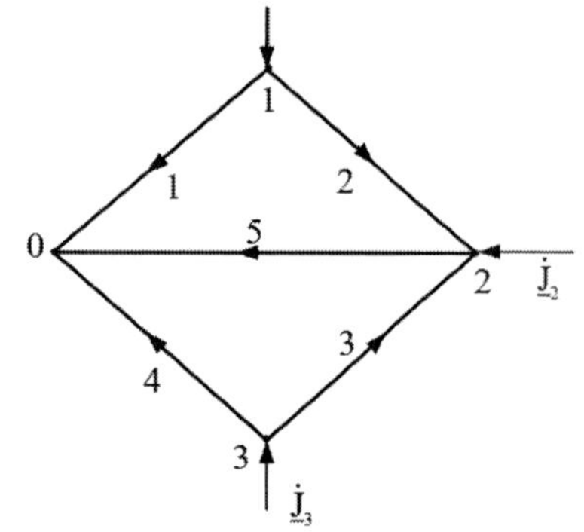

b

Fig. 2. The design diagram (a) and the directed graph of the 110 kV network (b).

As a result of cutting the nodes at points 1, 3, a branched, open network graph shown in Fig. 3 was obtained

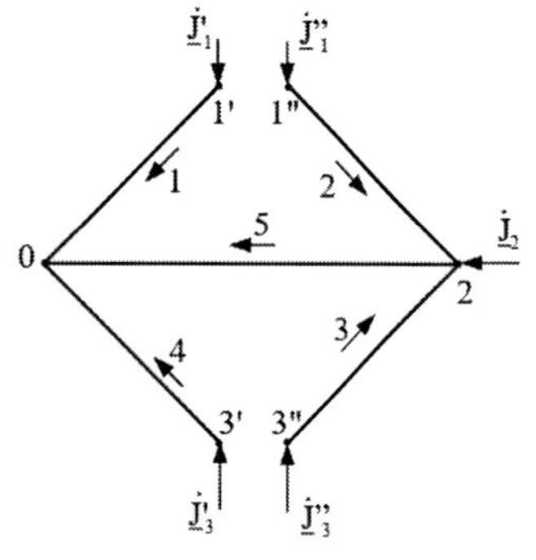

Fig. 3. Graph of open sub schemes (') and (").

Each cutting point is replaced by two fictitious points, provided with additional indices (' and "). The driving nodal currents at fictitious points are arbitrarily redistributed under the condition:

$$\left.\begin{aligned}\dot{J}'_1+\dot{J}''_1=\dot{J}_1\\ \dot{J}'_3+\dot{J}''_3=\dot{J}_3\end{aligned}\right\}, \quad (14)$$

The modes of the open circuits will be equivalent to the regime of a complex-closed circuit subject to identity:

$$\left.\begin{aligned}\dot{U}'_{\Delta 1}=\dot{U}''_{\Delta 1}\\ \dot{U}'_{\Delta 3}=\dot{U}''_{\Delta 3}\end{aligned}\right\}, \quad (15)$$

The values of the fictitious currents (14) ensuring the identity conditions (15) are determined on the basis of the matrix equation (7), which in the open form is written in the following form:

$$\left\|\begin{matrix}\dot{J}'_1\\ \dot{J}'_3\\ \dot{J}''_1\\ \dot{J}''_3\end{matrix}\right\| = \left\|\begin{matrix}1 & 0 & 1 & 0\\ 0 & 1 & 0 & 1\\ 6.6+19.9i & 0 & -(23+41.44i) & -(6.8+16.24i)\\ 0 & 13.5+21i & -(6.8+16.24i) & -(31.1+54.04i)\end{matrix}\right\|^{-1} \text{ч}$$

$$\text{ч}\left\|\begin{matrix}0.16+0,05i\\ 6+4,2i\\ 1.024+4.725i\\ 1.024+4.725i\end{matrix}\right\| = \left\|\begin{matrix}0.628+0.382i\\ 4.211+2.999i\\ -0.468-0.328i\\ 1.789+1.201i\end{matrix}\right\|,$$

where $\underline{\dot{E}}=\left\|\begin{matrix}1 & 0\\ 0 & 1\end{matrix}\right\|$ - diagonal unity matrix;

$J_p=\left\|\begin{matrix}\dot{J}_1\\ \dot{J}_3\end{matrix}\right\|$ - column matrix of driving currents;

$\dot{J}_1=0.16+0,05i$; $\dot{J}_3=6+4,2i$. - total currents of cut nodes;

$$\underline{C}^t_p\underline{Z}_b\underline{C}\ ' = \left(\left\|1 \quad 1\right\| \cdot \left\|\begin{matrix}Z_1 & 0\\ 0 & Z_4\end{matrix}\right\| \cdot \left\|\begin{matrix}1 & 0\\ 0 & 1\end{matrix}\right\|\right)' = \left(\left\|\begin{matrix}6.6+19.9i & 0\\ 0 & 13.5+21i\end{matrix}\right\|\right)' -$$

matrix of node resistance of the scheme with respect to the cut nodes with indices (');

$$\underline{C}^t_p\underline{Z}_b\underline{C}\ '' = \left(\left\|\begin{matrix}1 & 0 & 1\\ 0 & 1 & 1\end{matrix}\right\| \text{ч} \left\|\begin{matrix}Z_2 & 0 & 0\\ 0 & Z_3 & 0\\ 0 & 0 & Z_5\end{matrix}\right\| \text{ч} \left\|\begin{matrix}1 & 0\\ 0 & 1\\ 1 & 1\end{matrix}\right\|\right)'' =$$

$$= \left(\left\|\begin{matrix}23+41.44i & 6.8+16.24i\\ 6.8+16.24i & 31.1+54.04i\end{matrix}\right\|\right)''$$

matrix of node resistance of the scheme with respect to the cut nodes with indices (").

$\dot{J}_2=0.27+0,05i$ -uncut node driving current;

$$\dot{U}_{\Delta\Sigma} = \dot{U}''_{\Delta n} - \dot{U}'_{\Delta n} = \left(\left\|\begin{matrix}1.024+4.725i\\ 1.024+4.725i\end{matrix}\right\| - \left\|\begin{matrix}0\\ 0\end{matrix}\right\|\right) = \left(\left\|\begin{matrix}1.024+4.725i\\ 1.024+4.725i\end{matrix}\right\|\right)$$

- algebraic sum of nodal voltage drops from currents of uncut nodes;

$$\dot{U}''_{\Delta n} = \underline{C}^t_p\underline{Z}_b\underline{C}_n\ ''\text{ч}\dot{J}''_n = \left(\left\|\begin{matrix}1 & 0 & 1\\ 0 & 1 & 1\end{matrix}\right\| \text{ч} \left\|\begin{matrix}Z_2 & 0 & 0\\ 0 & Z_3 & 0\\ 0 & 0 & Z_5\end{matrix}\right\| \text{ч} \left\|\begin{matrix}0\\ 0\\ \dot{J}_2\end{matrix}\right\|\right)'' =$$

$$= \left(\left\|\begin{matrix}1.024+4.725i\\ 1.024+4.725i\end{matrix}\right\|\right)'';$$

$$\dot{U}'_{\Delta n} = \underline{C}^t_p\underline{Z}_b\underline{C}_n\ '\text{ч}\dot{J}'_n =$$

$$= \left(\left\|\begin{matrix}1 & 0\\ 0 & 1\end{matrix}\right\| \text{ч} \left\|\begin{matrix}Z_1 & 0\\ 0 & Z_4\end{matrix}\right\| \text{ч} \left\|\begin{matrix}1 & 0\\ 0 & 1\end{matrix}\right\| \text{ч} \left\|\begin{matrix}0\\ 0\end{matrix}\right\|\right) = \left\|\begin{matrix}0\\ 0\end{matrix}\right\|;$$

$\underline{C}'_p=\left\|1 \quad 1\right\|$, $\underline{C}''_p=\left\|\begin{matrix}1 & 0 & 1\\ 0 & 1 & 1\end{matrix}\right\|$ - matrix of open circuits fictitious currents distribution coefficients;

 https://doi.org/10.1051/e3sconf/20185802022

$\underline{C}'_n=\begin{Vmatrix}1 & 0\\ 0 & 1\end{Vmatrix}$, $\underline{C}''_n=\begin{Vmatrix}1 & 0\\ 0 & 1\\ 1 & 1\end{Vmatrix}$ - matrixes of distribution coefficients of currents of uncut nodes of subsystems;

$\underline{\dot{J}}'_n=\begin{Vmatrix}0\\ 0\end{Vmatrix}$, $\dot{J}''_n=\begin{Vmatrix}0\\ 0\\ \dot{J}_2\end{Vmatrix}$ - matrixes of the driving currents of uncut nodes of the corresponding subsystems.

If we use the matrix equation (10), then to determine the fictitious currents $\underline{\dot{J}}''$, we write the formula, which in expanded form looks as:

$$\begin{Vmatrix}\dot{J}''_1\\ \dot{J}''_3\end{Vmatrix}=-\begin{Vmatrix}-29.6\text{-}61.34i & -6.8\text{-}16.24i\\ -6.8\text{-}16.24i & -44.6\text{-}75.04i\end{Vmatrix}^{-1}\text{ч}$$

$$\text{ч}\begin{Vmatrix}1.043+1.184i\\ 8.224\text{-}177.975i\end{Vmatrix}=\begin{Vmatrix}-0.468\text{-}0.328i\\ 1.789+1.201i\end{Vmatrix},$$

where

$$\underline{Z}=-\left(\begin{Vmatrix}\underline{C}^t_p\underline{Z}_b\underline{C}_p\end{Vmatrix}''-\begin{Vmatrix}\underline{C}^t_p\underline{Z}_b\underline{C}_p\end{Vmatrix}'\right)=-\left(\begin{Vmatrix}\underline{Z}_2+\underline{Z}_5 & \underline{Z}_5\\ \underline{Z}_5 & \underline{Z}_3+\underline{Z}_5\end{Vmatrix}+\begin{Vmatrix}\underline{Z}_1 & 0\\ 0 & \underline{Z}_4\end{Vmatrix}\right)=$$

$$=\left(\begin{Vmatrix}-29.6\text{-}61.34i & -6.8\text{-}16.24i\\ -6.8\text{-}16.24i & -44.6\text{-}75.04i\end{Vmatrix}\right);$$

,

$$\dot{U}_{\Delta\Sigma}=\dot{U}''_{\Delta n}-\begin{Vmatrix}\underline{C}^t_p\underline{Z}_b\underline{C}_p\end{Vmatrix}'\text{ч}\dot{J}_p=$$

$$=\left(\begin{Vmatrix}1.024+4.725i\\ 1.024+4.725i\end{Vmatrix}-\begin{Vmatrix}1 & 0\\ 0 & 1\end{Vmatrix}\text{ч}\begin{Vmatrix}\underline{Z}_1 & 0\\ 0 & \underline{Z}_4\end{Vmatrix}\text{ч}\begin{Vmatrix}1 & 0\\ 0 & 1\end{Vmatrix}\text{ч}\begin{Vmatrix}0.16+0,05i\\ 6+4,2i\end{Vmatrix}\right)=$$

$$=\left(\begin{Vmatrix}1.043+1.184i\\ 8.224\text{-}177.975i\end{Vmatrix}\right).$$

For the remaining unknown currents, the following expression is true:

$$\dot{J}'_p=\dot{J}_p-\dot{J}''_p=\begin{Vmatrix}\dot{J}_1-\dot{J}''_1\\ \dot{J}_3-\dot{J}''_3\end{Vmatrix}=$$

$$=\begin{Vmatrix}0.16+0,05i\\ 6+4,2i\end{Vmatrix}-\begin{Vmatrix}-0.468\text{-}0.328i\\ 1.789+1.201i\end{Vmatrix}=\begin{Vmatrix}0.628+0382i\\ 4.211+2.999i\end{Vmatrix}.$$

As can be seen, the values of fictitious nodal currents obtained by different methods turned out to be equal among themselves.

The analysis of the modes of the source scheme after the determination of fictitious currents at the subsystems intersections nodes, is performed on the basis of calculations of open circuits which matrix of current distribution coefficients are equal to:

$$\underline{C}'_p=(\underline{C}'_p\underline{C}'_n)=\begin{Vmatrix}1 & 0 & 0\\ 0 & 0 & 1\\ 1 & 1 & 1\end{Vmatrix};\quad \underline{C}''=(\underline{C}''_p\underline{C}''_n)=\begin{Vmatrix}1 & 0\\ 0 & 1\end{Vmatrix}$$

then, the node voltage drops of the scheme of the individual subsystems provided with indices ('and') are defined as:

$$\begin{Vmatrix}\dot{U}'_{\Delta 1}\\ \dot{U}'_{\Delta 3}\end{Vmatrix}=\begin{Vmatrix}6.6+19.9i & 0\\ 0 & 13.5+21i\end{Vmatrix}\text{ч}\begin{Vmatrix}0.628+0382i\\ 4.211+2.999i\end{Vmatrix}=\begin{Vmatrix}-3.468+15.012i\\ -6.118+128.92i\end{Vmatrix};$$

$$\begin{Vmatrix}\dot{U}''_{\Delta 1}\\ \dot{U}''_{\Delta 2}\\ \dot{U}''_{\Delta 3}\end{Vmatrix}=\begin{Vmatrix}1 & 0 & 1\\ 0 & 0 & 1\\ 0 & 1 & 1\end{Vmatrix}\text{ч}\begin{Vmatrix}16.2+25.2i & 0 & 0\\ 0 & 24.3+37.8i & 0\\ 0 & 0 & 6.8+16.24i\end{Vmatrix}\text{ч}$$

$$\text{ч}\begin{Vmatrix}1 & 0 & 1\\ 0 & 0 & 1\\ 1 & 1 & 1\end{Vmatrix}\text{ч}\begin{Vmatrix}0.628+0382i\\ 0.27+0,05i\\ 4.211+2.999i\end{Vmatrix}=\begin{Vmatrix}-3.468+15.012i\\ -4.17+32.115i\\ -6.118+128.92i\end{Vmatrix},$$

where

$$\begin{Vmatrix}\dot{I}'_1\\ \dot{I}'_4\end{Vmatrix}=\begin{Vmatrix}1 & 0\\ 0 & 1\end{Vmatrix}\text{ч}\begin{Vmatrix}0.628+0382i\\ 4.211+2.999i\end{Vmatrix}=\begin{Vmatrix}0.628+0382i\\ 4.211+2.999i\end{Vmatrix};$$

$$\begin{Vmatrix}\dot{I}''_2\\ \dot{I}''_3\\ \dot{I}''_5\end{Vmatrix}=\begin{Vmatrix}1 & 0 & 0\\ 0 & 0 & 1\\ 1 & 1 & 1\end{Vmatrix}\text{ч}\begin{Vmatrix}0.628+0382i\\ 0.27+0,05i\\ 4.211+2.999i\end{Vmatrix}=\begin{Vmatrix}-0.468\text{-}0.328i\\ 1.789+1.201i\\ 1.591+0.923i\end{Vmatrix}.$$

The correctness of the calculations can be verified by the identity conditions for the intersection points:

$$\left.\begin{array}{l}\dot{U}'_{\Delta 1}=\dot{U}''_{\Delta 1}\\ \dot{U}'_{\Delta 3}=\dot{U}''_{\Delta 3}\end{array}\right\}$$

5 Conclusions

1. A method was developed for dividing a system based on the equality of the voltage of the nodes of the subsystems intersection circuits;
2. Nodal voltages of the subsystems intersection circuits are formed directly according to the scheme.
3. The reduction of a complex closed network to an equivalent open network greatly simplifies the calculation algorithms of its steady-state modes.

References

1. Kron G. Investigation of complex systems by parts – diakoptiks. Nauka, 542 (1972).
2. Gamm A.Z., Glinevich G.A., Krumm L.A. and others. The use of decomposition to solve certain problems and the management of large energy systems. Irkutsk, SOAN SSR, Volume 1. 172-192 (1970)
3. Geraskin O.T. Diakoptics and sparsity of tasks and calculations of large electric energy systems. News of higher educational institutions of the USSR. Power engineering. No.1. 7-12 (1982).
4. Kuzmin Y.F., Makhnitko A.Y. Calculation of electrical grids and given powers of the nodes of the methods of partitioning into subschemes. Electricity. No.5. 16-20 (1972)
5. Makhnitko A.Y. The method of calculating the steady-state mode of the power system in individual subsystems. Electricity. No. 4. 26-30 (1972)

6. Griben R.M. Topological method of decomposing graphs into parts. Izvestiya/ Energy and transport. 7-10 (1975)
7. Gridin V.N., Anissimov V.I., Atu A.S. Application of the method of diakoptics for modeling and calculating large systems. Control problems. No. 4 (2014)
8. Geraskin O.T. The problem of calculating modes in large electric energy systems and its solution by methods of diakoptics and decomposition, Problems of Energy. No. 1-2. 63-77 (2000)
9. Savina N.V., Tagirov S.V. Decomposition of electrical networks as a means of simplifying the calculation of steady-state modes under market energy conditions. Bulletin of the Amur State University. Series: Natural and economic sciences. No. 43. 60-64 (2008)
10. Khachatryan V.S., Badalyan N.P. Calculation of the steady-state mode of a large electric power system by the method of diakoptics. Electricity. No. 6. 13-17. (2003)
11. Dauren S. Akhmetbayev, Daurenbek A. Aubakir, Yermek Zh. Sarsikeyev, Bakhtybek A. Bainiyazov, Mikhail A. Surkov, Vitaliy I. Rozhkov, Gulbahit N. Ansabekova, Assel S. Yerbolova, Azamat T. Suleimenov, Miramgul S. Tokasheva. Development Of Topological Method For Calculating Current Distribution Coefficients In Complex Power Networks, Results in Physics, 7, 1644-1649 (2017)
12. Akhmetbayev D. Method of calculating the steady-state modes of electrical networks based on the current distribution coefficients. Electricity, No. 11, 23-27, (2010)
13. Dauren Akhmetbayev, Arman Akhmetbayev, Azamat Suleymenov, Michal Kolcun. Modeling the Set Mode of Complex Power Grid, Based on Infeed Coefficients. Proceedings of the 9th International Scientific Symposium, StaráLesná, SlovakRepublic, ELEKTROENERGETIKA, 12.-14, 9 (2017)
14. Akhmetbayev D., Akhmetbayev A., Aidarova A. Determination of the rational transformations distribution networks distribution. E3S Web of conferences 25.04003 (2018).
15. Arman Akhmetbayev, Dauren Akhmetbayev, Serik Zhumazhanov, Bauyrzhan Zhakishev. New Modeling of Steady-State Modes of Complex Electrical Grids of Power Systems MATEC Web of Conferences 155, 01043 (2018)
16. Akhmetbayev D., Akhmetbaev A., Berdygozhin A.S. Topological method for the formation of an inverted form of the node equations of electric energy systems. Electricity, No. 5, (2018)
17. Kholmsky V.G. Calculation and optimization of modes of electrical networks/ High school, 280 (1975)

Semi-Markov and hidden semi-Markov models of energy systems

Yuriy E. Obzherin[1*]

[1]Sevastopol State University, Higher Mathematics Department, 299053 Sevastopol, Russia

Abstract. The problem of information control systems creation for energy systems and transition to intelligent control and engineering is one of the important problems of reliability and efficiency theory for energy systems. The solution of this problem is possible based on construction of mathematical models concerning different aspects of these systems structure and operation. The possibilities of application of semi-Markov processes with common phase space of states, hidden Markov and semi-Markov models for energy system modeling are considered in the paper.

1 Introduction

According to the road map of National Technology Initiative "EnergyNet" between the main purposes of this project are:
1."Digitalization" of energy systems, creation of digital networks and digital substations.
2.Creation of information control systems for energy systems and transition to intelligent control and engineering.

The solution of these problems is possible based on mathematical model construction concerning different aspects of energy system structure and operation and development of algorithms and information systems for decision support during design and maintenance of the systems as well as forecast of their states. When constructing adequate models of energy systems it is necessary to use existing possibilities of mathematical apparatus existing.

The possibilities of application of semi-Markov processes with common phase space of states, hidden Markov and semi-Markov models for energy system modeling are considered in the paper.

2 Semi-Markov processes, hidden semi-Markov models, their application for energy system modelling

Semi-Markov processes are widely used to construct models and analyse systems of different purposes: technical, production, energy, information, economic, biology etc.

Research in this field is actively carried out in France, the USA, Italy, Great Britain, People Republic of China, India, Russian Federation, Ukraine, Sweden, Japan and some other countries. Under the results of this research there are published several monographs and a large number of papers published. Thus, the leading world publishing houses Elsevier, Springer, World Scientific have recently published the next monographs on this thematic [1 - 7].

In most of these papers semi-Markov models with finite space of states were used. When constructing the models of specific systems it was assumed that most of the random variables characterizing the system have exponential distribution what narrows noticeably the application possibilities of the results obtained.

A large contribution to the development of semi-Markov processes theory and applications was made by works of scientists from the CIS countries V.S. Koroluk, A.F. Turbin, I.A. Ushakov, A.V. Swischuk, V.A. Kashtanov, V.M. Shurenkov. In their works [8 - 13] the algorithms of asymptotic and stationary phase merging are worked out, the models of evolutionary stochastic systems and algorithms of their averaging are built. These papers are actual currently, they provide large opportunities to model systems of different purposes including energy systems.

An effective instrument of system models construction and analysis is the apparatus of theory of semi-Markov processes with common phase space of states [2],[5],[8],[9],[11]. Using of this class of random processes let:
- to get rid of a number of restrictions when model construction, in particulars from the hypothesis on exponential distribution law for random variables characterizing the system;
- to obtain analytic expressions for system's characteristics which can be used for engineering calculations;
- to construct the models for a number of energy systems.

To solve the problem of the dimension of the model one can use the algorithms of asymptotic and stationary phase merging developed by V.S. Koroluk, A.F. Turbin,

* Corresponding author: objsev@mail.ru

A.V. Swischuk [8-11]. This approach opens large opportunities for system modeling.

Since the 80s hidden Markov and semi-Markov models have been widely used for system modeling and analysis. In these models there is supposed that observed signal sequence is connected with Markov (semi-Markov) process with non-observable (hidden) states and the task is set to evaluate the structure and characteristics of unobservable process from observations of the signals. A large number of works including recently printed monographs [14-16] is devoted to hidden semi-Markov models.

Hidden semi-Markov models are applied in the next areas: equipment diagnostics and maintenance; reliability of systems, system operation forecasting, network performance evaluation, speech recognition and synthesis, machine translation, pattern recognition, FMRT brain mapping, pathologic events early determination, human activity recognition, protein structure forecasting, financial time series analysis modeling and in some other areas. It is noted in the literature that this approach is used in more than thirty areas.

Semi-Markov processes with common phase space of states and hidden semi-Markov models can be effectively used to construct the models, concerning various aspects of structure and operation of energy systems: reliability, efficiency, control, diagnostics, maintenance and forecasting.

On this basis, the following problems can be solved:

1. To build semi-Markov and hidden semi-Markov models of:
   - reliability of energy systems;
   - energy systems diagnostics and control;
   - energy systems maintenance;
   - energy systems forecasting.
2. To find systems technical and economical characteristics suitable for engineering calculations;
3. To solve the problems of energy systems optimum value finding on the basis of objective functions obtained with the aid of the models.
4. To create the software for information systems of energy systems reliability evaluation, control, diagnostic, maintenance and forecasting of states based on the results obtained.

3 Semi-Markov model of a single-stream line with intermediate storage devices

We consider a multiphase, single-stream system consisting of serving devices and intermediate storage devices, the connections between which are depicted in Fig.1. Systems of this kind play an important role in the modeling of energy systems [17].

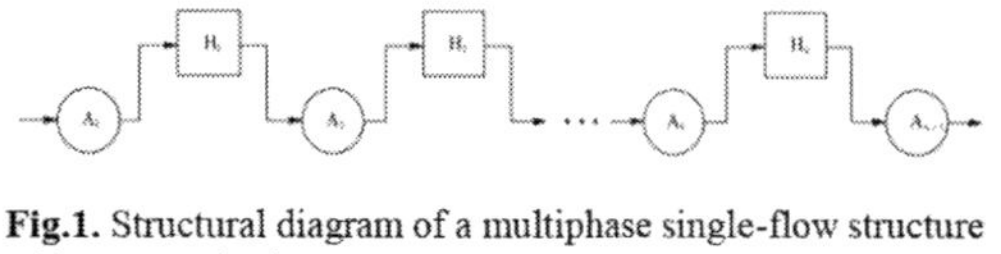

Fig.1. Structural diagram of a multiphase single-flow structure with storage devices.

The following notation is used in the scheme: A_i, $i=\overline{1,n+1}$ – service devices; H_i, $i=\overline{1,n-1}$ – intermediate storage devices. The model is being constructed under the following assumptions.

1. The possible states of each of the serving devices A_i are: operable, recovery and shutdown.

2. Time to failure (renewal time) for the device A_i is a random variable (RV) $\alpha_i^{(0)}(\alpha_i^{(1)})$ with distribution function $F_i^{(0)}(x)\,(F_i^{(1)}(x))$. RV $\alpha_i^{(0)}(\alpha_i^{(1)})$ are independent and have finite expectations; there are distribution densities $f_i^{(0)}(x)(f_i^{(1)}(x))$ for $F_i^{(0)}(x)\,(F_i^{(1)}(x))$.

3. The storage devices H_i are absolutely reliable devices that have limited capacity $h_i \geq 0$ (The storage capacity of the device H_i is expressed in units of time that device A_{i+1} will need to fully free this storage device).

4. An operable device A_i disconnects,being in operable state, if the storage device H_{i-1} is empty or the storage device H_{i+1} is full.

5.The productivity of device A_i is constant and equal to c_i, herewith $c_i \geq c_{i+1}$.

6. The system is in failed state if the output device A_{n+1} does not produce products; the renewal of the device A_i is considered as infinite.

To describe the operation of the system let us introduce the next discrete-continuous space of semi-Markov states:

$$\mathrm{E} = \overline{d}\overline{x}\overline{z} : i=\overline{1,n+1}, \overline{d}=(d_1,...,d_{n+1}),$$

$$\overline{x}=(x_1,...,x_{n+1}), \overline{z}=(z_1,...,z_n) \quad ,$$

where i is the number of device A_i that failed or was recovered by the last. The element d_k of the vector $\overline{d}$ fixes the state of the device: operable ($d_k = 0$), recovery ($d_k = 1$), shut-down ($d_k = 2$). The value of the element x_k of the vector $\overline{x}$ is the time lasted from the instance of device A_k change of state, it should be noted that $x_i = 0$. The component z_k of the vector $\overline{z}$ defines the time within which the storage device H_k can supply the device A_{k+1} with production, $0 \leq z_k \leq h_k$, $k=\overline{1,n}$.

To find approximately the reliability characteristics of the system considered let us use the phase merging algorythm [11] which consists in the following.

Let us suppose that stochastic kernel of embedded Markov chain (EMC) $\{\xi_n; n \geq 0\}$ for the initial system semi-Markov process $\xi(t)$ is close to the stochastic kernel of EMC $\{\xi_n^{(0)}; n \geq 0\}$ for some supporting system S_0 with the single stationary distribution $\rho(dx)$. Then the next approximate formulas represented in [11] can be used for approximate finding of the average stationary time to failure T_+, the average stationary renewal time T_- and the stationary availability factor K_a:

$$T_+ \approx \frac{(\rho, \overline{\mathrm{m}}_1)}{(\rho, P^{(r)}\overline{1}_0)}, \; T_- \approx \frac{(\rho, P^{(r)}\overline{\mathrm{m}}_0)}{(\rho, P^{(r)}\overline{1}_0)}, K_r = \frac{T_+}{T_+ + T_-}, \quad (1)$$

where

$$\bar{m}_1(x)=\begin{cases} m(x), x\in E_-,\\ 0, x\in E_-,\end{cases}\quad \bar{m}_0(x)=\begin{cases} 0, x\in E_-,\\ m(x), x\in E_-,\end{cases}$$

$$\bar{1}_0(x)=\begin{cases} 0, x\in E_+,\\ 1, x\in E_-,\end{cases}\quad (\rho,f)=\int_E f(x)\rho(dx),$$

$\rho(dx)$ is supporting EMC $\{\xi_n^{\varepsilon(0)}; n\geq 0\}$ stationary distribution; $m(x)$ are the average sojourn times for the states $x\in E$ of the initial system; $P^{(r)}(x,B)$ are the initial system EMC $\{\xi_n; n\geq 0\}$ transition probabilities; r is the minimum number of steps through which the system can transit to the failure states subset E_- from the operable states subset E_+ comprised to the ergodic class E^0.

An important moment of using of the method applied is the choice of the supporting system S_0.Let us suppose that devices $A_i, i=\overline{1,n}$ have fast renewal, i.e. their renewal times $\alpha_i^{(1)}$ depends on small positive parameter ε in such a manner that

$$\lim_{\varepsilon\to 0} E\alpha_i^{(1,\varepsilon)}=0,\quad i=\overline{1,n}, \tag{2}$$

and output device A_{n+1} has fixed time to failure and recovery time. This leads to the fact that recovery system S_0 will be the system with instant recovery of the devices $A_i, i=\overline{1,n}$ and completely filled storage devices H_i. Using the formulas (1) one can show that stationary characteristics of the system considered: stationary time to failure $\mathrm{T}_+^{(h_1,\ldots,h_n)}$ and average stationary renewal time $\mathrm{T}_-^{(h_1,\ldots,h_n)}$ are approximately calculated by the formulas:

$$\mathrm{T}_+^{(h_1,\ldots,h_n)}\approx \prod_{k=1}^{n+1}E\alpha_k^{(0)}+\sum_{i=1}^{n}\prod_{\substack{k=1,\\k\neq i}}^{n+1}E\alpha_k^{(0)}\int_0^{h_i}\bar{F}_i^{(1,\varepsilon)}(t)dt \Big/ \prod_{k=1}^{n}E\alpha_k^{(0)}+$$

$$+\sum_{i=1}^{n}\prod_{\substack{k=1,\\k\neq i}}^{n}E\alpha_k^{(0)}\int_0^{h_i}\bar{F}_i^{(1,\varepsilon)}(t)dt+\sum_{i=1}^{n}\prod_{r=1}^{i-1}E\alpha_r^{(0)}\bar{F}_i^{(1,\varepsilon)}\Big(\sum_{k=i}^{n}h_k\Big)\times \tag{3}$$

$$\times\prod_{m=i-1}^{n+1}\int_{\sum_{l=i}^{m-1}h_l}^{\infty}\bar{F}_m^{(0)}(x_m)dx_m\ .$$

$$\mathrm{T}_-^{(h_1,\ldots,h_n)}\approx\Big[E\alpha_{n+1}^{(1)}\prod_{i=1}^{n}E\alpha_i^{(0)}+E\alpha_{n-1}^{(1)}\sum_{i=1}^{n}\prod_{\substack{k=1,\\k\neq i}}^{n}E\alpha_k^{(0)}\int_0^{h_i}\bar{F}_i^{(1,\varepsilon)}(t)dt+$$

$$+\sum_{i=1}^{n}\prod_{r=1}^{i-1}E\alpha_r^{(0)}\int_{\sum_{k=i}^{n}h_k}^{\infty}\bar{F}_i^{(1,\varepsilon)}(t)dt\prod_{m=i+1}^{n+1}\int_{\sum_{l=i}^{m}h_l}^{\infty}\bar{F}_m^{(0)}(x_m)dx_m \Big/$$

$$\Big/\Big[\prod_{k=1}^{n}E\alpha_k^{(0)}+\sum_{i=1}^{n}\prod_{\substack{k=1,\\k\neq i}}^{n}E\alpha_k^{(0)}\int_0^{h_i}\bar{F}_i^{(1,\varepsilon)}(t)dt+ \tag{4}$$

$$+\sum_{i=1}^{n}\bar{F}_i^{(1,\varepsilon)}\Big(\sum_{k=i}^{n}h_k\Big)\prod_{r=1}^{i-1}E\alpha_r^{(0)}\prod_{m=i+1}^{n-1}\int_{\sum_{l=i}^{m}h_l}^{\infty}\bar{F}_m^{(0)}(x_m)dx_m\ .$$

Knowing the expressions for $\mathrm{T}_+^{(h_1,\ldots,h_n)}$ and $\mathrm{T}_-^{(h_1,\ldots,h_n)}$, one can find the stationary availability factor $K_a^{(h_1,\ldots,h_n)}$ by the formula (1).

The Table 1 represents the results of four-phase information system stationary characteristics finding using the formulas (3), (4) and using the simulation modeling. During the modeling it was assumed that all RV (operation and renewal times for all the devices) are distributed by the exponential law.

Table 1. Modeling results for the four-phase system.

$E\alpha_1^{(0)}=E\alpha_2^{(0)}=E\alpha_3^{(0)}=E\alpha_4^{(0)}=17\ h,$
$E\alpha_1^{(1)}=E\alpha_2^{(1)}=0{,}6\ h,\ E\alpha_3^{(1)}=E\alpha_3^{(1)}=0{,}8\ h,$
$h_1=h_2=h_3=h,\ processing\ time\ -\ 0{,}2\ h.$

Time reserve	Analytic modeling			Simulation modeling
h	$\mathrm{T}_+^{(h_1,h_2,h_3)}$, h	$\mathrm{T}_-^{(h_1,h_2,h_3)}$, h	$\mathrm{K}_\Gamma^{(h_1,h_2,h_3)}$	$\mathrm{K}_\Gamma^{(h_1,h_2,h_3)}$
0	4.250	0.700	0.859	0.879
0.1	6.518	0.736	0.899	0.929
0.2	8.693	0.736	0.919	0.932
0.3	10.507	0.780	0.931	0.923
0.4	11.929	0.790	0.938	0.933
0.5	13.029	0.795	0.943	0.927
0.6	13.885	0.797	0.946	0.937
0.7	14.556	0.799	0.948	0.940
0.8	15.084	0.799	0.950	0.939
0.9	15.501	0.800	0.951	0.935
1.0	15.830	0.800	0.952	0.941
1.1	16.089	0.800	0.953	0.943
1.2	16.292	0.800	0.953	0.945
1.3	16.451	0.800	0.954	0.946
1.4	16.575	0.800	0.954	0.947
1.5	16.671	0.800	0.954	0.945

4 Conclusion

The paper shows the possibilities of semi-Markov processes with common phase space of states and hidden semi-Markov models application to model construction and operation analysis of energy systems. To solve the dimension problem it is proposed to use the algorithms of asymptotic and stationary phase merging of the systems.

The research was carried out within the state assignment of the Minobrnauki of Russia (№ 1.10513.2018/11.12), with financial support by RFBR (project № 18-01-00392a).

References

1. F.Grabski, *Semi-Markov Processes: Applications in System Reliability and Maintenance* (Elsevier, 2015)
2. Yu. E. Obzherin, E.G. Boyko, *Semi-Markov Models: Control of Restorable Systems with Latent Failures* (Elsevier, Academic Press, 2015)
3. J. Janssen, R. Manca, *Applied Semi-Markov Processes* (Springer Science + Business Media, 2006)
4. V.S. Korolyuk, N. Limnios, *Stochastic Systems in Merging Phase Space* (World Scientific, Imperial Coledge Press, 2005)
5. N. Limnios, G. Oprisan, *Semi-Markov Processes and Reliability* (Springer Science + Business Media, 2001)
6. J. Janssen, N. Limnios (Eds.), *Semi-Markov Models and Applications* (Kluwer Academic Publishers, 1999)
7. D. Silvestrov, S. Silvestrov, *Nonlinearly Perturbed Semi-Markov Processes* (Springer, 2017)
8.V. S. Koroluk, A. F. Turbin, *Markovian Restoration Processes in the Problems of System Reliability* (Naukova Dumka, Kiev, 1982)
9.V. S. Koroluk, *Stochastic System Models* Naukova Dumka, Kiev (1989)
10. V.S.Korolyuk, A.V. Swishchuk, *Evolutionary Stochastic Systems. Algorithms of Averaging and Di usion Approximation. Mathematics and its Applications* (Institute of Mathematics, National Academy of Sciences of Ukraine, Kyiv,2000)
11. A. N. Korlat, V. N. Kuznetsov, M. M. Novikov, A. F. Turbin, *Semi-Markov Models of Recoverable Systems and Queuing Systems* (Shtiinta, Chisinau, 1991)
12. V. M. Shurenkov, *Ergodic Markovian Processes* (Nauka, Moscow, 1989)
13. V. A. Kashtanov, A. I. Medvedev, *Reliability Theory of the Complex Systems (Theory and Practice)* (Fizmatlit, Moscow, 2010)
14. Shun-Zheng Yu, *Hidden Semi-Markov Models: Theory, Algorithms and Applications* (Elsevier, 2015)
15. V.S. Barbu, N. Limnios, *Semi-Markov Chains and Hidden Semi-Markov Models toward Applications: their use in Reliability and DNA Analysis* (Springer , 2008)
16. J. Hoek, R.J. Elliott, *Introduction to Hidden Semi-Markov Models* (Cambridge University Press, 2018)
17.Y. N. Rudenko, I. A. Ushakov, *Reliability of power systems*, 2nd ed (Novosibirsk, Nauka, 1989)

Analysis of reliability of systems with component-wise storages

Yuriy E. Obzherin[1], *Stanislav M.* Sidorov[1,*], and *Mikhail M.* Nikitin[1]

[1]Sevastopol State University, Higher mathematics Department, Sevastopol, Russia

Abstract. Time redundancy is a method of increasing the reliability and efficiency of the operation of systems for various purposes, in particular, energy systems. A system with time redundancy is given additional time (a time reserve) for restoring characteristics. In this paper, based on the theory of semi-Markov processes with a common phase space of states, a semi-Markov model of a two-component system with a component-wise instantly replenished time reserve is constructed. The stationary reliability characteristics of the system under consideration are determined.

1 Introduction

Most modern large systems have high reliability, but they are characterized by a significant impact of operator errors (the human factor) on the functioning and efficiency of the system. Time redundancy reduces the impact of such errors by providing additional time (a time reserve) to eliminate them without affecting the operation of the system.

Time redundancy [1-9] is a method of improving the reliability and efficiency of systems, in which the system in the process of functioning is given the opportunity to spend some additional time (time reserve) for restoring characteristics. For systems with time redundancy, a malfunction of the system is not necessarily accompanied by a system failure, since it is possible to restore the system's operability for a standby time.

A time reserve is called instantly replenished if the same time is allocated for the restoration of operability after the failure of any element, regardless of the number of previous failures and the time spent on their elimination. At the end of the recovery, such a time reserve is immediately replenished to the original level [1-3].

Time reserve in the systems of a power engineering [10-12] can be created by increase in power (efficiency, a channel capacity) the generating inventory extracting inventories, subsystems of transport of energy resources, electricity transmissions and other constituents of systems of a power engineering, by creation of internal stocks of the made or transported production, introduction of parallel devices for increase in total capacity, use of the functional inertance of systems and restricted speed of development of the processes caused by adverse effects of various physical nature.

Time redundancy is used in gas transmission systems in which underground gas storage is the source of the time reserve; in the electric power industry, the time reserve is realized at the expense of high-capacity energy storage devices [10, 11].

In connection with this, problems arise in determining the capacities of storage devices and their locations.

In this paper, based on the theory of semi-Markov processes with a common phase space of states [13-18], a semi-Markov model of a two-component system with a component-wise instantly replenished time reserve is constructed. Stationary reliability characteristics of the system are found, the effect of the time reserve on the characteristics obtained is analysed.

2 Description of the system functioning

The system S, consisting of two components, time to failure of which are random variables (RVs) α_i with the distribution functions (DFs) $F_i(x)$, a restoration times are RVs β_i with DFs $G_i(x)$, $i=\overline{1,2}$. Each component of the system has a random instantly replenished time reserve τ_i with DF $R_i(x)$. RVs α_i, β_i, τ_i are assumed to be independent in aggregate, having finite mathematical expectations; DFs $F_i(x)$, $G_i(x)$, $R_i(x)$ have distribution densities $f_i(x)$, $g_i(x)$, $r_i(x)$.

The time reserve starts to be used at the time the component begins to recover. The failure of system S occurs when both elements are restored and the time reserve for each element is completely spent. It continues until the restoration of one of the failed elements, while the time reserve of the restored element is instantly replenished to the level τ_i (instantly replenished time reserve).

* Corresponding author: xaevec@mail.ru

https://doi.org/10.1051/e3sconf/20185802024

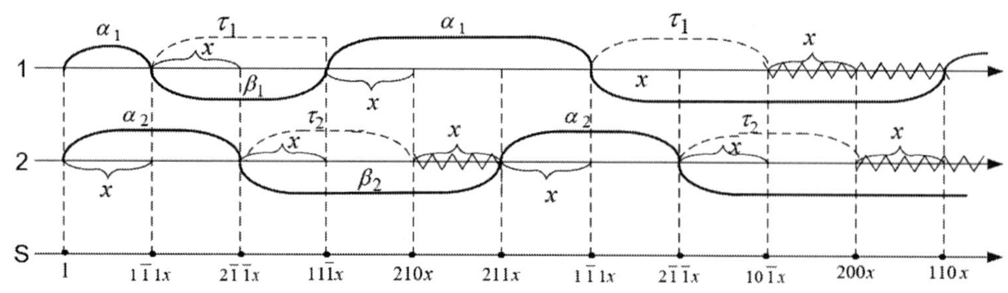

Fig. 1. Time diagram of the functioning of the system S.

3 Semi-Markov model building

To describe the functioning of the system S, we use the semi-Markov process $\xi(t)$ [7-10]. We introduce the space of states of the form:

$$E=\{1,\ i\bar{d}x:\bar{d}=(d_1,d_2),\ x>0\}, \tag{1}$$

where $i=\overline{1,2}$ is the number of the component in which the state change occurred. Component d_k of the vector $\bar{d}$ describes the physical state of the element with the number k:

$$d_k=\begin{cases}0, \text{ if k}-th\text{ component is in failure,}\\ 1, \text{ if k}-th\text{ component is operational,}\\ \bar{1}, \text{ if k}-th\text{ component is restored and}\\ \quad\text{functions due to the time reserve.}\end{cases}$$

The continuous component x indicates the elapsed time since the last change in the system state. The time diagram of the functioning of the described system is shown in Figure 1.

We construct the Markov renewal process (MRP) $\{\xi_n,\theta_n:\text{n}\geq 0\}$ that describes the functioning of the system S.

We define the transition probabilities of the embedded Markov chain (EMC) $\{\xi_n:\text{n}\geq 0\}$. Introduce the notation:

$$\bar{F}(t)=1-F(t)\,,\ \bar{i}=\begin{cases}2,\ i=1,\\ 1,\ i=2.\end{cases}$$

1. Let us consider the case of states id_1d_2x, $d_i=\bar{1}$, $d_{\bar{i}}=1$.

In this case, the following transitions are possible:

a) $id_1d_2x\rightarrow id_1'd_2'y$ in conditions: $d_k'=d_k$ at $k\neq i$, $d_i'=1$, $y>x$. The transition probability density in this case is calculated from the formula:

$$p_{id_1d_2x}^{id_1'd_2'y}=\frac{g_i(y-x)\bar{R}_i(y-x)\bar{F}_{\bar{i}}(y)}{\bar{F}_{\bar{i}}(x)}, \tag{2}$$

b) $id_1d_2x\rightarrow id_1'd_2'y$ in conditions: $d_k'=d_k$ at $k\neq i$, $d_i'=0$, $y>x$. The transition probability density in this case has the form:

$$p_{id_1d_2x}^{id_1'd_2'y}=\frac{r_i(y-x)\bar{G}_i(y-x)\bar{F}_{\bar{i}}(y)}{\bar{F}_{\bar{i}}(x)}, \tag{3}$$

c) $id_1d_2x\rightarrow \bar{i}d_1'd_2'y$ in conditions: $d_k'=d_k$ at $k\neq i$, $d_{\bar{i}}'=\bar{1}$, $y>0$. In this case, the probability density of the transition is

$$p_{id_1d_2x}^{\bar{i}d_1'd_2'y}=\frac{f_{\bar{i}}(y+x)\bar{G}_i(y)\bar{R}_i(y)}{\bar{F}_{\bar{i}}(x)}. \tag{4}$$

2. Let us consider the case of states id_1d_2x, $d_1=d_2=0$.

In this case, the following transitions are possible:

a) $id_1d_2x\rightarrow id_1'd_2'y$ in conditions: $d_k'=d_k$ at $k\neq i$, $d_i'=1$, $y>x$. The transition probability density is calculated by the formula:

$$p_{id_1d_2x}^{id_1'd_2'y}=\frac{\int_0^\infty r_i(t)g_i(y-x+t)dt\int_0^\infty r_{\bar{i}}(t)\bar{G}_{\bar{i}}(y+t)dt}{P(\beta_i>\tau_i)\int_0^\infty r_{\bar{i}}(t)\bar{G}_{\bar{i}}(x+t)dt}, \tag{5}$$

b) $id_1d_2x\rightarrow \bar{i}d_1'd_2'y$ in conditions: $d_k'=d_k$ at $k\neq i$, $d_{\bar{i}}'=1$, $y>0$. The transition probability density will be:

$$p_{id_1d_2x}^{\bar{i}d_1'd_2'y}=\frac{\int_0^\infty r_{\bar{i}}(t)g_{\bar{i}}(y-x+t)dt\int_0^\infty r_i(t)\bar{G}_i(y+t)dt}{P(\beta_i>\tau_i)\int_0^\infty r_{\bar{i}}(t)\bar{G}_{\bar{i}}(x+t)dt}. \tag{6}$$

3. Consider the case of states id_1d_2x, $d_1=d_2=\bar{1}$.

a) $id_1d_2x \to id_1'd_2'y$ in conditions: $d_k' = d_k$ at $k \neq i$, $d_i' = 1$, $y > x$. The transition probability density in this case is calculated from the formula:

$$p_{id_1d_2x}^{id_1'd_2'y} = \frac{g_i(y-x)\bar{R}_i(y-x)\bar{G}_{\bar{i}}(y)\bar{R}_{\bar{i}}(y)}{\bar{G}_{\bar{i}}(x)\bar{R}_{\bar{i}}(x)}, \qquad (7)$$

b) $id_1d_2x \to id_1'd_2'y$ in conditions: $d_k' = d_k$ at $k \neq i$, $d_i' = 0$, $y > x$. In this case, the transition probability density is calculated by the formula:

$$p_{id_1d_2x}^{id_1'd_2'y} = \frac{r_i(y-x)\bar{G}_i(y-x)\bar{G}_{\bar{i}}(y)\bar{R}_{\bar{i}}(y)}{\bar{G}_{\bar{i}}(x)\bar{R}_{\bar{i}}(x)}, \qquad (8)$$

c) $id_1d_2x \to \bar{i}d_1'd_2'y$ in conditions: $d_k' = d_k$ at $k \neq i$, $d_{\bar{i}}' = 1$, $y > 0$. The transition probability density is

$$p_{id_1d_2x}^{\bar{i}d_1'd_2'y} = \frac{g_{\bar{i}}(y+x)\bar{R}_{\bar{i}}(y+x)\bar{G}_i(y)\bar{R}_i(y)}{\bar{G}_{\bar{i}}(x)\bar{R}_{\bar{i}}(x)}, \qquad (9)$$

d) $id_1d_2x \to \bar{i}d_1'd_2'y$ in conditions: $d_k' = d_k$ at $k \neq i$, $d_{\bar{i}}' = 0$, $y > 0$. The transition probability density is found from formula:

$$p_{id_1d_2x}^{\bar{i}d_1'd_2'y} = \frac{r_{\bar{i}}(y+x)\bar{G}_{\bar{i}}(y+x)\bar{G}_i(y)\bar{R}_i(y)}{\bar{G}_{\bar{i}}(x)\bar{R}_{\bar{i}}(x)}. \qquad (10)$$

For the remaining states of the system, the transition probabilities are determined in a similar manner.

4 Finding the characteristics of the system

To determine the stationary reliability characteristics of the system, we find the stationary distribution of the EMC $\{\xi_n : n \geq 0\}$.

Suppose that for a stationary distribution of EMC $\{\xi_n : n \geq 0\}$ exist densities $\rho(id_1d_2x)$.

We introduce the following substitutions:

$$\tilde{\rho}(id_1d_2x) = \frac{\rho(id_1d_2x)}{\bar{F}_{\bar{i}}(x)}, \quad d_{\bar{i}} = 1,$$

$$\tilde{\rho}(id_1d_2x) = \frac{\rho(id_1d_2x)}{\bar{R}_{\bar{i}}(x)\bar{G}_{\bar{i}}(x)}, \quad d_{\bar{i}} = \bar{1},$$

$$\tilde{\rho}(id_1d_2x) = \frac{\rho(id_1d_2x)}{\dfrac{\int_0^\infty r_{\bar{i}}(t)\bar{G}_{\bar{i}}(x+t)dt}{P(\beta_{\bar{i}} > \tau_{\bar{i}})}}, \quad d_{\bar{i}} = 0.$$

The system of integral equations for the stationary distribution of the EMC $\{\xi_n : n \geq 0\}$ has the following form:

1. If $d_i' = 1$, $d_{\bar{i}}' = d_{\bar{i}}$, then

$$\tilde{\rho}(id_1d_2x) = \int_0^x \tilde{\rho}(id_1'd_2'y)f_i(x-y)dy + \int_0^\infty \tilde{\rho}(id_1'd_2'y)f_i(x+y)dy, \qquad (11)$$

2. If $d_i' = \bar{1}$, $d_{\bar{i}}' = d_{\bar{i}}$, then

$$\tilde{\rho}(id_1d_2x) = \int_0^x \tilde{\rho}(id_1'd_2'y)r_i(x-y)\bar{G}_i(x-y)dy + \int_0^\infty \tilde{\rho}(id_1'd_2'y)r_i(x+y)\bar{G}_i(x+y)dy, \qquad (12)$$

3. If $d_i' = \bar{1}$, $d_{\bar{i}}' = d_{\bar{i}} = 0$, then

$$\tilde{\rho}(id_1d_2x) = \int_0^x \tilde{\rho}(id_1'd_2'y)g_i(x-y)\bar{R}_i(x-y)dy + \int_0^\infty \tilde{\rho}(id_1'd_2'y)g_i(x+y)\bar{R}_i(x+y)dy, \qquad (13)$$

4. If $d_i = 1$, $d_i' = \bar{1}$, $d_i'' = 0$, $d_{\bar{i}}' = d_{\bar{i}}'' = d_{\bar{i}} \neq 0$, then

$$\tilde{\rho}(id_1d_2x) = \int_0^x \tilde{\rho}(id_1'd_2'y)g_i(x-y)\bar{R}_i(x-y)dy + \int_0^\infty \tilde{\rho}(id_1'd_2'y)g_i(x+y)\bar{R}_i(x+y)dy + \int_0^x \tilde{\rho}(id_1''d_2''y)dy\frac{\int_0^\infty r_i(t)g_i(x-y+t)dt}{P(\beta_i > \tau_i)} + \int_0^\infty \tilde{\rho}(id_1''d_2''y)dy\frac{\int_0^\infty r_i(t)g_i(x+y+t)dt}{P(\beta_i > \tau_i)}, \qquad (14)$$

5. $\int_E \rho(i\bar{d}x)dx = 1$. (normalization condition). $\qquad (15)$

By substitution, one can verify that the solution of the system of equations for the stationary distribution of the EMC has the form:

$$\rho(id_1d_2x) = c\rho_i\rho_{\bar{i}}\bar{F}_{\bar{i}}(x), \quad d_{\bar{i}} = 1, \qquad (16)$$

$$\rho(id_1d_2x) = c\rho_i\rho_{\bar{i}}p_i\bar{R}_{\bar{i}}(x)\bar{G}_{\bar{i}}(x), \quad d_{\bar{i}} = \bar{1}, \qquad (17)$$

$$\rho(id_1d_2x) = \frac{c\rho_i\rho_{\bar{i}}p_{\bar{i}}\int_0^\infty r_{\bar{i}}(t)\bar{G}_{\bar{i}}(x+t)dt}{P(\beta_{\bar{i}} > \tau_{\bar{i}})}, \quad d_{\bar{i}} = 0, \qquad (18)$$

$$\rho(id_1d_2x) = \frac{c\rho_i\rho_{\bar{i}}p_ip_{\bar{i}}\int_0^\infty r_{\bar{i}}(t)\bar{G}_{\bar{i}}(x+t)dt}{P(\beta_{\bar{i}} > \tau_{\bar{i}})}, d_i = d_{\bar{i}} = 0, \qquad (19)$$

where $p_i = P(\beta_i > \tau_i) = \int_0^\infty \bar{G}_i(t) r_i(t)dt$, $\rho_i = const$, which is equal to $\rho_i = \dfrac{1}{2+p_i}$, the constant c is found from the normalization condition.

We calculate the mean residence times in the states of the EMC.

1. When $d_i = d_{\bar{i}} = 1$,

$$\theta_{id_1d_2x} = \alpha_i \wedge [\alpha_{\bar{i}} - x]^+,\ E\theta_{id_1d_2x} = \int_0^\infty \frac{\bar{F}_i(t)\bar{F}_{\bar{i}}(x+t)}{\bar{F}_{\bar{i}}(x)}dt.$$

2. In case of $d_i = \bar{1}$, $d_{\bar{i}} = 1$,

$$\theta_{id_1d_2x} = \beta_i \wedge \tau_i \wedge [\alpha_{\bar{i}} - x]^+,$$

$$E\theta_{id_1d_2x} = \int_0^\infty \frac{\bar{G}_i(t)\bar{R}_i(t)\bar{F}_{\bar{i}}(x+t)}{\bar{F}_{\bar{i}}(x)}dt.$$

3. If $d_i = d_{\bar{i}} = \bar{1}$, then

$$\theta_{id_1d_2x} = \beta_i \wedge \tau_i \wedge [\beta_{\bar{i}} - x]^+ \wedge [\tau_{\bar{i}} - x]^+,$$

$$E\theta_{id_1d_2x} = \int_0^\infty \frac{\bar{G}_i(t)\bar{R}_i(t)\bar{G}_{\bar{i}}(x+t)\bar{R}_{\bar{i}}(x+t)}{\bar{G}_{\bar{i}}(x)\bar{R}_{\bar{i}}(x)}dt.$$

4. When $d_i = 0$, $d_{\bar{i}} = 1$,

$$\theta_{id_1d_2x} = [\beta_i - \tau_i]^+ \wedge [\alpha_{\bar{i}} - x]^+,$$

$$E\theta_{id_1d_2x} = \int_0^\infty \frac{\bar{F}_{\bar{i}}(x+t)}{\bar{F}_{\bar{i}}(x)P(\beta_i > \tau_i)}dt\int_0^\infty r_i(z)\bar{G}_i(t+z)dz.$$

5. If $d_i = 1$, $d_{\bar{i}} = \bar{1}$, then

$$\theta_{id_1d_2x} = \alpha_i \wedge [\beta_{\bar{i}} - x]^+ \wedge [\tau_{\bar{i}} - x]^+,$$

$$E\theta_{id_1d_2x} = \int_0^\infty \frac{\bar{F}_i(t)\bar{G}_{\bar{i}}(x+t)\bar{R}_{\bar{i}}(x+t)}{\bar{G}_{\bar{i}}(x)\bar{R}_{\bar{i}}(x)}dt.$$

6. In case of $d_i = 0$, $d_{\bar{i}} = \bar{1}$,

$$\theta_{id_1d_2x} = [\beta_i - \tau_i]^+ \wedge [\beta_{\bar{i}} - x]^+ \wedge [\tau_{\bar{i}} - x]^+,$$

$$E\theta_{id_1d_2x} = \int_0^\infty \frac{\bar{G}_{\bar{i}}(x+t)\bar{R}_{\bar{i}}(x+t)}{\bar{G}_{\bar{i}}(x)\bar{R}_{\bar{i}}(x)P(\beta_i > \tau_i)}dt\int_0^\infty r_i(z)\bar{G}_i(t+z)dz.$$

7. When $d_i = \bar{1}$, $d_{\bar{i}} = 0$,

$$\theta_{id_1d_2x} = \tau_i \wedge \beta_i \wedge \left[[\beta_{\bar{i}} - \tau_{\bar{i}}]^+ - x\right]^+,$$

$$E\theta_{id_1d_2x} = \frac{\int_0^\infty \bar{G}_i(t)\bar{R}_i(t)dt\int_0^\infty r_{\bar{i}}(z)\bar{G}_{\bar{i}}(x+t+z)dz}{\int_0^\infty r_{\bar{i}}(z)\bar{G}_{\bar{i}}(x+z)dz}.$$

8. If $d_i = 1$, $d_{\bar{i}} = 0$, then

$$\theta_{id_1d_2x} = \alpha_i \wedge \left[[\beta_{\bar{i}} - \tau_{\bar{i}}]^+ - x\right]^+,$$

$$E\theta_{id_1d_2x} = \frac{\int_0^\infty \bar{F}_i(t)dt\int_0^\infty r_{\bar{i}}(z)\bar{G}_{\bar{i}}(x+t+z)dz}{\int_0^\infty r_{\bar{i}}(z)\bar{G}_{\bar{i}}(x+z)dz}.$$

9. In case of $d_i = d_{\bar{i}} = 0$,

$$\theta_{id_1d_2x} = [\beta_i - \tau_i]^+ \wedge \left[[\beta_{\bar{i}} - \tau_{\bar{i}}]^+ - x\right]^+,$$

$$E\theta_{id_1d_2x} = \frac{\int_0^\infty dt\int_0^\infty r_i(y)\bar{G}_i(t+y)dy\int_0^\infty r_{\bar{i}}(z)\bar{G}_{\bar{i}}(x+t+z)dz}{P(\beta_i > \tau_i)\int_0^\infty r_{\bar{i}}(z)\bar{G}_{\bar{i}}(x+z)dz},$$

where $\wedge$ is the minimum sign.

RV $[\alpha - x]^+$ is the residual residence time of the semi-Markov process $\xi(t)$ in the state, provided that the residence time in this state exceeded the value x.

Then $P\{[\alpha - x]^+ > t\} = \dfrac{\bar{F}(x+t)}{\bar{F}(x)}$ and

$$P\{[\alpha - x]^+ \in dt\} = \frac{f(x+t)}{\bar{F}(x)}.$$

RV $[\beta - \tau]^+$ is given by the following probabilities:

$$P\{[\beta - \tau]^+ > t\} = \frac{\int_0^\infty r(z)\bar{G}(t+z)dz}{P(\beta > \tau)},$$

$$P\{[\beta - \tau]^+ \in dt\} = \frac{\int_0^\infty r(z)g(t+z)dz}{P(\beta > \tau)}.$$

Define a random variable $\left[[\beta_{\bar{i}} - \tau_{\bar{i}}]^+ - x\right]^+$ by

$$P\{\left[[\beta - \tau]^+ - x\right]^+ > t\} = \frac{\int_0^\infty r(z)\bar{G}(t+x+z)dz}{\int_0^\infty r(z)\bar{G}(x+z)dz},$$

$$P\{[[\beta-\tau]^+-x]^+\in dt\}=\frac{\int_0^\infty r(z)g(t+x+z)dz}{\int_0^\infty r(z)\bar{G}(x+z)dz}.$$

Let us turn to the determination of the stationary reliability characteristics of the system S: mean stationary operating time of the system to failure T_+, mean stationary restoration time T_- and stationary availability factor K_a of the system.

To find the characteristics, we use the following formulas presented in [18]:

$$T_+=\frac{\int_{E_+} m(e)\rho(de)}{\int_{E_+} P(e,E_-)\rho(de)},\quad T_-=\frac{\int_{E_-} m(e)\rho(de)}{\int_{E_+} P(e,E_-)\rho(de)},$$

$$K_a=\frac{T_+}{T_++T_-}, \tag{20}$$

where $\rho(de)$ is stationary distribution of EMC $\{\xi_n;n\geq 0\}$, $P(e,E_-)$ – transition probabilities of the EMC $\{\xi_n;n\geq 0\}$ to a subset of fault states, $m(e)$ is mean residence time of the semi-Markov process in the state $e\in E$.

Consider the parallel connection of system components. In this case

$$E_-=\{100x,\ 200x\},$$

$$E_+=\{i\bar{d}x:\bar{d}=(d_1,d_2),\ \bar{d}\neq(0,0),\ x>0\}.$$

We calculate the components of formulas (20) using solutions of the system of integral equations (16) - (19), the transition probabilities of the embedded Markov chain (2) – (10) and average residence times in the states obtained above.

In the transformations, we will use the following formula, proved in [19]:

$$\sum_{j=1}^{n}\int_0^\infty\ldots\int_0^\infty \bar{F}_j(t)\left(\prod_{\substack{k=1,\\k\neq j}}^{n}\bar{F}_k(t+y_k)dy_k\right)dt=\prod_{j=1}^{n}E\alpha_j.$$

where $F_k(t)$ are distribution functions of independent nonnegative random variables α_k with mathematical expectations $E\alpha_k$.

We write out expressions for the numerator and denominator for mean stationary operating time of the system to failure and mean stationary restoration time.

$$\int_{E_-} m(e)\rho(de)=$$

$$=c\rho_1\rho_2\int_0^\infty dx\int_0^\infty dt\int_0^\infty r_2(y)\bar{G}_2(y+t)dy\int_0^\infty r_1(z)\bar{G}_1(y+x+z)dz+$$

$$+c\rho_1\rho_2\int_0^\infty dx\int_0^\infty dt\int_0^\infty r_1(y)\bar{G}_1(y+t)dy\int_0^\infty r_2(z)\bar{G}_2(y+x+z)dz=$$

$$=c\rho_1\rho_2 p_1p_2E([\beta_1-\tau_1]^+)E([\beta_2-\tau_2]^+),$$

$$\int_{E_+} P(e,E_-)\rho(de)=$$

$$=c\rho_1\rho_2\int_0^\infty dx\int_0^\infty r_2(x+y)\bar{G}_2(x+y)dy\int_0^\infty r_1(t)\bar{G}_1(y+t)dt+$$

$$+c\rho_1\rho_2\int_0^\infty dx\int_0^\infty r_1(x+y)\bar{G}_1(x+y)dy\int_0^\infty r_2(t)\bar{G}_2(y+t)dt+$$

$$+c\rho_1\rho_2\int_0^\infty dx\int_0^\infty r_2(y)\bar{G}_2(y)dy\int_0^\infty r_1(t)\bar{G}_1(y+x+t)dt+$$

$$+c\rho_1\rho_2\int_0^\infty dx\int_0^\infty r_1(y)\bar{G}_1(y)dy\int_0^\infty r_2(t)\bar{G}_2(y+x+t)dt=$$

$$=c\rho_1\rho_2\left[p_1\int_0^\infty\bar{G}_2(x)R_2(x)dx+p_2\int_0^\infty\bar{G}_1(x)R_1(x)dx\right],$$

$$\int_{E_+} m(e)\rho(de)=$$

$$=c\rho_1\rho_2\big[p_1E([\beta_1-\tau_1]^+)(E\alpha_2+E(\beta_2\wedge\tau_2))+$$
$$+p_2E([\beta_2-\tau_2]^+)(E\alpha_1+E(\beta_1\wedge\tau_1))+$$
$$+(E\alpha_1+E(\beta_1\wedge\tau_1))(E\alpha_2+E(\beta_2\wedge\tau_2))\big].$$

Let's write down formulas for finding of stationary characteristics of reliability:

$$T_-=\frac{c\rho_1\rho_2p_1p_2E([\beta_1-\tau_1]^+)E([\beta_2-\tau_2]^+)}{c\rho_1\rho_2\left[p_1\int_0^\infty\bar{G}_2(x)R_2(x)dx+p_2\int_0^\infty\bar{G}_1(x)R_1(x)dx\right]}= \tag{21}$$
$$=\frac{E([\beta_1-\tau_1]^+)E([\beta_2-\tau_2]^+)}{E([\beta_1-\tau_1]^+)+E([\beta_2-\tau_2]^+)},$$

$$T_+=\frac{p_1E([\beta_1-\tau_1]^+)(E\alpha_2+E(\beta_2\wedge\tau_2))}{p_1p_2\left(E([\beta_1-\tau_1]^+)+E([\beta_2-\tau_2]^+)\right)}+$$
$$+\frac{p_2E([\beta_2-\tau_2]^+)(E\alpha_1+E(\beta_1\wedge\tau_1))}{p_1p_2\left(E([\beta_1-\tau_1]^+)+E([\beta_2-\tau_2]^+)\right)}+ \tag{22}$$
$$+\frac{(E\alpha_1+E(\beta_1\wedge\tau_1))(E\alpha_2+E(\beta_2\wedge\tau_2))}{p_1p_2\left(E([\beta_1-\tau_1]^+)+E([\beta_2-\tau_2]^+)\right)}.$$

As an example of the use of formulas (21), (22), consider a system in which K_1 operating time $E\alpha_1$ = 8 h, K_2 operating time $E\alpha_2$ = 6 h, K_1 recovery time $E\beta_1$ = 0.71 h, K_2 recovery time $E\beta_2$ = 0.83 h, RV α_1, α_2,

β_1, β_2 have 5th order Erlang distribution. Each component has a non-random time reserve ($R_i(t) = 1(t-h_i)$), which varies from 0 to 0.7 hours in 0.1 increments. The corresponding values of mean stationary operating time of the system to failure $T_+(h_1,h_2)$, mean stationary restoration time $T_-(h_1,h_2)$ and stationary availability factor $K_a(h_1,h_2)$ of the system for the specified distribution were calculated. The results are presented in Table 1, calculated on the condition that $h_1 + h_2 = 0.7$.

Table 1. The influence of the time reserve on the system reliability characteristics.

h_1	h_2	$T_-(h_1,h_2)$	$T_+(h_1,h_2)$	$K_a(h_1,h_2)$
0	0	0.385	49.551	0.9923
0	0.7	0.241	102.274	0.99765
0.1	0.6	0.241	91.035	0.99736
0.2	0.5	0.238	84.365	0.99719
0.3	0.4	0.233	82.291	0.99717
0.4	0.3	0.228	85.172	0.99733
0.5	0.2	0.224	93.617	0.99762
0.6	0.1	0.219	108.69	0.99799
0.7	0	0.214	132.3	0.99839

Analysis of the data in the table shows the significant effect of the time reserve on reliability characteristics.

5 Conclusion

In this paper we construct a semi-Markov model of a two-component system with a component-wise random instantly replenished time reserve. On a concrete example, influence of capacities of component-wise stores on stationary characteristics of reliability of system is shown. The effect of the time reserve on the reliability characteristics obtained is analysed.

In the future, it is planned to build semi-Markov models of multicomponent systems with a component-wise time reserve.

The results of this work can be used to construct semi-Markov models of systems with different types and strategies for using the time reserve, engineering calculations and solving optimization problems associated with the use of a time reserve.

The research was carried out within the state assignment of the Minobrnauki of Russia (No. 1.10513.2018/11.12), with financial support by RFBR (project No. 18-01-00392a).

References

1. G.N. Cherkesov, *Reliability of Technical Systems with Time Redundancy* (Sovietskoe Radio, Moscow, 1974)
2. B.P. Kredentser, *Prediction of Reliability of Systems with Time Redundancy* (Naukova Dumka, Kiev, 1978)
3. V.Y. Kopp, Yu.E. Obzherin, A.I. Peschanskiy, *Stochastic models of automized system with time reservation* (Publishing of Sevastopol State Technical University, Sevastopol, 2000)
4. I.A. Ushakov, *Probabilistic Reliability Models* (Wiley, San Diego, Calif., 2012)
5. G.N. Cherkesov, *Reliability of Hardware and Software Systems* (Textbook) (St. Petersburg, Piter, 2005)
6. A.M. Polovko, S.V. Gurov, *Basics of Reliability Theory* (St. Petersburg, BHV-Peterburg, 2008)
7. N.A. Severcev, *Reliability theory of the composite systems in working off and operation* (Tutorial) (Moscow, Urait Publishing House, 2018)
8. G.V. Druzhinin, *Reliability of the automated production systems* (Moscow, Energoatomizdat, 1986)
9. Yu.E. Obzherin, S.M. Sidorov, S.N. Fedorenko, MATEC Web of Conferences **129**, 03009. (2017)
10. Yu.N. Rudenko, I.A. Ushakov, *Reliability of Energy Systems* (Nauka, Novosibirsk, 1989)
11. G.N. Antonov et. al., *Reliability of energy systems and their equipment* (Handbook: in 4 volumes) (Moscow, Energoatomizdat, 1994)
12. G.N. Cherkesov, Yu.N. Rudenko, I.A. Ushakov, *Scientific and engineering problems of energy system reliability* (New York, Harwood academic publishers, 1987)
13. N. Limnios, G. Oprisan, *Semi-Markov Processes and Reliability* (Springer Science+Business Media, New York, 2001)
14. V.S. Koroluk, A.F. Turbin, *Markovian restoration processes in the problems of system reliability* (Naukova Dumka, Kiev, 1982)
15. V.S. Koroluk, *Stochastic System Models* (Naukova Dumka, Kiev, 1989) (in Russian)
16. Y.E. Obzherin, E.G. Boyko, *Semi-Markov Models: Control of Restorable Systems with Latent Failures* (Elsevier Academic Press, London, 2015)
17. V.A. Kashtanov, A. Medvedev, *Reliability Theory of the Composite Systems* (Theory and Practice) (Moscow, Fizmatlit, 2010)
18. A.N. Korlat, V.N. Kuznetsov, M.M. Novikov, A.F. Turbin, *Semi-Markov Models of Recoverable Systems and Queuing Systems* (Stiinta, Chisinau, 1991)
19. V.N. Kuznetsov, A.F. Turbin, G.Zh. Tsaturyan, *Semi-Markov model of recoverable systems* (In-t matematiki AN USSR, Kiev, 1981)

Formation of the list of gas industry critical objects from the standpoint of energy security of the country

Sergey Senderov[1,*], *Viktor Rabchuk*[1], and *Sergey Vorobev*[1]

[1]Energy Systems Institute SB RAS, 130 Lermontov str., Irkutsk, Russia

Abstract. The paper is devoted to the search for critical objects of the Russian gas industry in terms of energy security of the state and regions. The task of identifying critical objects of the gas industry is presented and an algorithm for their distribution according to the lists of federal and regional levels is shown. A list of critical facilities of the federal gas industry is presented. This list includes, along with gas transmission network objects, head compressor stations at the exits from gas fields and underground gas storages.

1 Introduction

When managing the development of the country's fuel and energy complex (FEC), it is necessary to take into account the requirements of energy security. In large, there are two such requirements. The first requirement is a long-term, non-deficit provision of domestic consumers with all necessary types of fuel and energy resources (FER) and fulfillment of compulsory export deliveries of Russian fuel and energy resources in the operation of the fuel and energy complex under normal conditions. The second requirement is to create conditions for ensuring domestic consumers of the country with all kinds of fuel and energy resources and export deliveries of Russian fuel and energy resources in the minimum required quantities in the event of emergency and crisis situations, without allowing significant social and economic damage from non-delivery of fuel and energy resources (both to domestic consumers and to exports).

Here and below in the text, an emergency means a partial or complete failure of a limited number of objects at the same time. The critical situation means situation, when the objects of the FEC (or a some energy system (ES)) practically all simultaneously have to function in an abnormal mode for example, especially cold days in a large part of the country or any other conditions that cause ESs to work in an abnormal regime). Accounting for the second requirement is associated with the need to carefully identify the critical objects (COs) of the FEC and ESs, i.e. those objects, partial or complete failure of which in different conditions can significantly reduce the production capacity of ESs and FEC and lead to significant shortages in the supply of appropriate types of energy. FEC and ESs can and should have different lists of COs at different times. This is due to the fact that the configuration of energy transport communications and the loading of nodes, both consumption and production, gradually change over time. Some objects may lose their significance over time, but new more significant objects may appear, the failure of which will have a critical impact on the production capacities of the ESs or the FEC of the country. The study of the interconnected work of the different ESs within the framework of a single FEC will make it possible to understand which COs from the lists of such objects of some ES can be included in the list of the COs of FEC. The analysis of the interconnected work of different energy systems within the framework of a single FEC will make it possible to understand which COs from the lists of such objects of individual energy systems can be included in the list of the COs of the FEC level. The criterion of such selection can be the level of negative consequences for consumers because of failure of a specific CO of some ES, taking into account the compensating abilities of the FEC to reduce the negative consequences (interchangeability of FER, diversification of their sources, etc.).

The study of critical infrastructures in the energy sector has become one of the important areas of scientific research. Many works are devoted to this question [1, 2 & oth.]. This topic also concerns the materials of the European Commission [3] and the US Department of Energy [4]. Unfortunately, any estimates and model estimates in these documents are not present. Basically, there are discussions about the policy and organization of energy supplies, as well as the interaction between the elements of energy systems. In such works there is no access to specific objects due to the fragmentation of energy systems and their work not in the framework of a single complex, but on the basis of the principles of market trade.

In general, the identification of the COs in the FEC can effectively solve the problems of identifying and minimizing the consequences of implementing various types of threats that are fraught with crisis and

* Corresponding author: ssm@isem.irk.ru

emergency situations at the FEC, as well as the tasks of preparing these objects in advance for the duration of the specified situations. Equally important is the solution of the problem of concentration of material, monetary and human resources in solving the problems of increasing the sustainability of the functioning of the identified COs of FEC without sputtering such, as a rule, limited resources.

2 Algorithm for the formation of the COs lists

Today, the main role in the fuel and energy complex of Russia belongs to gas. Its share in the balance of boiler-furnace fuel (BFF) is 74% in Russia as a whole. However, for the regions of the European part and the Urals (where 88% of the Russian Federation's population live), this share in a number of regions reaches 90-95%, and in some regions of the Russian Federation - up to 98-99%. Given the dominant role of gas in the fuel and energy complex, at the first stage we will focus on the Unified Gas Supply System (UGSS) of the country (including export outlets of Russian gas beyond its borders and imports of imported gas to Russia). The example of the UGSS should address the following issues:

- development of a general algorithm for the identification COs of a some system;
- evaluation of negative consequences for the some system from partial or complete failure of various COs in the event of emergency on these objects;
- assessment of the role of specific COs in ensuring the operability of the some system under the large-scale emergency situations;
- formation of a list of specific measures to minimize the negative consequences from a decrease in the level of efficiency of each CO of the some ES;
- substantiation of the list of invariant measures to minimize the negative consequences of emergency situations on the COs of the ES, taking into account possible simultaneous combinations of emergencies on different objects.

From the standpoint of energy security, two types of objects can be assigned to critical objects of ES:

- objects whose shutdown can cause significant shortages of energy resources in the whole country (deficit in the relative amount of δ_{sum} and more of the total demand for this energy resource), such objects can be considered as the ES COs of the federal level;
- objects that are not included in the list of federal COs for the some ES, but the termination of their work, can cause significant shortfall of energy resource in at least one of the regions (deficit in relative amount of δ_{reg} or more of the total demand of the region for this energy resource). Such objects can be considered as the ES COs of the regional level.

As δ_{sum} for the gas industry, a value of 5% was assumed earlier in [5]. As δ_{reg}, in the first approximation, a value of 30% can be used. It should be understood that these values are still very arbitrary, for their comprehensive justification, special studies are needed.

The algorithm for the formation of the COs lists of the regional and federal levels for the some ES is presented in Fig. 1.

3 Characteristic of the settlement network

Let's consider the real situation in the Russian gas industry. In 2017, Russia produced 690 billion m^3 of gas. 8 billion m^3 of natural gas (from Central Asian countries and Azerbaijan) was imported. Domestic consumption (together with the gas sector's own needs) amounted to 472 billion m^3, and exports amounted to 226 billion m^3, including slightly more than 192 billion m^3 to non-CIS

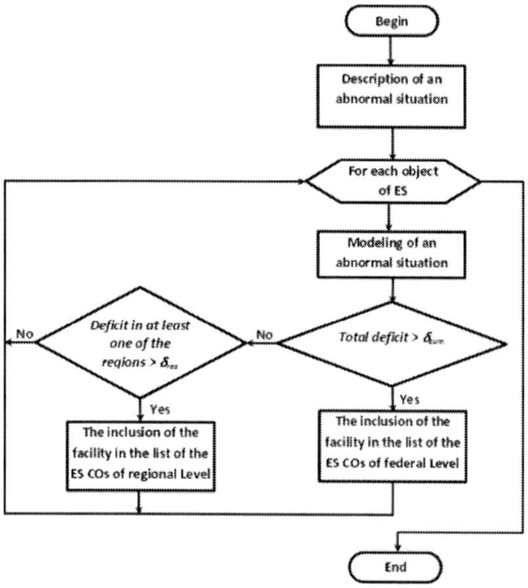

Fig. 1. Algorithm for the formation of the ES CFs lists of the regional and federal levels

countries. The country has an unified gas supply system (UGSS) with an extensive system of main gas pipelines covering most of the european part of Russia's territory. The existing territorial structure of the Russian gas supply system has a number of significant shortcomings. In fact, the main gas consumer within the country (the parts of its European part) does not have its own gas resources. All gas consumed here is mainly produced in the north areas of the Tyumen region (NATR), located in 2-2.5 thousand km from places of intensive gas consumption. Today, the NATR produces more than 85% of the total Russian gas, i.e. practically all Russian gas has to be transported for long distances by main gas pipelines with a colossal concentration of gas flows in one corridor. These corridors have a large number of crossings and bridges. The gas pipelines in one corridor can be located from each other at a very small distance. Currently, more than 20 potential crossings of main gas pipelines potentially dangerous for the operation of the UGSS can be identified in Russia. Violation of some of them may lead to a restriction of up to 85% of gas supply to domestic consumers in the country as a whole or to almost complete cessation of gas exports (while retaining a 50% restriction of gas supplies to domestic consumers).

In the studies carried out earlier [5, 6] facilities (20 crossings of main gas pipelines) characterized as the COs of gas transmission network of Russia have already been shown. At the same time, the issue of the remaining facilities of the gas industry for their inclusion in the list of COs from the energy security standpoint has not yet been raised. Along with the major intersections of the main gas pipelines at nodal compressor stations and outside, the COs of the gas industry from the standpoint of its operability are the main compressor stations on the outputs from the gas fields and underground gas storages (UGS). At present, there are 22 UGSs in the Russian gas transmission network, 5 more UGSs are operating outside the Russian Federation (3 in Belarus, one in Armenia and Germany). Another 7 UGSs (in which Gazprom participates as a co-investor) operate in the gas pipeline network on the territory of the European states [7]. All these UGSs incorporated into a specially designed flow model gas industry (within the software "Oil and Gas in Russia") [5, 6, 8]. This software allows imitating all aspects of the functioning of not only the UGSS of Russia, but also the technologically related gas networks of European countries. The settlement scheme of the model contains 382 nodes, including the above UGSs, 28 gas sources (in the model these are the main compressor stations (CS), 64 gas consumers, 268 nodal compressor stations, and 628 arcs representing the corridors of main gas pipelines and single main gas pipelines, as well as outlets to distribution networks.

4 Mathematical statement of the solution of the problem

Mathematically, Russia's unified gas supply system is represented as a network that varies with time, whose nodes have the facilities for production, conversion and consumption of material flows relating the enterprises. When solving the problem of state estimation of the system after a disturbance the criterion of flow distribution optimality is represented by the minimum shortage of the energy resource for consumer at the minimum costs of its delivery.

A change in the state of system facilities requires solving the problem of flow distribution in the system in order to supply energy carriers to the maximum extent possible, i.e. the model can be formalized as a problem about the maximum flow [8-10]. Then it is necessary to add two fictitious nodes to the graph scheme: O – total source, S – total sink, and introduce additional sections connecting node O with all sources, and all consumers with node S. Mathematically, the stated problem has the following form:

$$\max f \qquad (1)$$

$$\sum_{i\in N_j^+} x_{ij} - \sum_{i\in N_j^-} x_{ji} = \begin{cases} -f, j=O \\ 0, j\neq O,S \\ f, j=S \end{cases} \qquad (2)$$

$$0 \leq x_{ij} \leq d_{ij} \qquad (3)$$

Here N_j^+ - a subset of edges incoming to node j; N_j^- - a subset of edges outgoing from node j; f - a value of the total flow in the network; x_{ij} - a flow in edge (i, j); d_{ij} - constraints on the flow in edge (i, j).

Problem (1)-(3) about the maximum flow in a general case has a non-unique solution. The next step suggests solving the problem of maximum flow of the minimum cost, i.e. the cost functional is minimized:

$$\sum_{(i,j)} C_{ij} x_{ij} \rightarrow \min \qquad (4)$$

where C_{ij} - price or specific costs of the energy resource transportation.

The comprehensive approach to solving the stated problems throughout the entire technological chain from receiving gas by the network to its transportation to the distribution network or export makes it possible to obtain the total estimate of the production capabilities of the entire system in the extreme conditions. Solution to the problem is represented by the values of gas shortage at the consumption nodes under emergency conditions. Based on these results, it becomes possible to build a list of all the facilities in the industry whose shutdown will lead to a potential gas shortage in the network. Rank this list by the relative size of the resource deficit in the network. By cutting off objects whose shutdown will lead to a potential gas deficit in the grid that is less than the previously assigned value, for example, at 5%, we can get a list of the gas industry's COs. Such a list will also be ranked by the value of the impact on network capacity.

5 The results of the research

Appropriate studies have been carried out on the model of the gas industry of Russia described above. The initial conditions for the calculations were as follows: the average day of the maximum gas consumption in the network, based on statistics on gas consumption by regions in January 2017. In such days the network can be considered extremely tense with maximum load. The total flow of gas through the network on such days, taking into account export deliveries, amounted to 2250 mln m^3 approximately. The results of these studies showed that the potential gas shortage for consumers will be observed when each of the 441 gas objects in Russia is shut down (242 nodes and 199 arcs of the network settlement graph). From this list, 61 objects should be included in the list of gas industry COs of the federal level. Among them there are 25 arcs between nodal CSs and 36 nodes, including 30 nodal CSs, 5 main CSs on the outlets from large gas fields and one UGS. Information on the estimated values of the relative gas deficits in the network with the disconnection of specific nodes and arcs in ranked by the degree of reduction of the gas deficit is presented in Table 1 (the real names of the UGSS objects in this paper were replaced by conditional numbers).

From the data of Table 1 it can be seen that if each of the first eight objects of the ranked list of federal UGSS COs is disconnected, the relative gas deficit in the system may amount to about 20% of the required total supply. Disconnecting each of the following 15 objects can result in a gas flow restriction within 10-16%. Disconnecting all other objects from the list of CFs can lead to a relative gas deficit in the system in the range of 5-9%.

Table 1. Estimated relative gas deficits in the network at the most intense day of January, 2017, when federal UGSS COs are cut off

The ordinal number of the ranked list of COs in the calculated graph	Object type	Relative gas deficit in the system when the CO is disconnected,%
1, 2, 3, 4	Node	21
5, 6, 7	Arc	21
8	Node	19
9, 13, 14	Arc	16
10[a], 11, 12, 15	Node	16
16	Arc	12
17, 18, 19, 22, 23	Node	10
20, 21	Arc	10
24	Node	9
25, 26, 28[a]	Node	8
27	Arc	8
29, 31, 33, 35, 37, 39, 41	Arc	7
30[a], 32, 34, 36, 38, 40	Node	7
42, 48, 50	Arc	6
43[a], 44[a], 45, 46[b], 47, 49, 51	Node	6
52, 55, 56, 59, 60	Arc	5
53, 54, 57, 58, 61	Node	5

[a] The object of gas extraction, i.e. to the maim CS at exits from gas fields

[b] Underground gas storage

6 Conclusion

From the standpoint of ensuring the survivability of the gas industry and, in general, from the standpoint of ensuring the energy security of the country and its regions, it is first of all necessary to pay attention to the objects found in the result of the submitted studies from the list of the UGSS COs). Appropriate organizational measures should be taken to prevent emergencies in the first place at these facilities. A strategic task in the development of the industry may be the task of forming directions and concrete ways to reduce the critical importance of such COs for the potential productivity of the UGSS. With the development of the experience of identifying the COs in the gas industry for different time slices, the work can be continued in the part of identifying COs in other energy systems and in the fuel and energy complex as a whole.

The work was carried out within the framework of a scientific project III.17.5.1 of program of fundamental research of the SB RAS, reg. number AAAA-A17-117030310451-0.

References

1. Critical Infrastructure in the Energy Sector / https://www.me.government.bg/en/themes/critical-infrastructure-in-the-energy-sector-336-300.html
2. I.E. Anastasakis, Energy-critical infrastructure protection: "The case of Greece", https://www.google.ru/url?sa=t&rct=j&q=&esrc=s&source=web&cd=4&ved=0ahUKEwjxsKe0vJvcAhVlGZoKHX_SClMQFghGMAM&url=https%3A%2F%2Fhrcak.srce.hr%2Ffile%2F251955&usg=AOvVaw23toJ3XkfktOeJTwxmVxlk"
3. Protection of critical infrastructure / https://ec.europa.eu/energy/en/topics/infrastructure/protection-critical-infrastructure
4. Energy. Critical Infrastructure and Key Resources Sector-Specific Plan as input to the National Infrastructure Protection Plan (Redacted). May 2007. https://www.energy.gov/sites/prod/files/oeprod/DocumentsandMedia/Energy_SSP_Public.pdf
5. Senderov S.M., Rabchuk V.I., Edelev A.V., Izv. RAN. Energetika [Proc. of the Russ. Acad. of Sciences. Pow. Engin.], **1**, 70-78 (2016)
6. Senderov S., Edelev A.. Energy (2017). doi:10.1016/J.ENERGY.2017.11.063
7. http://www.gazprom.ru/about/production/underground-storage/
8. Voropai N.I., Senderov S.M., Edelev A.V., Energy (2012). doi:10.1016/j.energy.2011.07.038
9. Khramov A.V., Enikeyeva S.M., Khrustaleva N.M. & oth., in *Methods and models of research into the survivability of energy systems* (Nauka Sib. Branch Novosibirsk, 1990, 86-91)
10. L.R. Ford Jr., D.R. Fulkerson, *Flows in Networks* (The Rand Corporation, Santa Monica, California, 1962)

Search of critically important combinations of objects of the gas industry from the positions of the system operability

Sergey Vorobev[1*], *Sergey* Senderov[1], *Alexey* Edelev[1]

[1]Melentiev Energy Systems Institute, 130 Lermontov str., Irkutsk, Russia

Abstract. The paper is devoted to the search for critical combinations of gas industry objects from the standpoint of system operability. Critically important is the object partial or complete failure which can cause significant damage to the country from the fuel and energy complex in general or within the framework of a separate energy system. It is possible to failure of a critical and other object of the system. The paper presents the main major critical combinations of gas industry objects. Their significance and impact on the efficiency of the gas industry in Russia is shown. Conclusions are made about the expediency of searching for critical combinations of gas industry objects from the standpoint of system operability.

1 Introduction

One of the main aspects of ensuring energy security is the creation of conditions for the greatest possible degree of satisfaction of consumers with energy resources in emergency situations [1]. The study of this aspect of energy security requires the identification of critical objects for the fuel and energy complex in general and for individual energy systems. Critically important from the standpoint of energy security, it is necessary to consider an object whose partial or complete failure can cause significant damage to the country from the fuel and energy complex due to shortages to consumers of the necessary fuel and energy resources.

The increase in the number of major accidents in energy systems in recent years is due to the significant wear and tear of fixed assets, and the lack of significant financial investment in their reconstruction. Large-scale accidents in energy systems caused by the failure of various critical objects entail significant, sometimes irreparable, damage to consumers in the form of large short-term deliveries of final types of energy.

Current calculations show that in case of accidents at the major intersections of main gas pipelines, consumers of final types of energy resources in the regions will suffer significantly because of their deficiencies. In some cases, such deficits can amount to 30-50% of their need for certain periods of the year that are tense for consumption of energy resources. Along with the intersections of the main gas pipelines, potentially large-scale nodal compressor stations, as well as linear sections of separate main gas transmission corridors, may be potentially dangerous for the operation of the system. From all this list of potentially dangerous facilities for the functioning of the system, the most significant, that is, critically important objects are currently being identified, followed by the development of measures aimed at improving the reliability of these objects to maintain uninterrupted fuel and energy supply to consumers.

Previously, a number of studies were conducted on the identification of critical objects in the gas transportation network. The list of crossings of main gas pipelines in the Unified Gas Supply System of Russia has been determined, the disruption of which will result in a relative shortage of daily gas supplies for the system as a whole of 5% or more [2, 3]. Investigations were carried out to find and identify combinations of separate sections of main gas pipelines, the simultaneous disruption of which may lead to a significant shortage of daily gas supplies through the system (5% or more) [4, 5].

At this stage of research of critical gas industry objects, critical combinations of gas industry objects are considered, that is, such facilities are identified whose failure, together with one of the critical objects, will lead to a much larger gas deficit in consumers than if they fail these objects separately. After identifying possible critical combinations of objects in each specific case, the possibility of bypassing bottlenecks is analyzed by short-term increase in the capacity of individual sections of main gas pipelines. This is done with a view to minimizing gas shortages among consumers. As a result of such measures, the significance of a number of potential critical combinations of objects can be reduced. Thus, all possible critical combinations of gas industry objects will be identified and their list, ranked according to the degree of significance, will be formed.

* Corresponding author: seregavorobev@isem.irk.ru

2 Determination of critical elements in energy systems

In the ongoing study, both for determining the critical objects themselves and for searching for critical combinations of objects, a flow model was used, which is the core of the software complex "Oil and Gas of Russia" [6, 7]. Its application makes it possible to determine the degree of satisfaction of gas needs within the country and to ensure export supplies. In addition, the "Oil and Gas of Russia" allows identifying "bottlenecks" - sections of the gas transmission network, which in some cases limit the production capacity of the system.

The flow distribution model in the Unified Gas Supply System of Russia in the "Oil and Gas of Russia" is designed to assess the production capabilities of the Unified Gas Supply System of Russia in conditions of various kinds of disturbances. The purpose of such studies is to minimize gas deficits at the consumption sites. The Unified Gas Supply System of Russia in the model is represented as a set of three subsystems: gas sources, main gas transport network and consumers.

When solving the problem of estimating the state of a system after a perturbation, the criterion of the optimality of the distribution of flows is the minimum gas deficit in the consumer with minimum costs for delivering gas to consumers. This problem can be solved by finding the maximum flow over the network. The problem of the maximal flow was investigated in [8] and is formulated as follows:

$$f \to max, \tag{1}$$

with follow conditions

$$\sum_{i\in N_j^+} x_{ij} - \sum_{i\in N_j^-} x_{ji} = \begin{cases} -f, j=O \\ 0, j\neq O, S \\ f, j=S \end{cases} \tag{2}$$

$$0 \le x_{ij} \le d_{ij}, \text{ for all of (i, j).} \tag{3}$$

In this formulation, f is the maximized variable corresponding to the maximum flow.

With complex schemes, which are the Russian and European gas transmission networks, there may be several solutions, that is, several possible maximum flows. Then it is advisable to talk about minimizing the cost of delivering gas to consumers and using the Basaker-Gowen algorithm [8]:

$$\sum_{(i,j)} C_{ij} x_{ij} \to min, \tag{4}$$

$$x \in X^*. \tag{5}$$

Its application allows us to determine the maximum flow of the resource at its minimum cost or the optimal volumes of daily gas from the underground gas storages, which maximally provide the given volumes of gas supply to consumers at the minimum costs for extraction, transportation of gas and its withdrawal from underground gas storages.

O - the node number of the common source; S - is the number of the common drain node; N^+_j – is a subset of edges entering into the node j; N_j – is the subset of outgoing edges from the node j; f – is the value of the total flow over the network; x_{ij} – is the flow along the edge (i, j); d_{ij} – restrictions on the flow along the edge (i, j); X^* - is the set of solutions of problem (4)-(6); C_{ij} – specific costs for gas transportation.

The node O is connected by fictitious edges with all real sources of energy resources, and node S - with all consumers.

In the flow distribution model, the Basaker-Gowen algorithm is used to calculate the maximum flow of minimum cost, which as a result allows us to determine the possible level of satisfaction of consumers with gas. As a result of implementation of various contingencies, there may be a gas shortage in consumers caused by a lack of capacity of certain sections of the gas pipelines. Bypassing such narrow or limiting the production capacity of the system's sites, in permissible amounts, will reduce the gas shortage that has arisen in the situation in question among consumers.

If there is a gas shortage in consumers caused by a shortage of capacity of the gas pipelines, other branches of the main gas pipelines, which are not affected by the violation in question, assume increased volumes of gas. In such a situation, the network load is changing and there may be a lack of capacity on other sections of the main gas pipelines.

The subsequent expansion of bottlenecks in the gas transportation network will allow to minimize gas shortages among consumers and makes an assessment and definition of possible critical combinations of gas industry objects as much as possible.

"Oil and Gas of Russia" has graphic capabilities, which allow the researcher to identify many potential bottlenecks in the analysis of calculation results, that is, objects that do not have a reserve of production capabilities. Most often, such facilities are either, or may become, in the future the reason for the shortage of the necessary amount of gas to consumers.

Even with the use of graphic capabilities of "Oil and Gas of Russia" it is impossible without additional modelling to find real bottlenecks, which at the moment really limit the ability of the gas transportation network to supply gas to consumers.

The result of solving the problem (1) - (5) is the value of the maximum flow in the system and the gas deficit in consumers. It is clear that in order to compensate for the gas deficit in consumers, the amount of flow should become more than calculated. Then the problem arises of increasing the carrying capacity of edges, in order to obtain a flow of the required value.

This problem of optimal network transformation [8] can be written in the following form:

$$f \to max, \tag{6}$$

with follow conditions

$$\sum_{i\in N_j^+} (x_{ij} + y_{ij}) - \sum_{i\in N_j^-} (x_{ji} + y_{ji}) = \begin{cases} -f, j = O \\ 0, j \neq O, S, \\ f, j = S \end{cases} \tag{7}$$

$$0 \le x_{ij} \le d_{ij}, \tag{8}$$

$$0 \le y_{ij} \le b_{ij} \quad (9)$$

In this situation, there may also be several solutions, that is, several possible maximum flows. Then it is advisable to talk about minimizing the cost of delivering gas to consumers:

$$\sum_{(i,j)} C_{ij} x_{ij} + \sum_{(i,j)} A_{ij} y_{ij} \to min, \quad (10)$$

$$x \in X^*. \quad (11)$$

f – maximal flow, y_{ij} –edge's capacity increment (i, j); A_{ij} – price or unit costs for transport of gas by increment y_{ij}; b_{ij}– edge-increment limits (i, j).

At the same time, the unit costs for gas transportation by the increment y_{ij} should be many times higher than the unit costs for transporting gas through the grid.

The required bottlenecks are those objects for which, after solving the problem (6) - (11), y_{ij} will be greater than zero.

To solve the problem of bypassing the bottlenecks found in the flow distribution model, the possibility of incrementing the gas flow along the edges within 10% of the capacity of the flow was laid. Such a short-term increase in the capacity of the main gas pipeline section is possible with an increase in the working capacity of compressors at large main compressor stations [9]. As a result of increasing the working pressure in the gas pipeline, an increase in the throughput capacity of the main gas pipeline section is achieved up to 10%. As a result (through the use of the technical capabilities of the gas transportation network), the problem of minimizing gas shortages among consumers is being solved.

3 Definition of critical combinations of gas industry objects

The scheme of the Unified Gas Supply System of Russia used for the calculations in this paper on the flow model takes into account all the main features of the operation of the Unified Gas Supply System of Russia and contains:

- 382 nodes, including:
  - 28 gas sources;
  - 64 gas consumers (subjects of the Russian Federation);
  - 24 underground gas storage facilities;
  - 266 nodal compressor stations;
- 486 edges representing main gas pipelines and outlets to distribution gas networks.

Initial data, such as daily consumption, extraction, export and import of gas, throughput capacity of existing gas pipelines are adopted in accordance with official statistics [10-12] for 2017. In a specially conducted study [13], an appropriate analysis was carried out, as a result of which 61 objects of the gas industry were assigned to critical gas industry objects. Among these objects there are 25 edges between the node compressor stations and 36 nodes, including 30 nodal compressor stations, 5 head compressor stations at the outlets from large gas fields and a compressor station on one underground gas storage.

Received the above-mentioned 61 objects, calculations were carried out in this settlement scheme to determine critical combinations of objects for each individual critical object. With the subsequent "expansion" of bottlenecks - the implementation of measures aimed at minimizing gas shortage among consumers.

The criterion for assigning a pair to potential critically important combinations was to consider the difference in the ΔQ deficit.

$$\Delta Q = Q_{pair} - Q_{obj}, \quad (12)$$

$$\Delta Q \ge \delta \quad (13)$$

Q_{pair} – total gas deficit in the system caused by failure of a couple of objects, Q_{obj} – gas deficit in case of failure of a critically important object, δ – restriction of gas deficit increment.

Calculations show that more than 15,000 combinations of other objects with identified critical objects lead to an increase in gas deficit in consumers. Based on considerations of the possibility of expert analysis, we limit the value of δ to within 5%. As a result of the calculation in pairs of all 61 critical gas industry objects with the rest objects of the system (in total it was analyzed for their simultaneous disconnection of more than 61,000 pairs), 630 pairs were obtained, consisting of one critical object and one gas object, whose failure can lead to the occurrence of ΔQ more than 5%.

In table. 1 (the real names of objects of the Unified gas supply system in this article are replaced by conditional serial numbers), the average number of pairs for all objects is clearly represented, as well as the possible total volumes of gas deficit in the system both in the event of critical objects failure, combinations of other gas industry objects with this critically important object.

Table 1. Critical combinations of objects.

Critical object	The number of critical combinations of objects (pairs)	Gas deficit, %		ΔQ, %	After traversing bottlenecks,%	
		Single object	Pair		Gas deficit (pair)	Decrease of deficit
1, 6	10	21	28-32	7-11	25-30	2-3
2, 5, 7	11	21	28-32	7-11	25-30	2-3
3, 4	8	21	28-29	7-8	25-26	2-3
8	7	19	25-30	6-11	24-29	2-3
9-11	5	16	22-25	6-9	22-25	0
12-15	7	16	22-27	6-11	21-25	2-3
16	5	12	18-21	6-9	18-19	0-2
17	14	10	17-20	7-10	13-18	1-4
18-22	5	10	16-19	6-9	13-15	3-4
23	11	10	15-17	5-7	11-16	0-4
24	11	9	15-17	6-8	11-16	0-4
25	16	8	14-20	6-12	12-18	1-4
26, 27	11	8	13-16	5-8	9-14	0-4
28	1	8	14	6	10	4
29	11	7	13-15	6-8	9-13	0-4
32, 33	11	7	12-15	5-8	9-13	0-4
34, 35	16	7	12-19	5-12	10-16	1-4
36-39	10	7	13-18	6-11	10-14	3-4
40	14	7	13-22	6-11	9-18	3-4
41	14	7	13-21	6-14	9-18	3-4
42	16	6	12-18	6-12	10-15	1-4
43	1	6	12	6	11	1
45	11	6	11-13	5-7	10-12	1
46	1	6	11	5	7	4
47	11	6	11-12	5-6	7-12	0-4
48	11	6	11-14	5-8	7-12	0-4
49, 50	16	6	11-17	5-11	9-15	1-4
51	24	6	11-18	5-12	9-15	1-4
52	9	5	12-14	7-9	9-10	3-4
53	19	5	10-25	5-20	8-22	4-5
54	17	5	11-25	6-20	6-22	3-5
55	13	5	10-25	5-20	6-22	3-5
56	15	5	10-22	5-17	6-18	3-5
57	21	5	10-19	5-14	6-16	3-5
58, 59	17	5	10-19	5-14	6-16	3-5
60, 61	20	5	10-19	5-14	6-16	3-5

Table. 1 show that apart from some critically important objects (28, 43, 46) having only one critically important combination of objects (one pair each), from 5 to 24 critically important combinations of objects were obtained for the remaining critical objects. So the first seven critical objects are present in 8 to 11 critical combinations of objects, the failure of which will entail a total daily gas deficit for consumers in the system as a whole in the amount of 27 - 32%. The difference in the deficit ΔQ for these pairs is 7-11%.

In general, the following should be noted from the results of the study. Violation of the functioning of a critical combination of objects can lead to a greater gas deficit in consumers by an average of 8-10% than in the case of a violation of the functioning of a particular critical object.

In possible situations with the failure of the critical combinations of objects considered, measures to bypass bottlenecks result in an insignificant decrease in the gas deficit (on average by 2-3%), which also confirms the high importance of identifying these combinations. Here it is worth noting the critical combination of objects with critical objects 9-11, to which measures to bypass bottlenecks have no effect, that is, that before, that after the bypassing of bottlenecks the same volumes of gas deficit remain.

It is necessary to highlight 4 objects - 2 nodal compressor stations and 2 edges between the nodal compressors stations, which are present in 60% of the critical combinations of objects obtained. Violation of the normal functioning of these facilities together with a number of critical objects can significantly increase the gas deficit in consumers.

4 Conclusion

As a result of this research, the issues of searching for critical combinations of gas industry objects were considered. The definition of such objects in the gas industry, failure of which, coupled with one of the critical objects, will lead to a much larger gas deficit in consumers than in the event of failure of these objects separately. To search for critical combinations of gas industry objects, the flow distribution model laid in the software "Oil and Gas of Russia" was used. This model is designed to assess the production capabilities of the gas transmission network in conditions of various kinds of disturbances. A detailed analysis of the major critical combinations of gas industry objects is given.

The conducted studies showed that in the modern configuration of the Unified Gas Supply System, situations are possible when, in the event of a critical combination of objects failing, the total gas deficit in consumers can reach 32% of the total gas demand. Research and accounting of such objects and their combinations is necessary to improve the reliability of the Unified Gas Supply System in its development and reconstruction works. The main efforts in this case should be aimed at reducing the conditional significance of the identified specific critical combinations of objects for the operability of the entire system.

The work was carried out within the framework of a scientific project III.17.5.1 of program of fundamental research of the SB RAS, reg. number AAAA-A17-117030310451-0.

References

1. N.I. Pyatkova, [and others], *Energy security of Russia: problems and solutions*, 198 p. (2011).
2. S.M. Senderov, V.I. Rabchuk, A.V. Edelev, Izv. Ros. Akad. Nauk. *Energetika*. № **1**. p. 70-78. (2016).
3. S. Senderov, A. Edelev, *Energy*. doi: 10.1016 / J.ENERGY.2017.11.063. (2017).
4. S. Vorobev, A. Edelev, *MLSD 2017*, IEEE, 2017. DOI 10.1109 / MLSD.2017.8109707. (2017).
5. S. Vorobev, A. Edelev, E. Smirnova, *RSES 2017*. E3S Web Conf. Vol. **25**, DOI 10.1051 / e3sconf / 20172501004. (2017).
6. A.V. Edelev, S.M. Enikeeva, S.M. Senderov, *Computational technologies*. – Vol. 4, **№ 5**. – p. 30 – 35. (1999).
7. S.V. Vorobev, A.V. Edelev, *Software products and systems*. – **№3**. – p. 174 – 177. (2014).
8. L.R. Ford, D.R. Fulkerson. *Flows in networks*. Princeton University Press. 276p. (1966).
9. Yu.P. Korotaeva, R.D. Margulova, *Extraction, preparation and transport of natural gas and condensate. Reference manual in 2 volumes. Volume II*. Nedra, 288 p., (1984).
10. *Export of the Russian Federation of the most important goods in 2011 - 2016 (according to the Federal Customs Service of Russia)* http://customs.ru/ index.php?option=com_ newsfts&view =category&id= 52&Itemid=1978&limitstart=60.
11. *InfoTEK Monthly oil and gas magazine*. **№1**, – P.154. (2017).
12. *Ministry of Energy of the Russian Federation. Statistics*. http://minenergo.gov.ru/activity/statistic.J. Johansson, H. Hassel, E. Zio, *Reliability Engineering & System Safety*. **No. 120.** P.27-38. (2013).
13. S.M. Senderov, V.I. Rabchuk, S.V. Vorobev, *Proc. of the col. of rep. Methodological issues of reliability research of large energy systems. 90th meeting "Reliability of developing energy systems"*. July 1-7, Irkutsk. (2018).

Electromagnetic safety enhancing in railway electric supply systems

Natal'ya Buyakova[1, *], *Vasiliy* Zakaryukin[2], *Andrey* Kryukov[2, 3], *Tu* Nguyen [3]

[1] Angarsk State Technical University, Angarsk, Russia , 665835;
[2] Irkutsk State Transport University, Irkutsk, Russia, 664074;
[3] Irkutsk National Research Technical University, Irkutsk, Russia, 664074

Abstract. The methods of simulation modeling of traction power supply systems is developed, allowing analyzing the electromagnetic safety conditions in complex traction power networks equipped with sucking transformers. The effectiveness of the application of the sucking transformers (ST) connected in the rail cut with a distance of 3 km between transformers is small due to rails' currents. The sucking transformers with return wires do not have this drawback. The use of ST with a return wire allows significantly improving the electromagnetic safety conditions: the level of magnetic field strength is reduced by tens of percent.

1 Introduction

Railway traction power networks of alternating current are electromagnetically unbalanced and therefore can cause dangerous voltages in disconnected adjacent power lines and communications [1–7]. The sucking transformers (ST) are used to reduce the voltage of magnetic influence in traction power supply systems (TPSS), increasing the interrelation between the overhead contact system and rails. The traction current, instead of returning through the ground, flows along the rails or through a special return wire (Fig. 1).

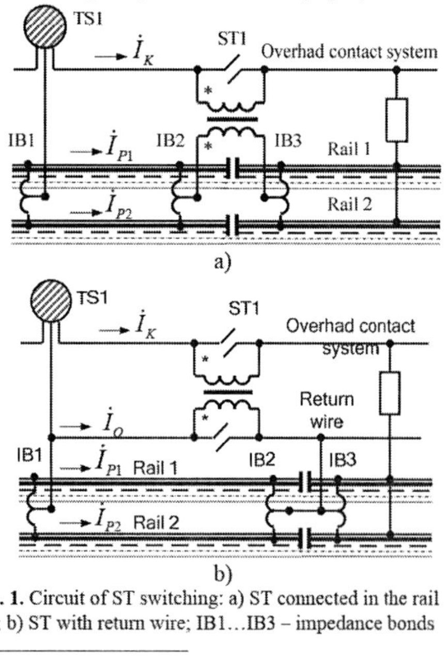

Fig. 1. Circuit of ST switching: a) ST connected in the rail cut; b) ST with return wire; IB1...IB3 – impedance bonds

Reducing the distance between the overhead contact system and return current circuit leads to a decrease in the magnetic field of traction power network and reduction induced voltages to adjacent lines. In practice, two variants of ST switching are used: in the first variant, the transformers are connected to the rail network (Fig. 1, a), in the second variant, the return wire is used (Fig. 1, b).

ST transformation coefficient equals to one. The operating mode of ST is close to short circuit, since the secondary winding is connected either to the return wire with the grounding of the latter or directly to the rails [5, 6]. In both circuits, the currents in the transformer windings are almost antiphase, and the voltages on the windings are relatively small. The return wire for efficient use should be mounted as close as possible to the contact suspension. As indicated in [5], the optimal distance between the sucking transformers connected to the rail cut is 3 km, and for ST with return wire this distance is 4.5 km.

2 Modeling of sucking transformers and impedance bonds

From the point of view of the magnetic field generated by the traction power network, the inter-substation zone with ST can be divided into sections of three types:

- a section of local currents from the traction load to the nearest sucking transformers;
- sections of transit currents on the remaining part of the inter-substation zone, except for the sections from traction substation (TS) to closest ST;

* Corresponding author: bn_900@mail.ru

- sections from substations to ST; they require separate consideration, since in these sections the sucking transformers are located only on one side, and the return of traction currents, which largely depends on the resistance of the substation earth lead and the transient resistance of the rail – ground, occurs through the ground.

The article presents the results of the analysis of the electromagnetic field created by the currents of section near the traction substation. The remaining two variants are characterized by other regularities partially considered in [5], and require additional research.

Modeling of ST is not difficult, since the manufacturing plant gives data in the passport of sucking transformer OMO-800/35 which is similar to the power transformer. The OMO-800/35 single-phase double-winding transformer has the following nominal parameters: power 800 kV·A, winding voltage 1,05 kV, short-circuit voltage 8.5%.

A more complicated situation arises when modeling impedance bonds (IB), since the manuals show the A-parameters of IB as a four-pole for a frequency of 25 Hz, as well as the resistance of the primary winding at 50 Hz and its nominal current.

The algorithm for determining the IB parameters which are suitable for modeling in the program Fazonord includes the following steps [8].

1. According to given four-pole A-parameters $\underline{A}$, $\underline{B}$, $\underline{C}$, $\underline{D}$ input impedances of idling and short circuit are determined $\underline{Z}_K$. The impedance $\underline{Z}_K$ is half divided to obtain active resistances, as well as two coils inductance, both primary and secondary, assuming their equality:

$$\underline{Z}_{1S} = \frac{\underline{Z}_K}{2} = \frac{1}{2}\frac{\underline{B}}{\underline{D}}.$$

2. The magnetization branch impedance is determined from the idling equation of the four-pole:

$$\underline{Z}_\mu = \frac{\underline{A}}{\underline{C}} - \underline{Z}_{1S}.$$

3. Imaginary parts of impedances $\underline{Z}_{1S}$ and $\underline{Z}_\mu$, which are determined for 25 Hz frequency, are positive and substantially higher than real ones, vary in proportion to the frequency and are recalculated at 50 Hz frequency.

4. For modeling, it is more convenient to consider impedance bonds as an autotransformer whose nominal voltage is equal to its 50 Hz impedance multiplied by the nominal current of the primary winding composed of two half-windings. For the purposes of analyzing the modes of the traction power network, the secondary winding cannot be considered, since the loading of this winding at 50 Hz frequency is small. The autotransformer model parameters are calculated through the nominal current I_n according to the following formulas:

– losses and voltage of short circuit

$$P_\kappa = 2\,\mathrm{Re}(Z_{1S})I_n^2;\; U_\kappa = \frac{2\cdot|\underline{Z}_{1S}|\,I_n}{U_n}\cdot 100;$$

– losses and current of idling

$$P_x = \mathrm{Re}\{\dot{U}_n[\mathrm{Re}(\dot{I}_x) - j\,\mathrm{Im}(\dot{I}_x)]\};\; \dot{I}_x = \frac{\dot{U}_n}{\underline{Z}_{1S} + \underline{Z}_\mu}.$$

The DT-1-300 impedance bond, most common on the alternating current road network, is characterized by the following parameters:

- rated current of primary winding 300 A;
- impedance at 50 Hz frequency 1.0 Ohm;
- primary winding contains 5 + 5, and the secondary winding contains 30 turns;
- A-parameters at a frequency of 25 Hz are equal to $\underline{A} = 0.33e^{-j4°}$, $\underline{B} = 0.062e^{j14°}$, $\underline{C} = 0.37e^{-j57°}$, $\underline{D} = 3.0e^{j4°}$.

With these initial data, the IB model in the form of an autotransformer has the following parameters:

- nominal power 300 kV·A;
- nominal voltage of the primary winding 150+150 V;
- short-circuit voltage 13.5%;
- short circuit losses 4.65 kW;
- idling current 30%;
- idling losses 17.25 kW.

When checking the model in idle and short circuit conditions, the regime parameters differ from the initial ones by not more than 2%.

3 Methods and results of modeling of TPSS modes

For a detailed study of sucking transformer protective action, a typical inter-substation zone of a double-track section of 25 kV alternating current railway with a length of 45 km with a contact suspension PBSM95+MF100 of two tracks and R-65 rails was considered. The A-185 return wires were supposed to be located above the contact suspensions of each track. The modeling was carried out according to the methods of [9–14] using the Fazonord software [10].

The model of the multi-wire system contained four wires of a contact suspension, two return wires and four rail threads. The return wires were not used in the model without sucking transformers and in the model with connection of ST to the rails.

For option of connecting the sucking transformers to the rail cut, model included 15 elements of traction power network, each of which had 3 km length. For the variant with the return wire, the model included 20 elements of the traction power network with 2.25 km length. The two-way supply of the inter-substation zone was simulated with loads of 5 + *j*5 MV·A applied at the end of the second section of the traction power network between the overhead contact system and the rails of each track.

For analysis of sucking transformers effectiveness, the operation of TPSS has been simulated in three versions:

- absence of sucking transformers;
- ST with return wire located at intervals of 4.5 km;
- ST connected to the rail cut through 3 km; rail – ground transient resistance is taken to be 2 Ohm·km.

Fig. 2 shows dependences of the amplitudes of both the electric and magnetic field strengths, as well as both the horizontal and vertical components effective values, from the X coordinate at height of 1,8 m. Strength values were calculated after TPSS modes defining based on technique described in work [14].

Horizontal and vertical components of the electrical field strength created by a system consisting of N wires, were defined by the following ratios:

$$\dot{E}_Y = -\frac{1}{\pi\varepsilon_0}\sum_{i=1}^{N}\dot{\tau}_i\frac{y_i[(x-x_i)^2 - y^2 + y_i^2]}{\zeta_i};$$

$$\dot{E}_X = \frac{2}{\pi\varepsilon_0}\sum_{i=1}^{N}\dot{\tau}_i\frac{(x-x_i)yy_i}{\zeta_i},$$

where $\zeta_i = [(x-x_i)^2 + (y+y_i)^2][(x-x_i)^2 + (y-y_i)^2]$; x, y – the coordinates of the point where the strength is defined; $\dot{\tau}_i$ – i wire charge per length unit.

The charges values were calculated from the Maxwell's formulas first group

$$\dot{\mathbf{T}} = \mathbf{A}^{-1}\cdot\dot{\mathbf{U}},$$

where $\dot{\mathbf{U}} = [\dot{U}_1 \;\; \ldots \;\; \dot{U}_N]^T$ – the wires voltages relative to the zero potential point (the earth); $\dot{\mathbf{T}} = [\dot{\tau}_1 \;\; \ldots \;\; \dot{\tau}_N]^T$ – the wires charges per length unit; $\mathbf{A}$ – matrix whose components are potential coefficients

$$\alpha_{ii} = \frac{1}{2\pi\varepsilon_0}\ln\frac{2y_i}{r_i};$$

$$\alpha_{ij} = \frac{1}{2\pi\varepsilon_0}\ln\frac{\sqrt{(x_i-x_j)^2 + (y_i+y_j)^2}}{\sqrt{(x_i-x_j)^2 + (y_i-y_j)^2}},$$

where x_i, y_i – the wires coordinates; r_i – the wires radiuses; ε_0 – electrical constant.

Equations of the strength vector hodograph were represented as

$$E_x(t) = \sqrt{2}\,E_X\sin(\omega t + \varphi_X);$$
$$E_y(t) = \sqrt{2}\,E_Y\sin(\omega t + \varphi_Y).$$

Maximum strength E_{MAX} was reached at moments which were defined using formula

$$t_{max} = \frac{1}{2\omega}\text{Arctg}(\varsigma),$$

where $\varsigma = \dfrac{E_X^2\sin 2\varphi_X + E_Y^2\sin 2\varphi_Y}{E_X^2\cos 2\varphi_X + E_Y^2\cos 2\varphi_Y}$.

The choice of the needed arctangent value was affected by condition

$$E_X^2\cos 2(\omega\, t_{max} + \varphi_X) + E_Y^2\cos 2(\omega\, t_{max} + \varphi_Y) < 0$$

The following expressions were used to calculate vertical and horizontal strength components of the magnetic field:

$$\dot{H}_X = \frac{1}{2\pi}\sum_{i=1}^{N}\dot{I}_i\frac{y-y_i}{(x_i-x)^2 + (y_i-y)^2};$$

$$\dot{H}_Y = -\frac{1}{2\pi}\sum_{i=1}^{N}\dot{I}_i\frac{x-x_i}{(x_i-x)^2 + (y_i-y)^2}.$$

The presented technique implemented in Fazonord software allows to evaluate of electromagnetic safety in traction networks taking into account Carson Solution for EMF middle and distant zones.

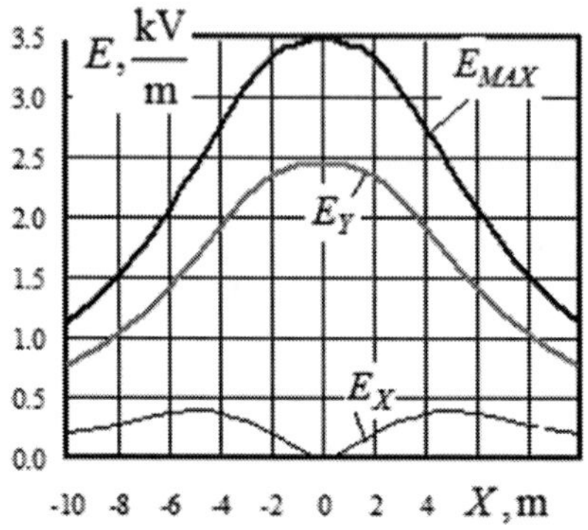

Fig. 2,a. Electric field strengths dependences from X coordinate, absence of sucking transformers

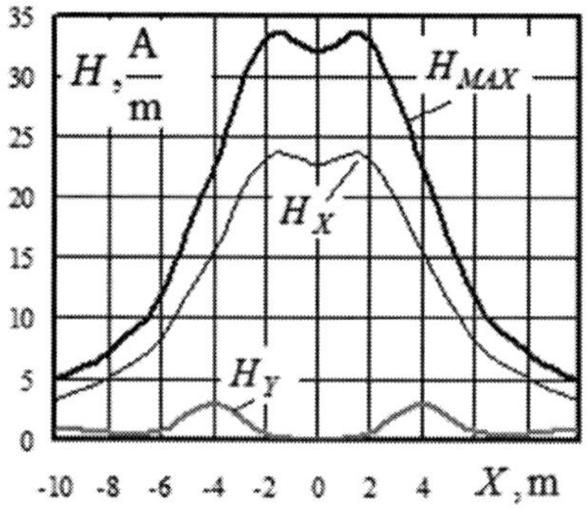

Fig. 2,b. Magnetic field strengths dependences from X coordinate, absence of sucking transformers

 https://doi.org/10.1051/e3sconf/20185803003

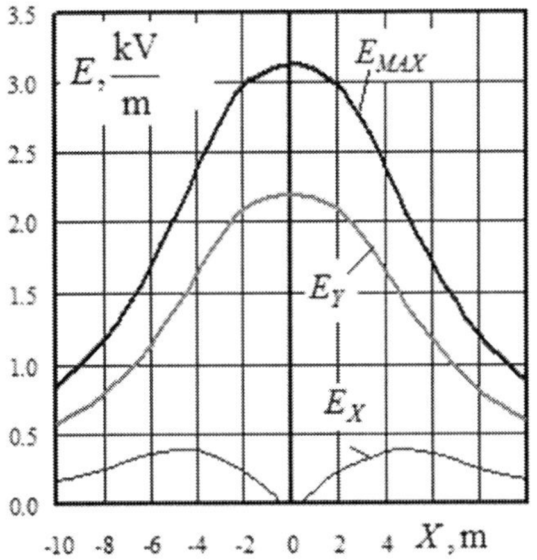

Fig. 2,c. Electric field strengths dependences from X coordinate, sucking transformers with return wire

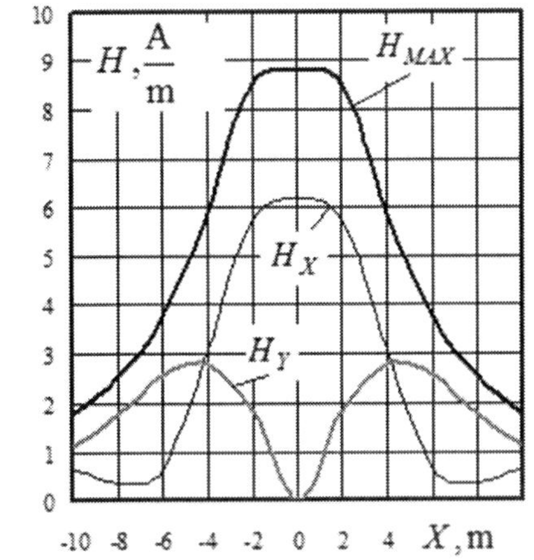

Fig. 2,d. Magnetic field strengths dependences from X coordinate, sucking transformers with return wire

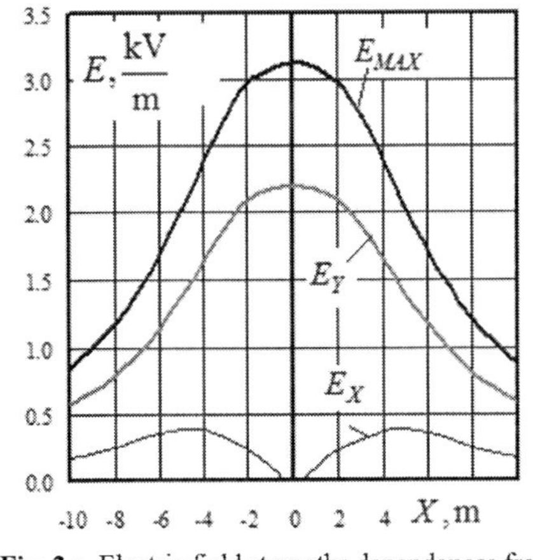

Fig. 2,e. Electric field strengths dependences from X coordinate, sucking transformers in rail cut

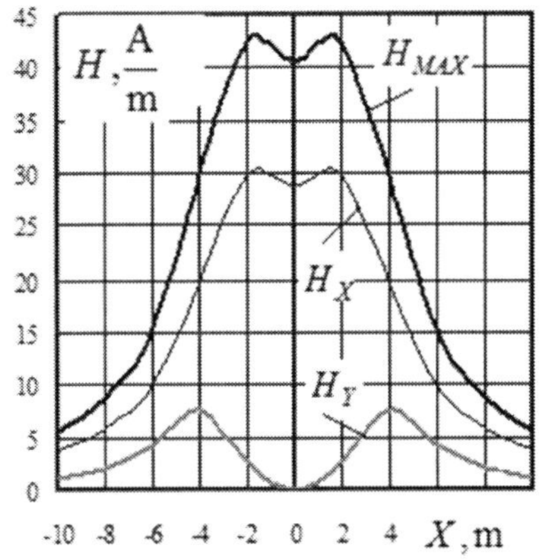

Fig. 2,f. Magnetic field strengths dependences from X coordinate, sucking transformers in rail cut

Comparative strengths are shown in Fig. 3, 4. Summary indicators of electromagnetic field amplitudes are given in Table 1.

Table 1 Summary indicators of strengths amplitudes

Parameter	Electric field, kV/m			Magnetic field, A/m		
	1	2	3	1	2	3
Minimum	1,13	0,83	1,13	5,00	1,79	5,70
Average	2,28	1,94	2,28	17,4	4,89	22,2
Maximum	3,46	3,12	3,46	33,0	8,78	42,4

Notes. 1 – absence of sucking transformers; 2 – ST with return wire; 3 – ST connected to rail cut

Fig. 5 shows dependence of electromagnetic field strengths on distance from left TS to observation point along the railway at the above-mentioned load for the inter-track space ($X = 0$) at a height of 1.8 m

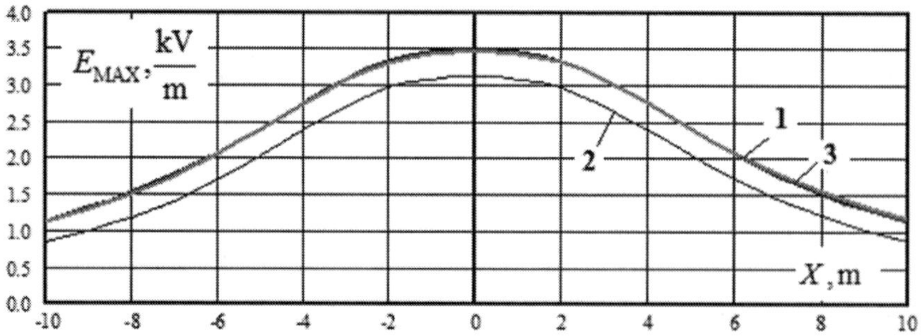

Fig. 3. Comparative dependences of electric field strengths from X coordinate:
1 – absence of sucking transformers; 2 – sucking transformers with return wire; 3 – sucking transformers in rail cut

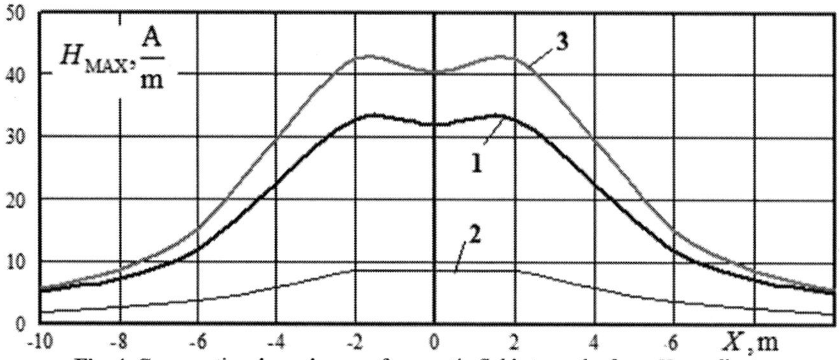

Fig. 4. Comparative dependences of magnetic field strengths from X coordinate

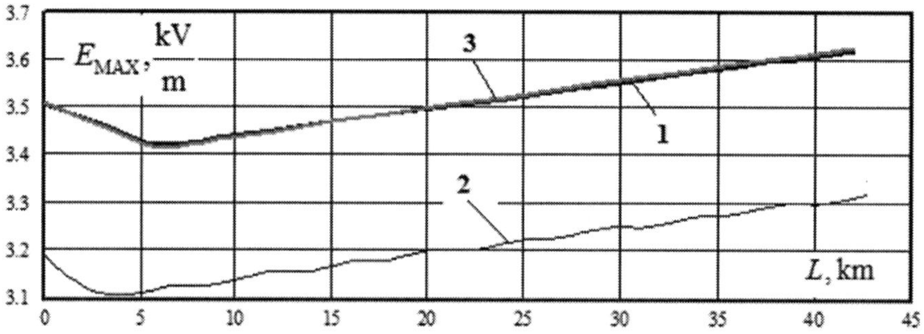

a)

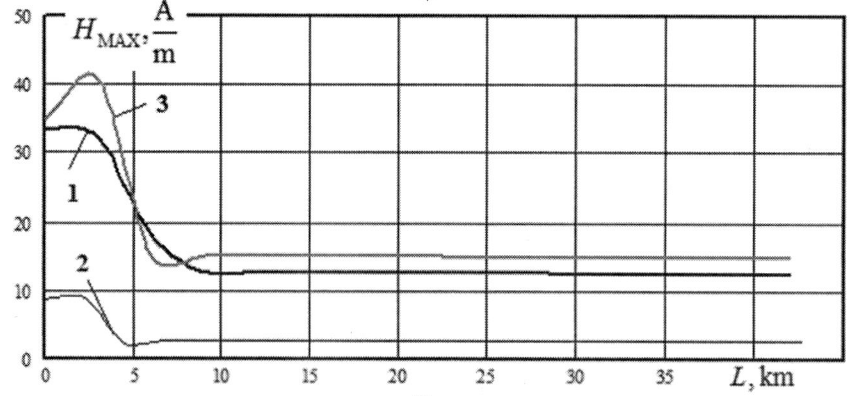

b)

Fig. 5. Dependences of field strengths on distance from left TS: a) electric field; b) magnetic field

4 Conclusions

The final conclusions on the results presented above can be formulated as follows.

1. The developed methods of traction power supply system's simulation allow analyzing the electromagnetic safety conditions in complex traction power networks equipped with sucking transformers.

2. The sucking transformers connected to the rail cut practically do not affect the levels of electric field strengths under the overhead contact system near traction substation. The magnetic field is extended due to the increase in the current of rails: the maximum value of strength is increased by 28%.

3. In the presence of sucking transformers with return wire, the electromagnetic safety conditions are significantly improved: the electric field strength is reduced by 10% and the magnetic field by 73% at the level of the person's head in the inter-track space.

References

1. Apollonskiy S.M. Problems of electromagnetic safety on electrified railway. In 2 volumes. T. II. Electromagnetic safety on railway with alternating current in the traction power network. Moscow, 2017. 414 p.
2. Ogunsola A., Mariscotti A. Electromagnetic Compatibility in Railways. London: Springer, 2013, 529 p.
3. Energieversorgung elektrischer bannen / H. Biesenack, E. Braun, G. George, etc. / Wiesbaden: B.G. Teubner Verlag. 2006. 732 p.
4. Steimel A. Electric traction motive power and energy supply. Basics and practical experience. Munchen: Oldenbourg Industrieverlag, 2008. 334 p.
5. Pavlov I.V. Sucking transformers in the traction power networks of alternating current. M.: Transport, 1965. 204 p.
6. Bader M.P. Electromagnetic compatibility. Moscow, 2002. 638 p.
7. Bardushko V.D., Zakaryukin V.P., Kryukov A.V. Principles for the construction of the power supply systems for railway transport. Moscow, 2014. 166 p.
8. Zakaryukin V.P., Kryukov A.V., Alekseenko V.A. Modeling of track impedance bonds in the calculation of rail track circuit modes // Transport infrastructure of the Siberian region. Irkutsk, 2011. Vol. 1. P. 551-555.
9. Zakaryukin V.P., Kryukov A.V. Complex asymmetric modes of electric systems. Irkutsk: Irkutsk University, 2005. 273 p.
10. Kryukov A.V., Zakaryukin V.P. Methods of joint modeling of both traction and external power supply systems for alternating current railways. Irkutsk, 2011. 170 p.
11. Zakaryukin V., Kryukov A., Cherepanov A. Intelligent Traction Power Supply System // International Scientific Conference Energy Management of Municipal Transportation Facilities and Transport. EMMFT 2017. Advances in Intelligent Systems and Computing, Vol 692. Springer, Cham. P. 91-99.
12. Zakaryukin V.P., Kryukov A.V. Clarified methods for determining mutual electromagnetic influences of adjacent transmission lines // News of higher educational institutions. Problems of power production. No. 3-4. 2015. P. 29-35.
13. Zakaryukin V.P., Kryukov A.V. Determination of the induced voltages when nonparallel power lines are adjacent to one another // Power Technology and Engineering. 2015. Vol. 49. No. 4. P. 304–309.
14. Kryukov A.V., Zakaryukin V.P., Buyakova N.V. Electromagnetic safety in the systems of power supply of the railroads: modeling and management. Angarsk, 2018. 382 p.

https://doi.org/10.1051/e3sconf/20185803004

Choosing of the FEC critical objects in model researches of energy security problems and its features

Natalia Beresneva[1,*], *Natalia* Pyatkova[1]

[1]Melentiev Energy Systems Institute, 130 Lermontov str., Irkutsk, Russia

Abstract. The overall objective of the proposed work is to formulate the main provisions of the approach to the determination of critical objects for the fuel and energy complex within the framework of hierarchical system studies of the energy security problems for the country and its regions. This approach involves the use of a hierarchical system of models, presupposes a preliminary choice of critical objects for energy systems. The approach is focused on the analysis and evaluation of interrelated systems functioning when implementing threats to energy security (in the form of contingencies), requires the accounting for changes in the work of critical objects for industries. To assess the criticality of these objects for the energy complex as a whole, a criterion is proposed for the significance of objects by the most important model indicators.

1 The features of the critical objects choosing in energy security problems researches

Energy security (ES) of the country and its regions is essentially a balance of supply and demand for energy resources [1]. The main content of research tasks to ensure energy security in fact reduced to:
- forecasting the conditions for the functioning and development of energy systems (SE) and the fuel and energy complex as a whole in case of emergency situations caused by ES threats;
- assessment of the energy systems condition in these conditions, the selection of possible directions and measures to prevent or reduce the negative impact of contingencies, taking into account specialized safety measures (reserves and storage).

The most important component of these studies is the identification of energy objects critical to the industries (CO of SE) or FEC (CO of FEC), determination the degree of their criticality in the context of the situations studying. The critical objects are most vulnerable objects, whose disruption of work can cause significant failures in reliable satisfaction of needs [2].

The critical objects is determined within the framework of studies (sectorial or general energy) scenarios for implementing contingencies and possible compensatory measures aimed at adjusting the performance of energy facilities, or on their restructuring. In the case of an abnormal situation, the task of selecting objects whose performance may decrease the efficiency of the system as a whole is solved. When introducing compensatory measures, objects are determined, despite the inoperability of which, the state of the system as a whole is able to improve. If these objects are classified as critical, they can be deprived of this status. The abnormal situations can be classified into object and territorial, technogenic and functional (by type of disturbance), point and group (by the number of objects of application of disturbances). Compensatory measures, technogenic or functional, can also be applied to individual objects or to their groups.

Energy systems and the fuel and energy complex as complex technical infrastructures in the search for COs can be viewed from three positions. Energy systems in research can be analyzed from the structural and functional point of view, the FEC - from the functional and taking into account the factor of ensuring ES, which is possible only in the case of an analysis of the interrelated work of energy systems. Structural aspect, is connected with the situations of system restructuring, functional aspect is connected with changes in the production characteristics of the objects. The factor of ensuring ES, in turn, is respected with respect to the modeled variants in general.

Accordingly, the following criteria can be used as the selection criteria of the CO:
- structural indicators characterizing changes in the efficiency of the systems functioning as a whole in modeled situations in comparison with their normal mode of operation;
- functional indicators that characterize the effectiveness of the participation of facilities in covering the total demand for energy resources;

- indicators of energy security, whose threshold values can become a criterion for determining the admissibility of modeled situations in terms of energy security providing.

In this case, there is a factor of consistency of the CO (Figure 1) in the sense that the industry CO in the case of determining its criticality for the FEC as a whole

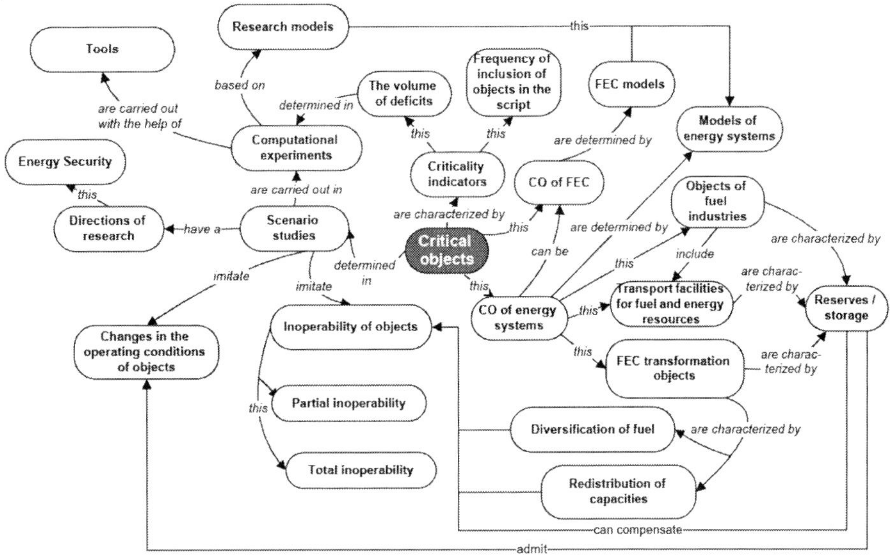

Fig. 1 -. Scheme for the coordination of the selection of CO in energy security researches

can be selected as the COs of the FEC. The possibility of such a choice can be justified by the fact of interconnected work of the branch systems that are the basis of the whole energy system, within the fuel and energy sector. The necessity of proving such a choice is explained by the structural redundancy within the FEC in the form of:

- fuel reserves (coal, gas, fuel oil, diesel fuel);
- volumes of fuel and energy interchangeability among consumers;
- the interchangeability of energy resources in the electric power industry (both by distributing loads between power stations operating on different energy resources, and by changing fuel consumption in dual-fuel power plants;
- reserves of capacity and productivity at the production and transport facilities of coal, gas, fuel oil, electricity and heat;
- reserves (stocks) of gas and oil pipelines throughput.

Such a coherence of the multilevel COs gives the right to assert that:

the importance of the same CO for the industry and for the fuel and energy complex can be different, since the branch CO has a more technological (structural) character, the CO of FEC have more system and functional character;

- a list of branch COs can be considered as potential elements of the fuel and energy COs, can be used for calculated conditions in scenario researches of ES;
- the list of the COs of the fuel and energy complex can be more optimal (rational) in quantity and quality in comparison with the list of branch COs.

All this justifies the task of selecting the CO of the fuel and energy complex on the basis of branch COs (regardless of the type of energy systems), the ideology of this task is reflected in Fig. 2.

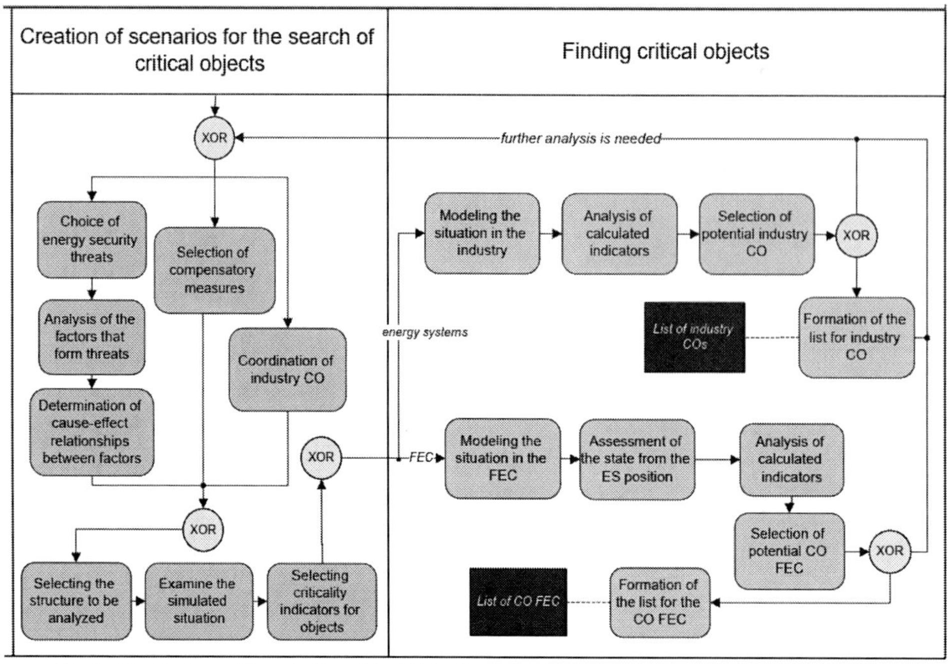

Fig. 2 - The ideology of the selection of the CO for the fuel and energy complex on the basis of branch COs

2 Methodical approach to the determination of critical objects for FEC on the basis of critical objects for industry

An approach is proposed for obtaining the CO of the fuel and energy complex on the basis of a pre-formed set of branch COs, taking into account the requirements of the ES. This approach is focused on the use of diurnal models of fuel and energy complex functioning. At the same time, the list of Cos of energy systems may not coincide with the list of fuel and energy complex objects depending on the degree of structural detailing of sectorial models and the degree of aggregation of objects in the fuel and energy complex models. This is a feature of the implementation of the proposed approach.

Wherein:

1. The CO of the FEC can become:

- either directly by the sectorial COs in the case of proving their significance for the fuel and energy complex as a whole, provided that the degree of detail of the objects at the sectorial and complex level is identical;

or aggregated fuel and energy complex objects, including dedicated sectorial CO, also under condition that this object is significant for the functioning of the fuel and energy complex as a whole.

2. The choice of the CO of the fuel and energy complex should be carried out on the basis of the permissible from the standpoint of energy security positions. Only in this way is it possible to take into account the requirements of the ES. It is provided by modeling contingencies caused by ES threats, as well as through the use of ES indicators to evaluate and select the calculated conditions of the fuel and energy complex. At the same time, the assessment should be made on a large scale at the level of the country, which is mainly due to the need to obtain a single complex assessment of the state, regardless of the strategy of the analysis (states can be assessed by one indicator, step by step , or by a generalized integral indicator).For the evaluation of the ES level, the following indicators can be used [3]:

- share of the dominant type of fuel in the structure of the consumed boiler fuel;
- the ratio of the volume of shortage of energy resources in the country as a whole to the total demand for them;
- the ratio of the daily total reserves of boiler-furnace fuel at the beginning of the heating period to daily consumption;
- in case of realization of the threat of cooling, an indicator of the level of potential provision of consumers under the conditions of growing demand for energy resources.

3. The assessment of the importance of sectorial COs for the fuel and energy sector as a whole should be based on a system indicator that characterizes the contribution of the sectorial CO to the relative assessment of states according to one model indicator. In the case of the analysis of states over a set of model indicators, a convolution of the numerical values of the set of system indices formed in this case should be provided. Model indicators should be relative at the same time, be considered at the country level. It can be considered:

- shortage of energy resources (deficits);

- the level of the storages using;
- the level of production capacity using in the fuel and energy complex.

4. The selection of CO for the FEC from the list of pre-assessed by the above-mentioned indicator of sectorial COs should be made by experts taking into account the specifics of the tasks to be performed.

Thus, the actual analyzed element in the selection of the COs of the fuel and energy complex is the calculated states of the fuel and energy complex, in the formation of which changes are made in the characteristics of the functioning of the objects. In this case, the search for the CO includes the following steps. First, using the ES indicators, a set of FEC states is defined. Then, the significancy of the selected FEC states is determined by the relatively model parameters taking into account the priority of the sectorial systems. Further, on the basis of the obtained state estimates, the importance of sectorial COs for the fuel and energy complex as a whole is determined using a special system indicator. This indicator characterizes the relative importance of a particular CO for a particular model parameter. The coefficient of significance for the i-[th] object is determined from the j-[th] index by the following formula

$$ZO_i^j = \frac{\sum_{k=1}^{n} ZS_k^j}{\sum_{l=1}^{m} ZS_l^j} \qquad (1)$$

$j = \overline{1,p}$,*where*

ZO_i^j – the importance of the i-[th] object on the j-[th] indicator,

ZS_k^j – estimations of states with inclusion of the i-[th] object on the j-[th] indicator,

ZS_l^j – estimates of all states with respect to the j-[th] indicator,

n – the number of states evaluated with respect to the j-[th] state with the inclusion of the i-[th] object;

m – the total number of states, estimated by the j-[th] index,

p – number of analyzed indicators.

A generalized scheme for searching critical objects for the fuel and energy sector within the framework of the proposed approach is shown in Fig. 3.

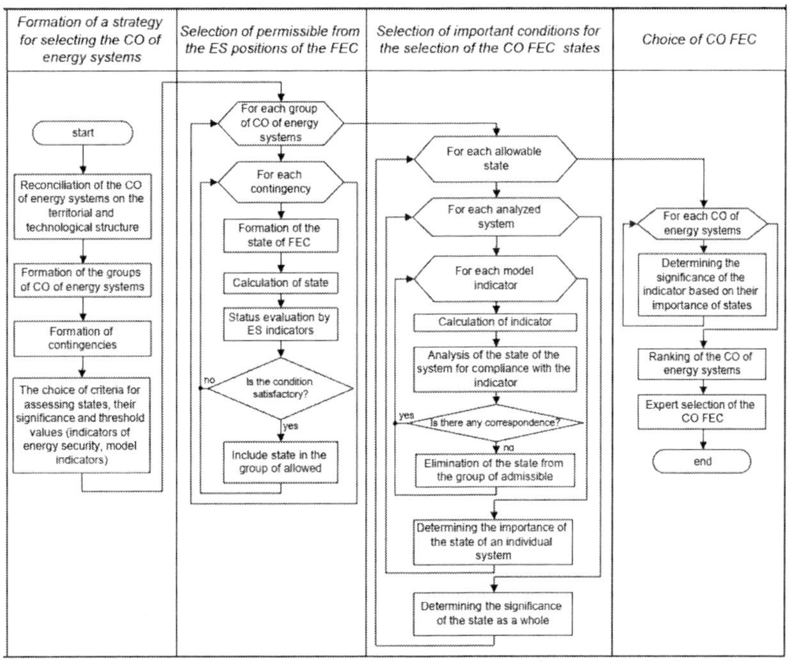

Fig. 3 - Scheme of selection of the COs of the fuel and energy complex on the basis of sectorial COs.

It identifies four main stages of the selection of the CO of the fuel and energy complex on the basis of sectorial COs:

1. The stage of forming a strategy for selecting the CO of the fuel and energy complex, within the framework of which at the expert level:

- the technological and territorial level of representation of sectorial COs in the modeled territorial-production structure of the fuel and energy complex is coordinated;
- groups of coordinated COs of energy systems are formed for their subsequent inclusion in modeled situations;
- analyzable threats of energy security are selected, for them non-standard situations are formed;

- a set of criteria for assessing the states (indicators of ES, relative model indicators) is formed, the significance of the criteria, their threshold values is determined.

2. The stage of permissible from the ES positions of the fuel and energy complex states selection, at which for each group of industrial objects in the conditions of realization of emergency situations, possible states of the fuel and energy complex are formed and calculated. These states are analyzed for compliance with the level of the country's EBS in general with the help of previously selected ES indicators.

3. The stage of significant FEC states selection, consisting in the evaluation of previously selected states for each analyzed relative model index.

4. Stage of the CO formation for the fuel and energy complex, which includes an assessment of the significance for the facilities based on the analyzed model indicators, the subsequent expert selection of the sectorial COs, which are important for the fuel and energy complex.

Detailing of the work on the selection of the COs of the fuel and energy sector on the basis of sectorial COs in the framework of the model calculations of the fuel and energy complex is considered in the next section of the article.

3 Technique of fuel and energy complex functioning models application in the selection of the COs for FEC

Complex scenario calculations in determining the CO of the fuel and energy complex within the framework of the proposed approach should be based on the territorial-production models of the fuel and energy complex functioning. These models include the main industry blocks that imitate the interrelated work of industry objects in varying degrees of territorial and technological detail. In the temporal aspect, as applied to the task of selecting the COs of the fuel and energy complex, they are oriented to the daily cut, since, in the case of more significant time intervals, the consequences of the failure of the operation of the sectorial facilities for the fuel and energy complex are generally leveled.

Technically conducted on the models the experiments are traditionally represented by the processes of data preparation and verification, by the procedures for the formation and calculation of various options for the operation of the fuel and energy complex, by the analysis of calculated situations, possibly using by the additional calculated indicators. At the same time, the starting point in the studies is the annual balanced version of the model, on the basis of which daily variants are subsequently formed (the average daily variant and diurnal variants with perturbations). Such a chain of work with model variants is justified by the composition and level of presentation of statistical and analytical industry information, especially with regard to external supply of energy resources, the relative ease of balancing the annual indicators. Correction coefficients used for obtaining daily variants are either formed by experts from the specifics of the situations analyzed, or are calculated by presence of the monthly reports from industry companies.

With regard to the task of selecting the COs taking into account the factor of ES providing, three types of diurnal versions of the model must be consistently formed: the average daily balanced version, the maximum version of the load of energy facilities, and the options for implementing contingencies with maximum utilization of industrial objects. In the latter two cases, when deficits arise, activation of model mechanisms for reserving and diversifying energy resources is required. In case of formation of variants with contingencies, additional accounting of changes in the work of groups of critical industrial objects is necessary, which allows to determine the criticality of these facilities for the fuel and energy complex as a whole in conditions of ensuring the requirements of the ES.

The conceptual scheme of working with models of fuel and energy complex functioning in the selection of the COs of the fuel and energy complex within the framework of the main stages of the proposed approach is shown in Fig. 4.

The details important for this scheme, debugging the basic annual version of the model, are represented by the following sequence.

1. Debugging of the electricity and heat block, including preparation of data on stations (or their groups) and boiler houses (or their groups), namely:
- conversion of data on fuel consumption ;
- definition of technological characteristics of equipment of stations and boiler-houses (their capacities, volumes of heat and electricity supply on separate types of fuel);
- balance control of the release of converted resources and fuel consumption (verification of specific fuel consumption) within stations and boiler houses;
- adjustment of external electricity flows;
- balance heat control in entities, electricity in the country as a whole.

2. Debugging of the fuel supply block, including:
- determination of mandatory fuel demand for the territories based on the total and energy needs of the regions;
- if necessary, the breakdown of the compulsory fuel demand by entities for specific categories (for example, for the needs of the population and industry)
- possible correction of the data in the event of a lack of balance on fuel consumption in the subjects;
- correction of external supplies of fuel;
- analysis of transmission capacities of transport subsystems for fuel.

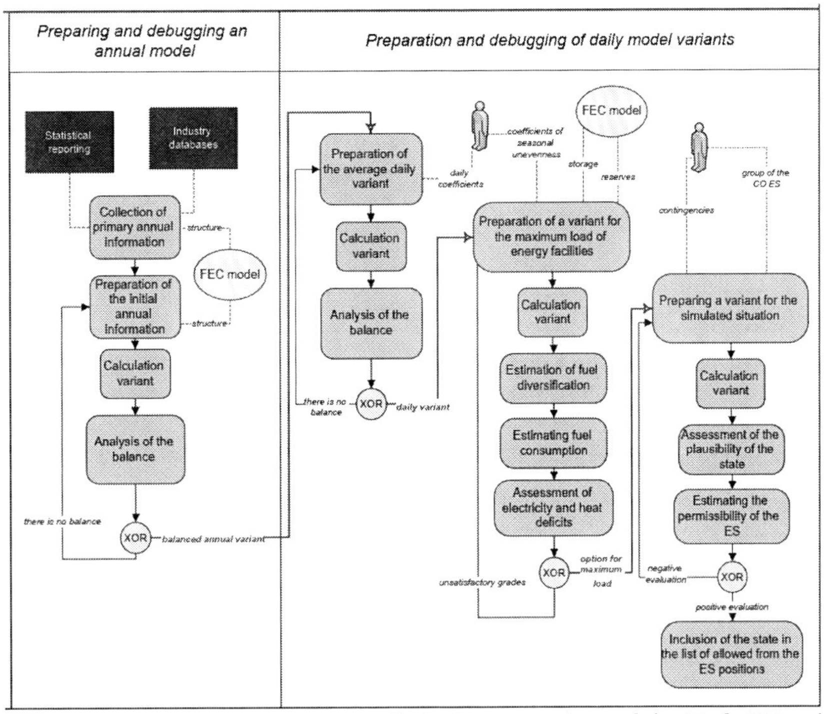

Fig. 4 - The scheme of work with models of functioning of fuel and energy complex at a choice COs for FEC on the basis of industrial COs

3. The balance estimate for all types of resources without taking into account their stocks.

4. Carrying out model calculations, during which should be analyzing:

- correctness of the initial information;
- "locked-in" energy resources, determination the causes of their occurrence;
- fuel deficits in the areas associated with functional insufficiency of transport subsystems, or with imbalance of own production capacities and needs in case of isolation of the territory;
- deficits of final types of energy resources, the reasons for their occurrence (non delivery of primary fuel and energy resources, technological insufficiency of transportation facilities);
- the degree of occupancy of production facilities in the fuel and energy complex.

This analysis is carried out in stages for each of the industries. The results of this analysis are the annual version of the model, balanced for all types of energy resources.

4 Conclusion

The paper outlines the features of the selection of critical facilities in the fuel and energy complex on the basis of the list of the COs of energy systems in researches of the ES problems. The methodological provisions for determining the COs of the fuel and energy complex have been developed, and the criteria for their selection have been presented. The scheme of the coordinated selection of the COs is justified, the indicators and the scheme for determining their significance with the use of models for the functioning of the fuel and energy complex are proposed. In the future, it is planned to carry out experimental calculations on the proposed methodological provisions for the formation of the list of COs for FEC.

References

1. N.I. Pyatkova, S.M. Senderov and others, *Energy security of Russia: problems and solutions* /Ed. By N.I. Voropay, M.B. Cheltsov (SB RAS Publishing House, Novosibirsk, 2011).
2. S.M. Senderov, V.I. Rabchuk, A.V. Edelev, Izvestiya RAN. Energetika. [Proc. of the Russ. Acad. of Sciences. Pow. Engin.], **1**, 9 (2016)
3. S.M. Senderov and others, *Methodology for monitoring the state of Russia's energy security at the regional level*. (Irkutsk: ISEM SB RAS, 2014)
4. S.V. Vorobiev, A.V. Edelev, Energy Policy, **1**, 7 (2018).

 https://doi.org/10.1051/e3sconf/20185803005

The future of Russian LNG on the world gas markets

Natalia Pyatkova [1]*

[1]Melentiev Energy Systems Institute, 130 Lermontov str., Irkutsk, Russia

Abstract. The main purpose of the article is to analyze the situation on the world markets of liquefied natural gas (LNG) and Russia's ability to enter this market. This article continues the work carried out 15 years ago on possible ways of diversifying the forms and directions of Russian gas exports [1]. An analysis of the state of the LNG market for this period was made, prospects for the development of world trade in LNG and promising Russian projects were considered.

1 The current state of the LNG markets

In the 90 years of the last century, as a result of the use of the latest LNG technologies, the export of natural gas in liquefied form has significantly increased. Now this is a large-scale business, providing more than a quarter of international gas trade (Table 1). LNG technologies are increasingly competing the traditional for Russia gas exports using pipelines. In the long term future growth of production and consumption of liquefied natural gas is forecasted, and an increasing part of international gas trade will occur as LNG supplies. According to the available forecasts, the LNG trade in 2020 may amount to more than 450 billion m3, and by 2030 the deliveries can reach almost 700 billion m3 and will account for 45% of the world gas trade.

Table 1. The share of liquefied natural gas in world gas exports [2-5].

	2001 г.	2005 г.	2010 г.	2015 г.	2016 г.
World gas exports, bn. m^3	554,2	721,4	975,2	1034,5	1084,1
including pipeline gas	411,2	532,6	677,6	709,1	737,5
LNG	143,0	188,8	297,6	325,5	346,6
Share in total exports, %					
pipeline gas	74,2	73,8	69,5	68,5	68,0
LNG	25,8	26,2	30,5	31,5	32,0
Russia, billion m^3	126,9	151,3	199,9	207,5	204,8
pipeline gas	126,9	151,3	186,5	193,0	190,9
LNG	-	-	13,4	14,5	14,0
The Russia share in total world exports, %					
pipeline gas	30,9	28,4	27,5	27,2	25,9
LNG	-		4,5	4,3	4,0

Compared with gas deliveries by pipeline, deliveries using LNG technologies have a number of important advantages:
- "mobility" of LNG - if necessary, it can be delivered from any liquefaction plant to any regasification terminal of any country. Manufacturers and buyers of liquefied natural gas are more free to choose a trading partner than those who focus only on pipeline gas;
- LNG technologies allow to promptly change volumes of production and gas supplies depending on market conditions, and gas transportation is not connected with transit through the territories of third countries and, consequently, there is the possibility of rapid response to fluctuations in demand. The opportunities for market trade in LNG contracts are particularly attractive for the largest importers, namely, the largest consumer of Russian gas, the European Union, which is reforming its gas market in this direction;
- LNG is the only possible technology for the organization of trans-oceanic transport, which is especially important for the countries of the Asia-Pacific region.

LNG technologies are become constantly cheaper (Figure 1). Unit capital costs for the construction of gas liquefaction plants, for example, decreased to $ 250 / ton. Due to a significant increase in the carrying capacity

* Corresponding author: nata@isem.irk.ru

https://doi.org/10.1051/e3sconf/20185803005

of tankers, the cost of transporting LNG is also reduced. As a result, the cost of LNG at the location of major gas markets is approaching the cost of pipeline gas.

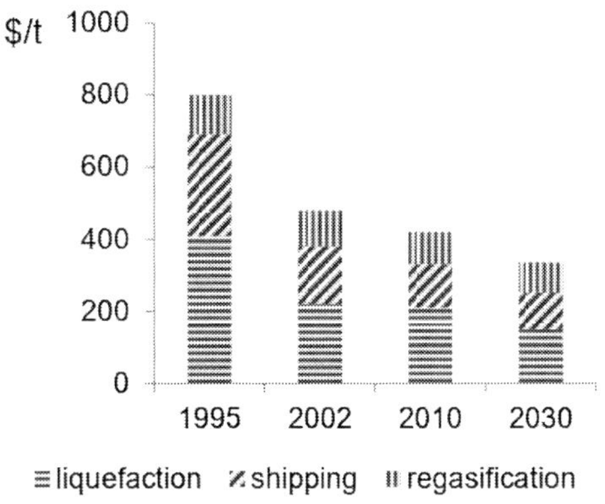

Fig. 1. Capital costs for LNG technology

2 The main consumers and suppliers of LNG

In the 1960s and 1970s, LNG was supplied mainly to the Atlantic market - to countries in Europe and the USA, but beginning in the 1970s and 1980s, the emphasis shifted towards the Pacific market. Japan became the first LNG importing country in the Asia-Pacific region (APR), which, in an effort to reduce dependence on oil supplies from the Middle East, established regular LNG supplies from Indonesia.

In the mid-80s and early 90s, South Korea and Taiwan started importing LNG, in 2004, India became one of the Asian importers of LNG, in 2006 - China, which began importing LNG from Australia.

Currently, about 70% of the world's LNG volumes are shipped to the Asia-Pacific region, which will continue to dominate the future as an importer of gas. Experts predict that China, India, Pakistan, Vietnam, Indonesia, Malaysia and Thailand will be the main "points of growth" in consumption.

In Europe, in recent years, there has been a slowdown in the growth in demand for natural gas, and there is a noticeable tendency to increase the specific weight of LNG in gas imports. The fact that the European LNG market will gain momentum is indicated by the rapid growth of regasification capacities in the region, which have already reached half of the pipeline capacity for gas imports and continue to grow despite the low level of loading. In the future, experts believe, when choosing between pipeline and LNG deliveries, European countries will often opt for the latter. According to the forecast [6], the demand for LNG in Europe by 2030 will grow from the current 47 million to 150 million tons

In the United States, the recent "shale revolution" radically changed the situation in the country. An increase in domestic gas production by 30% led to its excess in the domestic market, and the States ceased to need imported LNG volumes.

Currently, the USA and Canada, where shale deposits are also being actively developed, are preparing to become major world exporters of LNG, their extracting companies are investing heavily in the conversion of receiving terminals into export and construction of industrial and transport infrastructure.

In 2016, the following structure of LNG exporters was formed and the next demand for LNG is forecasted for the future (Fig. 2).

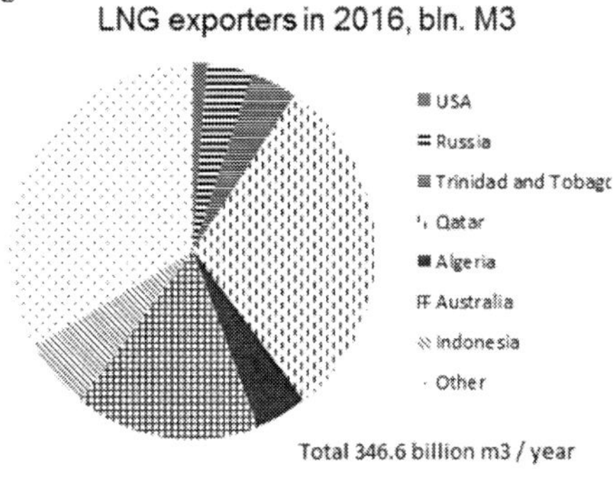

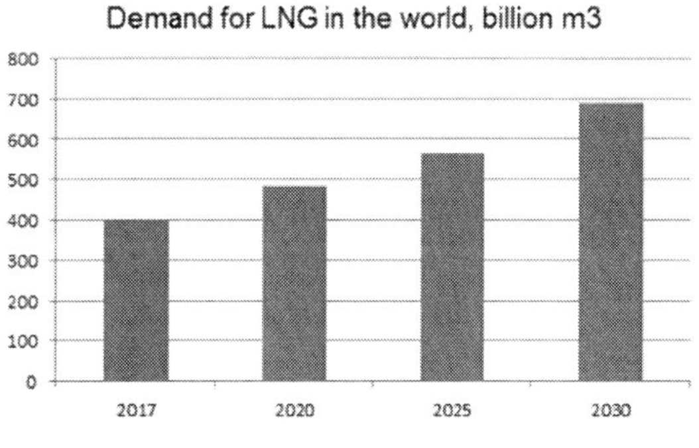

Fig 2. LNG exporters and prospects for LNG demand in the world [7].

At present, the main consumers of natural gas in the Asia-Pacific region are Japan, the Republic of Korea, China and India. With insignificant natural gas resources, the gas supply to the economy of Japan, as to South Korea, is based on imports of liquefied natural gas (Figure3).

https://doi.org/10.1051/e3sconf/20185803005

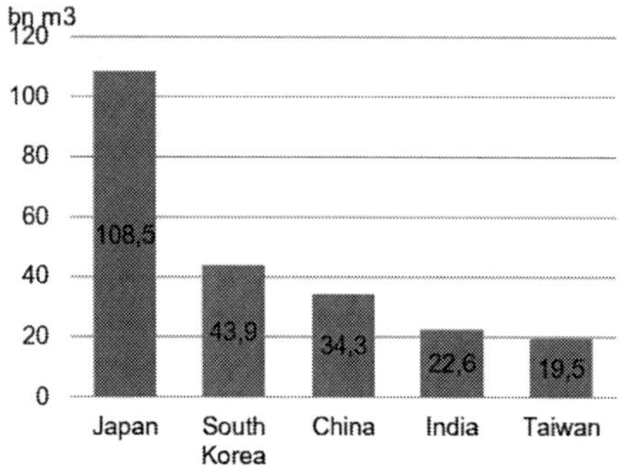

Fig. 3. Largest importers of LNG (2016 г.) [5]

The main exporters are Qatar, Australia, Malaysia, Nigeria and Indonesia (Figure 4).

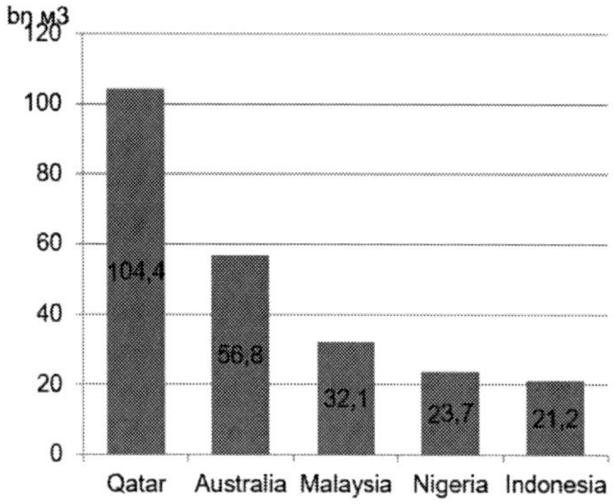

Fig. 4. The largest LNG exporters (2016 г.) [5]

3 Forecasts of growth in LNG production and Russia's capabilities

In the long term, from 2016 to 2040, the import of natural gas in the form of LNG will increase according to one of the IEA scenarios around the world (Fig. 5).

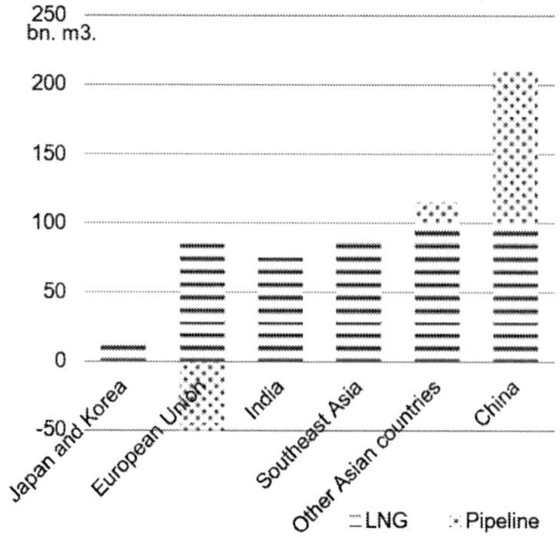

Fig. 5. Change in gas imports by countries of the world, billion m3 / year [8].

Russia should be interested in diversifying its gas consumers and delivery means. To do this, there is every opportunity: liquefied natural gas from the Yamal fields and from the shelf of the western part of the Arctic seas is convenient to supply to Northern Europe, and the gas reserves of the shelf of Eastern Siberia and the island of Sakhalin to consumers of the rapidly developing Asian LNG market. To strengthen Russia's position in the gas markets of the world, it is important to realize this potential.

The first plant for large-tonnage liquefaction of natural gas in Russia appeared in 2009 as part of the Sakhalin-2 project, which operates under the terms of the Production Sharing Agreement. Its initial capacity was 9.6 million tons per year (two lines in 4.8 million tons), after modernization of production capacity their capacity increased to 10.8 million tons.

The second project began work only after 8 years. In December 2013, the final investment decision was accepted for the Yamal LNG project, and in December 2017 the first line of the plant with a capacity of 5.5 million tons was launched (Table 2). Thus, at the moment Russian capacity is 16.3 million tons.

Table 2. The main Russian LNG projects (large-capacity production) [9,10]

Plant	Capacity, million tons / year	Year launching	Status	Shareholders
Sakhalin-2 T1, T2	10,8 (after modernization)	2009	Action.	Gazprom (50%), Shell (27,5%), Mitsui (12,5%), Mitsubishi (10%)
Sakhalin-2 T3	5,4	After 2023	Project	Gazprom» (50%), Shell (27,5%), Mitsui (12,5%), Mitsubishi (10%)
Yamal LNG T1	5,5	2017	Action.	«Novatek» (50,1%), «Total» (20%), CNPC (20%), Silk Road Foundation (9,9%)
Yamal LNG T2, T3 + T4	5,5*2+0,9	2018, 2019	Under construction	«Novatek» (50,1%), «Total» (20%), CNPC (20%), Silk Road Foundation и (9,9%)
Arctic LNG-2 T1-T3	6,6*3	After 2023	Project	«Novatek» and partners
Baltic LNG	10	After 2023	Project	«Gazprom», Shell
Far East. LNG	5	After 2030	Project	«Rosneft», ExxonMobil
Pechora LNG	<10	After 2030	Project	«Rosneft», «Alltek»

After the launch of the second, third and fourth Yamal LNG lines under construction (approximately, by the end of 2019), the total capacity of Russian large-tonnage LNG production will reach 28.2 mil-lion tons, which is already comparable to the existing capacities in Malaysia or Indonesia.

However, further plans of companies to expand capacity are associated with great uncertainties. Gazprom has plans for two large-scale projects: within the framework of Sakhalin-2, the expansion of the project is being discussed for many years due to the construction of a third line with a capacity of 5.4 million tons. This production has the lowest prime cost among the new LNG plants in Russia. However, the unresolved problem with the resource base hinders implementation with this project. In addition, the project "Baltic LNG", which assumes gas from the Unified Gas Supply System as a source of gas, is being discussed.

NK Rosneft has plans for the construction of its own LNG production (Far Eastern LNG), but un-der current price conditions its implementation seems unlikely: it is a relatively small (5 million tonnes) greenfield project, which makes the LNG cost is very high. The company is also considering the Pechora LNG project on the basis of the Kumzhinsky and Korovinskoye fields, while certainties for this project are just as few.

Among the most likely new productions - the project "Arctic LNG 2" ("Novatek") - three lines of 6.6 million tons with a resource base of the Morning deposit of the Gydan Peninsula.

The project "Arctic LNG 2" involves the construction of an LNG plant not on land but on gravity-type grounds (300 m in length and 150 m in width), which is expected to reduce capital costs by 30%. The target cost of a unit of power is in the range of 650-750 dollars per ton. The construction of an LNG plant on platforms will allow to reduce logistics costs and install the main equipment "remotely" - LNG lines will be built at the Center for Large-Scale Marine Construction in Murmansk.

Moreover, based on the existing resource base on the Yamal and Gydan Peninsulas, "Novatek" is considering the possibility of implementing the projects "Arctic LNG 1" and "Arctic LNG 3", also for 19.8 million tons each, which implies an increase of the total capacity of the company's factories in the region up to 76.8 million tons (taking into account the construction and operating facilities), which is comparable to the existing capacities of Qatar. But it is clear that it takes to make a long way to realize these ambitious plans.

Based on the current situation and statements of company representatives, the possible total LNG capacity in Russia by the beginning of the 2030s. can be estimated at 60-80 million tons (Figure 6). Some of the more complex projects may not be realized, and new, yet unannounced productions (first of all, in the context of monetization of the gas reserves of "Novatek" on Yamal and Gydan Peninsulas) is possible.

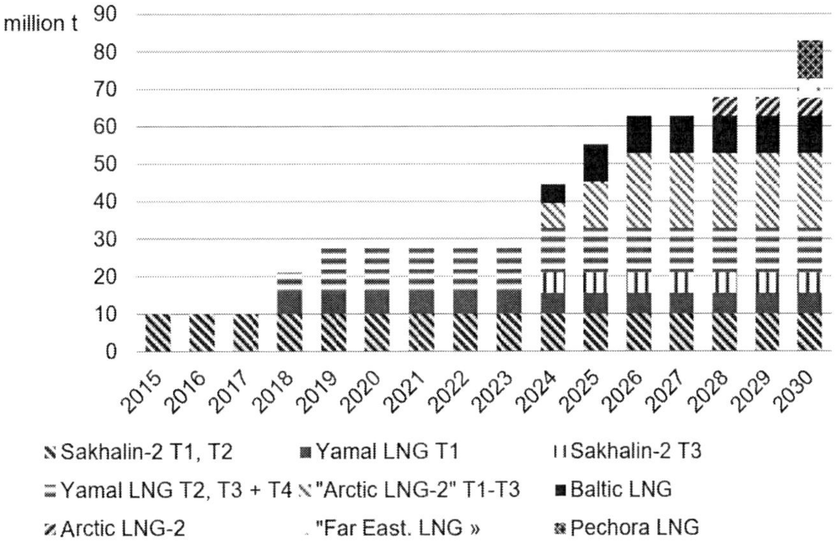

Fig. 6. Possible dynamics of capacity for large-capacity liquefaction in Russia [9]

4 Conclusion

The emergence of new Russian LNG projects on world markets occurs in a difficult price environment. In [9] it is shown that Russian LNG occupies an average position among competitors. Estimating the full cost of production and delivery to North-West Europe and China in the perspective of 2025 shows that Russian LNG is not a closing supplier: new Australian projects and US projects have higher costs [9,10]. The unit costs at the Yamal-LNG plant for deliveries to the APR countries are more than 30% less than the Australian ones and 10% cheaper than the American projects. Projects Sakhalin-2 cheaper in a similar situation is cheaper than more than 2 times.

In favor of Russia, such factors as the availability of large gas resources with a fairly low cost, the de-veloped transportation infrastructure, proximity to potential markets (to European and Asian) are acting.

It is possible to develop low-tonnage LNG production, which can be used in difficult situations on transit export corridors, as well as when demand for Russian gas is reduced by external consumers.

According to all forecasts, the world market of LNG will grow rapidly in the nearest decades. But Russia's slow entry into the LNG market can take place in a tough competition. In addition, the biggest negative factor in the implementation of LNG projects in Russia will be the main threat to energy security - a shortage of investments. When all possibilities are realized, the share of Russian LNG by 2030 may account for about 10-15% of the world trade in liquefied natural gas.

References

1. N.I.Pyatkova, M.B.Cheltsov, Perspectives of energetic, 7, 8 (2003)
2. BP Statistical Review of World Energy June 2001.
3. BP Statistical Review of World Energy June 2011.
4. BP Statistical Review of World Energy June 2016.
5. BP Statistical Review of World Energy June 2017.
6. *http://www.morvesti.ru/tems/detail.php?ID=53333*
7. A.Toporkov, Vedomosti, 14.02.2018
8. World Energy Outlook-2017
9. Transforming global LNG market: how Russia cannot lose the window of opportunity/ Energy Center LNA SKOLKOVO. 2018, 59 p.
10. A.Sobko, Oil and gas vertical, **11**, 7 (2018).

 https://doi.org/10.1051/e3sconf/20185803006

Renewable energy sources in future energy balance of the republic of Kazakhstan

Bekzhan Mukatov[1,*], *Ravil* Khabibullin[1]

[1]KEGOC JSC, National Dispatch Center of System operator, Astana, Kazakhstan

Abstract. The article describes the main factors determining the development of renewable energy sources in the world. The assessment of the applicability of foreign RES development strategies to Kazakhstan's energy system has been made. The main tasks facing Kazakhstan's energy system with large-scale implementation of renewable energy were formulated. On the basis of the analysis and performed calculations recommendations and basic principles have been made on development strategy of renewable energy sources in the Republic of Kazakhstan.

1 Introduction

Renewable energy is one of the fastest growing sectors, attracting more than $ 250 billion of capital investments every year worldwide. Many countries, including developing ones, set ambitious long-term goals for the development of renewable energy sources (RES). According to the UN, by the middle of the century, perhaps more than half of humanity's energy needs will be provided through renewable sources [1].

According to the report [2], in 2015, investments in the production of electricity from coal and natural gas amounted to about $ 130 billion, which is less than half of capital investments in renewable energy sources, amounting to $ 286 billion.

It is noteworthy that the amount of investment in RES was higher in developing countries than in developed countries. In Brazil, South Africa, Mexico, Chile, Honduras, Morocco, Pakistan, the Philippines and Uruguay – investment ranged from $500 million up to $7 billion. Investment in developing economies rose 19% to $156 billion, while in developed countries decreased by 8%, to $130 billion (for example, in Germany investments totaled $8.5 billion).

Kazakhstan also creates favorable conditions for the development of renewable energy sources. According to the plans of the government of the Republic of Kazakhstan, the share of renewable and alternative energy sources in total electricity production should be brought to 3% by 2020, 30% by 2030 and 50% by 2050 [3].

The purpose of this paper is to analyze the factors affecting the development and optimal level of renewable energy sources in the country's energy balance.

2 The factors in the development of renewable energy sources and the current situation in the world

One of the main factors stimulating RES development is to reduce greenhouse gas emissions in energy production, to combat global warming. Currently, it is believed that the probability that a large part of the temperature change caused by increases in greenhouse gas concentrations due to human activities is 90 %. The following estimate is given in the reports [2,4]: "It is extremely likely that human influence was the main cause of warming observed from the middle of the 20th century".

However, a significant difference in the proportionality between investments in renewable energy sources and the level of CO2 emissions in different countries shows different strategies for increasing the volume of RES despite environmental issues.

Another important factor stimulating the development of renewable energy sources is the desire of countries to reduce their dependence on external energy supplies. For this, energy saving and more complete use of own energy resources are used as the main methods. In modern conditions, renewable energy sources also contribute to the solution of this problem.

As a main measure of energy conservation, the direction was chosen to structural reorganization of the economy with a reduction in energy-intensive industrial production and an increase in the share of services. However, energy conservation, which is very effective at the national level, does not have a significant impact on the world energy consumption level, as the reduction in the production of energy-intensive products in developed countries causes an increase in energy consumption for

* Corresponding author: mukatov@kegoc.kz

the production of relevant products in other countries of the world.

Substitution of fossil fuel-based generation by nuclear energy is not always possible, as the use of nuclear power plants is directly connected with the policy implemented by the State and often depends on public opinion.

It is expected that at the global level, interest in renewable energy sources will continue to increase due to the growing demand for energy, the reduction of proven fossil fuel reserves, as well as environmental problems. Currently, more than 138 countries identified RES as development goals at the national level. The increase in the share of RES occurring despite falling global energy prices and the allocation of fossil fuel subsidies.

At the current stage of development, renewable energy sources (excluding large hydropower stations and geothermal power plants) in most cases are difficult to compete economically with traditional energy industry. Therefore, in all countries, developing renewable energy sources, there are a variety of measures to support them, artificially increasing their competitiveness. The most popular measure of support is the establishment of special high tariffs for electric power produced by renewable energy sources, which guarantee the payback of the project. Also, it is often practicable to compensate for the cost of connecting RES to electric grids.

Currently, preferential tariffs for RES power plants are stipulated in the legislation of many countries, including not only developed European countries, but also China, India, poor African countries like Kenya and Tanzania, as well as Iran and Algeria, rich in hydrocarbons [5].

The result of all these incentive measures is the most active development of renewable energy sources in the world over the past decade.

3 Features of the energy system of Kazakhstan. Limitations and risks for large-scale penetration of renewable energy sources

The development of the energy infrastructure in Kazakhstan was significantly determined by the presence of large reserves of cheap coal and gas, which makes the share of electricity generation using fossil fuels at a high level.

First of all, increasing the share of RES brings the following technical problems to the energy system:

Complication of dispatching control due to the increase in the number of generation nodes and the transformation of previously passive distribution networks into active ones [6].

Negative impact of RES on the reliability of energy system operation, the complexity of the UPS operation modes forecasting.

The need for reconstruction of networks with mass input of renewable energy sources.

Large volumes of variable generation of RES are able to cause power leaps in interfaces due to a drastic change in the production mode. This requires an analysis of the effect of RES on irregular power fluctuations, as well as the consideration of possible power leaps in determining the power transfer limits in the interfaces. There is a need to maintain large reserves of maneuverable generation or reserve transmission capacity in interfaces.

When using foreign experience of introducing RES, it is necessary to take into account the peculiarities of Kazakhstan's energy system.

First, in the conditions of a sharply continental climate, the main part of the energy demand is made up of thermal energy, which exceeds the demand for electricity several times. A significant part of the energy (about 40%) is generated by CHP plants that operate on a thermal schedule and can not be used to cover unbalances in the generation of RES. Renewable generation based on solar and wind energy, partially covering the demand for electricity, is not able to provide heat in the required volumes. Since in most Western countries there is no centralized heat supply system, it is unacceptable to directly distribute the results obtained in European countries to Kazakhstan.

Secondly, for concentrated energy systems of Western countries, the key limitation is the current loading, while for Kazakhstan energy system the determinative limitation for the transmission of electricity through the backbone network is steady state stability. Considering that the total capacity of RES, which is supposed to integrate into the UPS of Kazakhstan, significantly exceeds the natural transmission capacity of transmission lines (even 500 kV), this will be one of the main problems in ensuring the sustainability of the UPS of Kazakhstan in the development of RES. Definitely this will require a fundamentally new approach to load flow management, primarily automatic.

In IEA report there are counter-arguments against challenges described above [7]. However, statements are not accompanied by strong evidence. In the conditions of the full load of power transmission lines according to steady state stability limits due to the highly echeloned and complex automation, it is impossible to state that the deviations of RES are safe for the power system.

An uncontrolled increase in the share of RES due to state support can lead to a significant increase in electricity tariffs, which at some point may push consumers to avoid centralized power supply and switch to cheaper small generation, including operating in isolation from the UPS. With sufficiently high tariffs, large consumers can start installing own power plants, refusing from centralized power supply services. This is especially true for areas with significant hydrocarbon reserves with a large-scale development of the gas transportation system.

The creation by consumers of own generation will not pass without a trace for the UPS, which will lose the large and strong entities of the wholesale market. As a result, the remaining participants of the grid will increase their workload to maintain the conditions for the operation of the entire power system.

In addition, the priority given to renewable energy leads to substitution of existing generation and a decline in the economic returns of traditional power plants, whose construction becomes less attractive for investors. While RES is not able to be created and operate without the support of traditional maneuverable generation.

Renewable generation, due to its uncertainty, requires reserve capacity to cover unbalances. Theoretically, the variability of renewable sources can be compensated for by the flexibility and modulation of the fleet of stations. However, in practice, this is difficult to implement, because these stations have own technical and economic limitations. Power storage significantly increases the capital investment in the construction of RES.

Countries with predominantly coal generation, which are most interested in reducing CO2 emissions, are now facing a shortage of maneuverable power plants. In the absence of a sufficient volume of regulatory capacity, it will be necessary to conclude contracts for regulating services with adjacent power systems, the costs of which are difficult to forecast and will be included in the tariff.

For the energy system of Kazakhstan, the contradictions between the desire to introduce large volumes of renewable energy sources and ensure the reliability of the UPS become obvious. Therefore, it can be argued that the spontaneous growth of renewable energy without a scientific study is the main problem.

The opinion expressed in 1975 by academician P. Kapitsa regarding the prospect of using renewable energy as the main source of electricity is also relevant today [8]. As noted by P. Kapitsa, the generation of electricity from RES in practice for high-power engineering is limited by the limited value of the energy flux density (1).

The density of incoming energy is limited by the physical properties of the medium through which it is transmitted. In a material medium, the energy flux density U is limited by the expression:

$$\mathbf{U} < \mathbf{v}F, \qquad (1)$$

where **v** is the speed of propagation of energy, F-energy density. **U** is the vector value. When stationary processes **div U** determines the amount of energy conversion in another form.

Direct conversion of solar energy into electricity for high-power engineering is associated with a limited value of the energy flux density. In this case the speed of propagation of energy **v** is almost equal to the speed of light, however, the energy density is low.

4 Impact of large-scale RES penetration. Modeling of influence with respect to UPS of Kazakhstan

Important results were obtained by the authors in [9]. In the course of the research, a mathematical model was created, the input parameters for which were: material capital, fossil fuel resources and infrastructure required for RES.

In this model, the growth of material capital is possible in the presence of a surplus of electricity, and the resources of fossil fuels take into account the initial energy costs for the creation of RES and the cost of production over time. Parameters: fossil fuel resources and the cost of production are necessary to define a limit for the extraction of fossil fuels. In this paper, for the parameter "infrastructure required for RES", the most important characteristic is "renew return" (RE), which characterizes the return of RES energy relative to the spent energy for RES creation.

Of the more than 100 simulated scenarios for more than a quarter have shown the unstable state of the system when you enter RES, i.e. the economy of the country was unable to meet demand for electricity as the depletion of fossil fuel reserves. The equilibrium was ensured by the introduction of renewable energy sources in the event of signals about a shortage of electricity as the resources of fossil fuels were depleted. Simultaneously, the simulation results showed a significant effect on the result of the dynamics of the introduction of renewable energy as the deficit approaches. With the slow introduction of renewable energy sources, the country's economy was not able to provide material resources and energy for the creation of renewable energy infrastructure because of the decline in electricity production from the traditional sources. Based on a series of sustainable scenarios, it can be concluded that the maximum amount of renewable energy is determined not by the technical limitations of the energy system, but by the dynamics of the use of the existing infrastructure and depletion of fossil fuels.

In the case of Kazakhstan, the following time contours of RES development can be formulated. Based on the forecasted balance of the UPS of Kazakhstan for a seven-year period (until 2024) [10], in the absence of input of generating capacities, a power shortage of 180 MW occurs by 2022 (excluding reserves). With scheduled inputs of traditional generation, there is no electricity deficit in the next seven-year period. Given the competitiveness of RES, it is possible to replace part of the traditional generation with renewable energy sources.

Given the expected power deficit and the timing of the implementation of renewable energy projects in 2022, the deadline for the beginning of active renewable energy input is 2021. However, given the availability of fossil fuels, the type of energy source (for example, nuclear power plants) should be carefully determined, and in case of non-competitiveness of renewable energy sources, it is necessary to postpone the mass entry of renewable energy sources at a later date.

The introduction of renewable energy sources on a competitive basis (without the creation of benefits) can be carried out at any time, as with the competitiveness of renewable energy sources one can speak of the inefficiency of traditional generation.

Kazakhstan power system consists of three zones: North, South and West (Figure 1).

https://doi.org/10.1051/e3sconf/20185803006

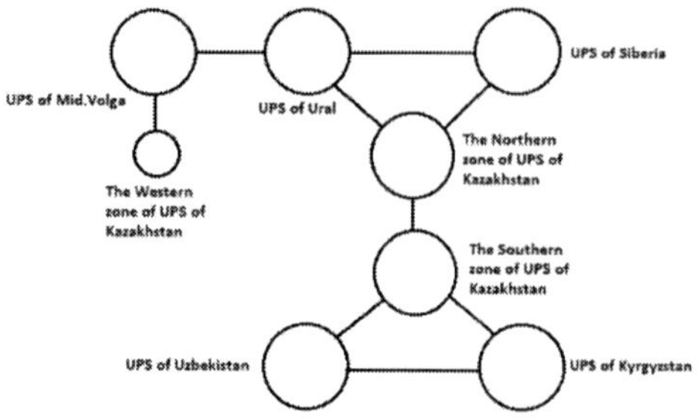

Fig. 1. The simplified scheme of the Unified Power System of Kazakhstan.

The Southern zone is energy-deficient by about 1500 MW, and covers the deficit from the Northern zone. The main problems with stability in UPS of Kazakhstan are observed in interface, connecting the Northern and Southern zones. The planned significant increase in renewable energy in the South zone has the potential to exacerbate existing problems with stability. While maintaining the current dynamics of input of wind farms and PVs, by 2021 the maximum instantaneous output of PVs and wind farms will be about 7% of the total generation of the country and 45% in the Southern zone. The increase in the volume of renewable energy requires a greater maneuverable capacity of traditional stations to cover the net load, since solar generation changes the form of the daily consumption schedule in the direction of decreasing the noontime demand. Net load is power consumption, excluding generation from renewable energy sources, i.e. it is a demand that must be covered by traditional stations. By 2021, on the day of maximum consumption at maximum output from PVs, the required capacity of traditional stations to cover the evening peak of the net load will increase by 800 MW, i.e. by 65% (Figure 2).

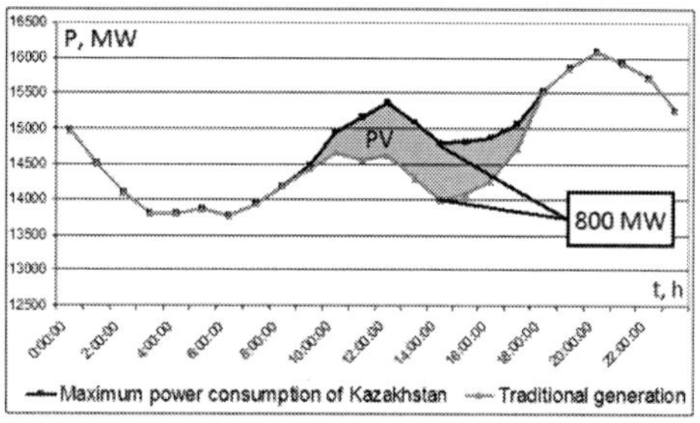

Fig. 2. Daily schedule on the day of maximum load, taking into account the maximum output of PVs (2021).

A situation was simulated in which the growing power consumption in the South zone is covered exclusively by introducing RES, without introducing traditional capacities, and the effect of reducing parts of the variable generation on the stability is considered.

In power systems of developed countries with a high degree of introduction of RES, the accuracy of predicting variable generation is about 10%. The accuracy of the forecast strongly depends on the amount of historical data and modern means of climate monitoring. At the initial stages of penetration of renewable energy in Kazakhstan it will be difficult to achieve high accuracy of forecasting.

The loss of a part of RES generation in the South zone causes a power leap in the North-South interface. At a consumption level of 2021, the transmission capacity of the North-South interface allows reserving possible losses of generation in the Southern zone due to the flow from the North Zone.

If the annual growth of consumption in the South zone is maintained at 5%, the planned increase of the RES is not able to cover the deficit. The growing deficit will lead to an increase in power flow across the North-South interface and a decrease in interfaces stability margin. At the same time, the increasing share of renewable energy in the Southern zone leads to an increase in the possible power leaps in the North-South interface.

With the increase in consumption and the increase in the share of renewable energy sources, critical power leaps can occur with a smaller percentage of the switched-off generation of RES. In 2031, the loss of 80% of renewable energy will lead to a violation of transient stability, due to power leaps on the North-South interface. In 2035, the loss of 50% of renewable energy will lead to a violation of transient stability (Figure3).

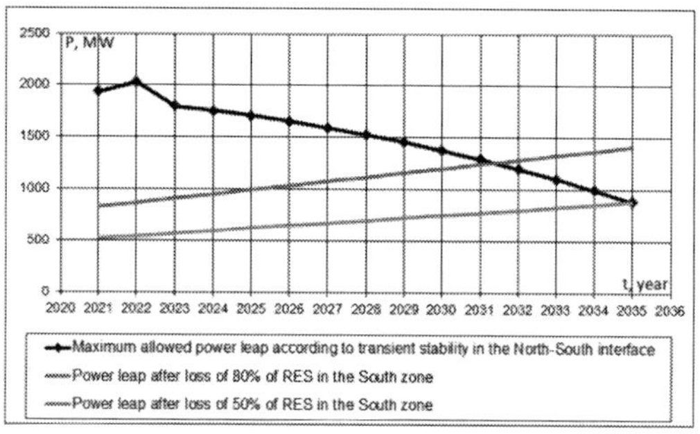

Fig. 3. Power leaps at various %% off of RES in the Southern zone.

Even if the entire deficit growth in the South zone is covered by introducing RES, the problems with stability will remain relevant. In this case, it will be necessary to introduce a larger amount of RES, which will also lead to an increase in the possible power leaps (Figure 4).

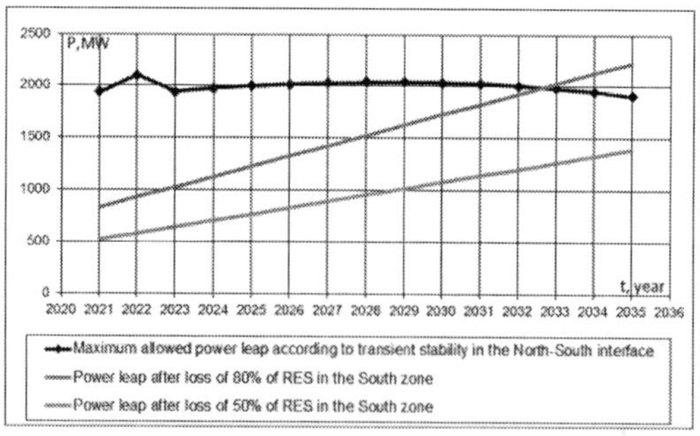

Fig. 4. Power leaps at various %% off of RES in the Southern zone.

The emergence of a powerful stable generation in the southern zone is capable of relieving North-South interface and providing a larger reserve of capacity, which will ensure the safe operation of the Southern power district when introducing large volumes of RES.

Obviously, coverage of imbalances of energy-deficient areas exclusively at the expense of renewable energy is impossible. Proceeding from this, in parallel with the large-scale commissioning of renewable energy in the Southern zone of the UPS of Kazakhstan, it is necessary to strengthen the links or enter synchronous generation of commensurate capacity for unloading interfaces.

5 Conclusion

Many countries embarked on a large-scale implementation of renewable generation. However, when planning the development of the energy system of Kazakhstan, it is inadmissible to copy the energy strategies of other countries. When planning a model for the development of renewable energy in a particular country or region, the features of the power system configuration should be taken into account.

On the basis of the analysis, the following basic principles can be formulated when introducing renewable energy sources into the energy system of the Republic of Kazakhstan:

• The transmission capacity of the interfaces must be selected taking into account the possible power leaps caused by variable generation.

• When planning the operation modes, it is necessary to take into account the volume and rate of change of the maneuverable generation of regulating stations.

• Coverage of imbalances of energy-deficient areas exclusively at the expense of renewable energy is impossible. In parallel with the large-scale commissioning of renewable energy in energy-deficient areas, it is necessary to strengthen the links or introduce synchronous generation of commensurable capacity for unloading interfaces.

• Increasing volume of variable generation will require involvement of additional regulation stations or contracting regulation with neighboring countries. In Kazakhstan currently there is no sufficient maneuverable generation. Therefore, the development of renewable energy requires the modernization of the market model in order to create conditions for the development of maneuverable power plants.

• The penetration of renewable energy in isolated parts of the UPS or in any parts on a competitive basis (without the creation of benefits) can be carried out at any time, since with the competitiveness of renewable energy sources with traditional energy, one can speak of the extreme inefficiency of the latter.

Taking into account the peculiarities of Kazakhstan power system, it becomes obvious that an unreasonable, constant increase in the share of renewable energy in the production of electricity is not an optimal option for the development of the energy system. Insufficiently thought-out policy of constant increase of the share of RES in the power system in pursuit of nominal indicators can disrupt the entire electric power sector. Until the capital costs for the creation of renewable energy can be justified by the energy received for converting solar or wind energy, RES will not contribute to the sustainable development of the energy system and the economy of the country as a whole.

To achieve maximum reduction of harmful emissions with minimal negative impact on the energy system, the use of combinations of different technologies, primarily nuclear power, should now be considered.

Thus, the penetration of renewable energy will require a significant change in the entire energy industry of the country, both in terms of technology, and in terms of economic and legal relations.

References

1. V. Slivyak. "Energy turnaround": who and why? // Ecology and law, No. **55**, pp. 8-11 (2014).
2. Frankfurt School-UNEP Centre/BNEF.2016. Global Trends in Renewable Energy Investment 2016.
3. The concept of Kazakhstan on transition to "green economy": The decree of the President of the Republic of Kazakhstan dated May 30, 2013 № 577
4. IPCC: Climate Change 2013: The Physical Science Basis. Contribution of Working Group I to the Fifth Assessment Report of the Intergovernmental Panel on Climate Change [Stocker, T.F., D. Qin, G.-K. Plattner, M. Tignor, V. Bex and P.M. Midgley (eds.)]. Cambridge University Press, Cambridge, United Kingdom and New York, NY, USA, 1535 pp (2013).
5. Press-service of JSC "RusHydro". Support of renewable energy: global experience // the Blog of JSC RusHydro, 03 November (2011).
6. B.B. Mukatov, N.A. Karjaubaev, A.G. Fishov. Features of Ensuring the Reliability of Power Supply in Isolated Power Grids with Small Generation // Reports of the Academy of Sciences of the Higher School of the Russian Federation, No. **4**, p. 94-104 (2015).
7. OECD/IEA, Getting Wind and Sun onto the Grid. A Manual for Policy Makers, International Energy Agency, Paris (2017).
8. P.L. Kapitsa. Energy and physics. Report on the scientific session devoted to 250-anniversary of the Academy of Sciences of the USSR // Vestnik SSSR, no. **1**, pp. 34-43 (1976).
9. Mediavilla M, Miguel LJ, Castro C. From fossil fuels to renewable energies // In: 26th international systems dynamics conference, Athens, pp. 1972 - 1986 (2008).
10. On approval of the forecasted balances of electric power and capacity for 2018-2024: Order of the Minister of Energy of the Republic of Kazakhstan dated 25.12.2017, No. 475

E3S Web of Conferences **58**, 03007 (2018) https://doi.org/10.1051/e3sconf/20185803007
RSES 2018

Prospects of using new technologies in Russia's electric power industry to comply with international commitments to reduce CO_2 emissions

Alexey Tchemezov [1,2*], *Elena* Chemezova[3], *Anton* Syromyatnikov[2]
[1]Melentiev Energy Systems Institute of Siberian Branch of Russian Academy of Sciences, 664033 Irkutsk, Lermontov str., 130, Russia
[2]Irkutsk National Research Technical University, 664072 Irkutsk, Lermontov str., 83, Russia
[3] Exploration College of Irkutsk National Research Technical University, 664072 Irkutsk, Lermontov str., 104, Russia

Abstract. This study addresses the methodology of projecting the electric power industry developments, taking into account environmental constraints. I obtained quantitative assessments of long-term electric power industry development in a Russian region, determined the emission of greenhouse gases from fuel combustion at thermal power plants (TPPs), and the efficiency of technologies to reduce greenhouse gas emissions in the electric power industry.

1 Introduction

Presently, when planning and programming economy development of the countries and regions, the global climate change problem related to emissions of greenhouse gases – carbon dioxide, CO2 being the main among the latter – at fuel combustion is regarded the most acute [1]. The constraints on its emission imposed by the Paris Agreement, in fact, define the boundaries impeding the power engineering development, and, consequently, they restrain the economy development, especially in developing countries. At present, it is difficult to estimate how much these constraints are justified, because investigations into the anthropogenous emission effect on the climate and its change have not been completed yet, with the preliminary conclusions being ambiguous and even contradictory [2].

In Russia, a number of Federal Laws, according to which greenhouse gas emissions should be reduced to 75% of the 1990 level by 2020, and to 70% by 2030 [3], have been passed.

There is no real alternative to fossil hydrocarbon fuel as a primary energy carrier over a long perspective. Reducing the carbon dioxide emissions into the atmosphere may be achieved by:

- active promotion of energy-saving technologies;
- increase in electric power production efficiency through improving thermodynamic cycles, equipment, fuel combustion techniques, use of fuel elements, etc.;
- introduction of multi-purpose power technological installations that, besides electric power, produce additional marketable products (for example, through the catalytic synthesis methods, etc.) increasing the fuel usage coefficient as compared with separate production of electric power and those products;
- switching to fuels containing smaller carbon amount (replacing coal with natural gas, or oil motor fuel with hydrogen);
- CO_2 recovery in the power installation cycle with its subsequent durable disposal (sequestering).

2 Modelling long-term development of Russia's electric power industry

Figures and tables, as originals of good quality and well contrasted, are to be in their final form, ready for reproduction, pasted in the appropriate place in the text. Try to ensure that the size of the text in your figures is approximately the same size as the main text (10 point). Try to ensure that lines are no thinner than 0.25 point.

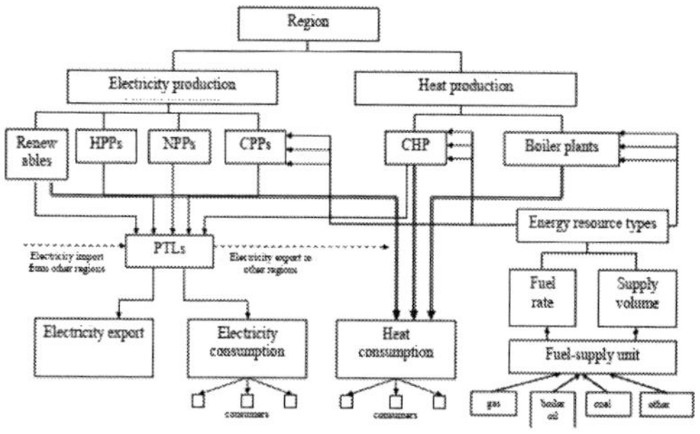

Fig. 1. Structure of the model for production, distribution, and consumption of electric and district heating

As per the above model construction principle, each energy carrier produced at this or that stage is appropriated by the corresponding equation in the model, whereas each power plant is appropriated by a variable:

* Corresponding author: tchemezov@isem.irk.ru, atchemezov@mail.ru

$$I^t_{e,r} + \sum_k R^t_{e,k,r} + \sum_{r'} Y^t_{e,r'r} + \sum_{p \in r}(p_e \times X^t_{p,r}) - \sum_{r'}(b^t_{e,r,r'} \times Y^t_{e,r,r'}) + \sum_d D^t_{e,d,r} + Ex^t_{e,r} \quad (1)$$

for all: $e \in 1, ..., E; r \in 1, ..., r'..., R; t \in 1, ..., T$,
where: $I^t_{e,r}$ is the energy carrier e import to the region r in year t; $R^t_{e,k,r}$ is the extraction of the given energy carrier e by all plants k in the territory of the region r $(k \in r)$ in year t; $Y^t_{e,r'r}$ is possible import of the energy carrier e from other regions r'; $X^t_{p,r}$ is the production capacity of the power plant p in the region r $(p \in r)$ in year t; p_e is the coefficient determining the output (consumption) of the energy carrier e at the plant p (p_e $=-1/f_p$, if the energy carrier is consumed, $p_e=1$, if the energy carrier is produced); $Y^t_{e,r,r'}$ are possible deliveries of the energy carrier e to other regions r' considering transport losses $b^t_{e,r,r'}$ in year t; $D^t_{e,d,r}$ is the consumption of the end energy carrier e by customers' categories d in the region r in year t; $Ex^t_{e,r}$ is the energy carrier e export from the region r in year t.

All the power plants are split into operating $(RESX^t_{p,r})_{t=1}$ by the settling period start, and newly-built plants $(X^t_{p,r})_{t>1}$. For the operating plants, I specify the dynamics of their possible retirement by stages of the settling period $(RESX^{t+1}_{p,r} \leq RESX^t_{p,r})$, and creation of new production capacities. Newly-commissioned capacities at each temporal stage may be restrained and should retire as the standard life comes to an end. To consider the dynamics factor, I introduce the capacity transport equations for each power plant by stages t of the settling period T. For a separate plant p in the region r, this equation looks like:

$$X^t_{p,r} = RESX^t_{p,r} + \sum_{b<t} \min(RL^{b,t}_p / n_t, 1) \times X^b_{p,r} \quad (2)$$

for all $p \in 1,...,P$,
where: $X^t_{p,r}$ is the power plant p capacity in the region r in year t; $RESX^t_{p,r}$ is the residual capacity of the plant p in the region r at the stage t, that was commissioned by the settling period start; $RL^{b,t}_p$ is the number of years, over which the power plant p capacity built at the stage b persists (proceeding from the standard life) at the stage t; n_t is the number of years at the stage t; $X^b_{p,r}$ is the power plant p capacity introduced at the stage b in the region r;

In general case, the technological aspect of the power plant (technology) is described by the following metrics: coefficients determining the consumption and production of energy carriers by the given plant (technology); power installation efficiency coefficient; operation mode; plant life.

As the economic metrics of power plants, I use: specific constant and variable costs related to maintaining power plant capacities and to producing power resources and energy carriers; specific holdings necessary for building (renovating) power plant capacities, energy-saving technologies and emission reduction technologies.

3 Assessing the efficiency of using new technologies in Russia

Effective (in terms of the CO_2 emission reduction, included) is perfecting the TPP energy production based on [4, 7, 8]:

- introducing coal power-generating units at supercritical (41% eff.) and supercritical (46% eff.) steam parameters;
- introducing steam-gas plants (55-60% eff.);
- using boilers with circulating fluidized bed at low-grade-fuel combustion;
- using fuels with an increased calorific value, and natural gas;
- using technologies of CO_2 recovery and disposal.

To assess the performance of the offered methodical approach, I address two scenarios for the Irkutsk Oblast electric power industry development: optimistic and basic [5, 6].

The power consumption was projected by considering the economy development and the population perspective metrics. Inasmuch as the planned long-term regional major energy-intensive investment projects, the projection for the need in electric power was performed separately for the industries and services (developed for today) and for new industries (projects) [7-10].

Table 1. Projection of the electric power and heat production in Irkutsk Oblast

	1990	2000	2010	2015	2020		2025		2030	
					o.s.	b.s.	o.s.	b.s.	o.s.	b.s.
Electricity production, bil. kW·h	**67,1**	**54,5**	**62,4**	**48,0**	**64,3**	**56,9**	**73**	**57,5**	**77**	**60,6**
Hydro Power Plants	47,5	47,3	49,3	35,9	46,3	46,3	46,3	46,3	48,6	48,6
Thermal Power Plants	19,6	7,2	13,1	12,1	18,0	10,6	26,7	11,2	28,4	12,0
Thermal energy production, mln. GCal	**72,6**	**54,7**	**49,1**	**41,2**	**46**	**44**	**50**	**46**	**59**	**54**

As per the electric power industry development scenarios, we calculated the CO_2 emissions from fuel combustion at TPPs (Table 2).

Table 2. CO_2 emission volume at boiler-furnace fuel (BFF) combustion, mln. tonnes*

	1990	2000	2010	2015	2020		2025		2030	
					o.s.	b.s.	o.s.	b.s.	o.s.	b.s.
Thermal power plants	**35,4**	**22,4**	**22,7**	**20,7**	**26,4**	**23,3**	**36,3**	**33,3**	**32,7**	**30,9**
including coal	32,0	19,6	19,3	17,9	22,9	19,9	32,8	29,8	22,6	21,8
gas	0	0	0	0	0	0	0	0	6,6	5,7
oil	0,6	0,3	0	0	0,09	0,09	0,09	0,09	0,1	0,1
other, etc	2,8	2,52	3,4	2,8	3,4	3,4	3,4	3,4	3,4	3,4

* - *Author's calculations*

The CO_2 emissions from TPP fuel combustion will exceed the 1990 level by 2025, and reach 36,3 and 33,3 mln. t. in the optimistic and basic scenarios, respectively. By 2030, they decrease provided new power stations are commissioned.

However, if one considers further decrease in the CO_2 emissions, it is necessary to change the structure of the consumed fuel (to use natural gas), or to use new technologies of steam-gas plants (integrated gasification combined cycle (IGCC), etc.).

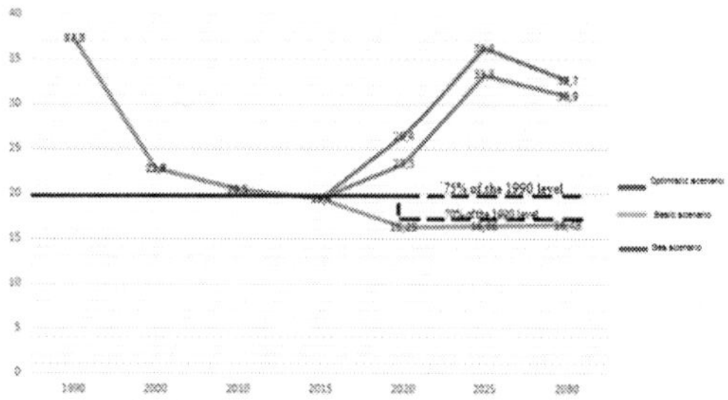

Fig. 2 Projection for the CO_2 emissions from TPP fuel combustion, mln. tonnes

To implement President's Decrees on the CO_2 emission reduction in any scenario for the Irkutsk Oblast electric power industry development, it is necessary to apply additional measures. Considering the capital investments and operational costs, TPP switching to natural gas is the most inexpensive way (of all the above processes) to reduce CO_2 emissions. Herewith, the CO_2 reduction may reach 20-22 mln. tonnes (Figure 2).

4 Conclusion

The proposed methodical approach enables to select an optimal version for Russia's (regions') electric power industry development, considering environmental constraints.

I provided an example of assessing the CO_2 emission reduction technology efficiency in Irkutsk Oblast, and showed the necessity to change the structure of the BFF consumption and TPP switching to natural gas.

References

1. A. Tchemezov, B. Saneev, A. Lagerev Outlooks of Russia's power industry development in the 21st century and greenhouse gas emissions // International Journal of Energy Technology and Policy. 2003, Vol. 20, No. 4, pp. 382-392. doi: 10.1504/IJGEI.2003.004411.
2. National system of greenhouse gas cap-and-trade. - Moscow, 2011, P. 40, in Russian.
3. 2013 September 30 Russian Federation President's Decree No. 752 «On reducing greenhouse gas emissions» // Official website of Russia's President, 2013 September 30 [e-resource]. URL: <http://static.kremlin.ru/media/acts/files/0001201310010043.pdf> (access date: 2018 May 15).
4. B. Saneev , A. Lagerev, V. Khanaeva, and A. Tchemezov Kyoto mechanisms as a tool of expanding the international cooperation to provide energy security of Russia // Izvestiya RAN (Proceedings of the Russian Academy of Sciences. Power Engineering, 2006, No. 3, pp.121-126, in Russian.

5. On approving the scheme and program for the Irkutsk Oblast electric power industry development over 2017-2021 // Official website of the "Kodeks" LLC, 2016 December 22 [e-resource]. URL: <http://docs.cntd.ru/document/441789760> (access date: 2018 May 15).
6. 2010 October 18 Irkutsk Oblast Government Order No. 266-pp «Long-term target program Gasification of Irkutsk Oblast over 2011-2015» // Website of Irkutsk Oblast Government, 2010 November 10 [e-resource]. URL: http://irkobl.ru/
7. Prutkovsky E.N., V.S. Varvarskiy, V.I. Gritsenko, et al. Ecological and Energy Aspects of introducing 3rd-generation SGPs with ICG to Power Engineering // Teploenergetika (Thermal Energy Engineering), 1993, No. 11, pp.18-22, in Russian.
8. Stennikov V.A., and S.V. Zharkov. Method of assessing production efficiency by CO2 emission minimization with the focus on energy supply systems // Al'ternativnaya energetika i ekologiya (Alternative power engineering and ecology), 2017, Nos.16-18, pp.118-132, in Russian.
9. Irkutsk Oblast Fuel and Energy Complex: Current State and Development Perspectives / Ed. by B.G. Saneev and N.A. Voronin. - Moscow: "Energiya" Publishing House, 2013, P. 304, in Russian.
10. Baikal Region Fuel and Energy Complex: Current state, Development Perspectives // Ed. by B.G. Saneev; Russian Academy of Sciences, Siberian Branch, Melentiev Energy Systems Institute. - Novosibirsk: GEO Academic Publishers. 2015, P. 176, in Russian.

Ensuring energy security in the Siberian Federal District: analysis of the situation and the main trends

Elena Smirnova[1,*]

[1]Melentiev Energy Systems Institute, 130 Lermontov str., Irkutsk, Russia

Abstract. This paper presents an assessment of the level of energy security of the subjects of the Russian Federation located on the territory of the Siberian Federal District. The approach to the assessment is based on the use of the monitoring and indicative analysis of energy security developed at ESI SB RAS. As a result, the main problems were identified with ensuring energy security in the regions, with a view to further developing ways to improve the situation in them.

1 Introduction

This paper presents the process of assessing the level of energy security of Russia subjects located on the territory of the Siberian Federal District. This assessment was carried out on the basis of the use of the monitoring apparatus and indicative analysis of energy security developed at ESI SB RAS. The main problems with ensuring energy security for individual regions are reflected, which is the basis for forming directions for improving the existing situation in the Siberian Federal District as a whole.

The tasks monitoring of Russia's ES and its regions are to identify the observed and expected processes, phenomena and parameters that determine the level and threats to energy security. At the same time, the identification process is based on a system of indicators that adequately describe the situation in one or another aspect of securing ES. Thus, the meaning and essence of monitoring and indicative analysis consist in displaying information on the degree of implementation of ES threats using an indicator system when comparing the numerical values of these indicators with their threshold values.

Energy security - the state of protection of citizens, society, the state, the economy from the threat of a deficit in providing their energy needs with economically accessible energy resources of acceptable quality, from threats of disrupting the continuity of energy supply. In fact, at any level, energy security is the balance of energy supply and demand or the lack of energy deficit. The reasons for the imbalance can be found both in the state of the objects of the fuel and energy complex and in adverse changes in natural, political and economic conditions. Comprehensive assessment of potential threats of ES, which are imminent within the fuel and energy complex and arising from outside due to changes in external energy conditions, is rather difficult to obtain. At the same time, an assessment of the level of the region's ES, determined by the state of the fuel and energy complex objects in this region and related federal energy systems, can be found using numerically evaluated indicative indicators, using the data [1] on the relevant parameters of the objects and the processes recorded.

An indicative assessment of the level of energy security of a specific region of the country is proposed to implement three, largely interrelated, blocks of indicators: production and resource provision of the region's fuel and energy supply system; reliability of the fuel and energy supply system in the region; state of the basic production assets energy systems in the region.

On the first two blocks, specific indicators can be distributed quite arbitrarily, since they are directly or indirectly related to each other.

The composition of the most important indicative indicators of energy security at the regional level, structured by blocs, is shown in Table. 1.

With the use of a methodology specially developed at ESI SB RAS, which allows taking into account the peculiarities of energy supply in individual regions [1-3], threshold values of indicative indicators for all RF subjects were expertly determined.

2 Results of the conducted indicative analysis

When performing the study, the values of indicators by region were quantitatively correlated with their threshold values adopted in [3]. This allowed us to assess the qualitative state of each indicator in the relevant subject of analysis. As an example, the results of such work for the regions of the Siberian Federal District are presented below (Tables 2-5).

* Corresponding author: smirnova.e.m@isem.irk.ru

Table 1. Composition of the most important indicators of regional energy security.

1. Block of production and resource provision of the region's fuel and energy supply system
1.1. The ratio of the total available capacity of the region's power plants to the maximum electric load of consumers on its territory. 1.2. The ratio of the amount of available capacity of power plants and the throughput of interconnections between the region and neighboring consumers to the maximum electric load on its territory. 1.3. Possibilities to meet the needs for primary energy from the region's own sources.
2. The block of reliability of fuel and energy supply of the region
2.1. The share of the dominant resource in the total primary energy consumption in the region. 2.2. The share of the largest power plant in the installed electric capacity of the region. 2.3. The level of potential supply of demand for fuel in the conditions of a sharp cooling (10% consumption of consumption) in the region.
3. Block of the state of basic production assets energy systems in the territory of the region
3.1. Degree of depreciation of the basic production assets in the energy sector of the region. 3.2. The ratio of the average annual input of installed capacity and reconstruction of power plants in the region over the last 5-year period to the established capacity of the region.

Table 2. Characteristics of the status of indicators in the territory of the subjects of the Siberian Federal District for the block of production and resource provision of the fuel and energy supply system for 2012, 2016

Indicator	Dimension	The threshold values of the indicator		The meaning and status of the indicator, year			
		N	C	2012		2016	
1	2	3	4	5	6	7	8
Altai region							
1.1	ед.	0,5	0,3	0,74	N	0,72	N
1.2	ед.	1,5	1,2	3,03	N	3,1	N
1.3	%	60	40	1,55	C	1,75	C
Kemerovo Region							
1.1	ед.	0,5	0,3	0,85	N	1,01	N
1.2	ед.	1,5	1,2	2,27	N	2,55	N
1.3	%	40	20	5,88	N	1052	N
Novosibirsk region							
1.1	ед.	0,5	0,3	1,03	N	1,03	N
1.2	ед.	1,5	1,2	2,18	N	2,18	N
1.3	%	40	20	33,7	N	40,4	N
Omsk Region							
1.1	ед.	0,5	0,3	0,84	N	0,86	N
1.2	ед.	1,5	1,2	2,44	N	2,46	N
1.3	%	40	20	48,3	N	36,9	PC
Tomsk Region							
1.1	ед.	0,5	0,3	0,81	N	0,77	N
1.2	ед.	1,5	1,2	2,76	N	2,77	N
1.3	%	60	40	121	N	294	N
Krasnoyarsk region							
1.1	ед.	0,7	0,5	1,65	N	2	N
1.2	ед.	1,5	1,2	2,39	N	2,74	N
1.3	%	60	40	140	N	142	N
Irkutsk region							
1.1	ед.	0,7	0,5	1,58	N	1,6	N
1.2	ед.	1,5	1,2	2,11	N	2,13	N
1.3	%	60	40	122	N	122	N
The Republic of Khakassia							
1.1	ед.	0,7	0,5	2,85	N	2,42	N
1.2	ед.	1,5	1,2	4,67	N	4,28	N
1.3	%	60	40	557	N	740	N
Transbaikal region							
1.1	ед.	1,0	0,8	1,22	N	1,19	N
1.2	ед.	1,5	1,2	1,6	N	1,57	N
1.3	%	60	40	266	N	233	N
The Republic of Buryatia							
1.1	ед.	0,5	0,3	1,31	N	1,28	N
1.2	ед.	1,5	1,2	3,35	N	3,19	N
1.3	%	60	40	49,6	PC	76,1	N
Tyva Republic							
1.1	ед.	0,5	0,3	0,38	PC	0,22	C
1.2	ед.	1,5	1,2	1,18	C	1,78	N
1.3	%	60	40	199	N	240	N
Altai Republic							
1.1	ед.	0,5	0,3	0	C	0,13	C
1.2	ед.	1,5	1,2	0	C	0,13	C
1.3	%	60	40	7,69	C	12,2	C

Along with the qualitative assessment of the energy security level of the regions in 2016, in order to show some trends, information is provided on the qualitative assessment of the relevant indicators in 2012. As a result of analysing the data of the corresponding tables for the subjects of the Siberian Federal District, it is possible to briefly characterize the trends inherent in the energy sector from positions of requirements of power safety.

According to the first block of indicators, the situation is acceptable from the EB's point of view in practically all subjects of the Siberian Federal District.

The exception is the Republic of Altai, where the values of the indicators are in a crisis state (Table 2). Here you can note a low level of the maximum electrical load of consumers. But the situation slowly began to improve and for the period 2014-2016. Three solar power stations were commissioned with a total capacity of 15 MW (Kosh-Agach SES-1, 2, Ust-Kansky SES), and further new capacity inputs are planned.

In the Republic of Tuva in comparison with 2012. by 2016 the values of indicator 1.2 went into the range of acceptable values. In 2014. the throughput of intersystem connections was increased due to the modernization of substations (PS Kyzylskaya, SS Chadan) to 280 MW of capacity, of which up to 100 MW can be transferred to Western Mongolia. However, in general, the problem of energy deficit of the republic is not solved.

In addition, for the first block of indicators, in particular for indicator 1.3, one can note the Altai Territory, the Republic of Altai with crisis values. Omsk region, where by 2016. The situation has worsened due to a 20% decrease in the production of fuel oil over the past five years. In the Republic of Buryatia, on the contrary, the values of the indicators have moved into the region of acceptable, as a result of the development of a number of small coal deposits for local needs and an increase in the level of coal production by 30% over the past five years.

Satisfactory situation for the first block of indicators in the Republic of Khakassia. With sufficient margin, the maximum electrical load is ensured. Extraction of significant volumes of coal provides positive values of the indicator of supply with own primary energy. According to the second set of indicators, the situation in most of the regions is aggravated by an excessively high share of dominance of one of the imported resources (indicator 2.1, Table 3).

Table 3. Characteristic of the status of indicators in the territory of the subjects of the Siberian Federal District for the fuel and energy supply reliability block

Indicator (dimension - %)	The threshold values of the indicator			The meaning and status of the indicator, year			
	N	PC	C	2012		2016	
1	2	3	4	5	6	7	8
Altai region							
2.1	40		70	92,4	C	88,5	C
2.2	50		70	22,8	N	24,2	N
Kemerovo Region							
2.1	90	>90		83,5	N	96,0	N
2.2	50		70	25,8	N	23,7	N
Novosibirsk region							
2.1	40		70	72,6	C	68,9	PC
2.2	50		70	39,4	N	39,4	N
Omsk Region							
2.1	40		70	50,6	PC	54,3	PC
2.2	50		70	43,4	N	42,8	N
Tomsk Region							
2.1	90	>90		68,7	N	62,9	N
2.2	50		70	24,3	N	27,7	N
Krasnoyarsk region							
2.1	90	>90		73,5	N	75,4	N
2.2	40		50	39,2	N	32,8	N
Irkutsk region							
2.1	90	>90		76,5	N	70,6	N
2.2	40		50	34,1	N	33,9	N
The Republic of Khakassia							
2.1	90	>90		96,5	PC	96,5	PC
2.2	40		50	90,9	C	89,5	C
Transbaikal region							
2.1	90	>90		96,4	PC	98,1	PC
2.2	30		40	29,3	N	28,3	N
The Republic of Buryatia							
2.1	40		70	95,1	C	94,2	C
2.2	50		70	83,3	C	79,1	C
Tyva Republic							
2.1	90	>90		97,2	PC	100	PC
2.2	50		70	36,3	N	59,5	PC
Altai Republic							
2.1	40		70	53,8	PC	43,8	PC
2.2	50		70	100	C	64,1	PC

In the Republic of Buryatia and the Altai Territory, the share of coal exceeded in 2012-2016 - 90%. The pre-crisis situation, with a share of the dominant resource, is observed in the Novosibirsk, Omsk regions, the Republic of Khakassia, Tyva, as well as in the Transbaikal Territory and the Altai Republic. In Irkutsk, Kemerovo regions and the Krasnoyarsk Territory, the situation on this indicator can be considered acceptable, due to the dominance of own fuel and energy resources in these regions. At the same time, regarding the discussion of the threshold values of the indicators, the qualitative assessment of the "pre-crisis" in the self-reliant regions (belonging to group 1) indicates the desirability of greater species diversification of fuel and energy supply in order to increase the systems' readiness for potential changes in the structure of the country's TEBs and its regions for various reasons, including for reasons related to the prices of primary fuel and energy resources. According to indicator 2.2 (the share of the largest power plant in the installed electric capacity of the territory), the most acute situation is observed in the Republic of Buryatia (Gusinoozerskaya GRES - 79% of the total installed capacity) and Khakassia (Sayano-Shushenskaya HPP - 89% of the total installed capacity). Since in the event of an accident, such a high share of a single source is very dangerous due to possible problems in the electricity supply of consumers. The situation in the Altai Republic has improved over this indicator in recent years, moving from crisis to pre-crisis by introducing new capacities and redistributing the load.

In the state of pre-crisis, i.e., where it is desirable to develop the growth trend of generating capacities in order to cover the growing demand for electric power, it is the Republic of Tyva (Mobile GTPP - 59% of the installed capacity of the region).

Another important aspect affecting the provision of energy security in the regions is the state of the main production assets of the energy sector. The average data on the wear and tear of the basic production assets energy industries in the territories, correlated with the book value of these industries in these territories allow approximately to estimate the average wear and tear in the energy sector of the territories.

Analysis of the data in Table 4 shows that in recent years, the worsening of energy equipment in the Novosibirsk Region, the Altai Republic and the Altai Territory has worsened (and, most importantly, continues to deteriorate at a fairly rapid pace).

Table 4. Characteristics of the status of indicators in the territory of the subjects of the Siberian Federal District by the state block of the basic production assets of energy systems

Indicator (dimension - %)	The threshold values of the indicator		The meaning and status of the indicator, year			
	N	C	2012		2016	
1	2	3	4	5	6	7
Altai region						
3.1	40	60	58,5	PC	60	C
3.2	2	1	1,6	PC	1,7	PC
Kemerovo Region						
3.1	40	60	47,5	PC	43,7	PC
3.2	2	1	3,7	N	1,9	N

Novosibirsk region						
3.1	40	60	55	PC	61	C
3.2	2	1	1	C	0,7	C
Omsk Region						
3.1	40	60	38,3	N	36,7	N
3.2	2	1	9	N	7,3	N
Tomsk Region						
3.1	40	60	33,6	N	39,7	N
3.2	2	1	12,2	N	0,9	C
Krasnoyarsk region						
3.1	40	60	44	PC	38	N
3.2	2	1	2,9	N	4,9	N
Irkutsk region						
3.1	40	60	54,2	PC	53,7	PC
3.2	2	1	1,8	PC	0,6	C
The Republic of Khakassia						
3.1	40	60	39,5	N	38,7	N
3.2	2	1	7,7	N	4,4	N
Transbaikal region						
3.1	40	60	48,3	PC	47,7	PC
3.2	2	1	2,9	N	3,9	N
The Republic of Buryatia						
3.1	40	60	45,9	PC	43,4	PC
3.2	2	1	2	PC	5	N
Tyva Republic						
3.1	40	60	50,1	PC	58,4	PC
3.2	2	1	29,2	N	7,3	N
Altai Republic						
3.1	40	60	62	C	70	C
3.2	2	1	0	C	33,7	N

The situation on the average in the energy sectors and in most other districts of the district continues to deteriorate, while remaining in the middle of the pre-crisis range in such regions as Irkutsk Region, Republic of Tyva, Republic of Buryatia, Kemerovo Region. Positive trends by 2016 with the renewal and modernization of the basic production assets energy sector are observed in Khakassia, due to the active restoration and modernization of the Sayano-Shushenskaya HPP, as well as modernization at the Abakanskaya CHPP with the commissioning of two power units with a total capacity of 256 MW. In the Kemerovo Region, Kuzbass-Energo carried out a major overhaul of 11 turbine units, and power units were put into operation at Novokuznetskaya GTPP. In the Omsk region and the Republic of Buryatia, acceptable indicators for the indicator are also associated with an active policy for capital repairs and reconstruction of power generating capacities. To a large extent, the value of this indicator is due to the commissioning of new capacities, major repairs and technical re-equipment of existing power generating sources. In the crisis situation in the aspect reflected by indicator 3.2 are: Novosibirsk, Tomsk, Irkutsk regions and the Republic of Altai, where for the analysis preceding the 5-year period there was insufficient commissioning of new capacities and practically no serious work was done to modernize the installed equipment, which in turn led to a decrease in the level of energy security of regions in the aspect described by this indicator.

Smaller inputs and some equipment upgrades were carried out in the Altai Republic and the Altai Territory (commissioning 130 MW at Barnaulskaya CHP-2 in 2016), but in insufficient (in terms of energy security) volumes to reverse the negative trends with the aging of the basic production assets. As for the indicator of the renewal of power generating equipment in the Irkutsk Region and the Trans-Baikal Territory, there were not enough new power inputs.

In the Irkutsk region in 2012, a turbine unit with a capacity of 50 MW was launched at Novo-Irkutsk TPP. In the Trans-Baikal Region, the commissioning of a hydroelectric unit at Kharanorskaya HPP in the same year, with a capacity of 225 MW. In addition, major repairs of equipment and reconstruction were carried out.

These actions brought the situation on this indicator in an acceptable state from the EB's position in the Trans-Baikal Territory, but they were insufficient for the withdrawal of the Irkutsk region from the crisis.

Above were presented and analyzed the values of the main indicators, which form the basis for an integrated assessment of the level of energy security in the territory of the subjects of the Siberian Federal District. To obtain such an assessment, an approach based on the convolution of indicator values was used, taking into account their specific weights. Qualitative characteristics of the state of all the indicators discussed in Table. 2-3, were collected according to the respective territories and processed according to a special methodology. As a result, a qualitative final assessment of the energy security status of the territories of the subjects of the Siberian Federal District was presented. 5.

Based on the analysis of data in Table 5, the best condition for energy security is observed in the Tomsk region and the Krasnoyarsk Territory. The level of energy security close to acceptable can be considered in the Kemerovo, Omsk, Irkutsk regions and the Trans-Baikal Territory. It is necessary to pay serious attention to indicators, whose values in these territories are located in the zones of "crisis" and "pre-crisis" values. This signals serious problems in the field of fuel and energy supply of territories in the part described by the values of the respective indicators. In addition, the negative state of the indicators characterizing the condition of the OPF and the renewal of energy in these territories is, above all, hindering the achievement of a better position.

The most acute situation with ensuring energy security should be noted in the Republic of Altai. Here, the crisis situation of most of the monitored indicators is evident, this concerns both the degree of the maximum electrical load and the share of the largest power generating source in the installed capacity in the territory, and the wear and tear of the energy sector.

 https://doi.org/10.1051/e3sconf/20185803008

Table 5. Assessment of the qualitative state of certain regions of the Russian Federation

Year	The order numbers of the estimated ES indicators								The sum of the specific weights by state			Quality condition ES
	1.1	1.2	1.3	2.1	2.2	2.3	3.1	3.2	Boundaries of states			
	Specific weights of indicators								C	PC	N	
	0,104	0,138	0,133	0,120	0,079	0,170	0,127	0,129				
1	2	3	4	5	6	7	8	9	10	11	12	13
Altai region												
2012	N	N	C	C	N	N	PC	PC	0,253	0,256	0,491	PC
2016	N	N	C	C	N	N	PC	PC	0,253	0,256	0,491	PC
Kemerovo Region												
2012	N	N	N	N	N	N	PC	N	0	0,127	0,873	N
2016	N	N	N	N	N	N	PC	N	0	0,127	0,873	N
Novosibirsk region												
2012	N	N	PC	C	N	N	PC	C	0,249	0,260	0,491	PC
2016	N	N	N	C	N	N	C	C	0,376	0	0,624	PC
Omsk Region												
2012	N	N	N	PC	N	N	N	N	0	0,120	0,880	N
2016	N	N	PC	PC	N	N	N	N	0	0,253	0,747	N
Tomsk Region												
2012	N	N	N	N	N	N	N	N	0	0	1	N
2016	N	N	N	N	N	N	N	C	0,129	0	0,871	N
Krasnoyarsk region												
2012	N	N	N	N	N	N	PC	N	0	0,127	0,873	N
2016	N	N	N	N	N	N	N	N	0	0	1	N
Irkutsk region												
2012	N	N	N	N	N	N	PC	PC	0	0,256	0,744	N
2016	N	N	N	N	N	N	PC	C	0,129	0,127	0,744	N
The Republic of Khakassia												
2012	N	N	N	PC	C	N	N	N	0,079	0,120	0,801	N
2016	N	N	N	PC	C	N	N	N	0,079	0,120	0,801	N
Transbaikal region												
2012	N	N	N	PC	N	N	PC	N	0	0,247	0,753	N
2016	N	N	N	PC	N	N	PC	N	0	0,247	0,753	N
The Republic of Buryatia												
2012	N	N	PC	C	C	N	PC	PC	0,199	0,389	0,412	PC
2016	N	N	N	C	C	N	PC	N	0,199	0,129	0,674	PC
Tyva Republic												
2012	PC	C	N	PC	N	N	PC	N	0,138	0,351	0,511	PC
2016	N	C	N	PC	PC	N	PC	N	0,242	0,326	0,432	PC
Altai Republic												
2012	C	C	C	PC	C	C	C	C	0,880	0,120	0	C
2016	C	C	C	PC	PC	C	C	N	0,672	0,199	0,129	C

3 Conclusion

Analysis of the dynamics of qualitative assessments of energy security in the regions and quantitative indicators of the weights of indicators that are in different states from 2012 to 2016 allows to draw a conclusion that the situation with ensuring energy security during this period on average had positive trends: for all subjects of the Siberian Federal District, with the exception of the Republic of Altai.

References

1. N. Pyatkova, S. Senderov, M. Cheltsov, *"Application of a two-level technology for research in solving energy security problems,"* Izvestiya RAN, Power Engineering, **6**, (2000)
2. V. Bushuyev, N. Voropai, A. Masterpanov, Yu. Shafranik, *"Energy security of Russia,"* Novosibirsk, Science, Siberian Publishing Company RAS (1998)

3. S. Senderov, N. Pyatkova, V. Rabchuk, G. Slavin, S. Vorobev, E. Smirnova, *"Methodology for monitoring the state of energy security of Russia at the regional level,"* ISEM SB RAS, (2014)

4. S. Senderov, *"Assessment of the level of energy security in the regions of Russia and the basic principles for the establishment of a system for monitoring energy security,"* Safety of objects of the fuel and energy complex, **1**,(2012)

5. Statistical form of Rosstat, *"Information on balances, receipt and consumption of fuel and heat, collection and use of used oil products"*, (2012-2016)

6. Statistical form of Rosstat, *"Information on the use of fuel and energy resources"*, (2012-2016)

7. Statistical form of Rosstat, *"Technical and economic indicators of power plants, district boilers"*, (2012-2016).

Determination of Critically Objects of Electric Power Systems from the Position of Energy Security

Dmitry Krupenev[1,*]

[1]Melentiev Energy Systems Institute of Siberian Branch of the Russian Academy of Sciences, ESI SB RAS, Irkutsk, Russia

Abstract. The paper deal the problem of definition and ranking of critical objects (CO) in electric power systems (EPS). The identification of the CO is necessary for the timely adoption of measures to organize the provision of the required level of energy security of the region where the object is location. The adequacy model is used for determine the critical objects of EPS, within the framework of which the simulation of the functioning of EPS during the given time interval is carried out, taking into account the main random factors affecting its operation. Approbation of the proposed approach is presented on the interconnected power system.

1 Introduction

The energy security of Russia and its regions includes two main aspects:
- we need to provide for long-time no-deficit with the required types of energy resources of consumers in the operation of energy in standard conditions;
- we need to create of conditions for provision of energy resources to consumers in emergency situations (ES).

When we considerate of the second aspect we have to define of critical objects (CO) of the fuel and energy complex (FEC), i.e. those facilities, partial or complete failure of which can cause significant damage to the economy of the country. In this case, the reasons for the failure can be different from the effects of natural disasters before terrorist activities.

The CO FEC are objects whose violation or termination of their functioning will lead to a loss of economic management of the Russian Federation, a constituent entity of the Russian Federation or an administrative-territorial unit, its irreversible negative change (destruction) or a significant decrease in the safety of the population [1]. In this paper the problem of determining the CO focuses on the electric power industry, namely, on the electric power system (EPS).

The definition of the CO FEC should be based on the following main possible negative consequences for society and the economy of the country when it is damaged:
1. Economic consequences: damage to electricity consumers due to short supply and energy companies in case of damage to the power plant, as well as in case of damage to the thermal power plant, damage from shortage of heat energy to consumers.
2. Environmental consequences: possible negative processes on the environment in the event of an accident at the power industry site.
3. Social consequences: the impact of a possible accident on the health and lives of people.

The importance of the object is determined on the basis of a comprehensive evaluation of the submitted criteria. In this paper, the focus is on assessing the economic consequences of accidents at power facilities. Assessment of economic consequences and ranking of electric power facilities on the importance of this indicator will allow making priorities in the al-location of funds for ensuring energy security and the development of the energy system and, thereby, increasing the economic efficiency of the operation of EPS. The identification and ranking of the CO of EPS will allow focusing on selected objects to increase their security against different types of threats, including from terrorist acts (including cyber terrorism).

The definition of the CO is an important task not only for the electric power industry, there are already approaches to the definition of the CO in the gas transportation system [2, 3]. In these sources, it is suggested that, based on the analysis of the operation of the transport system, the CO should be assessed at the most intense gas consumption per day, by determining the resulting gas deficiencies, by disconnecting in turn the elements from the pre-established list.

EPS is a complex technical facility that is fundamental for the successful development of the country's economy. EPS has a number of characteristics related to the definition of CO:
- there are many states of electricity and power consumption characteristic of different seasons, under these conditions, in determining the CO, it is necessary to analyse the maximum number of states, since the degree of importance of the object can manifest itself not only in the state of maximum power consumption. In different regions the maximum power consumption can be for different months;

* Corresponding author: krupenev@isem.irk.ru

- we need to take into account the scheduled repairs of power equipment in addition to the consumption regime in determining the CO, since their conduct is mandatory and, in the final analysis, the equipment that is under planned repair will have an additional impact on the possible power shortage and underexposure - power outage upon failure of the analysed object;
- there are some equipment in the EPS that can go out of operation during the calculated period in addition to the failure of the analysed object, and thereby aggravate the current situation.

2 Statement of the problem of determining and ranking the CO EPS and the methodology for its solution

We have to use a model that simulates the operation of EPS for a year, and in which all the factors affecting the power shortage and under-supply of the factors stated above for the determination and ranking of EPS CO. The model of estimating the adequacy of EPS is reasonable to take as a basis [4]. In this model based on the Monte Carlo method, many EPS operating modes are simulated over a one-year period, while scheduled repairs, unscheduled repairs, regular and irregular load fluctuations are recorded. The model consists of three computational stages:

1. Stage of formation of design conditions of EPS.
2. Stage of determining the capacity deficits of the generated EPS states. The mathematical formulation of this problem is as follows [5]:

for assessment of the power shortage in the k-th power system state, $k = 1, \ldots, N$, it is necessary to determine:

$$\sum_{i=1}^{n} (\bar{y}_i^k - y_i) \to \min \qquad (1)$$

subject to the balance constraints

$$x_i - y_i + \sum_{j=1}^{n} (1 - z_{ji} a_{ji}) z_{ji} - \sum_{j=1}^{n} z_{ij} = 0,\ i = 1, \ldots, n,\ i \neq j, \qquad (2)$$

$$y_i \leq \bar{y}_i^k,\ i = 1, \ldots, n, \qquad (3)$$

$$x_i \leq \bar{x}_i^k,\ i = 1, \ldots, n, \qquad (4)$$

$$z_{ij} \leq \bar{z}_{ij}^k,\ i = 1, \ldots, n,\ j = 1, \ldots, n,\ i \neq j, \qquad (5)$$

$$y_i \geq 0,\ x_i \geq 0,\ z_{ij} \geq 0, \qquad (6)$$

where: x_i is the usable capacity at node i, $\bar{x}_i^k$ is the available generating capacity at node i, y_i is the served load at node i, $\bar{y}_i^k$ is the load value at node i, z_{ij} - is the power flow from node i to node j, $\bar{z}_{ij}^k$ is the transfer capability of transmission lines between nodes i and j, a_{ij} - are the given positive coefficients of specific power losses at its transmission from node i to node j, $i \neq j$, $i = 1, \ldots, n$, $j = 1, \ldots, n$, $i \neq j$, $k = 1, \ldots, N$.

3. Stage of determining mathematical expectation (m.e.) of undersupply of the electricity to consumers and m.e. of the capacity deficits.

We used this sequence of actions to determine and rank the CO:

1. Formation of the list of objects of EPS for the definition of m.e. capacity shortage and m.e. of undersupply of the electricity to consumers in the event of their failure. Formation can be carried out in several ways:
- alternate search of power stations and power lines for one object;
- by sequences sort out of expertly assigned objects of EPS.

2. Assessment of the adequacy of all assigned states, depending on the chosen method in the first step. The definition for each variant is m.e. of the capacity deficits and m.e. of undersupply of the electricity to consumers. The assessment can be carried out for any a priori of the appointed time interval: year, month, day, and hour.

3. Ranking of the results. Identifying the objects most strongly affecting m.e. of the capacity deficits and m.e. of undersupply of the electricity to consumers, determination and ranking of the CO of EPS.

3 Experimental investigation

We will show the definition of CO on the interconnected power system (IPS) of Siberia. IPS of Siberia is a large power association that is part of the Unified Power System of Russia. Large scale thermal and hydraulic power plants operate in the UPS of Siberia, as well as transmission lines (TL) with a voltage of 220 and 500 kV. In fig. 1 shows the scheme of the IPS of Siberia [5] and the transformed model of the IPS of Siberia for determining the CO. Clustering in the zones of reliability of the IPS of Siberia was carried out on the basis of the division into the constituent entities of the Russian Federation, the exception is the Bodaibo download center, formally belonging to the Irkutsk region, but actually connected with the Buryatia energy system.

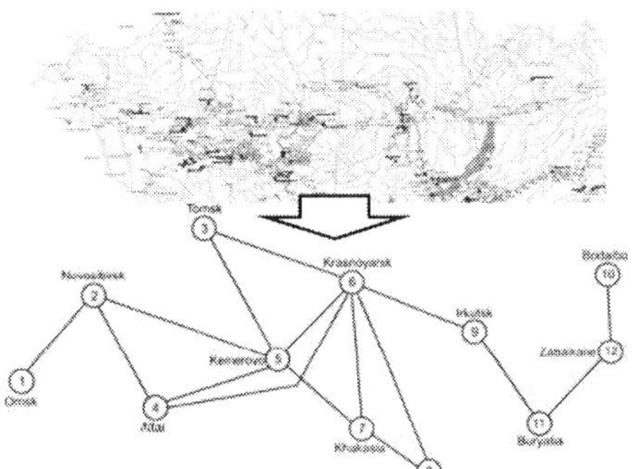

Fig. 1. Model of the IPS of Siberia for the assessment of the CO.

Table 1 presents the characteristics of the reliability zones of the calculated model of the IPS of Siberia.

Table 1. Characteristics of reliability zones of the IPS of Siberia.

№ zone	Zone name	Annual peak load, MW	Available capacity, MW
1	Omsk	1782	1479
2	Novosibirsk	2690	2730
3	Tomsk	1302	918
4	Altai	1884	1444
5	Kemerovo	4535	4462
6	Krasnoyarsk	6235	6848
7	Khakasia	2155	5025
8	Tyva	152	40
9	Irkutsk	7570	9048
10	Bodaibo	90	20
11	Buryatia	945	898
12	Zabaikalie	1260	1156

Table 2 shows the throughput capacity of tie lines connections of the calculation model of the IPS of Siberia.

Table 2. Capacity of tie lines of IPS of Siberia.

№ tie line	Connected zones	Capability of tie line, MW
1	1. Omsk – 2. Novosibirsk	1305
2	2. Novosibirsk – 4. Altai	1440
3	2. Novosibirsk – 5. Kemerovo	950
4	3. Tomsk – 5. Kemerovo	1170
5	3. Tomsk – 6. Krasnoyarsk	780
6	4. Altai – 5. Kemerovo	950
7	4. Altai – 6. Krasnoyarsk	850
8	5. Kemerovo – 6. Krasnoyarsk	1560
9	5. Kemerovo – 7. Khakass	1650
10	6. Krasnoyarsk – 7. Khakass	3400
11	6. Krasnoyarsk – 8. Tyva	135
12	6. Krasnoyarsk – 9. Irkutsk	3630
13	7. Khakasia – 8. Tyva	135
14	9. Irkutsk – 11. Buryatia	885
15	10. Bodaibo – 11. Buraytia	66
16	11. Burayta – 12. Zabaikalie	410

We need to determine the CO of the IPS of Siberia, at the first stage, we will expertly identify a number of power plants in the reliability zones and TL in tie lines connections, which will be outage off during the research. Reliability zone is a concentrated node, inside of which there are no restrictions on the transmitted power, the most significant contribution to the value of the m.e. of the capacity deficits and m.e. of undersupply of the electricity to consumers will be made by the largest stations in each zone, therefore during the implementation of the experiment at the first stage we will consistently disconnect the largest power stations in each zone.

Table 3. List of power plants to be switched off.

Zone name	Plant name, MW	Available capacity of plant, GW
Omsk	TEP 5	0,73
Novosibirsk	TEP 5	1,2
Tomsk	AO SHK	0,43
Altai	Biisk TEP	0,508
Kemerovo	Tom-Usinskaia GDEP	1,34
Krasnoyarsk	Krasnoyarsk GEP	5,762

Khakasia	Sayano-Shushensk GEP	5,326
Tyva	Kisil TEC	0,17
Irkutsk	Bratsk GEP	4,2184
Bodaibo	Mamakan GEP	0,086
Buryatia	Gusinoozersk GDEP	1,16
Zabaikalie	Haranorsk GDEP	0,665

Simulation of the operation of the IPS Siberia was carried out during the day corresponding to January 15. In table 4 shows the results of calculations after disabling the generating objects from table 3.

Table 4. The effect of outage off power plants in the zones of reliability of the UPS of Siberia.

Zone name	M.e. of undersupply of the electricity to consumers in the UPS without plant outage, kWh	M.e. of under-supply of the electricity to consumers in the UPS with plant outage, kWh	M.e. of under-supply of the electricity to consumers reliability zone, in which the stations were outage, kWh
Omsk	3	29	0
Novosibirsk	0	30	0
Tomsk	0	26	0
Altai	0	23	0
Kemerovo	0	20	0
Krasnoyarsk	0	23	0
Khakassia	0	32	0
Tuva	0	24	0
Irkutsk	0	32	0
Bodaibo	0	401	358
Buryatia	0	30	0
Zabaykalsk	24	30	27
Sistem	27		

We can see in table 4 the current state of the IPS of Siberia from the standpoint of providing electricity consumers to the CO can only be attributed to the Mamakanskaya GEP in the Bodaibo. In other reliability zones, the withdrawal of the largest power plant from work did not lead to an increase in the under-supply of electricity either in this zone or in the IPS of Siberia, i.e. in the IPS of Siberia there is a sufficient amount of reserve generating capacity to meet the power requirements for the failure of the largest station in each node of the system under consideration.

We made similar outages with electric transmission lines in all link connections of the IPS of Siberia, i.e. we will conduct a series of calculations with the sequential withdrawal of the largest TL in each tie line. Table 5 shows the list of power lines that are switched off during the experiment.

Table 5. List of outage power TL.

№ tie line	Substation of start TL	Substation of finish TL	Voltage TL, kV
1	Tavricheskay	Barabinskay	500
2	Zarya	Altay	500
3	Zarya	Urga	500
4	Tomskaya	Novo-Ansherskaya	500
5	Tomskaya	Itatskaya	500
6	Barnaulskaya	Novokyznetskaya	500
7	Altay	Itatskaya	500
8	Novo-Ansherskaya	Nazarovskaya GREP	500
9	Novokyznetskaya	Sayano-Sushenskaya GEP	500
10	Abakanskaya	Itatskaya	500
11	Ergaki	Turan	220
12	Kamala	Taishet	500
13	Abaza	Ak-Dovurak	220

14	Kluchi	Gusinoozerskaya GREP	500
15	Tacsimo	Mamacan	220
16	Gusinoozerskaya GREP	Petrovsk-Zabaikalsk	220

After carrying out a number of calculations, the values of m.e. of undersupply of the electricity to consumers in the IPS, formed in connection with the disconnection indicated in table. 5 transmission lines. In Table. 6 shows these results.

Table 6. M.e. of undersupply of the electricity to consumers in the IPS of Siberia during the link of power transmission lines in inter-zone communications, kWh and the ratio to the required electricity generation,% (the zone numbers correspond to the numbers from the previous tables)

№ zone	2	4	5	6	7	8	9	11	12
1	3027/0,45	-	-	-	-	-	-	-	-
2	-	29/0	32/0	-	-	-	-	-	-
3	-	-	22/0	31/0	-	-	-	-	-
4	-	-	40/0	46/0	-	-	-	-	-
5	-	-	-	24/0	33/0	-	-	-	-
6	-	-	-	-	28/0	46/0	35/0	-	-
7	-	-	-	-	-	42/0	-	-	-
9	-	-	-	-	-	-	-	47/0	-
10	-	-	-	-	-	-	-	602/0	-
11	-	-	-	-	-	-	-	-	295/0

We can see in table 6 expectation of undersupply of the electricity to consumers in the IPS of Siberia has changed slightly, only when the largest TL in the tie line Omsk-Novosibirsk, Bodaibo - Buryatia and Buryat - Zabaykalsk was cut off the values assumed large values. With respect to the required generation of electricity that amounts to 663171 million kWh during the entire disconnection of the above-mentioned power lines, there is no significant change, although locally for the connected reliability zones the above-mentioned power lines can lead to a significant under-release and damage.

Thus, after the analysis of the IPS of Siberia on the subject of CO from the standpoint of energy security, it was revealed that in the current conditions, the CO can include: Mamakanskaya GEP and TL 500 kV Tavricheskaya-Barabinskaya; TL 220 Taksimo-Mamakan; TL 220 Gusinoozerskaya GREP - Petrovsk-Zabaikalsk.

4 Conclusions

The article considers the problem of determining critical objects in electric power systems. The methodology for determining the CO EPS is proposed. The situation with CO has been analysed on the example of the IPS of Siberia, it is clarified that in the current conditions of operation of the IPS, Siberia has rather large reserves for both generating facilities and the network part, but several objects are critically important which should be given close attention when planning the development of the IPS of Siberia.

References

1. *On the safety of objects of the fuel and energy complex* (Federal Law N 256, 2011)
2. Senderov S.M., Rabchuk V.I., Edelev A.V. Features of the formation of the list of critical objects of the Russian gas transmission network, taking into account the requirements of energy security and possible measures to minimize the negative consequences from emergency situations at such facilities. Izvestiya. Power Engineering, No. 1, (2016)
3. Senderov S., Edelev A. Formation of a List of Critical Facilities in the Gas Transportation System of Russia in Terms of Energy Security. Energy 2017 (2017)
4. Kovalev G.F., Lebedeva L.M. *Reliability of power systems*. (Novosibirsk: Science, 2015)
5. On the approval of the scheme and program for the development of the Unified Energy System of Russia for 2016-2022 (Order of the Ministry of Energy of Russia No. 147 2016)

https://doi.org/10.1051/e3sconf/20185803010

Methods of linear and nonlinear state estimation of distribution network

Irina Golub[1,*] and *Evgeny* Boloev[1]

[1] Energy Systems Institute of Siberian Branch of the Russian Academy of Sciences, Department of Electric Power System , 664033 Lermontov str., 130, Irkutsk, Russia

Abstract. The paper proposes a new approach to the problem of state estimation of a low voltage distribution network by the measurements coming from smart meters. The problem of nonlinear state estimation based on the measurements of nodal powers and voltages is solved by the method of simple iteration which minimizes the quadratic function of the residues with and without the consideration of the constraint on the zero currents in the transit nodes. The same algorithms are proposed to use for linear state estimation based on the measurements of nodal currents and voltages. The effectiveness of the proposed methods for linear and nonlinear state estimation is illustrated on the 33 nodes three-phase four-wire low-voltage network.

1 Introduction

Methods of state estimation (SE) have found wide application in transmission systems designed to transport electrical energy from power plants to distribution networks. The most important event, which contributed to significant improvement of reliability of SE procedure for the transmission system is the introduction of synchronized phasor measurement unit (PMU). Such SE is instrumental in identification of the power passed to the primary medium-voltage distribution network, but it is unable to identify neither overall loads of secondary transformation substations, nor the loads of low voltage distribution network.

Traditional primary and secondary distribution networks operate as radial, with power flows in the feeders oriented from the primary distribution substation towards load nodes. Secondary distribution network is modeled as a three-phase network with neutral wire; its loads can be single-phase, two-phase and three-phase. The return current in neutral wire is due to the unbalanced load and the nonlinear characteristics of electrical equipment through the distribution feeder. The return current may be larger than the phase currents if three-phase loads are seriously unbalanced in some sections [1]. Imaginary components of the voltage drops in the feeders sections virtually equal zero, which is due to a large ratio of resistances to reactance and small lengths of individual sections. Secondary distribution network is modeled without considering mutual reactance between phase wires and without shunt capacitances.

In traditional distribution networks, measurements are usually performed only on primary substations from the medium voltage side, and the state of the secondary distribution substations is unknown. There are methods, in which such measurements are used for approximate estimation of load flow and energy losses in the distribution network [2], but these measurements are not sufficient for SE.

Transition of passive distribution network to active network, is first and foremost related to introduction into it of renewable energy generation sources of stochastic nature, as well as energy storages, charging and discharging of which can be regarded as additional load or generation, active or flexible loads, which allow to transfer the loads of peak period to non-peak within a planned span of time. Directions of power flows in active networks vary during the day, load nodes can become generator nodes, and voltage deviations may exceed admissible values. Voltage control with the help of results of load flow calculation based on the power arriving to the distribution network is impossible even for traditional networks, and even less so for active ones.

Load flow calculation will benefit from the use of average load values calculated on the basis of measurements of energy consumed and registered in the reports of modern automated system of commercial electrical energy accounting in terms of improving accuracy of defining node voltages, but even such accuracy is insufficient to monitor voltages of active distribution networks.

Information of the secondary distribution network state may be obtained by SE methods on the basis of measurements of smart meters (SM), which are installed at load and/or generator nodes. Modern measuring infrastructure Advanced Metering Infrastructure [3] combining an active distribution network with communication network and modern measurement system is responsible for the transfer of SM data to the system of data collection and back. Some SM can measure, besides

*Corresponding author: golub@isem.irk.ru

consumed energy, active and reactive power, active and reactive currents and voltage magnitudes.

Information on such SM measurements is handled by the SE algorithms, the results of which are used to analyze permissibility of voltages and currents, to calculate power losses and indicators of secondary network reliability. In addition, the estimates of power flows and/or currents arriving within each phase from the primary distribution network into the secondary network, as well as the estimates of phase voltages in the secondary network feeding points can be used as loads and voltages for calculation of load flow and/or three-phase state estimation of primary distribution network.

A list of tasks, which require information on the loads and load flow of primary distribution network, is extensive. It includes energy loss calculation, regulation of reactive power and voltage, optimization of capacitor placement, network reconfiguration, load forecasting, network strengthening.

However, the process of secondary distribution network SE should be preceded by taking into account the fact that the information on the phase, to which users are connected and for which consumption measuring SM are installed, is generally not available. Knowledge of measurement phase, as well as observability, is an indispensable necessary condition for SE of the secondary distribution network, and, consequently, loads of the primary distribution network.

An effective approach to the definition of the phases is a measurement experiment proposed in [4], which consists in assessment of cross-correlation between voltage profiles in two nodes within certain time period. The first one is the power distribution network node or the closest node with known phases, in each phase measurements are conducted with certain time intervals. The second node is sequentially represented by the nodes with SM, in which measurements of voltage magnitudes are performed at the same time, as the measurements in the phases of the first node. Definition of SM phase connection SM is based on maximum values of cross-correlation coefficients.

There is no single answer to the question on the possibility of synchronizing SM measurements. The first publications on SM [5] argued that there is a possibility to synchronize SM measurements of currents and voltages in the same distribution network through PLC communication technology, which helps to transmit measurements directly along power transmission lines. However, subsequent publications [6] analyzed the factors impeding synchronization, such as the delay in measurement transfer along communication channels, ranging from a few seconds to several minutes and even up to one day, while measurement upgrade is performed not more often than every 15 minutes.

Despite the existence of SM synchronization methods, such measurements are usually not synchronized and may not be used for SE as synchronized ones, since this would result in unacceptable decrease of estimation accuracy, [7] proposes to compensate for the lack of synchronization with the increase in measurement error with SE up to 10%.

2 Secondary distribution network state estimation

Below we analyze two variants of SM measurements in load and/or generator nodes for SE of secondary distribution network: 1) active $z_P^{a,b,c}$ and reactive $z_Q^{a,b,c}$ node power and voltage magnitudes $z_U{}^{a,b,c}$ in the phases *a*, *b*, *c*; 2) active $z_{Ja}^{a,b,c}$ and reactive $z_{Jp}^{a,b,c}$ component of node currents and voltage magnitudes $z_U{}^{a,b,c}$.

Additional measurements are zero values of active and reactive power and active and reactive components of node currents in transit nodes.

Elements of the state vector are represented by real $u_i'{}^{a,b,c}$ and imaginary $u_i''{}^{a,b,c}$ components of the vectors of phase voltages $u^{a,b,c} = u_i'{}^{a,b,c} + j u_i''{}^{a,b,c}$ in the nodes $i = 1,\ldots,n$ of the design scheme. Measured voltages can be set by their real components, which are equal to the measured voltage magnitudes. The validity of such replacement is related to the proximity of the imaginary voltage components in low-voltage network to zero. Imaginary component of the voltage in the power supply node of the secondary power distribution network is registered as zero.

For the first version of the measurements, it is proposed to use the SE procedure consisting of two iterative repetitive steps. At the first step of the first iteration, the measured values of active and reactive node powers and measurements of voltage magnitudes in the node *i* are used to determine the values of node currents, which we propose to call pseudo-measurements of node currents

$$z_{Jai}^{a,b,c} - j z_{Jpi}^{a,b,c} = \left(z_{iP}{}^{a,b,c} - j z_{iQ}{}^{a,b,c} \right) / z_{U'_i}^{a,b,c} . \quad (1)$$

At the second step by pseudo-measurements of currents and real voltage components from the solution of the linear system of measurement equations we define estimation of state variables, which are substituted to the denominator of the expression (1) in subsequent iterations

$$z_{Jai}^{a,b,c} - j z_{Jpi}^{a,b,c} = \left(z_{iP}{}^{a,b,c} - j z_{iQ}{}^{a,b,c} \right) / \left(u_i'{}^{a,b,c} + j u_i''{}^{a,b,c} \right) . (2)$$

Iterative process ends, when the maximum difference between the state variables obtained during adjacent iterations does not exceed predetermined calculation accuracy. Such a procedure for solution of nonlinear systems of measurement equations called simple iteration method is hereby proposed to be applied for secondary distribution network SE instead of traditionally used Newton's iterative method based on linearization of measurement equations through their expansion into Taylor series [8, 9]. The simple iteration method is known to work well provided the original values of node voltages are appropriate. In the present case, which presumes that SM measure both node powers and voltages, this condition is satisfied. The main advantage of such a solution of nonlinear measurement system is absence of necessity to form Jacobi matrix at iterations, which makes the solution of the nonlinear state estimation task as simple as that of linear task.

For the second variant of measurements, SE procedure includes solution only linear system equations of measurements of node currents and real voltage components.

Measurement linear equations corresponding to the measured in node i phase values of active and reactive components of the currents and real voltage component can be presented as

$$\begin{pmatrix} g_i^{a,b,c} & b_i^{a,b,c} \\ -b_i^{a,b,c} & g_i^{a,b,c} \\ \mathrm{I}^{a,b,c} & 0^{a,b,c} \end{pmatrix} \begin{pmatrix} u_i'^{a,b,c} \\ u_i''^{a,b,c} \end{pmatrix} = \begin{pmatrix} z_{Jai}^{a,b,c} \\ z_{Jpi}^{a,b,c} \\ z_{U'i}^{a,b,c} \end{pmatrix} \quad (3)$$

where $g_i^{a,b,c}$ and $b_i^{a,b,c}$ are of nodal conductance and susceptance matrices, $\mathrm{I}^{a,b,c}$ and $0^{a,b,c}$ – unit and the zero matrices.

In general terms, the system (3) can be presented as

$$H_z^{a,b,c} \cdot u^{a,b,c} = z_z^{a,b,c}. \quad (4)$$

Similarly, measurement equations for zero node currents in transit node i will appear as

$$\begin{pmatrix} g_i^{a,b,c} & b_i^{a,b,c} \\ -b_i^{a,b,c} & g_i^{a,b,c} \end{pmatrix} \begin{pmatrix} u_i'^{a,b,c} \\ u_i''^{a,b,c} \end{pmatrix} = \begin{pmatrix} z_{Jai}^{a,b,c} \\ z_{Jpi}^{a,b,c} \end{pmatrix} = \begin{pmatrix} 0 \\ 0 \end{pmatrix} \quad (5)$$

or

$$H_0^{a,b,c} \cdot u^{a,b,c} = 0. \quad (6)$$

We hereby present the combination of equation systems (4) and (6) as

$$H \cdot u = \begin{pmatrix} H_z^{a,b,c} \\ H_0^{a,b,c} \end{pmatrix} \cdot u^{a,b,c} = \begin{pmatrix} z_z^{a,b,c} \\ 0 \end{pmatrix} = z. \quad (7)$$

System (6) has a single solution if the number of its equations is equal to the number of variables – basic measurement system, and the rank of matrix H called the matrix of observability [10, 11], equals to the number of state variables. To smooth the error impact in individual measurements on the estimation of state variables, measurement equations are multiplied by weight coefficients $R^{-1/2}$

$$R^{-1/2} H \cdot u = R^{-1/2} z, \quad (8)$$

where R is measurement error variance.

Such weighing does not affect the solution of the basic measurement equation system. A necessary condition for weighing efficiency is measurement redundancy, which ensures absence of critical measurements, the loss of which leads to a loss of observability. Matrix $R^{-1/2}H$ in the presence of redundant measurements becomes redefined, as the number of its lines exceeds the number of columns. There is no classical solution for redefined system (8), but a solution vector u can be acquired, which would help to minimize the distance between the vectors of the right and left parts (8), to this end, the criterion of minimizing the sum of squares differences is used

$$J(u) = (z - Hu)^{\mathrm{T}} R^{-1} (z - Hu). \quad (9)$$

This method is called the method of weighted least squares (WLS) and the solution to the task of minimizing can be obtained from the normal system of equations with a square matrix

$$\left(H^{\mathrm{T}} R^{-1} H\right) u = H^{\mathrm{T}} R^{-1} z. \quad (10)$$

Estimation of state variables obtained from the solution (10)

$$u = \left(H^{\mathrm{T}} R^{-1} H\right)^{-1} H^{\mathrm{T}} R^{-1} \quad (11)$$

is used to calculate estimates of measured variables z, in our case – node currents and voltage magnitudes

$$z = H\left(H^{\mathrm{T}} R^{-1} H\right)^{-1} H^{\mathrm{T}} R^{-1} = Hu, \quad (12)$$

of residues

$$r = z - z = z - Hu. \quad (13)$$

Criterion (9) through the weighted residue can be presented as

$$J(u) = r^{\mathrm{T}} R^{-1} r. \quad (14)$$

Reduction of weight coefficients for measurements with large residues or their exclusion from the measurements can help to increase the accuracy of estimates of all variables. Another factor affecting estimates is conditionality of matrix $\left(H^{\mathrm{T}} R^{-1} H\right)$ characterizing the impact of measurement errors on the errors of state variables estimation. If the measurements include a large amount of node currents, the conditionality is lower than in the case of measuring line currents. Furthermore, there is an additional factor leading to deterioration of conditionality, which is the need to set very high weight coefficients for equations (6) associated with zero currents.

To obtain a solution ensuring equality to zero of currents in transit nodes, minimization (9) is replaced [12] by Lagrange function minimization, including the function $J_Z(u)$ recorded only for the measured variables and constraints of equations (6) with Lagrange multipliers λ

$$L = 0.5 \cdot (z - H_z u)^{\mathrm{T}} R_z^{-1} (z - H_z u) - \lambda^{\mathrm{T}} H_o u. \quad (15)$$

Solution of linear system of equations with a square matrix is instrumental in determination of state variables estimations u

$$\begin{pmatrix} H_z^{\mathrm{T}} R_z^{-1} H_z & H_0^{\mathrm{T}} \\ H_0 & 0 \end{pmatrix} \begin{pmatrix} u \\ \lambda \end{pmatrix} = \begin{pmatrix} H_z^{\mathrm{T}} R_z^{-1} z_z \\ 0 \end{pmatrix}. \quad (16)$$

The solution method in [12] is called WLS method with constraints.

To increase numerical stability, [13] proposes to express $J_Z(u)$ in Lagrange function through the residues (14) and, in addition to the constraint (5), to add the constraint on zero equality of the residues (13)

$$L = 0.5 \cdot r^{\mathrm{T}} R_z^{-1} r - \lambda^{\mathrm{T}} H_o u - \mu^{\mathrm{T}} (r - z + H_z u). \quad (17)$$

A solution on u, λ and μ in this case may be obtained from the linear system of equations with Hachtel matrix

$$\begin{bmatrix} R_z & H_z & 0 \\ H_z^{\mathrm{T}} & 0 & H_0^{\mathrm{T}} \\ 0 & H_0 & 0 \end{bmatrix} \begin{pmatrix} \mu \\ u \\ \lambda \end{pmatrix} = \begin{pmatrix} z \\ 0 \\ 0 \end{pmatrix}. \quad (18)$$

Only vector u is used to estimate the state.

It is obvious that the matrices in equations (10), (16), (18) recorded for the three phases are very large, which results in additional computing problems, compared to the case of calculation for single phase with balanced

 https://doi.org/10.1051/e3sconf/20185803010

loads. The experience of calculating load flow in low voltage four-wire distribution network with unbalanced loads [2] shows that in the case of zero mutual inductances between phase, neutral and earth, calculation of load flow for each phase can be carried out independently. Currents and voltages in a neutral wire and earth can be defined at the second stage of the calculation.

3 Results of the experiments

To illustrate the effectiveness of the use of these approaches for state estimation of the three-phase test network the topology data, Fig. 1, hour node powers made up from the loads and renewable generations – (photovoltaics) are taken from the work [14]. Only the results of nonlinear state estimation are illustrated on iterations of which a linear task solves using WLS method without constraints, with constraints and Hatchel method.

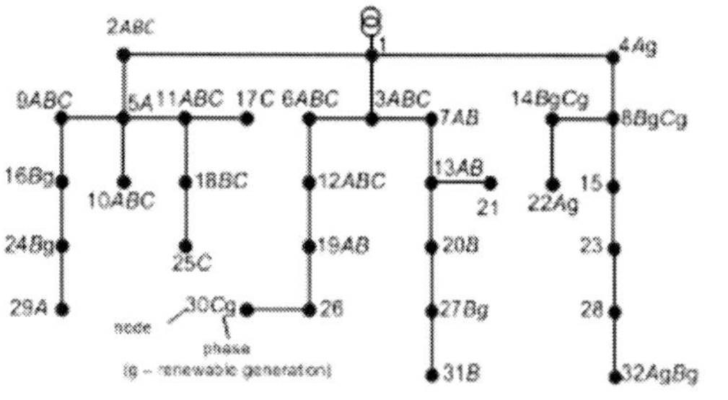

Fig. 1. Scheme of the secondary distribution network

The network includes 32 nodes; the node number 1 is a power supply node. In the phase A there are 15 transit nodes, 12 nodes with the load and 4 nodes with generation. Number of transit, load and generator nodes for phase B is equal to 13, 13 and 5, and for phase C – 17, 10 and 4.

Information on the loads and the generations was the basis for calculating load flow for each phase of the test network; for impedances of the phase and neural wires, the values, typical for low-voltage distribution networks, were taken. Results of calculation load flow called test results were used for the preparation for measurements of node power and voltage magnitudes including measurement errors. To model errors in measurements, in the test values there were entered errors, standard deviations of which were taken equal to 66 W and 0.4 V.

The minimum number of measurements providing topological observability of each test network phase is 64 – 32 measurements for the active model and 31 measurement for the reactive model, taking into account zero value of imaginary voltage component fixed in the power supply node of each phase. To ensure observability, SM should be installed in non-transit nodes.

In this case, measurements and pseudo-measurements of 31 active components of nodal currents and real voltage component in one of the nodes of each phase can be considered as base.

The number of redundant measurements in the active model will include real voltage components equal to measurements of voltage magnitudes, minus the voltage measurements included in base composition of measurements. In phase A there will be 16 redundant measurements, in phase B – 18, and in phase C – 14. The same measurement redundancy can be also achieved in the reactive model, if the number of redundant measurements at the nodes with the measured voltage magnitudes includes zero measurements of imaginary voltage components, but such approach requires additional research.

Fig. 2 shows daily graphs of absolute values of overall residuals of real voltage components and active node powers in A, B, C phases obtained at the iterations of simple iteration method by WLS methods without constraints – 1, with constraints – 2 and by Hachtel method – 3.

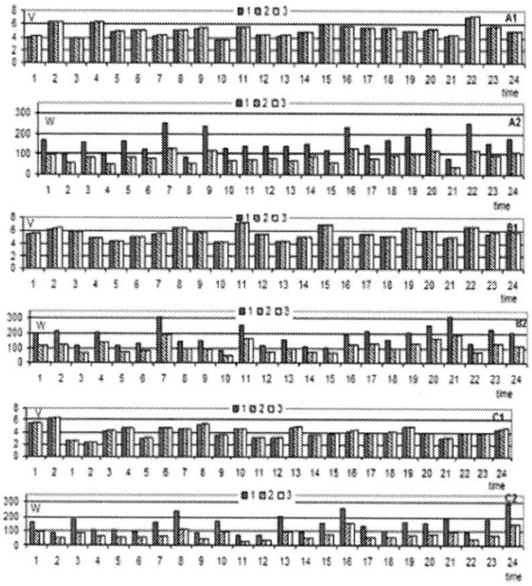

Fig. 2. Total absolute values of the residues in the phases A, B, C for the real components of the voltages (A1, B1, C1) and active node powers (A2, B2, C2) obtained by three methods

The analysis of the graphs shows that the residues of voltages are very close for the three methods, which does not make it possible to identify the best method. As regards the residues for active powers, in all phases, for all hours maximum overall are obtained for WLS method without constraints and the same minimum residues – for WLS method with constraints and Hatchel method.

Large residues in the first method appear despite the fact that the weight coefficients associated with zero power injections of transit nodes were included in the measurement equations with larger weights than weights to measured powers.

Let us show that with the gradual increase in weight coefficients for powers of transit nodes from 1 up to 40 times the overall residues for powers will approach the residues obtained by the least squares method with constraints, as shown in Fig. 3 for the 15th hour of daily node powers of phase A.

The advantage of using a simple iteration method in comparison with the widely used Newton's method is confirmed by Table 1 which presents the results of calculations for each phase for the 15th hour of daily node powers.

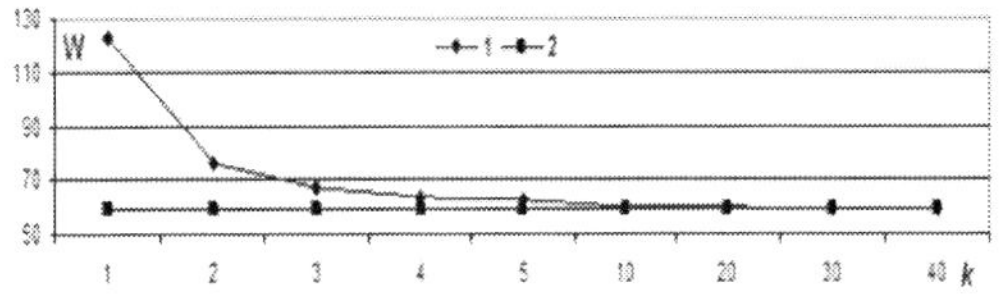

Fig. 3. Change of the total absolute values of nodal powers residues obtained by the WLS method, with an increase in the weight coefficients by k times for the powers of transit nodes – 1, the residues for the WLS method with constrains – 2

Table 1. Comparison methods of solving linear equations systems at the iterations of simple iteration method and Newton's method by the number of iterations (NI), the value of the criterion (VC) of the iteration completion and calculation time (CT)

Method	WLS method without constraints			WLS method with constraints			Hachtel method		
Phase	A	B	C	A	B	C	3A	3B	3C
Simple iteration method									
NI	5	5	5	5	4	5	5	4	5
VC	$3{,}88\cdot10^{-13}$	$6{,}21\cdot10^{-13}$	$1{,}52\cdot10^{-12}$	$3{,}93\cdot10^{-13}$	$6{,}94\cdot10^{-11}$	$2{,}99\cdot10^{-13}$	$3{,}94\cdot10^{-13}$	$6{,}94\cdot10^{-11}$	$2{,}98\cdot10^{-13}$
CT, s	0,042	0,041	0,034	0,039	0,027	0,03	0,04	0,028	0,034
Newton's method									
NI	7	7	8	7	7	8	7	7	8
VC	$4{,}48\cdot10^{-11}$	$4{,}85\cdot10^{-11}$	$7{,}25\cdot10^{-12}$	$4{,}48\cdot10^{-11}$	$4{,}83\cdot10^{-11}$	$7{,}26\cdot10^{-12}$	$4{,}48\cdot10^{-11}$	$4{,}83\cdot10^{-11}$	$7{,}26\cdot10^{-12}$
CT, s	0,12	0,059	0,061	0,161	0,052	0,06	0,064	0,061	0,057

4 Conclusion

1. The main result, the efficiency of which was confirmed by calculations, is the use of a simple iteration method to state estimation of the distribution network by SM measuring of nodal powers and voltage magnitudes.
2. The use of the simple iteration method makes the nonlinear state estimation procedure as trivial as the linear state estimation procedure
3. Experimental proof was presented to describe the advantage of using simple iteration method for solving nonlinear equations system as compared to Newton's method by the number of iterations and calculation time, as well as the advantage of using methods of solving measurement linear equations system with constraints for zero currents in transit nodes as compared to method of weighted least squares.

The work is done in the framework of the project III.17.4.2. program of fundamental research SB RAS, registration number AAAA-A17-117030310438-1

References

1. R.M. Ciric, A.P. Feltrin, L.F. Ochoa IEEE Trans. Power Syst. **18**, 1283 (2003)
2. O.N. Voitov, I.I. Golub, L.V. Semenova, Electricity **9**, 38 (2010).
3. M. Emmanuel, R. Rayudu, J NETW COMPUT APPL **74**, 133 (2016).
4. H. Pezeshki, P. J. Wolfs, ISGT Europe **3**, 1 (2012)
5. S. Santos, A. Llano, A. Arzuaga, T. Arzuaga, L. Marron, M. Zamalloa, CIGRE **44**, 8 (2012)
6. A. Alimardani, F. Therrien, D. Atanackovic, J. Jatskevich, E. Vaahedi, IEEE Trans. Smart Grid **6**, . 2919 (2015)
7. A. Primadianto, C. Lu, IEEE Trans. Power Syst. **32**, 3875 (2017)
8. F.C. Schweppe, J. Wildes, IEEE Trans. Power Appar. Syst. **89**, 120 (1970)
9. A.Z. Gamm, *Statistical methods of state estimation of electric power systems* (Nauka, Moskow, 1976)
10. A.Z. Gamm, I.I. Golub, D.Y. Keselman, Electricity **1**, 12 (1975).
11. A.Z. Gamm, I.I. Golub, *Observability of electric power systems* (Nauka, Moskow, 1990)
12. F. C. Aschmoneit, N.M. Peterson, E.C. Adrian PICA **10**, 427 (1977)
13. A. Gjelsvik, S. Aam, L. Holten IEEE Trans. Power Appar. Syst. **104**, 2987 (1985)
14. P.C. Olivala, A.G. Madureira, M. Matos Electric Power Systems Research **146**, 132 (2017)

Using LabView environment for the demonstration of generator dynamic behavior

Imanakunova Zh.S.[1]*, *Satarkulov K.*[1], *Baibagysova D.Zh.*[1], *Kadieva A.K.*[1]

[1] KSTU, Bishkek, Kyrgyzstan

Abstract. Under consideration is a virtual device (VD) in NI – LabVIEW environment for the demonstration of a generator dynamic behavior close to borders of its stability area. The generator is operating for the unlimited power system. Capabilities of VD are demonstrated on a specific example. It was established that dynamic modes of various complexity – from stationary mode to attractors of different types that correspond to different modes of the generator under study can be observed in the system under consideration. Identification of chaotic modes of generator operation is a topical task as deterministic chaos due to its unpredictability is certainly an inadmissible phenomenon from the point of view of dynamic stability of its operation.

1 Introduction

A system of equations used for the analysis of dynamical stability of generators (synchronous machines) normally relates to determinate systems of nonlinear differential equations (DSNDE).

The idea that solution of the given DSNDE with the known initial conditions is predictable at any time interval turned out to be wrong, as a chaotic solution [1-5] was found for such systems [6, 7].

There is some range of parameters in DSNDE, solutions in such a range although turn to be limited, but are noncyclic and unpredictable, i.e. gain random nature and therefore have non-discrete, but wideband continuous range [6, 8].

For electric energy systems (EES) identification of chaotic modes is a topical task, as deterministic chaos due to its unpredictability – is certainly inadmissible phenomenon from the point of view of dynamical stability in EES [9]. And besides, wideband range of chaotic oscillations will be a reason for appearing:

- large number of frequencies, some of them may match own frequencies of mechanical oscillations, for instance, shaft of turbo- or hydrogenerator, may result in their destruction due to occurrence of electromechanical resonance (EMR) [10];

- higher harmonics of current and voltage [9], that facilitate the impact upon the occurrence of additional loss of electricity in power grid elements [11].

The article shows the results of study with the help of VD-designed dependency of circular frequency deviation $\omega(t)$ from rated value $\omega_н$ for a generator operating for a system with unlimited power (SUP) with the purpose of identifying its chaotic behavior. Selection of $\omega(t)$ as a subject of research is explained by the fact that reliable operation of EES is related to value $\omega(t)$.

2 Task description

Nonlinear equations are solved by numerical techniques with computers, and the same approach is used while analyzing power systems stability. Conclusion of synchronous machines stability is made on the basis of change in angles δ_i their rotor in time [12].

When selecting numerical techniques, their numerical stability is of primary importance. Solution of a system of equations of generator rotor movement, i.e. identification of generator rotor angular displacement in time, is possible by the method of successive intervals. However, this method has a material weaknesses that is in accumulating a considerable error at comparatively large duration of intervals, and therefore its use is usually limited by the study of the first cycle of rotor swinging [13], which is unacceptable when studying a generator operation mode with the purpose of identifying chaotic modes or its operation for long time of simulation.

Simulation of large systems with nonlinear difference equations is possible on computers with parallel architecture with the use of respective numerical techniques based on parallel computations. There are many publications related to it, e.g., [14, 15, 16, 17].

In this paper, with the use of results of work [14] and equation of generator movement working for SUP, and also using mathematical possibilities of NI LabVIEW [18] environment, a task was set – to design a virtual device (VD) and demonstrate its work for numerical study and simulating modes of generator operation close to the borders of its stability area with the purpose of identifying a possibility of chaotic oscillation of its rotor in the given operation mode on specific examples.

* Corresponding author: jeniks.73@mail.ru

3 Numerical solution of generator nonlinear equation

Equation of generator movement working on SUP and its electric power are recorded as follows [12]:

$$\frac{\tau_J}{\omega_{ном}}\frac{d^2\delta}{dt^2} = \frac{2H}{\omega_{ном}}\frac{d\omega}{dt} = P_т - P_{эл}, \quad (1)$$

$$P_{эл} = P_m \sin(\delta), \quad (2)$$

where $\tau_J = 2H$ - in seconds; H - machine inertia constant; ω - angular frequency and its rated value $\omega_н$ – in radian/seconds; $d^2\delta/dt^2$ - angular acceleration in rad/s^2; $P_т$, $P_{эл}$ - power of primary engine and electromagnetic generator capacity in relative units respectively.

Equation of motion (1) in the initial mode recorded for a specific synchronous generator working on SUP via transformer and double-circuit power line, is with $P_т = 0{,}8$, $P_m = 2{,}222$ as follows [12]:

$$\frac{10}{377}\frac{d^2\delta}{dt^2} = (0{,}8 - 2{,}222\sin\delta). \quad (3)$$

Basic value of angle $\delta_0 = 21{,}09° = 0{,}367 rad$. Rotor initial acceleration:

$$\frac{d^2\delta}{dt^2} = 37{,}7[0{,}8 - (2{,}222\sin 0{,}367)] = 0{,}102\ {rad}/{s^2}. \quad (4)$$

Equation (3) was formed from [1] for system (fig.1,a) on the basis of the scheme of its replacement (fig.1,б) and the following baseline data:

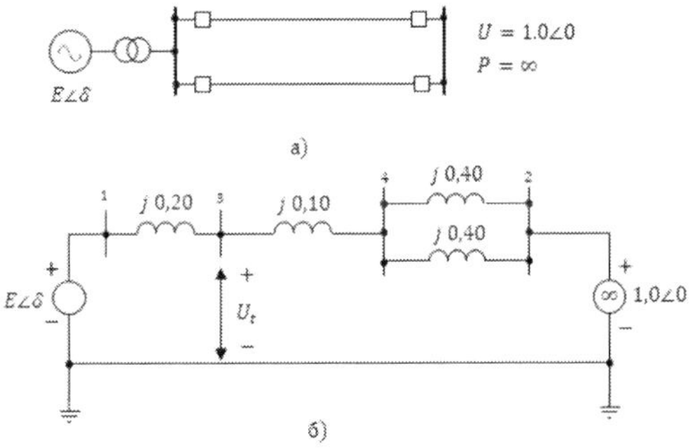

Fig.1. Scheme of the system a) and scheme of its replacement in the basic mode б)

voltage on system buses $U = 1$; generator reactive transient resistance along longitudinal axis – 0.2; transformer reactive impedance– 0.1; reactive impedance of each of two power transmission lines – 0.4, all values are related to baseline value of generator rated apparent power. In the basic mode, the machine gave active power of 0.8, voltage on terminals 1.05. Machine inertia constant $H = 5\,MJ/(MV*A) = 5s$. Active resistance is neglected.

Equations (3, 4) will be used for demonstrating dynamic behavior of generator close to borders of its stability area with the use of LabVIEW environment.

Let's replace state variables δ and $d\delta/dt = \omega(t)$ with x_1 and x_2 respectively, motion equation (1) – with two first-order difference equations:

$$\dot{x}_1 = x_2 - \omega_{НОМ}, \qquad \dot{x}_2 = -a\sin x_1 + b \quad (5)$$

where $a = (P_m\omega_{НОМ})/\tau_J = (P_m\omega_{НОМ})/2H$; $b = (P_т\omega_{НОМ})/\tau_J = (P_т\omega_{НОМ})/2H$; $\omega_{НОМ} = const$.

Let's note that while preparing a block diagram for the task to be solved in LabVIEW environment, it is necessary to remember that $\omega(t) = d\delta/dt + \omega_{НОМ}$.

Use of LabVIEW standard functions for integrating a system of equations determines representation of the right part of difference equation (6) in the form of a set of text constants which makes it difficult to make changes into the numerical values of certain parameters that should be made when studying the impact of, for instance, change in electric power upon generator dynamic stability. Therefore, this paper uses equation solver given in paper [14].

Authors of the paper [14] called the numerically stable and quick operating solver for parallel method a second-order method with diagonal correction (hereinafter - DC). Algorithm of method in solution of system (5) is as follows:

$$\left.\begin{aligned} x_{11} &= x_1[n] + x_2[n]h; \\ x_{21} &= x_2[n] + (b - \sin x_{11})h; \\ x_{22} &= x_2[n] + (b - \sin x_1[n])h; \\ x_{12} &= x_1[n] + x_{22}h; \\ x_1[n+1] &= (x_{11} + x_{12})/2; \\ x_2[n+1] &= (x_{21} + x_{22})/2, \end{aligned}\right\} \quad (6)$$

where h - integration step value.

Realize algorithm (6) as a virtual device of LabVIEW simulation environment. Front panel–interface of the virtual device–is shown on Fig.2:

1– placed are elements for entering simulation parameters (duration of simulation and integration step in seconds, baseline value of rotor angle in radians, mechanical and electric power in relative units, angular frequency in Hz); 2 – (XY Graph) graphical indicator for displaying of a phase pattern (along axis X–change of rotor angle in degrees; along axis Y–rotor angular speed in degr./s); 3 – virtual oscillograph for displaying dependency of rotor angle on time; 4 – graphical indicator for displaying a range of input signal; 5 – these elements display values of frequency and amplitude of the found harmonic motion; 6 – button, if pressed displayed are options of displayed output signals, such as baseline signal; baseline signal harmonic component; baseline signal residual component.

 https://doi.org/10.1051/e3sconf/20185803011

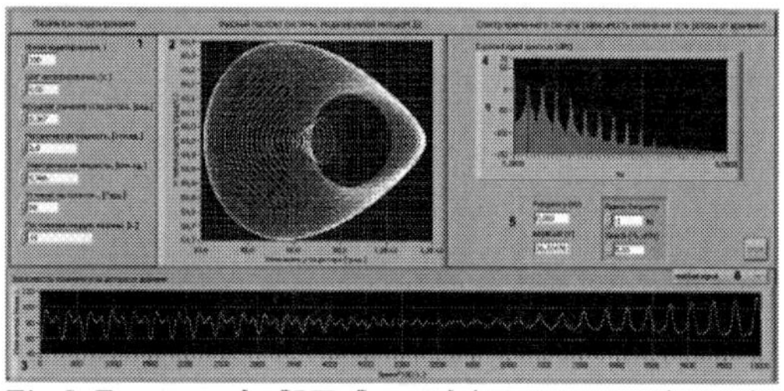

Fig.2. Front panel of VD for studying generator dynamic stability and its chaotic behavior

From among many functions of LabVIEW that were used in the designing VD, let's note only the function of measuring oscillograms' parameters "Extract Single Tone Information" ("Extract information about a single harmonic motion").

Thanks to Extract Single Tone Information function, a researcher can distinguish from rotor basic motion its components with VD, i.e. a single harmonic motion with the highest amplitude and residual signal, or search for certain range of frequencies and returns frequency, amplitude and phase of such oscillation.

Let's consider the change of rotor angle in time $\delta(t)$ and phase pattern of a discrete system, (6) on the example of equation of machine motion (2), simulated with VD numerical method, with the following baseline parameters of simulation: baseline value of rotor angle $\delta_0 = 21{,}09° = 0{,}367$ rad and its initial acceleration $0{,}102\ rad/s^2$ (4); mechanical capacity $P_т = 0{,}8$; electric power $P_m = 2{,}222$; rated angular frequency $\omega_{ном} = 377\ rad/s$; machine constant inertia $H = 5$ s; computations were made with DC method with time integration step of 0.0001.

The presented virtual device allows a researcher to simulate the work of generator and analyze the impact of all parameters of simulation upon its operating mode.

Let's set the task of identifying the degree of impact of P_m values parameters in generator motion equation (2) upon its operating mode. To solve the task, use the designed VD for performing a simulation exercise, varying the value of parameter P_m with unchanged values of others.

4 Results of simulation exercise

Results of simulation exercise are given in tab. (1, 2, 3, 4, 5). First line of each table1–5 includes a graph of rotor angle change $\delta(t)$ in time, second and third lines – its components, i.e. single harmonic oscillations with the highest amplitude and residual components. They were obtained, as was stated above, thanks to using Extract Single Tone Information function ("Extract information about a single harmonic motion") in the virtual device.

Right part of the table contains phase patterns and dependency ranges $\delta(t)$. Phase patterns represent phase trajectories. Coordinates of each trajectory point show values of angle δ or rotor in degrees (axis of X) and its angular speed in degr/s (Y axis) at the given moment of time t.

Table 1 Graphic expression of generator operating mode with $P_т = 0{,}8; P_m = 2{,}222$

	Law of rotor angle change in time (1) and its components (2,3)	Phase patterns and range
1 $\delta(t) = \delta_1(t) + \delta_2(t)$	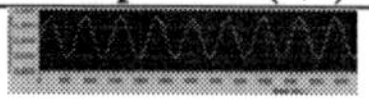	
2 $\delta_1(t)$		
3 $\delta_2(t)$		

Table 1 presents the results of simulation exercise under rated mode of generator operation.

Graphic expression of generator operating mode with $P_т = 0{,}8$; $P_m = 1{,}000$ is shown in tabl.2. In this mode, phase pattern is distorted, dependency $\delta(t)$ is other than sine curve, accordingly, additional harmonics appear and oscillation range is discrete.

Table 2 Graphic representation of generator operating mode with $P_т = 0{,}8$; $P_m = 1{,}000$

	Law of rotor angle change in time (1) and its components (2, 3)	Phase pattern and range
1 $\delta(t) = \delta_1(t) + \delta_2(t)$		
2 $\delta_1(t)$		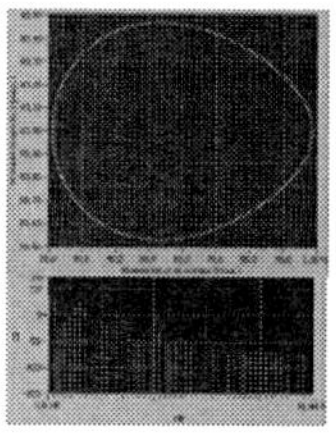
3 $\delta_2(t)$		

Tables 3, 4 show results of simulation exercise while simulating dynamic behavior of a generator close to borders of its stability area. The range of oscillations is non-discrete, but is becoming a continuous and wideband one.

In this mode, chaotic oscillations in the phase area of the trajectory develop continuously in time without

Table 3 Graphic representation of generator operating mode with $P_т = 0{,}8$; $P_m = 0{,}96337$

	Law of rotor angle change in time (1) and its components (2, 3)	Phase pattern and range
1 $\delta(t) = \delta_1(t) + \delta_2(t)$	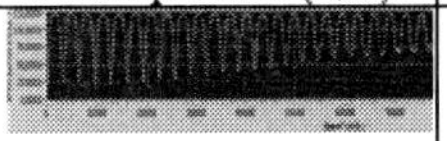	
2 $\delta_1(t)$	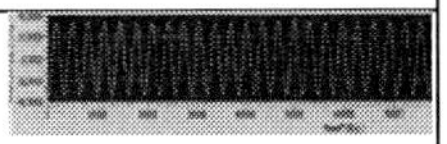	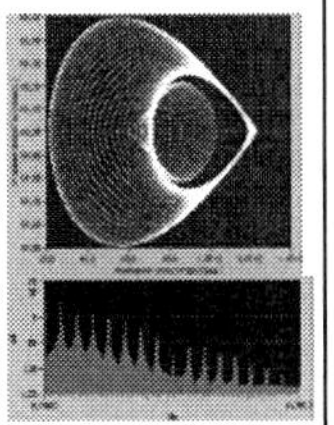
3 $\delta_2(t)$	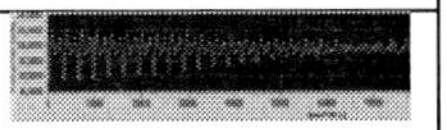	

self-crossing and gradually fill some of its area so that any indefinitely small area of phase pattern is crossed by infinitude of trajectory segments. This is what creates a random situation – chaos – in each area [6].

Range of frequencies changes from discrete into continuous, which proves the presence of large number of frequencies some of which may match with own frequency of generator shaft mechanical oscillation. Such matching in its turn may cause resonance oscillation that is extremely undesirable for the generator.

Table 4 Graphic representation of generator operating mode with $P_т = 0{,}8$; $P_m = 0{,}9633518548$

	Law of rotor angle change in time (1) and its components (2, 3)	Phase pattern and range
1 $\delta(t) = \delta_1(t) + \delta_2(t)$		
2 $\delta_1(t)$		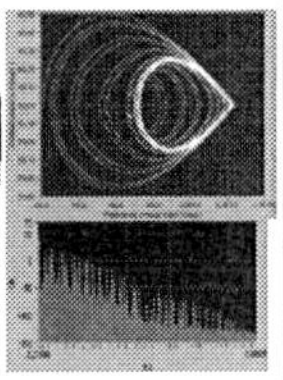
3 $\delta_2(t)$	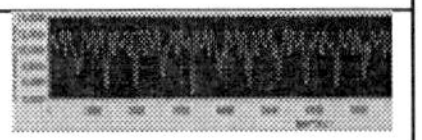	

Table 5 Graphic representation of generator operating mode with $P_т = 0{,}8$; $P_m = 0{,}9633518547$

	Law of rotor angle change in time (1) and its components (2, 3)	Phase pattern and range
1 $\delta(t) = \delta_1(t) + \delta_2(t)$	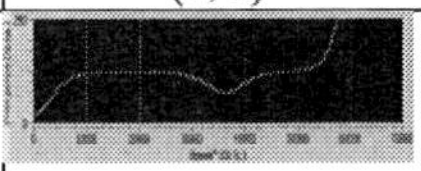	
2 $\delta_2(t)$	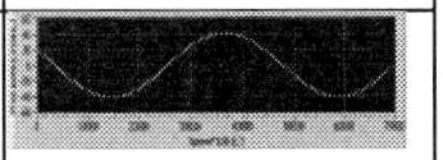	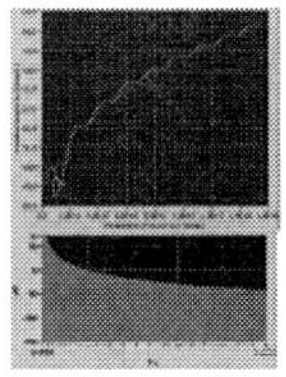
3 $\delta_2(t)$	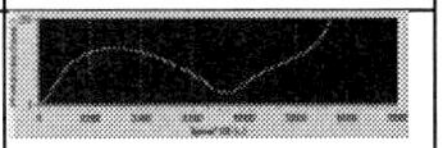	

Table 5 represents a case of occurrence of loss of generator operation stability.

5 Conclusion

The article presents the results of a study providing the picture of alteration of the dynamic behavior of simulation (3) of generator with change in parameter P_m. When the parameter was decreased from baseline value $P_m = 2{,}222$ and when approaching value $P_т = 0{,}8$, complex dynamics phenomena are observed which are related to occurrence of various attractors (tab.1-5). In the phase area (tab.1-2) stable mode is matched by the system state of equilibrium. Asynchronous mode (tab.3) is matched by a multitude of stable multiturn rotary cycles, where phase point jumps from one cycle to another. A chaotic attractor (tab.4) is matched by a mode when $\delta(t)$ is changed stochastically and the range comprises multitude of narrow lines corresponding to oscillations themselves, and a wide base corresponding to "random" oscillations. In this mode, it is possible for the highest harmonics of current and voltage to appear. It was found out that loss of mode stability is observed in the system (tab. 5) when $P_m = 0{,}963351857$.

So, the obtained results are of interest as concerns extending the horizon of knowledge about generator possible behavior. Especially important appears the found existence of areas of parameters values, when the system demonstrates chaotic attractors of various types.

Chaotic mode makes especially difficult the operation of synchronous generators, since chaotic modes have wideband range of frequencies and may induce harmonics of current and voltage that are dangerous for functioning of synchronous generators.

References

1. L.O. Chua, *Computerized analysis of electronic circuits: algorithms and computational method* / L.O.Chua – M.: Energiya, p. 640 (1980).

2. H.D.Chiang, *Chaos in simple power system* / H.D. Chiang. IEEE Trans. Power Syst., **Vol. 8(4)**, pp. 1407-1417 (1993).

3. S.Wiggins, *Introduction to Applied Nonlinear Dynamical Systems and Chaos*. Springer-Verlag, Telos, (1997).

4. H.G., Schuster, W. Just *Deterministic Chaos: An Introduction*, **4th edition**. WILEY-VCH Verlag GmbH & Co. KGaA, Weinheim, (2005).

5. J.C.Sprott, *Elegant Chaos: Algebraically Simple Chaotic Flows*. World Scientific, Singapore (2010).

6. V.K. Fedorov, *Deterministic chaos in nonlinear electric circuits and systems*. - Omsk: OmSTU, p. 130 (2006).

7. S.V.Sidorov, *Structure of solutions and dynamic chaos in nonlinear difference equations* // Vestnik RUDN. Series Math.. Inform.. Phys..– pp. 45-63, **2** (2013).

8. V.K. Fedorov, *Introduction to the theory of chaotic modes of nonlinear electric circuits and systems* / V.K. Fedorov. – Omsk: OmPU, p. 144, (1992).

9. V.K. Fedorov, *Peculiarities of energy dissipation in nonlinear electric circuits* / V.K. Fedorov, P.V. Rysev and E.Yu. Sveshnikova // Omsk. Nauch. Vest. **1(30)**. pp.131-135 (2005).

10. D.V.Rysev, *Stability of power system turbine–generator – load at electromechanical resonance occurrence* / D.V.Rysev and V.K. Fedorov // Omsk. Nauch. Vest. **3(103)**. pp. 194-198 (2011).

11. D.Zh. Baibagysova and K.Satarkulov, *Virtual device for examining spectral composition of voltage in overhead line in chaotic mode of operation of power grid and computation of additional loss* // KR HEIs News. **6**. p 20-29 (2016).

12. P. Anderson and A. Fuad A., *Operation of energy systems and stability* / Translation from Eng. under editorship of Ya.N.Luginskiy. M.: Energiya, p.568, illus. (1980)

13. I.M. Markovich, *Modes of energy systems*. M.; L.: Gosenergoizdat, p. 271 (1957)

14. D.N. Butusov, V.Yu. Ostrovskiy and A.V. Krasilnikov, *Simulation of nonlinear dynamic systems using parallel numerical techniques of integration* // Basic research (Engineering sciences). **№12**, pp.1873-1877 (2014).

15. D.N. Butusov, A.I. Karimov, T.I Karimov and G.K. Dolgushin, *Family of hardware-based methods of numerical integration* // Modern issues of science and education. **4** [Electronic resource] (2014)
URL: http://www.science-education.ru/ru/article/view?id=13480 (referral date: 27.05.2018).

16. K.G. Zhukov. *Algorithm of embedding parallel calculations by formulas of Runge-Kutta numerical integration* // Scientific and engineering bulletin of Saint-Petersburg State Polytechnic University. Informatics. Telecommunications. Managements. SPSPU, V. 6.2, **138**. pp. 143–149 (2011)

17. L.P. Feldman, *Parallel algorithms of simulation of dynamic systems described by simple difference equations*// Electronic simulation. V. 26, **1**. pp. 19–30 (2004).

18. A.O. Sevostyanov and V.V. Karataev. *Using LabVIEW environment in studies of transient phenomena in nonlinear circuits with piecewise linear characteristics* [Electronic resource]. URL: http://do.gendocs.ru/docs/239552.html
(date of referral: 07.08.2018)

Evaluation of the Influence of Non-sinusoidal Conditions on Power Transformers

Lidiia Kovernikova[1]*, *Ngo Van Cuong*[2]

[1]Melentiev Energy Systems Institute SB RAS, Irkutsk, Russia
[2]Irkutsk National Research Technical University, Irkutsk, Russia

Abstract. The electrical equipment operation is cost-effective and reliable when operating parameters of the electrical network correspond to the rated data of the equipment. The real operation conditions, however, differ from those required for electrical equipment, which negatively affects its efficiency. The non-sinusoidal conditions in electrical networks are currently very common. The paper provides an overview of the characteristics obtained from an analysis of publications, which are used to evaluate the effect of the non-sinusoidal conditions on power transformers. The results of the calculation of these characteristics for a transformer installed at a railway traction substation are presented. Parameters of the non-sinusoidal conditions are obtained as a result of measurements.

1 Introduction

The operating parameters of electrical networks should ensure their economically viable and reliable operation, which is largely determined by power quality. Power quality, in turn, affects the reliability of electrical equipment, its functioning, and service life. Electrical equipment works most efficiently, cost effectively, and has a long service life, when the operating parameters of the network correspond to the rated data of the equipment. In real conditions, the operating parameters of electrical networks often differ from those required for electrical equipment, in particular, they do not meet the requirements of the State Standard 32144-2013 [1].

At present, a major problem in electrical networks is the non-sinusoidal voltage, which is caused by loads with nonlinear voltage-current characteristic. The authors of [2] note that "the large values of indices K_U and $K_{U(n)}$, that characterize the voltage non-sinusoidality, are observed in the electrical networks that power AC railway, aluminum plants and large-scale metallurgical facilities. The problems related to providing the required levels of K_U and $K_{U(n)}$ ... are virtually never stated and solved". Non-sinusoidal conditions cause a large number of negative consequences, which have been reported in numerous publications for more than half a century [3-8]. Harmonic voltage and current appearing in non-sinusoidal conditions cause overheating of transformers, damage to capacitors, increased losses in induction motors, erroneous operation of protection and automatic systems, etc. According to [8], the costs associated with harmonics make up 5.4% of all costs related to low quality of electric energy.

Power transformers are an important component of electric power systems. In Russia, their total capacity due to several transformation stages is 6-6.5 times higher than the installed capacity of generators. The authors of [9] note that "in the event of an accidental disconnection of a power transformer, the losses are estimated at millions of Euros, excluding the cost of repairing the equipment or replacing it". In the presence of harmonic voltages and currents, the transformers are damaged due to an increase in the heat release in windings and other components. This accelerates the aging process of insulating paper, transformer oil, and magnetic system. Recently, there has been a growing interest in evaluating the influence of harmonics on power transformers, as evidenced by the papers of the international conferences [10-16] and the new edition of IEEE Standard C57.110-2008 [17].

This paper provides an overview of the characteristics used in publications to evaluate the influence of harmonics on transformers. The results of their calculations are presented. The calculations involved measured indices of power quality and parameters of non-sinusoidal conditions for the transformer TDTNZh-40000/110, installed at the Slyudyanka railway substation. The rated data of the transformer are: S_R=40000.0 kVA, U_R=115.0 kV, I_R=200.0 A, P_{NLR}=39.0 kW, P_{LLR}=200.0 kW, R_T=1.65 Ohm.

* Corresponding author: kovernikova@isem.irk.ru

https://doi.org/10.1051/e3sconf/20185803012

2 Analysis of non-sinusoidal conditions at the node connecting railway substation to supply network

Figure 1 shows a scheme for measuring the power quality indices and the operating parameters. Measurements were carried out for 24 hours with a 1-minute time interval between the measurements.

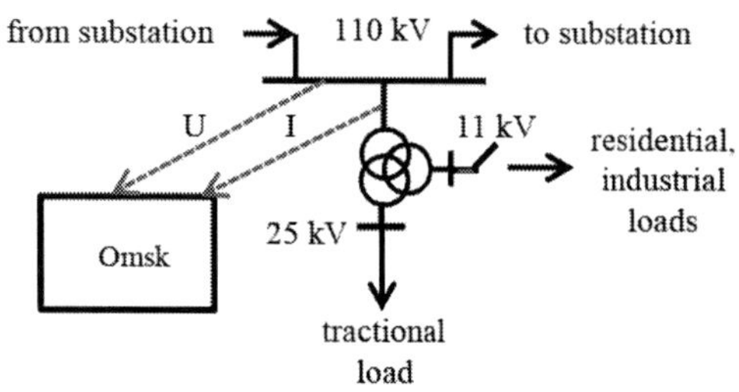

Fig. 1. Scheme for measuring the operating parameters.

Figure 2 presents the oscillograms of phase voltages and currents for one of the measurements, which show that the voltage and current waveforms are non-sinusoidal.

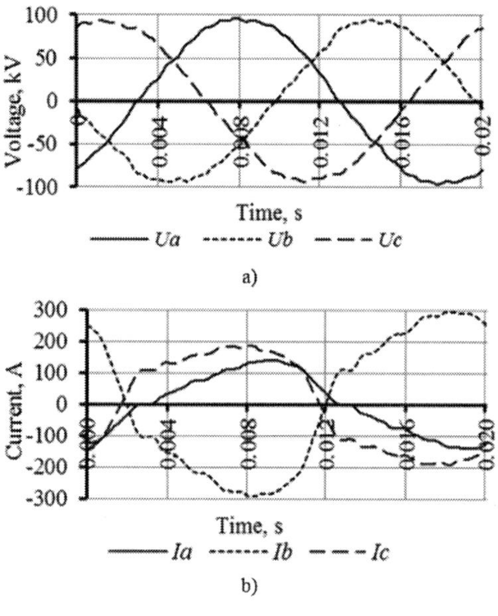

Fig. 2. Oscillograms of phase voltages a) and currents b).

Table 1 demonstrates for some harmonics the values of measured $K_{U(n)}$ and standard $K_{U(n)95\%}$ for the coefficients of the n-th harmonic voltage component [1] and the values of measured $K_{I(n)}$ for the coefficients of the n-th harmonic current component. The values exceeding the standard are shown in bold. As evidenced by the Table, the conditions are unbalanced. The values of harmonic voltages and currents are of random nature.

Table 2 gives statistical estimates for some harmonics of phase B. The estimates of the harmonic voltages and currents of three phases with a probability of 0.95 will be used to calculate the parameters characterizing their influence on transformer.

Table 1. $K_{U(n)}$ and $K_{I(n)}$, %.

n	3	5	7	9	11	23	25
$K_{U(n)A}$	0.9	**2.0**	1.0	**1.1**	**1.4**	**1.6**	**0.8**
$K_{U(n)B}$	**1.9**	**2.3**	**1.2**	**1.2**	**1.5**	**1.3**	**0.9**
$K_{U(n)C}$	1.2	**2.2**	0.9	**1.0**	**1.6**	**1.3**	**0.9**
$K_{U(n)95\%}$	1.5	1.5	1.0	0.4	1.0	0.4	0.4
$K_{I(n)A}$	12.0	9.9	5.1	2.7	2.9	1.9	1.6
$K_{I(n)B}$	1.6	7.4	3.4	2.7	2.5	1.8	0.9
$K_{I(n)C}$	36.4	9.3	7.1	5.0	2.7	3.0	1.7

Table 2. Statistical estimates of U_n and I_n in phase B.

Parameter	Value	n			
		3	5	7	11
U_n, V	min	15.4	145.0	4.8	4.1
	max	1635.7	1879.4	1238.5	1606.8
	evar	645.4	978.6	366.0	495.2
	0.95	1248.7	1538.4	777.5	1028.5
I_n, A	min	0.06	0.23	0.06	0.05
	max	33.2	13.9	10.0	5.5
	evar	14.6	7.0	2.5	1.5
	0.95	25.3	10.0	5.2	3.1

3 Active power losses in transformer under non-sinusoidal conditions

Harmonics of voltages and currents cause additional losses of active power in transformer. Figure 3 shows a classification diagram of transformer losses [14, 17].

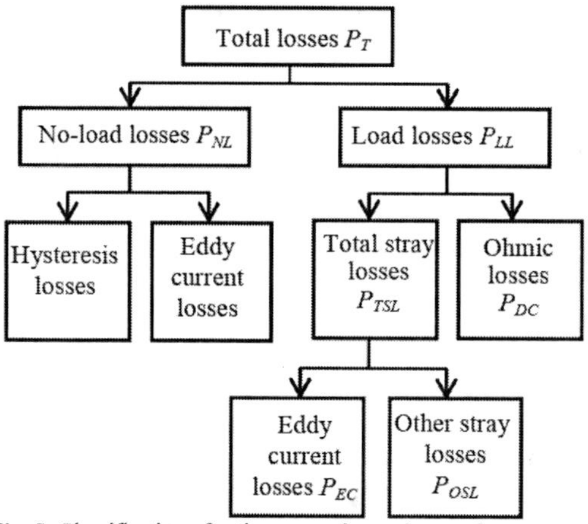

Fig. 3. Classification of active power losses in transformer.

3.1 No-load losses under non-sinusoidal conditions

In [14], the authors propose calculating the transformer no-load losses under non-sinusoidal conditions P_{NL-NS} using the expression

$$P_{NL-NS} = P_{NLR} \sum_{n=1}^{n=n_{max}} \left[1/n^{2.6}(U_n/U_1)^2\right], \quad (1)$$

where U_1 – effective value of voltage at fundamental frequency (the first harmonic), V; U_n – effective value of the n-th harmonic of voltage, V. Expression (1) can be transformed, considering $K_{U(n)} = 100U_n/U_1$. By substituting $K_{U(n)}$ in (1) we obtain

$$P_{NL-NS} = P_{NLR} \sum_{n=1}^{n_{max}} \left[1/n^{2.6}(K_{U(n)}/100)^2\right]. \quad (2)$$

Expression (2) was used to calculate the no-load losses for harmonics from 2 to 39 to 39 for the values of the index $K_{U(n)}$ with a probability of 0.95. The calculation results are presented in Table 3, where P_{TNL-NS} – total no-load losses. Losses P_{NL-NS} compared to the losses under sinusoidal conditions P_{NLR} vary insignificantly. The total excess in three phases is 2.4 W.

Table 3. No-load losses, W.

Parameter	Phase A	Phase B	Phase C
P_{NLR}	39000.0	39000.0	39000.0
P_{NL-NS}	39000.5	39001.2	39000.7
P_{TNL-NS}	117002.4		

3.2 Load losses under non-sinusoidal conditions

According to diagram in Fig. 3, load losses in transformer consist of two components [3, 14, 15, 17]: ohmic windings losses and total stray losses.

3.2.1. Ohmic losses

Ohmic losses in windings are determined by expression

$$P_{DC-NS} = P_{DC-S} \sum_{n=1}^{n_{max}} (I_n/I_1)^2, \quad (3)$$

where I_1 – effective value of current at fundamental frequency, A; I_n – effective value of the n-th harmonic current, A; P_{DC-S} – ohmic losses under sinusoidal current that are calculated as $P_{DC-S} = R_T I_R^2$, R_T – the transformer winding resistance calculated by rated data, Ω; I_R – effective value of the transformer winding rated current, A.

Expression (3) can be transformed, considering $K_{I(n)} = 100I_n/I_1$. After substitution in (3), we obtain

$$P_{DC-NS} = P_{R-S} \sum_{n=1}^{n_{max}} (K_{I(n)}/100)^2. \quad (4)$$

An increase in the active power losses under non-sinusoidal conditions is determined by expression

$$\Delta P_{DC-NS} = P_{DC-NS} - P_{DC-S}. \quad (5)$$

Table 4 presents ohmic losses in high and medium voltage windings that are calculated by (3)-(5). As seen from the Table, the non-sinusoidal current increases total ohmic losses by 14.5 kW.

Table 4. Ohmic losses, kW.

Parameter	Phase A	Phase B	Phase C
P_{DC-S}	66.0	66.0	66.0
P_{DC-NS}	68.1	68.1	76.3
ΔP_{DC-NS}	2.1	2.1	10.3
ΔP_{TDC-NS}	14.5		

3.2.2. Total stray losses

Total stray losses under non-sinusoidal conditions consist of two components: eddy current losses in windings and other stray losses in other components of the transformer (core, core lamps, tank wall and magnetic shields)

Eddy current losses in windings are calculated as follows

$$P_{EC-NS} = P_{EC-S} \left[\sum_{n=1}^{n_{max}} n^2 (I_n/I_1)^2\right], \quad (6)$$

where P_{EC-S} – eddy current losses in windings under sinusoidal conditions that are calculated as

$$P_{EC-S} = 0.33(P_{LLR} - P_{DC-S}). \quad (7)$$

By substituting $K_{I(n)} = 100I_n/I_1$ in (6) we obtain

$$P_{EC-NS} = P_{EC-S} \left[\sum_{n=1}^{n_{max}} n^2 (K_{I(n)}/100)^2\right]. \quad (8)$$

Other stray losses in the other transformer components under non-sinusoidal currents are calculated as follows

$$P_{OSL-NS} = P_{OSL-S} \left[\sum_{n=1}^{n_{max}} n^{0.8} (K_{I(n)}/100)^2\right], \quad (9)$$

where P_{OSL-S} – eddy current losses in other components of transformer under sinusoidal conditions that are calculated by

$$P_{OSL-S} = 0.67(P_{LLR} - P_{DC-S}). \quad (10)$$

Table 5 demonstrates eddy current losses in transformer windings and its other components that are calculated by (6-10). An increase in the eddy current losses under non-sinusoidal conditions compared to the sinusoidal ones is 399.0 kW.

Table 5. Total stray losses, kW.

Parameter	Phase A	Phase B	Phase C
P_{EC-S}	44.2	44.2	44.2
P_{OSL-S}	89.8	89.8	89.8
P_{EC-NS}	125.8	106.7	236.6
P_{OSL-NS}	100.7	100.5	130.7
ΔP_{EC-NS}	81.6	62.5	192.4
ΔP_{OSL-NS}	10.9	10.7	40.9
ΔP_{TSL-NS}	399.0		

3.2.3. Load losses

Load losses under non-sinusoidal conditions given (4), (8), (9) are calculated by expression

$$P_{LL} = P_{DC-NS} + P_{EC-NS} + P_{OSL-NS}. \quad (10)$$

The calculated load losses are presented in Table 6. An excess load loss under non-sinusoidal conditions compared to sinusoidal ones is 413.5 kW.

Table 6. Load losses, kW.

Parameter	Phase A	Phase B	Phase C
$P_{DC\text{-}NS}$	68.1	68.1	76.3
$P_{EC\text{-}NS}$	125.8	106.7	236.6
$P_{OSL\text{-}NS}$	100.7	100.5	130.7
$P_{LL\text{-}NS}$	294.6	275.3	443.6
$P_{TLL\text{-}NS}$	1013.5		
$\Delta P_{LL\text{-}NS}$	95.1	76.81	244.3
$\Delta P_{TLL\text{-}NS}$	413.5		

3.3 Reduction in the transformer service life

The authors of [8] propose determining the real service life of transformer (T_{RSL}) under non-sinusoidal conditions by expression

$$T_{RSL} = T_{SSL} / F_{AA}, \quad (12)$$

where T_{SSL} – a standard transformer service life, years; F_{AA} – transformer aging coefficient, p.u. It is calculated by

$$F_{AA} = e^{\left(\frac{15000}{383} - \frac{15000}{\Theta_H + 273}\right)}, \quad (13)$$

where Θ_H – hot spot temperature of the windings, °C.

Temperature Θ_H is calculated by

$$\Theta_H = \Theta_{TO} + \Theta_g + \Theta_A, \quad (14)$$

where Θ_{TO} – oil temperature rise with respect to ambient temperature, °C; Θ_g – conductor hot spot temperature rise relative to oil temperature, °C; Θ_A – ambient temperature, °C.

Temperature Θ_{TO} is calculated by

$$\Theta_{TO} = \Theta_{TO-R}\left[(P_{LL-NS} + P_{NLR})/(P_{LLR} + P_{NLR})\right]^{0.8}, \quad (15)$$

where Θ_{TO-R} – rated oil temperature rise with respect to the ambient temperature, °C.

Temperature Θ_g is calculated by

$$\Theta_g = \Theta_{g-R}\left[\begin{array}{l}(P_{DC-S} + P_{EC-S}K_{EC-NS})* \\ * \sum_{n=1}^{n_{max}}\left(\frac{K_{I(n)}}{100}\right)^2 /(P_{DC-S} + P_{EC-S})\end{array}\right]^{0.8}, \quad (16)$$

where $\Theta_{g-R} = \Theta_W - \Theta_{TO-R}$ – rated conductor temperature rise relative to oil temperature, °C; Θ_W – rated winding temperature rise over the ambient temperature, °C; $K_{EC-NS} = \sum_{n=1}^{n_{max}} K_{I(n)}^2 n^2 / \sum_{n=1}^{n_{max}} K_{I(n)}^2$ – coefficient of eddy current losses in transformer windings, p.u. [17].

Tables 7 and 8 show the initial data and results of the calculation of reduction in the transformer service life by (12) - (16). As evidenced by Table 8, the non-sinusoidal conditions in phase C can lead to a reduction in the standard transformer service life by 7.5 years.

Table 7. Initial information.

Parameter	Phase A	Phase B	Phase C
$P_{DC\text{-}S}$, kW	66.0	66.0	66.0
$P_{EC\text{-}S}$, kW	44.2	44.2	44.2
P_{NLR}, kW	39.0	39.0	39.0
$P_{LL\text{-}NS}$, kW	294.6	275.3	443.6
P_{LLR}, kW	200.0	200.0	200.0
Θ_{TO-R}, °C	25		
Θ_W, °C	40		
Θ_{g-R}, °C	15		
Θ_A, °C	35		
T_{SSL}, years	25		

Table 8. Calculation of the real service life of the transformer.

Parameter	Phase A	Phase B	Phase C
$K_{EC\text{-}NS}$	2.76	2.33	4.63
Θ_{TO}, °C	32.6	31.4	43.9
Θ_g, °C	23.6	21.8	34.7
Θ_H, °C	91.2	87.9	113.5
F_{AA}, p.u.	0.13	0.09	1.43
T_{RSL}, years			17.5

4 Conclusion

1.The calculation results indicate that the voltage and current harmonics lead to additional active power losses in transformer, which shorten the transformer service life.

2. At present, the systems for continuous monitoring of power quality indices are being developed. They can be supplemented with programs designed to calculate the characteristics that signal the effect of voltage harmonics and currents on electrical equipment.

This work is conducted in the framework of the research projects № AAAA-A17-117030310438-1 of the program of fundamental research of SB RAS III.17.4.

References

1. GOST 32144–2013. Electric energy. Electromagnetic compatibility of technical equipment. Power quality limits in public power supply systems, (2014) (in Russian)
2. L.I. Kovernikova, V.V. Sudnova, R.G. Shamonov et al., *Power quality: current state, problems and proposals for solving them* (Nauka, 2017) (in Russian)
3. E.A. Mankin, Electrichestvo **12**, 48-52 (1955) (in Russian)
4. Investigation of the influence of unbalanced and non-sinusoidal voltage on the operation of induction motors. Information materials **70** (Gosenergoizdat, 1963) (in Russian)
5. N.A. Melnikov, Electrichestvo **5**, 1-6 (1968) (in Russian)
6. I.V. Zhezhelenko, *High harmonics in the power supply systems of industrial plant* (Energoatomizdat, 1984) (in Russian)

7. V.N. Gromov, *Power quality control, Proceedings of the International Scientific and Practical Conference*, 85-92 (Moscow, 2014) (in Russian)
8. R. Targosz, D. Chapman, (2015), http://www.leonardo-energy.org/resources/297/the-cost-of-poor-power-quality-5800e490f1e14
9. Ernst Gockenbach, Xiang Zhang, Zhaolin Liu, Haibo Chen, Linghu Yan, Electric Power. Transmission and Distribution **4 (25)**, 82-88 (2014) (in Russian)
10. Saeed Mousavi, Bahman Tavan, Homayoun Bakhtiari, *The review and analysis the destructive effects of harmonics on a sample MV/LV transformer, Proceedings of 20-th International Conference on Electricity Distribution* (Prague, 2009)
11. Abla Gado, Hassan Abo Gad, Salah Radwan, *Effect of types of loads in rating of transformers supplying harmonic-rich loads, Proceedings of 21-st International Conference on Electricity Distribution*, (Frankfurt, 2011)
12. O.E. Gouda, G.M. Amer, W.A.A. Salem; Engineering, Technology & Applied Science Research Vol. **1,** No. **5,** 114-120 (2011)
13. Ahmed Abbas, Essam Abou El Zahab, Ahmed Elbendary, *Thermal modeling and ageing of transformer under harmonic currents, Proceedings of 23-rd International Conference on Electricity Distribution*, (Lyon, 2015)
14. Thinh Dao, H. Abdull Halim, Z. Liu, B.T. Phung, *Voltage Harmonic Effect on Losses in Distribution Transformers, Proceedings of International Conference on Smart Green Technology in Electrical and Information Systems*, (Bali, 2016)
15. Jaspreet Singh, Sanjeev Singh, Amanpreet Singh, *Effect of Harmonics on Distribution Transformer Losses and Capacity*, International Journal of Engineering Technology Science and Research **Vol. 4, Issue 6** (2017)
16. Emil Cazacu, Maria-Cătălina Petrescu, Valentin Ionită, Lucian Petrescu, *Nonsinusoidal load current effect on the electrical and thermal operating parameters of oil filled power distribution transformers, Proceedings of 18-th International Conference on Harmonics and Quality of Power*, (Ljubljana, 2018)
17. IEEE Std C57.110-2008, IEEE Recommended Practice for Establishing Liquid-Filled and Dry-Type Power and Distribution Transformer Capability When Supplying Nonsinusoidal Load Currents. IEEE Power Engineering Society, (2008)

Effect of asymmetry of supply voltages on asynchronous motor operation modes

Victoria Romanova[1]*, *Sergey* Khromov[1]

[1]Trans-Baikal State University, Department of Electric Power Generation and Electrical Engineering, Russia, Chita, 30, Aleksandro-Zavodskaya Str.

Abstract. Effect of asymmetry of supply voltages on asynchronous motor operation modes was investigated using the computer-aided simulation experiments, the software complex Matlab and the package program Simulink. Publication relevance is determined by the necessity for performance of comparative analysis of the level of effect of asymmetry of supply voltages on asynchronous motors of different series and power capacities. Six motors of three different power capacities (5.5, 55, 160 kW) and of two different series (4A, AI) were modeled for investigation. The value of asymmetry of supply voltages was changed from 0 to 5%. The motor shaft load was changed from 0 to 120% of the motor rated load. Comparison of dependencies of the phase currents on the voltage unbalance ratio by reverse sequence (K_{2U}) of asynchronous motors with short-circuited rotor series 4A and AI was made. The analysis and discussion of obtained investigation results were performed, recommendations on operation of asynchronous motors series 4A and AI in conditions of asymmetry of supply voltages were given.

1 Introduction

At present the three-phase AC asynchronous motors (AM) is the most common type of the electrical machines. As a rule, AM are produced by large series, of which the most widely spread are the general purpose machines of series 4A, AI. Series 4A, AI are the mass series of asynchronous motors designed for using in different industry areas. In accordance with [1,2], the asynchronous motors are designed for a life time 15 – 20 years without overhaul provided that they are operated properly. Proper operation implies their running under rated parameters specified in AM passport certificate. However, under actual operating conditions of motors, depending on different factors, there is significant deviation from the rated modes of operation.

As the investigations [3-6] and many years' experience of operation demonstrate, one of the dominant causes for breakdown of motors is the asymmetry of supply voltage. AM operation under conditions of asymmetry of supply voltages results in change of the maximum rotation moment, sliding, currents in the motor windings. As a result, due to excess of the phase current rated values [7], heating of AM active parts occurs leading to winding insulation overheating and reduction of asynchronous motor lifetime. Thus, with asymmetry of voltages making 2%, the asynchronous motor lifetime, due to additional real power loss, reduces by 10 % [8].

Currently, in Russia a number of regulatory documents exists, which establish the set of the electric energy indicators and standards of quality [9], as well as requirements for reliability electrical machines [10,11,12].

The regulatory base for formation of recommendations aimed at ensuring AM operational reliability, includes:

- GOST 32144-2013 "Electrical power. Electromagnetic compatibility of technical equipment. Electrical power quality standards in the general purpose power supply systems",
- GOST IEC 60034-26 – 2015. Part 26. Effect of unbalanced voltages on on operating characteristics of three-phase asynchronous motors,
- GOST P 52776-2007. Rotating electric machines. Rated data and features.

In accordance with GOST 32144 – 2013, the following standard values of the voltage unbalance ratio by reverse sequence K_{2U} are established: the values of the voltage unbalance ratio by reverse sequence K_{2U} in the electrical power delivery point, averaged at 10 minutes time intervals shall not exceed 2% during 95% and 4% during 100% of one-week time interval.

In accordance with GOST IEC 60034-26 – 2015. AM with K_{2U} above 1% (during continuous mode of operation) shall be unloaded "... to avoid its breakdown". And AM operation " with 5% of asymmetry excess is not recommended".

The analysis of modern publications [4,5,13] showed that these values do not always correspond to GOSTs and, in practice, can significantly exceed them.

* Corresponding author: romanova181@mail.ru

Within the framework of this publication, it is necessary to perform investigation of synchronous motor with short-circuited rotor, their most common series 4A and AI, to make analysis of the effect of asymmetry of voltages on AM operation modes based on the motor series and power capacity. To get a more complete picture of the effect of asymmetry of voltages on AM, it is necessary to compare dependencies of AM phase currents on the voltage unbalance ratio by reverse sequence (K_{2U}), under different motor shaft load values. It is obvious, that such investigation will make it possible to solve the issue of determining the level of effect of the asymmetry of voltages different series and power capacity AM, and besides, to identify the permissible rate of K_{2U} (under AM standard mode of operation) depending on AM load.

It is clearly understood that experimental investigations play significant part in provision of AM reliability. In their turn, they enable formulating the proposals aimed at improvement of AM operating reliability and, in addition, prevention of occurrence of situations determined by overheating of winding insulation.

2 Investigation materials and methods

During investigation of AM operation modes, the computer-aided simulation is taken as the basis for virtual model implemented in the visual and efficient means for interactive environment simulation modeling Simulink and programming Matlab [14,15], using T-shaped scheme for asynchronous motor replacement.

Earlier, the publication author developed the computer model enabling to make analysis of AM operation modes in conditions of asymmetry of supply voltages [16]. Besides, in the obtained computer model, there is an option for investigation of transition processes in the asynchronous motor, for reading operational and imitation mechanical characteristics, of supply grid voltage, currents and frequency.

The scheme for investigation of operation modes for synchronous motor with short-circuited rotor created by means of simulation modeling in Matlab/Simulink environment is given in (Fig.1).

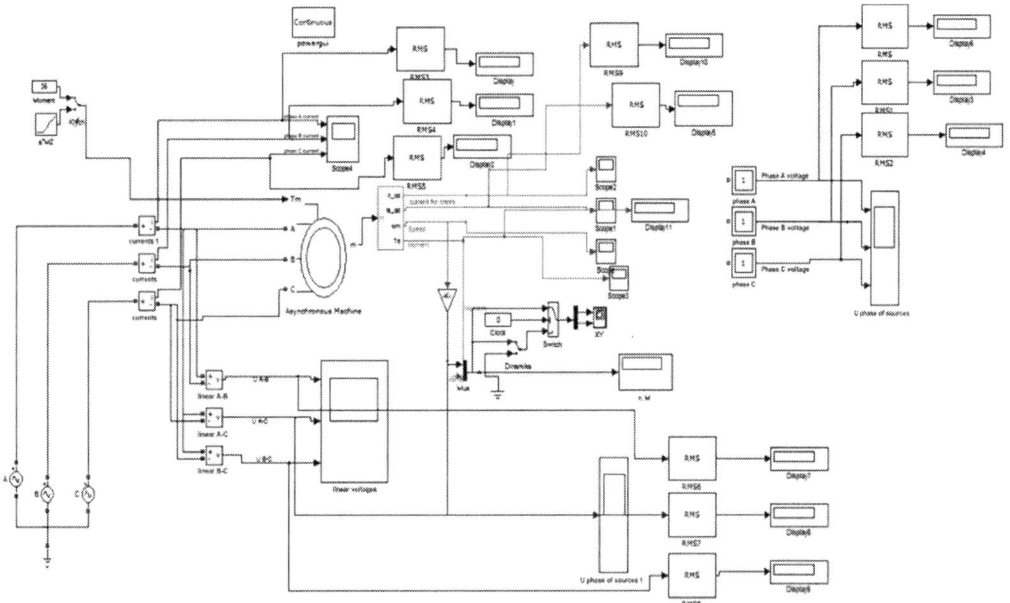

Fig.1. Simulation scheme implemented in Matlab/Simulink environment.

AM operation modes with account of below listed parameters were investigated using the developed model:
– the value of the motor shaft load was changed from 0 to 120 % with the pitch of 20 under different values of K_{2U},
– the value of the voltage unbalance ratio by reverse sequence – from 0 to 5 % with the pitch of 1 %.

The asynchronous motors series 4A, AI with the motor rated rotation frequency n=1500 rpm, with significant
scatter by power capacities from 5.5 kW to 160 kW were subjected to investigation.

The asynchronous motors with short-circuited rotor series 4A (4A112M4Y3 with Pn = 5.5 kW; 4A225M4Y3 with Pn = 55 kW; 4A315S4Y3 with Pn = 160 kW) and series AI (AIR112M4 with Pn = 5,5 kW; AIR225M4 with Pn = 55 kW; AIR315S4 with Pn = 160 kW) with the rated parameters [17,18] given in (Table1) were selected as examples.

Table 1. Cataloged data of investigated three-phase motors.

Motor power capacity, Pn, kW	Motor model	Rated rotation frequency, n, rpm	η	cosφ	Starting torque-to-nominal torque ratio, Kst	Starting current-to-rated current ratio, Ksc
5.5	4A112M4Y3	1500	85.5	0.85	2.2	7
55	4A225M4Y3	1500	92.5	0.9	2.5	7
160	4A315S4Y3	1500	93.5	0.91	2.2	6
5.5	AIR112M4	1500	85.5	0.86	2.5	7
55	AIR225M4	1500	93	0.89	2.6	7
160	AIR315S4	1500	93.5	0.91	2	5.5

Characteristic curves of phase currents from K_{2U} for AM series 4A, AI were obtained during computer-aided simulation. Current curves in the function from K_{2U}, under different values of motor shaft load were plotted for each phase of considered AM.

The main type of current change in the most loaded phase (phase "A") of AM series 4A, AI with Pn = 5.5; 55; 160 kW, under different values of K_{2U} and different values of motor shaft load is given in (Fig.2 - 4).

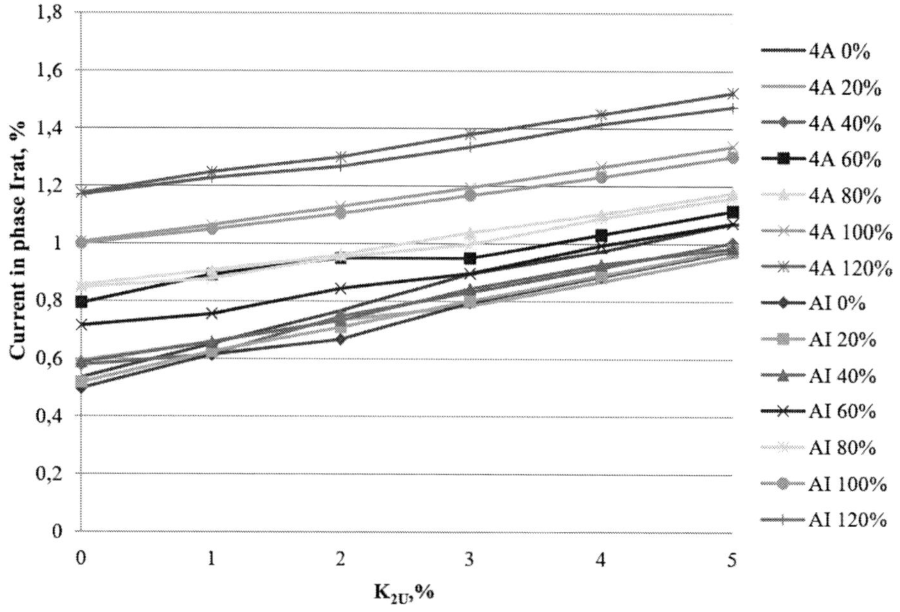

Fig.2. Curve of current in phase "A" AM series 4A, AI with Pn = 5.5 kW from K_{2U} under different permanent load factors.

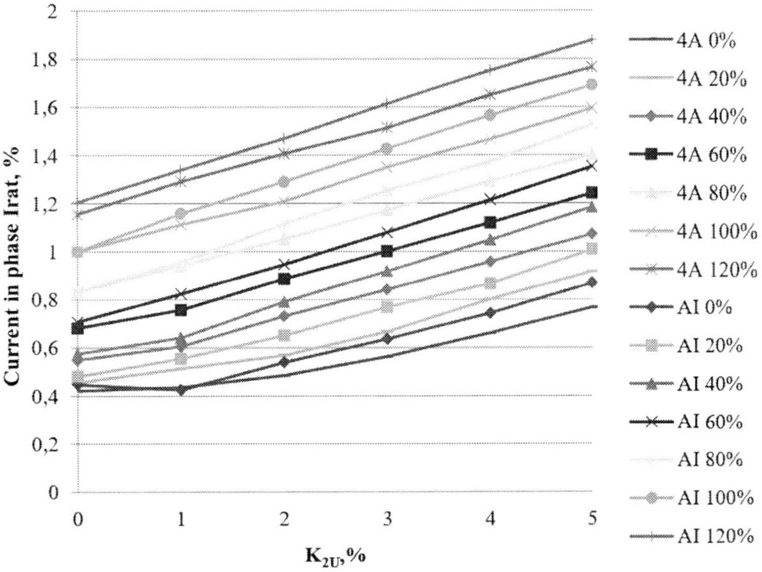

Fig.3. Curve of current in phase "A" AM series 4A, AI with Pn = 55 kW from K_{2U} under different permanent load factors.

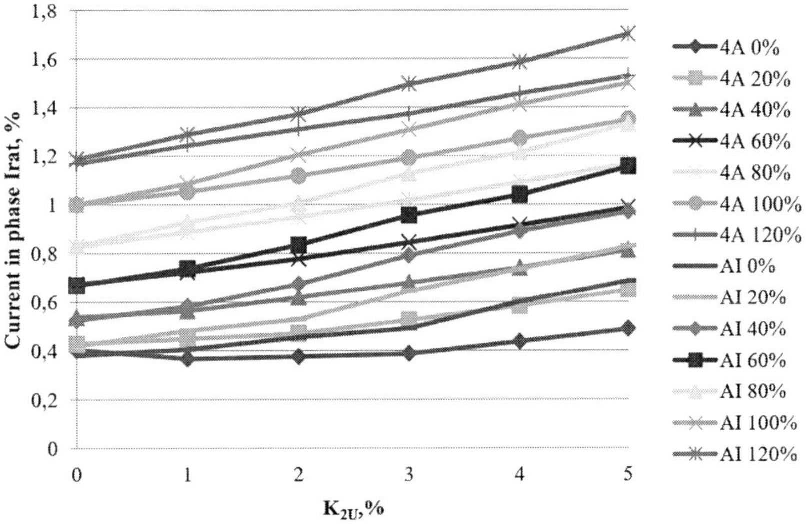

Fig.4. Curve of current in phase "A" AM series 4A, AI with Pn = 160 kW from K_{2U} under different permanent load factors.

3 Investigation results and their discussion

Analisys of computer-aided simulation (Fig.2-4) allowed revealing the following trends:

1) asynchronous motors with rated power capacity 5.5 kW:

- AM series 4A Pn = 5.5 kW, i.e. the specified series motors with small rated power, are subjected to greater effect of asymmetry of supply voltage and may breakdown even under standard GOST 32144–2013 values of K_{2U}. Thus, with the load value equal to 100%, the current in phase "A" exceeds the allowable over current accepted as equal to 1.1 of the rated current (this is the permissible excess value in accordance with relay protection requirement [19,20]), with K_{2U} = 1.3%, and for AM series AI Pn = 5.5 kW it occurs with the value K_{2U} = 2%. The motors series 4A overload under 80% load already working within the permissible range of K_{2U}, series AI under 100% load,

- overcurrent of motors series 4A Pn = 5.5 kW is observed with the value K_{2U} above 4% with K_Z = 60%, and for series AI with K_Z = 80%, under load factor increase up to 100%, the start of winding insulation heating is observed under K_{2U} from 1,3% to 2%,

2) asynchronous motors with rated power capacity 55 kW:

- the start of winding insulation heating is observed in the specified AM under K_{2U} = 0-4%, whereas this shall be outside that range limits. At that, this is observed with K_Z = 60-100%,
- with the load of the motor series AI up to 40%, and the motor series 4A up to 60 %, AM standard operation mode will be observed with the standard value K_{2U} from 0 to 4%,

3) asynchronous motors with the rated power capacity 160 kW:

- for the motor series AI Pn = 160 kW under K_Z up to 60% and the motor series 4A Pn = 160 kW under K_Z up to 80% the standard operation mode with K_{2U} from 0 to 4% will be observed, which corresponds to GOST 32144–2013.
- for both series, with K_Z above 80 % the start of winding insulation heating within the standard ranges of K_{2U} occurs,

4) the currents under motor full load are of ultimate interest. With the load value equal to 100%, the current in phase "A" for AM series 4A Pn = 5.5 kW exceeds the allowable overcurrent accepted as equal to 1.1 of the rated current under K_{2U} = 2% [19,20]. For the motor series 4A Pn = 55 kW, excess of the allowable overcurrent occurs upon reaching K_{2U} = 0.9 %. This attests to the fact that the motors have not only contrasting ranges of permissible values by K_{2U}, but also the values, which do not always correspond to the ones standardized by the GOST. Correspondingly, during operation of motors in locations with poor quality electric power, it is necessary to adjust the supply voltage guided not by the GOST for electric power quality, but by the motor operation mode.

Based on the objective set forth in the research study, the main conclusions can be represented:

1) the following indicators affect AM allowable operation mode under asymmetry of supply mode: motor series, power capacity, load factor (K_Z) and voltage unbalance ratio by reverse sequence (K_{2U}).

Thus, under K_Z within the range from 60% to 80% the start of majority of AM is observed under K_{2U} value below 4%. The range of AM series 4A operation modes is under K_Z from 0 % to 60%, the range of AM series AI is under K_Z from 0 % to 80% with K_{2U} from 0% to 4%.

The medium power capacity AM series 4A and AI are most susceptible to asymmetry of voltages. Under motor full load, they have the allowable K_{2U} value:

- AM series 4A Pn = 5.5 kW allowable value K_{2U} = 1.3%, AM series AI Pn = 5.5 kW K_{2U} = 2%,
- AM series 4A Pn = 55 kW allowable value K_{2U} = 0.9%, AM series AI Pn = 55 kW K_{2U} = 0.7%,
- AM series 4A Pn = 160 kW allowable value K_{2U} = 1.8%, AM series AI Pn = 160 kW K_{2U} = 1.4%.

Consequently, the allowable value K_{2U} differs not by AM series only, but also due to motor series, because of the motor power capacity immediately,

2) the allowable range of K_{2U} (under AM standard operation mode) with permanent asymmetry of voltages, full motor load:

- for the motor series 4A Pn = 5.5 kW, the allowable range K_{2U} from 0 to 1.3%, for the motor series AI Pn = 5.5 kW K_{2U} from 0 to 2%,
- for the motor series 4A Pn = 55 kW the allowable range K_{2U} from 0 to 0.9%, for the motor series AI Pn = 55 kW K_{2U} from 0 to 0.7%,
- for the motor series 4A Pn = 160 kW the allowable range K_{2U} from 0 to 1.8%, for the motor series AI Pn = 160 kW K_{2U} from 0 to 1.4%.

Consequently, under excess of ultimate permissible value K_{2U}, overheating of AM winding insulation will be observed,

3) AM reliability significant reduction occurs under K_{2U} exceeding 4%. To exclude overheating of both AM series 4A, and AI breakdown, it is necessary to reduce the load.

AM full load is possible only under K_{2U} maximum 1 % [11].

Thus, after performing the complex of investigations on effect of asymmetry of voltages on asynchronous motor operation modes, the recommendations on AM series 4A and AI operation under asymmetry of supply voltages were stated:

1) AM series 4A are subjected to the largest effect of asymmetry of voltages, therefore, at the stage of AM power supply system designing and modernization, this condition should be taken into account. Consequently, AM series AI are less exposed to effect of asymmetry of voltages, and have increased operation reliability,

2) to avoid the motor overheating and premature breakdown, it is necessary to reduce its load,

3) rate setting of the voltage unbalance ratio by reverse sequence should be carried out based on the motor permissible operation modes and their models produced in lots,

4) to make responsive control during UPQI (unified power quality index) deviations, for stable operation of AM, it is necessary to control the electric power parameters at the enterprise input or at the respective motors,

5) it is proposed to use the investigation results for correction of current standards in the area of electric power quality standards, as well as in specifications for particular motor series,

6) the investigation results will be necessary for development and designing of PSP (power supply system) containing AM. Particularly, it will be relevant

for regions with large amount of one-phase non-linear load,

7) the obtained values of allowable values K_{2U} for AM play the key role during symmetry facilities selection and implementation in the power supply system with the motor drive load,

8) it seems feasible to use the approach based on the method of computer-aided simulation in Matlab/Simulink environment for assessment of affect of voltage supply asymmetry on AM operation.

4 Conclusion

The effect of asymmetry of voltages on asynchronous motor operation modes was investigated using the computer-aided simulation experiments, the software complex Matlab and the package program Simulink. Following the investigation results, the significant effect of the asymmetry system of supply voltages on different series and power capacity AM operation modes was detected. Based on specific examples, change of characteristic curves, on the quantitative side, was demonstrated. Recommendations on operation of asynchronous motor series 4A and AI in conditions of asymmetry of supply voltages were presented.

During solution of issues for ensuring asynchronous motor operation reliability, the experimental investigations form the basis. Their practical implementation will allow for improving reliability of motors operated in conditions with available permanent asymmetry of voltages.

The work was executed within the framework of FSBEI HPE Trans-Baikal State University Council for Research and Innovative Activities Scientific Grant No. 247-GR of 30/01/2018 implementation.

References

1. GOST P 51137-98. Adjustable asynchronous electric drives for power generation facilities. General Specifications. – M.: IPC (Publishing and Printing Complex) Publishing House of Standards, **12** (1998).
2. M. Katsman, Electrical machines. – M.: Higher School, **463** (2000).
3. A. Gusarov, Electrical and Power Engineering. Scientific publications, Donetsk National Technical University. Issue 8 (140), 95 – 97 (2008).
4. I. Zhezhelenko, Yu. Sayenko, A. Gorpinich, I. Shvetsova, Priazovsk National Technical University Reporter: Collection of scientific articles / PNTU. – Mariupol Issue 15, P.2., 25 – 29 (2005).
5. M. Sorkind, Electric Engineering News. Information and Reference Publishing House No. 2 (32), (2005).
6. M. Fedorov, N. Ivchenko, A. Tkachenko, DNMBA (Donbass National Machine-Building Academy) Scientific Research Reporter. Collection of scientific papers No. 1 (11E) – DNMBA, Kramatorsk, 164–170 (2013).
7. O. Pipchuk, Electrical and Power Engineering. Scientific publications, Donetsk National Technical University. Issue 8 (140), 201 – 205 (2008).
8. I. Zhezhelenko, Electric power quality at industrial enterprises – M.: Energoatomizdat, – **261** (2005).
9. GOST 32144 – 2013. Electrical power. Electromagnetic compatibility of technical equipment. Electrical power quality standards in the general purpose power supply systems. – M.: Standartinform, **19** (2014).
10. GOST IEC 61000-4-27-2016 Electromagnetic compatibility (EMC). Part 4-27. Test and measurement methods. Test forimmunity to asymmetry of voltages for equipment with current consumption maximum 16 A per phase. – M.: Standartinform, **17** (2016).
11. GOST IEC 60034-26 – 2015. Part 26. Effect of unbalanced voltages on operating characteristics of three-phase asynchronous motors (IEC 60034-26:2006, IDT). – M.: Standartinform, **11** (2016).
12. GOST P 52776–2007 (IEC 60034–1–2004). Rotating electric machines. Rated data and features – M.:Standartinform, **68** (2008).
13. I. Suvorov, A. Sidorov, S. Khromov, Electrical safety. Quarterly Journal of Theory, Research and Practice – Chelyabinsk:SUSU (South Ural State University), No. 3, 17 – 30 (2014).
14. S. German-Galkin, G. Kardonov, Electrical machines: Laboratory works with PC. – SBR: Korona-print, **256** (2003).
15. I. Chernykh, Simulation of electrical devices using MATLAB, SimPowerSystems and Simulink. – SBR: Piter M.: DMK Press, **288** (2008).
16. V. Romanova, S. Khromov, Collection of scientific papers. – Voronezh: ITSRON (Russian Association Research Center), 35 – 41 (2015).
17. http:// principact.ru/content/view/166/72/
18. www.elektrikii.ru/publ/6-1-0-55
19. http://www.bmrzzakharov.narod.ru/raschet/overload.html
20. V. Kogorodsky, S. Kuzhekov, L. Paperno, Relay protection of motors with voltage above 1 kV. – M.: Energoatomizdat, **248** (1987).

Determination of optimal location for installation of symmetry facilities in 0.4 kV power supply systems with motor-drive load

Victoria Romanova[1]*

[1]Trans-Baikal State University, Department of Electric Power Generation and Electrical Engineering, Russia, Chita, 30, Aleksandro-Zavodskaya Str.

Abstract. This article considers the issue for creation of computer-aided program designed for determination of optimal location for installation of symmetry facilities in 0.4 kV power supply systems with motor-drive load in conditions of asymmetry of voltages. The developed program is implemented by means of C# programing language in Microsoft Visual Studio 2012 programming environment using the embedded cross-platform data base SQLite. The program is designed for making electrotechnical calculations during designing, modernization and operation of electrical grids with available motor drive load. It ensures plotting of schematic circuits for 0.4 kV power supply system sections with installed symmetry facilities. The program will enable quick and exact calculation of power capacity loss, determination of economic feasibility for taking special measures to eliminate the asymmetry of voltages, and to determine optimal application and location of symmetry facilities. Using the program in motor-drive load power supply systems will enhance the asynchronous motor reliability and efficiency.

1 Introduction

Consumers' operation under asymmetry of voltages is one of challenges faced in Russian electrical grids. The experimental investigations carried out in the current grids of of individual Russia regions [1,2] demonstrate that the levels of asymmetry of voltages exceed the established GOST 32144 – 2013 values [3]. Following performed investigations, there is sufficiently high level of asymmetry of voltages exceeding the value standardized by GOST 32144 – 2013by 2-3 times and is the permanent factor during operation of electrical grids. The voltage, which does not meet GOST 32144 – 2013 requirements, inflicts significant material damage, reduces technical and economical indicators of the electrical grid system operation on the whole. Different assignment types electrical grids widely use asynchronous electrical drive with large range of asynchronous motor (AM) rated power capacities. In AM field operation conditions, depending on various factors, there is significant deviation from the rated operation modes. The investigations [4-7], demonstrated that the asymmetry of voltages is one of causes for AM failures. The asymmetry of voltages results in accelerated ageing of insulation and breakdown of motors due to irreversible ionization processes. Approximately 75% of motors shut down due to insulation fault, and the economic damage can reach some tens of thousands of rubles [8].

Using of technical means and protection methods for motors against the emergency cases caused by reduction of quality and reliability of power supply systems, their efficient allocation in the power supply system is one of the options for solution of their reliable operation. The integrated solution of the specified problem is impossible without development and implementation of software computing facilities for power supply systems based on modern computer machines. Therefore, it is necessary to develop the procedure, the computing program for power supply systems, considering the allowable operation modes of the asynchronous motors in conditions of asymmetry of voltages.

The main objective of this research work is the development of the applied software program to determine the optimal installation location of symmetry facilities in 0.4 kV power supply systems (PSS) with the motor drive load.

The proposed software program is aimed at solution of practical tasks for enhancementof the asynchronous motor functional reliability by reducing the asymmetry of voltages. The above stated determines the problem relevance and implementation of the proposed software program.

* Corresponding author: romanova181@mail.ru

2 Task description

In accordance with the set objective, the task for development of the power supply system calculation procedure algorithm implementation will be based with account of using the the symmetry facilities and and their installation locations, on basis of which the software program for determining the optimal location of the symmetry facilities in 0.4 kV PSS with the motor drive load will be built.

Based on the data of experimental investigations of the main UPQI, the information on the power loss value in the elements of the power supply system under investigation, and besides, on the ground of consumers' allowable operation modes, it is possible to make selection of parameters for the symmetry facilities and determine optimal location of these devices in the power supply systems. Calculation of power capacity losses due to asymmetry of voltages will allow to justify the economical feasibility for the asymmetry elimination, to determine optimal location of the symmetry facilities.

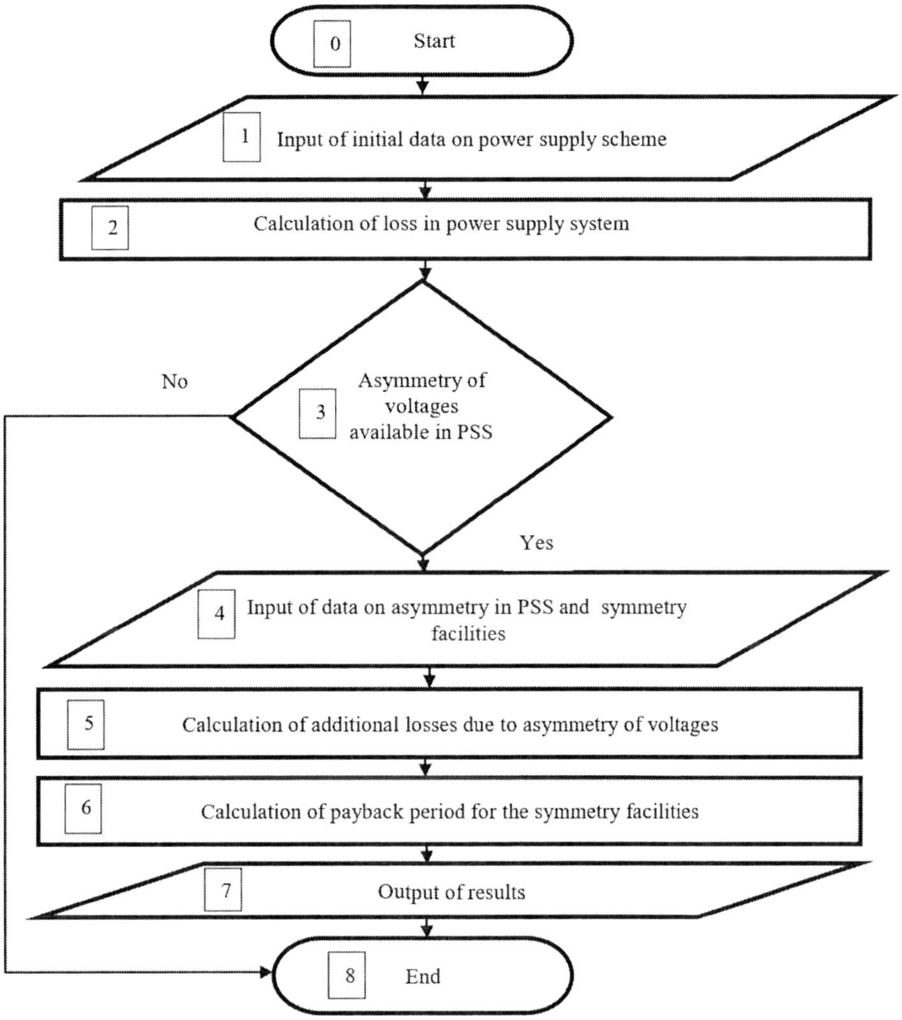

Fig. 1. Flow diagram for power supply system calculation algorithm with account of the symmetry facilities and and their installation locations.

3 Calculation procedure for power with account of the symmetry facilities and and their installation locations

The calculation procedure for power supply system calculation algorithm with account of the symmetry facilities and and their installation locations can be implemented according to the algorithm:

1) block 1 – input of initial data is performed in the software program.

The initial data includes:

- section diagram description (section general information);
- parameters of electricity transmission lines, line length; type of line wires and cables; line load);
- parameters of motors connected to electricity transmission line (line number, where the motor is located; quantity of motors in the line, motor type);
- parameters of transformer (quantity of transformers and types);
- the value of factors asymmetry of voltages by reversible and zero-phase-sequence in the calculated power supply system;
- cost of electric power.

2) block 2 – calculation of losses for symmetrical operation mode in the main elements of the power supply system is performed.

Symmetrical operation mode current (positive-sequence current) is determined based on the expression:

$$I_1 = S/U \qquad (1)$$

where S – substation full load.

Losses in electricity transmission lines are determined based on the expression:

$$\Delta P_{ETL} = 3I_1^2 R_{ph} \qquad (2)$$

where I_1 – symmetrical operation mode current (positive-sequence current);

R_{ph} – phase conductor resistance.

Substation full load (total of loads for individual line) is determined based on the expression:

$$S = {}_{i=1}^{n} P_i \qquad (3)$$

where P_i – line load.

Losses of active $\Delta P'$ and reactive $\Delta Q'$ power capacities for double-winding transformers are determined based on the expression:

$$\Delta P' = \frac{\Delta P_k}{n} \frac{S}{S_{rat}}^2 \qquad (4)$$

$$\Delta Q' = \frac{U_k}{100 \cdot n} \cdot \frac{S^2}{S_{rat}} \qquad (5)$$

where $\Delta P_k, U_k, S_{rat}$ – passport certificate data of individual transformer [9];

S – substation full load;

n – number of one-type transformers in substation operating in parallel.

3) block 3 – let us proceed to condition "the available asymmetry of voltages in the power supply system".

If "NO", calculation termination.

If "YES", proceed to block 4.

4) block 4 – let us proceed to special measures for elimination of asymmetry of voltages, for determination of optimal application and location of these facilities.

supply system calculation algorithm

Input of data of factors for asymmetry of voltages by reversible and zero-phase-sequence according to two calculation options, as well as selection of suitable symmetry facilities from the database is carried out.

a) option 1 – consists in installation of the symmetry facilities in the common node of lines with the motor drive load;

b) option 2 – consists in installation of the symmetry facilities in each line of the scheme section with asynchronous motor.

5) block 5 – calculation of additional losses for asymmetrical operation mode in the main elements of the power supply system is performed.

Available amplitude and angular asymmetry in the grid, when estimating increase of additional losses as compared with symmetrical operation load shall be determined using the factor:

$$K_{asym} = 1 + K_{2I}^2 + K_{0I}^2 \cdot \; 1 + 3 \cdot \frac{R_r}{R_{ph}} \qquad (6)$$

where K_{2I}, K_{0I} – factors of the asymmetry currents by reversible and zero-phase-sequence.

Based on results of processing of data of main UPQI in 0.4 kV grid nodes of the considered object, let us accept the values of factors.

R_r, R_{ph} – resistances of neutral and phase conductors.

Accordingly, with account of (6), the expression (2) will look as follows:

$$\Delta P_{AD.ETL} = \Delta P_{ETL} K_{asym} \qquad (7)$$

Additional losses of active power capacity in the asynchronous motor are determined based on the expression:

$$\Delta P_{AD.AM} = 2{,}41 K_{AM} K_{2U}^2 P_n \qquad (8)$$

where K_{AM} – the factor considering the parameters of the particular motor (rated power capacity, copper losses in stator, starting current-to-rated current ratio);

K_{2U} – factor of asymmetry of voltages by reverse sequence;

P_n – motor rated active power capacity.

The value of factor for industrial load on the whole is recommended to accept as equal to 1.85 [10].

Under long-term asymmetrical operation mode, due to running the reversible sequence currents in the power transformers, additional power capacity losses occur, which can be determined according to the following formula:

$$\Delta P_{AD.TR} = K_{2U}^2 \;\; \Delta P_{i.r.} + \frac{\Delta P_{SHC}}{U_k^2} \qquad (9)$$

where K_{2U} – the factor of asymmetry of voltages by reverse sequence;

$\Delta P_{i.r.}$ – losses in the mode of idle running;

ΔP_{SHC} – losses in the mode of short-circuit;

U_k – short-circuit voltage.

Calculation of overall losses is performed.

The losses for symmetrical operation mode are determined based on the expression:

$$\Delta P_{sym} = {}_{i=1}^{n\_line} \Delta P_{ETL} + \Delta P' \qquad (10)$$

where n_lines – quantity of lines;

i – line number.

The losses for asymmetrical operation mode are determined based on the expression:

$$\Delta P_{asym} = \Delta P_{sym} + \sum_{i=1}^{n\_line} \Delta P_{AD.ETL} + \sum_{j=1}^{n\_dv} \Delta P_{AD.AM} + \Delta P_{AD.TR} \quad (11)$$

where j – motor number;

n_dv – quantity of motors in the electricity transmission line.

6) block 6 – calculation of payback period for each option of installation for symmetry facilities is carried out and the shortest period is selected.

Additional losses without the symmetry facilities are determined based on the expression:

$$S_{AD.W} = \Delta P_{AD.W} + \Delta Q_{AD.W} = \sum_{i=1}^{n\_line} \Delta P_{AD.ETL} + \sum_{i=1}^{n\_line} \sum_{j=1}^{n\_dv} \Delta P_{AD.AM} + \Delta P_{AD.TR} \quad (12)$$

with the initial factors of asymmetry of voltages by reversible and zero-phase-sequence.

Additional losses with the symmetry facility are determined based on the expression:

$$S_{AD} = \Delta P_{AD} + \Delta Q_{AD} = \sum_{i=1}^{n\_line} \Delta P_{AD.ETL} + \sum_{j=1}^{n\_dv} \Delta P_{AD.AM} + \Delta P_{AD.TR} \quad (13)$$

with the initial factors for symmetry facilities of asymmetry of voltages by reversible and zero-phase-sequence.

Selection of parameters for symmetry facilities is determined based on the following conditions:

a) option 1 – power capacity of the symmetry facilities shall be higher or equal to total loads in the lines outbound of the central node:

$$P_{sym} \geq S \quad (14)$$

b) option 2 – power capacity of the symmetry facilities shall be higher or equal to the load in each line with the motor:

$$P_{sym} \geq P_i \quad (15)$$

Calculation of the total cost of the symmetry facilities is determined based on the expression:

$$S = \sum_{k=1}^{k\_d} C\_d \quad (16)$$

where S – the total cost of installed symmetry facilities;

C_d – the cost of each symmetry facility;

k_d – quantity of installed symmetry facilities;

k – symmetry facility number.

Calculation of payback period for the symmetry facilities can be determined according to the following formula:

PP

(17)

where PP – the payback period;

S – total cost of installed symmetry facilities;

$(\Delta P_{AD.W} + \Delta Q_{AD.W})$ – additional losses without symmetry facilities;

$(\Delta P_{AD} + \Delta Q_{AD})$ – additional losses with the symmetry facility;

C – electric power cost.

7) block 7 – output of the total data is made.

The output data includes:

- losses for symmetrical operation mode;
- additional losses under asymmetrical operation mode (without symmetry facilities; with installation of the symmetry facilities in the common node of lines, including the motor drive load – option 1; with installation of the symmetry facilities in all lines with asynchronous motors – option 2);
- cost of the symmetry facilities (cost of symmetry facility; total cost of the symmetry facilities - option 1; total cost of the symmetry facilities - option 2; payback period - option 1; payback period - option 2).

The above stated procedure for power supply system calculation will be the basis for implementation of the task for creation of the software program on optimal location of the symmetry facilities in 0.4 kV PSS with the motor drive load.

For the software program implementation and testing, let us consider the power supply system section of Taptugary village, Mogochinsky region of Trans-Baikal district.

The following initial information was used during this publication performance:

1) results of processing of data of main UPQI in 0.4 kV grid nodes in Mogochinsky region of Trans-Baikal district,

2) configuration of the power supply system of Taptugary village, description of electric power consumers.

Power supply system, where the software program was approbated, is given in (Fig.2).

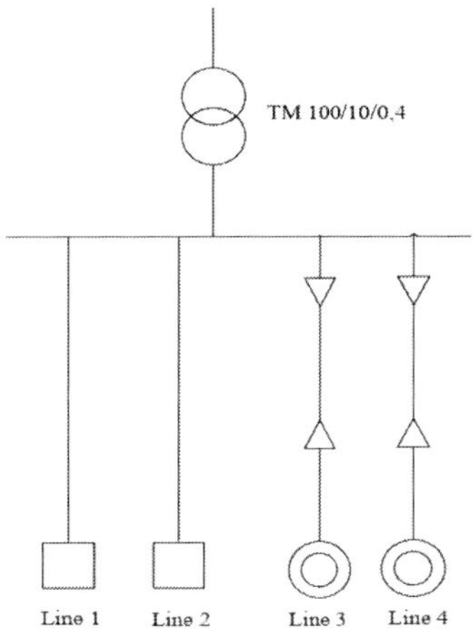

Fig.2. Scheme of Taptugary village power supply system section.

Scheme of power supply system section includes:

1) line 1 – is routed with conductor AS-25, line distance - 350 metres, feeds 7 private houses with the furnace heating,

2) line 2 – is routed with conductor AS-25, line distance - 50 metres, line distance double-storeyed house, the primary school, the kinder garden, the administration building, the library,

3) line 3 – is routed with conductor AVVshv 4*50, line distance - 70 metres, feeds the boiler house of two boilers with the electrical motors: type 4A132S4Y3 with Pn = 7.5 kW with 2 pieces, flue gas fan Pn = 3 kW, fan blower

Pn =1.4 kW,

4) line 4 – is routed with conductor SIP 4*25, line distance - 150 metres, feeds two band saws type AI132M4

Pn = 11 kW.

Table 1. Taptugary village grid section initial data.

Parameter	Line 1	Line 2	Line 3	Line 4	TS
Power capacity, kVA	30.4	49.4	20.9	25	100
Current, A	80	130	55	65.7	330.7

4 Program implementation

The software program is implemented by means of C# programing language in Microsoft Visual Studio 2012 programming environment using the embedded cross-platform data base SQLite [11-14].

The software program main functional capabilities:

1) input, editing of the initial parameters of the power supply system elements;

2) input, editing of the reference data of the power supply system elements;

3) input of additional data for calculation of option with installation the symmetry facilities;

4) plotting of power supply system section with the initial data entered into database;

5) calculation of options with installation of the power supply system symmetry facilities;

6) graphic presentation of schemes for options of calculations with particular location specification of the symmetry facilities in PSS.

The software program consists of the database storing all required information for calculation and designing of 0.4 kV power supply systems with the motor drive load; the computational calculation module with the algorithm for calculation of power capacity losses in the power supply system elements; the module for data input and output of 0.4 kV power supply system design calculation results. The software program carries out the required calculations automatically based on the set initial data. The initial data includes the data on 0.4 kV power supply system section with the motor drive load. The data is developed into data base (DB). Its structure allows for handling the data; it is flexible and eliminates superabundance and non-conforming dependences.

The data base includes the tables of the following nature: reference parameters of the power supply system elements, Table the symmetry facilities, tables of the power supply system input data, tables of tables of the power supply system output calculations. Each BD table has its unique attributes and its own designation in the system.

The most important specificity of the software program consists in plotting of the power supply system calculation model. At that, the accuracy of calculations will be mainly determined by the fact, how fully the calculation model plotted by means of special instrument set, the parameters of the designed or operated – real power supply system are considered. The software program interface is well understood and plain (Fig.3–4). The software program interface presents and displays the power supply system initial information, the output of the summary information on calculation results.

The proposed software essence consists in the fact, that, based on the obtained investigation results of of K_{2U} permissible values for AM [15] entered in the software program algorithm, the data of K_{2U} power supply system calculation experimental investigation, selection of parameters for the symmetry facilities and the optimal location of these devices in the power supply system are determined.

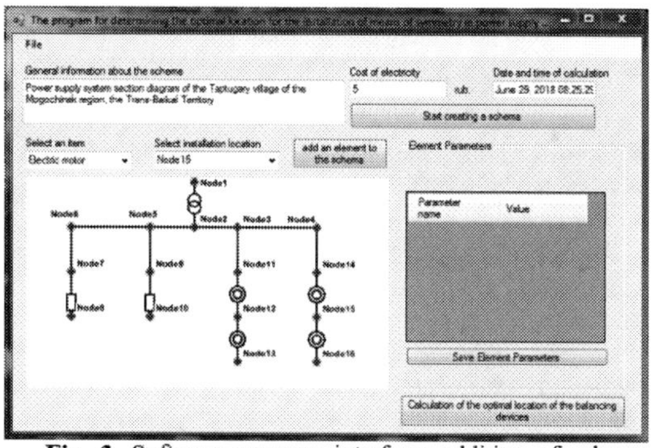

Fig. 3. Software program interface: addition of scheme elements by means of designing.

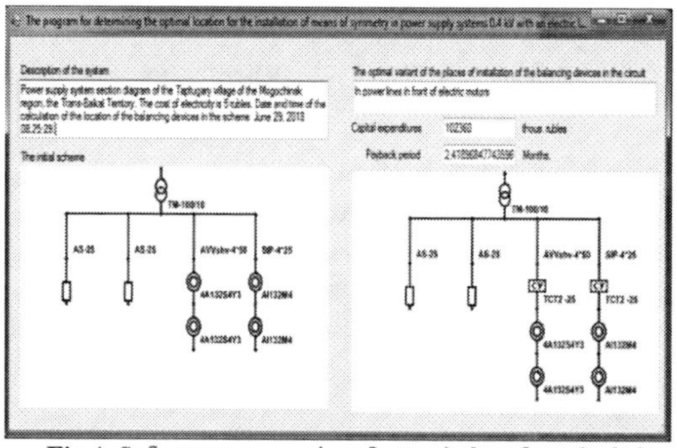

Fig.4. Software program interface: window for calculation results output.

When changing the calculation scheme parameters (for example, replacing the symmetry facilities and their quantity), it is possible to match them in such way that the sustainable and fail-free operation of asynchronous motors is ensured in the whole power supply system. The use of the software program will allow of significant increase of AM and the whole power supply system operation reliability and efficiency due to using the symmetry facilities (balance-unbalance transformers, frequency converters and filter-balancing devices) [16,17].

Based on the results of the software program operation according to the considered scheme (Fig.2), it is much in evidence that it is economically feasible to install the balance-unbalance transformers type TCT2 with power capacity 25 kW on line 3,4 having the motor drive load.

The necessity for using the symmetry devices for the industrial consumers in the investigated scheme was detected as a result of the software program implementation. Based on the capital expenditures and the payback period of actions aimed at asymmetry reduction, the most economically advantageous location of the symmetry facilities PSS was determined.

The proposed software program allows for the power supply system calculation procedure simplification without significant increase of labor costs due to using computer equipment. The software program operability was tested on different 0.4 kV PSS sections.

5 Conclusion

The program is designed for making electrotechnical calculations during designing, modernization and operation of power supply systems with available motor drive load. It ensures plotting of schematic circuits for 0.4 kV power supply system sections with installed symmetry facilities.

The software program provides for calculation of different symmetry options and for selection of the most economically feasible option. Such user options are provided for during the software program operation: replenishment, changing and removal of the reference parameters for the elements included in the power supply system; input of initial data in the power supply system section, and besides, input of additional data for procedure of the symmetry optimal option calculation and for payback period for each of them.

Practical significance of the developed software program consists in the fact that implementation of new research solutions in designing and operational practice ensures suppression of asymmetry of voltages in 0.4 kV PSS, and, at that, it improves the operation reliability of asynchronous motors. The developed software program is intended for the regions with available non-linear loads, where the actions directed for support of electric energy quality in the power grids are of the most priority.

The work was executed within the framework of FSBEI HPE Trans-Baikal State University Council for Research and Innovative Activities Scientific Grant No. 247-GR of 30/01/2018 implementation.

References

1. V. Romanova, S. Khromov, Ulan-Ude: BRC of RAS SB (Buryat Research Center of Siberian Branch of the Russian Academy of Sciences), 20 – 25 (2016).
2. L.I. Kovernikova, R.G. Shamonov, V.V. Sudnova. Electrical energy quality: modern state, problems, proposals on their solution. Monograph. Novosibirsk: Nauka Publishing House, **219** (2017).
3. GOST 32144 – 2013. Electrical power. Electromagnetic compatibility of technical equipment. Electrical power quality standards in the general-purpose power supply systems. – M.: Standartinform, **19** (2014).
4. A. Gusarov, Electrical and Power Engineering. Scientific publications, Donetsk National Technical University. Issue 8 (140), 95 – 97 (2008).
5. I. Zhezhelenko, Yu. Sayenko, A. Gorpinich, I. Shvetsova, Priazovsk National Technical University Reporter: Collection of scientific articles / PNTU. – Mariupol Issue 15, P. 2., 25 – 29 (2005).
6. M. Sorkind, Electric Engineering News. Information and Reference Publishing House No. 2 (32), (2005).
7. M. Fedorov, N. Ivchenko, A. Tkachenko, DNMBA (Donbass National Machine-Building Academy) Scientific Research Reporter. Collection of scientific papers No. 1 (11E) – DNMBA, Kramatorsk, 164–170 (2013).
8. http://tekhnosfera.com/sistema-povysheniyanadezhnosti-elektrodvigateley-v-selskom-hozyaystvena-osnove-kompleksnoy-diagnostiki-i-effektivnoyte#ixzz456v53hmH
9. O. Grunin, L. Savitsky, Electrical power systems and networks in examples and problems. Chita: Chita State University Publishing House, **209** (2010).
10. A. Ded, A. Parshukova, N. Khalitov, International Journal of Applied and Fundamental Research. Scientific Journal. Moscow: Academia yestestvoznania Publishing House, No. 10, P.3., 421 – 425 (2015).
11. http://www.w3ii.com/ru/sqlite/default.html.
12. http://er.nau.edu.ua/bitstream/NAU/10100/1/SQLite.Allow.pdf.
13. R. Frost, D. Day, K. Van Slike, Databases. Designing and development. Moscow: NT Press Publishing House, **570** (2007).
14. V. Medvedev, Specifics of object oriented programming on C++/CLI, C# and Java. Kazan: RIC Shkola (Russia Innovation Center), **444** (2010).
15. V. Romanova, S. Khromov, I. Suvorov, Modern technologies. System analysis. Modeling. Scientific Journal. Irkutsk: Irkutsk State University of Railways Publishing House, No. 1(53), 209 – 214 (2017).
16. https://www.liderteh.ru/katalog/transformatory/simmetriruyushie__transformatory
17. http://chistotnik.ru/chastotnye-preobrazovateli-dlya-asinhronnyh-dvigatelej.html

Analysis and modeling of harmonic distortions in consumer's distribution networks, containing a powerful non-linear load

Rakhmanov N.R.[1], Kurbatsky V.G.[2], Guliyev H.B.[1]*, Tomin N.V.[2], Rakhmanov R.N.[1]

[1] Azerbaijan Scientific–Research and Designed–Prospecting Institute of Power Engineering, Baku, Azerbaijan
[2] Melentyev Energy Systems Institute Siberian Branch of the Russian Academy of Sciences, Irkutsk, Russian

Abstract. The paper proposes to simulate a random variability of the nonlinear load in the form of probabilistic characteristics of the amplitudes of harmonic components of powers. The influence of the variability of distribution of amplitude values of the power harmonics on the load flow in the network, power losses and maximum power transmission in lines is studied. The results of design experiments for one of distribution networks of Azerbaijan power system, feeding the non-linear load, are given.

1 Introduction

In modern power grids, due to the wide use of pulse technique in the power consumer control, along with the improvement of operation efficiency, the proportion of distortion from harmonic currents in the network from these control devices increases. The increase of harmonics leads to the voltage curve distortion and additional-losses. The deterioration of the power quality has a negative impact on the quality of product [1-3].

At present methods of estimation of harmonic distortions in distribution systems from different types of non-linear loads are developed, in which it is supposed that the amplitude of power spectrum components of harmonic currents remains unchanged during the process of operation of consumers of non-linear loads (NL). The methodology of iterative decomposition on frequency components of NL currents is proposed in [1,2]. Distribution network up to the general connection point is calculated in the frequency range, while the power supply network of consumers is represented as a whole in the time interval. At that, the power (currents) of the power supply network of consumers is modeled using statistical methods. The analysis of probability of power changes of sources of harmonic distortions and assessment of interrelation between distortions in the power supply network and distribution networks is performed in [4-6]. In addition, in the same paper, the degree of influence of the voltage in the network on the harmonic distortions from each individual NL is estimated.

In this paper the analysis of an harmonicity of current and voltage in the distribution network, generated by random power changes of consumed NL, is performed. Arc steel furnaces (ASF) and power converters AC/DC installed for the connection of distributed generation system with an external system are considered as the nonlinear load. In addition, the problem of adaptive NL modeling using machine learning (ML) algorithms for the impact assessment of parallel connection of harmonic filters is considered.

2 Adaptive NL modeling for the harmonic filter connection impact assessment

Since the use of harmonic filter reduces the harmonic components of currents (HCC), such a device also changes the harmonic components of voltages (HCV) at the filter connection point [7]. Therefore, the filter connection will change the HCC NL. Usually, the presence of the filter reduces the voltage distortion, to which the NL responds with the increase of HCC. Such effect of positive feedback, called "harmonic amplification" [8], leads to the fact that the filter size calculated at the design stage becomes insufficient for the full-fledged reduction of HCC generated by NL. In turn, this can lead to damages of equipment of the non-linear consumer, for example, failure of ASF.

Theoretically, active harmonic filters (AHF) automatically adapt to load conditions. Based on the estimation of "harmonic amplification" during the AHF selection, the "harmonic amplification factor" parameter is used, which is often defined as the ratio of the nominal values of the total nonlinear distortions coefficient (NDC), K_U NL and actually measured. Depending on the load type, the requirements for the AFG installation, this factor can be accepted from 1.1 to 2 [9]. However, in practice, the choice of AFG parameters and the need for its connection is a non-trivial task, since often the neighboring loads and properties of NL are unknown.

To solve this problem, usually one resorts to the development of different models of the NL behavior, located below the connection point. Such models are rele-

* Corresponding author: huseyngulu@mail.ru

vant before the design and installation of the filter. Significant success in this direction was achieved by using ML algorithms that allow for creating adaptive training models capable of approximation of various nonlinear functions [7, 10].

This paper proposes a method for NL modeling using the ML model, to input of which the HCV values, measured at the filter connection point and the load factor of the system are entered. The output values of the model are the current of the main harmonic, as well as the subsequent most significant high-frequency HCC. In this case, the problem is represented as a multiple nonlinear regression recovery between the HCV and HCC. To solve this problem, two ML algorithms were studied in this paper in detail: extreme learning machine (ELM) [11] and random forest model for multiple regression (RF) [12].

3 Experimental measurements of the mode parameters and indices of power quality in the ASF connection mode

Currently, there is a sufficient range of equipment for the purpose of measuring and recording of the PQ parameters. Among these one can mention the devices of Siemens, LEM, S&I (UK), MARS and PARMA (Russia), instruments of Italian producers, etc. All these measuring devices are capable of monitoring and recording parameters of PQ in real time and comply with GOST-32144-2013 [13]. From this point of view, the Simeas Q measuring and recording device was chosen as satisfying the measurement objectives and taking into account that there is a significant amount of instrumentation and security equipment of Siemens company in the power grid.

Functional capabilities of the Simeas Q PQ meter are as follows: recording of electrical parameters for the analysis of electricity quality; ability to perform monitoring in single-phase networks, in 3- or 4-wire three-phase circuits; recording of exceeding limit values; transmission of recorded values through various communication channels (RS-232, RS-485, PROFIBUSDP).

4 Experimental analysis of harmonic distortions from NL

4.1 Experimental studies for HCV and HCC in the node with a powerful NL

As an example of NL, which creates stochastic changes in voltage and power in the network, the arc steel and aluminum melting furnaces, receiving power from the Azerenerji system, are considered. The 2x50 MW steel furnace is located in Baku city and connected to the power grid through a high-voltage distribution network. The diagram of the power supply network and electrical connections of this point with other load nodes of the power grid are shown in Fig.1. ASF is considered as a NL in the studied network. The load variability probabilistic nature level is shown in Figure 2 in the form of HCC amplitude dynamics curves in the ASF and HCV circuit (at the point of connection of the NL to the network in the operating melting mode). Change of NDC HCV at the connection point is shown in Fig. 3.

In the paper, based on the proposed method, the design and experimental studies on the assessment of the contribution of ASF operation mode to the voltage, current and power in the power supply network are performed. The spectral composition of the current and voltage at the general connection point of the steel plant was established experimentally on the basis of measurements carried out during the month. The average values of harmonic voltage and current are given in tables 1and 2, respectively.

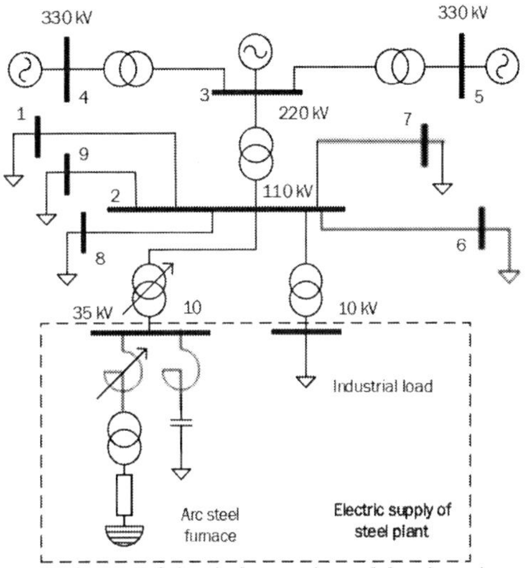

Fig.1. Diagram of electrical network supplying the Baku steel plant, taking into account the connections in the overall connection of ASF

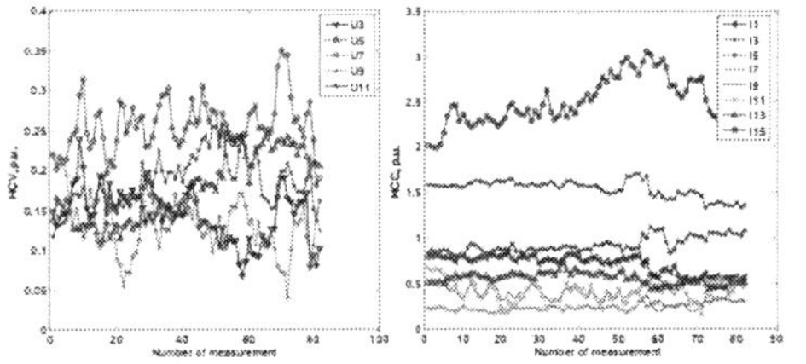

a) b)

Fig. 2. Change of HCV (U_n) circuit in the ASF connection point (a) and significant HCC (I_n) in ASF (b)

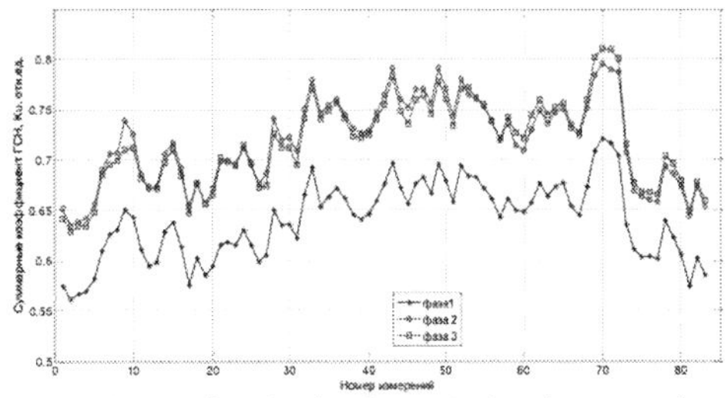

Fig. 3. Change of total ratio of HCV (K_U) at the connection point

As can be seen from tables 1 and 2, the most significant frequencies in the voltage and current spectrum are harmonics 2, 3, 4, 5 and 7. Taking this into account, we calculate the distortion power flow in the electrical network of the power supply system generated by electric furnace of the steel plant. The electrical network parameter values for these higher harmonics are determined using known relationships for transformer, reactor and capacitive branches, respectively [1, 3].

Table 1. Voltage harmonics spectrum (%).

Harmonics	Phase A $U = 21.8\ kV$	Phase B $U = 21.16\ kV$	Phase C $U = 21.14\ kV$
2	0,19	0,6	0,63
3	3,94	3,5	3,5
4	0,16	0,42	0,2
5	0,34	0,49	0,46
6	0,12	0,1	0,21
7	0,27	0,16	0,58
8	0,18	0,04	0,36
9	0,33	0,15	0,2
10	0,3	0,1	0,17
11	0,15	0,15	0,2
12	0,09	0,27	0,16
13	0,63	0,12	0,31
14	0,21	0,2	0,24
15	0,11	0,39	0,13
K_U	4,05	3,72	3,7

Table 2. Current harmonics spectrum (%).

Phase A Effective value $I = 161A$			
2	3,31	9	1,41
3	11,62	10	1,70
4	4,33	11	0,41
5	4,63	12	0,82
6	1,72	13	0,68
7	1,48	14	0,25
8	1,63	15	0,15
Phase C Effective value $I = 134A$			
2	8,25	2	1,72
3	10	3	1,32
4	9,48	4	0,85
5	10,60	5	1,43
6	1,98	6	1,24
7	3,58	7	0,02
8	0,38	8	0,89
Phase C Effective value $I = 152A$			
2	9,63	9	0,41
3	9,13	10	2,11
4	7,64	11	1,92
5	10,20	12	1,38
6	5,94	13	1,35
7	3,56	14	1,16
8	3,90	15	0,99

In accordance with this, using the electrical network parameters data for the fundamental frequency (Table 4), the values of these parameters for the frequencies of the selected harmonics are determined. Harmonic distortion powers at the input buses of 110 kV substation, supplying the steel plant (node 1 in Fig.1), and the distortion voltages for each phase are obtained as shown in Table 4.

Table 3. Total resistance of electric circuit of the furnace

Harmonic No.	Total resistance of electric circuit of the furnace, mOhm
2	10,26
3	15,39
4	20,52
5	25,6
7	35,91

Table 4. Harmonic distortion powers and voltages

Phase	Distortion power (kVA)	Distortion voltage (V)%
A	18,75	841/4
B	20,59	779/3,7
C	23,59	768/3,6

For the determination of the distortion power the following formula is used [15]:

$$D = \sqrt{S^2 + P^2 + Q^2} \quad (1)$$

where S - total power determined as follows:

$$S = \sqrt{\sum_{n=1}^{n_m} U_n^2 \cdot \sum_{n=1}^{n_m} I_n^2} \quad (2)$$

U_n and I_n – voltage and current harmonics respectively.

The total power of harmonic distortions is 64 kVA, which is 0.5% of the power of steady-state mode of the electric furnace – 13000 kVA. The energy of this power is imposed on the external network and is consumed in it in the form of losses in its elements and in the elements of nodal loads.

The amount of this energy for the period of one melting (in our case is about 2 hours) is $64 \cdot 0.9 \cdot 2 = 115\ kWt \cdot h$. At 10 – 12 meltings per day the harmonic distortion energy will be about 1380 kWh or ~ 41400 kW.h per month (accepted payable period for electricity consumed).

Designed distribution of this distortion energy in the electrical network is presented in Table 5. For the determination of the distortion energy the following formula is used [16]:

$$W_{\Sigma,D,P} = \sum_{n=2}^{40} W_{D,P,n} = \frac{T}{\pi} \sum_{n=2}^{40} P_n \quad (3)$$

$$W_{\Sigma,D,P} = \sum_{n=2}^{40} W_{D,Q,n} = \frac{T}{\pi} \sum_{n=2}^{40} Q_n \quad (4)$$

where $W_{D,P,n}$, $W_{D,Q,n}$ – distortion energy active and reactive components respectively; P_n, Q_n –active and reactive powers of harmonic components.

In the process of calculation of the harmonic content in the electrical network the distortion levels are determined not only in terms of the amount of energy emission in the form of additional power flows and losses of electricity, but also in magnitude (energy), and the voltages represent a quantitative measure of the contribution of each nonlinear load node to the deterioration of the PQ in the network of power supply organization.

Table 5. Load flows of harmonic distortion energy in the external network

Branch No.	Active energy flow (kW.h)	Energy losses in branches (kW.h)
1-2	3400	76
2-3	319,9	9,6
2-6	883,6	35,3
2-7	304,7	6,4
2-8	1009,3	32,3
2-9	285,6	9,1
2-10	300,5	9,0
3-4	137,5	2,9
3-5	179,2	5,6
Total:		186,2 кW.h

3.2 NL modeling on the basis of machine learning models

To create the ML models, the R software environment was used with attraction of "elmNN" and "MultivariateRandomForest" packages. The tested EML and RF models were trained on the basis of measurements received from SIMEAS Q and had the following structure: 9 input parameters (voltage of the fundamental harmonic and HCV harmonics 3, 5, 7, 9, 11, 13, 15, 17) and 5 output values (fundamental harmonic current and HCC 3, 5, 7, 9 harmonics).

At that, the ELM model had one hidden layer with 10 neurons with sinusoidal activation function (Fig. 4). RF model used the following parameters: number of trees, n_{tree}=5; the number of features for the selection of splitting, $m_{feature}$=5; minimum number of splitting objects min_{leaf}=5. At the intermediate stages of the study, the Bayesian ANN proposed for this task in [7] was also trained and tested. However, in our study of NL ASF modeling this neural network showed a low accuracy.

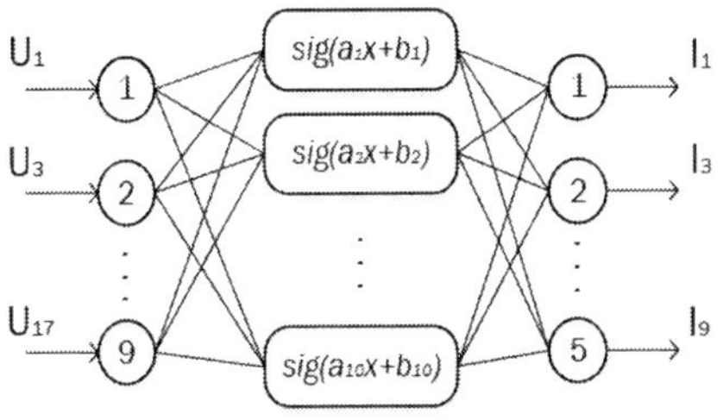

Fig. 4. Architecture of ELM used for the NL modeling

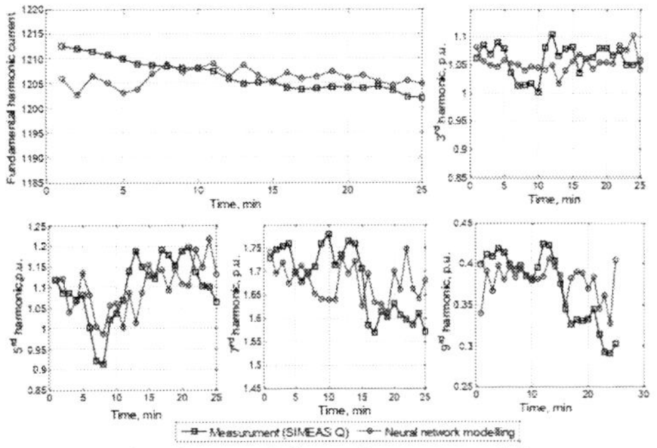

Fig. 5. Results of NL modeling using EML model

The results of testing of the trained models are presented in Fig. 5 and in Table 6. Standard regression metrics were used to evaluate the effectiveness: mean absolute error (MAE) and root- mean-square error (RMSE).

Table 6. Results of testing of machine learning models for the HCC ASF predict.

HCC	ELM model		RF model	
	MAE	*RMSE*	*MAE*	*RMSE*
Fundamental harmonic current	2.889	3.699	13.234	13.968
3rdharmonic	0.028	0.031	0.023	0.028
5thharmonic	0.056	0.068	0.065	0.083
7thharmonic	0.059	0.072	0.069	0.082
9thharmonic	0.031	0.040	0.042	0.051

As can be seen from Table 6, the best results are given by ELM, which, apparently, is determined by its best approximating properties in conditions of small samples [13]. However, it should be noted that a certain advantage of the RF model is that it does not require the analyzed data normalization. The choice of the optimal normalization method is a certain problem for neural network models, including the problem of NL behavior modeling [7].

As can be seen from Fig. 5, the obtained neural network model of the NL behavior on the example of ASF allows for the exact prediction of the HCC at different changes of HCV, for example, during the harmonic filter connection.

4 Conclusion

On the basis of experimental studies carried out on the 110 kV distribution network with the NL node in the form of a high-power ASF, the probability estimations of HCV in the nodes of the electric network from each harmonic are obtained and the total power distortion in the network is determined. The solution of the problem of adaptive NL modeling using ML algorithms to assess the effects of the inclusion of harmonic filters is presented.

This work was supported by the Science Development Foundation under the President of the Republic of Azerbaijan-Grant № EİF-BGM-4-RFTF-1/2017-21/09/1 and RFBR according to the research project №18-51- 06001.

References

1. Rakhmanov R.N., .Mamedyarov O.S., Bayramaliyeva E. M..*Identification of distortion of electrical network parameters with nonlinear loads*. Collection of reports of International conference. Identification and control problems. IPU, M. (2002)
2. R.N. Rahmanov, I.Z. Islamov. *Evaluation of the harmonic voltage distortion and its influence on power network insulation age*. in Trans. of CIGRE conference on Power system insulation coordination, Abu-Dhabi, 298-302 (1998)
3. Zhezhelenko I. V., Korotkevich M. A. *Electromagnetic compatibility in electrical networks*. - Minsk: Vysheyshaya school, 197 (2012)
4. Van den Keybus J. *Simulation of large numbers of non-linear loads in distribution systems* / J. Van den Keybus et al., in IEEE Ninth Int.Conf.on Harmonics and Quality of Power, **2**, (2000).
5. Rakhmanov R. N..Application of the neural network model to control the mode of electric power DSP // Energy problems.,No.1, 47-51 (2005)
6. Kovernikova L. I. Active powers of harmonics in the nodes of connection of nonlinear loads to the high voltage network / / Electricity, №3, 12-20 (2017)
7. Manuel L. *Modelling harmonics drawn by nonlinear loads* / L. Manuel, B. Josep, M. Juan, C. Montserrat and G. Eulalia, in 2015 9th International Conference on Compatibility and Power Electronics (CPE), Costa da Caparica, 93-97 (2015)
8. Sainz, L. Harmonic Interaction Influence Due to Current Source Shunt Filters in Networks Supplying Nonlinear Loads / L. Sainz, J. Balcells // IEEE Transactions on Power Delivery,**27**,No.3, 1385 - 1393 (2012)
9. Schipman, K. Harmonic amplification demystified [Electronic resource] / K.Schipman. ABB United Kingdom, (2011) – available from: http://www04.abb.com/global/seitp/seitp202.nsf/0/dba3255853945cfdc125783800573521?opendocument
10. Mazumdar J. RecurrentNeural Networks Trained With Backpropagation Through Time Algorithm to Estimate Nonlinear Load Harmonic Currents / J. Mazumdar and R. G. Harley // IEEE Transactions on Electronics, **55**, No.9, 3484-3491 (2018)
11. Huang G.-B. Extreme Learning Machine for Regression and Multiclass Classification / G.-B. Huang et al. // IEEE Transactions on Systems, Man, and Cybernetics,Part B, **42**, No.2, 513-529 (2011)
12. Segal, M. *Multivariate random forests* / M. Segal, and X. Yuanyuan // Wiley Interdisciplinary Reviews: Data Mining and Knowledge Discovery,**1**, No.1, 80-87 (2011)
13. Electric energy. Electromagnetic compatibility of technical means. Standards of quality of electric energy in power supply systems of general purpose. GOST 32144-2013-M.: Standardinform, 18 (2014)
14. Rakhmanov N. R., Kurbatsky V. G., Guliyev G. B., Tomin N. B. *Short-term forecasting of power output of wind power plants to ensure the reliability of electrical networks* / The 87-th Int. scientific seminar "Methodological issues of research of reliability of large energy systems", Minsk, Belarus, (2015)
15. Kopylov I. P. Definition of the dynamic energy indices at non-sinusoidal asymmetric voltage / I. P. Kopylov // Energetika, No.2, 31-38 (2001)
16. Farkhadzadeh E.M., Guliyev H.B. Bases of the Theory of Distortion in Electrical Circuits with Nonlinear Rower // Physics, №2, 29-33 (2003)

Modeling of three-phase electric motor operation by the MATLAB system with deteriorated power quality in the 0.38 kV distribution networks

I.V. Naumov[1]*, N.V.Savina[2] , M.V. Shevchenko[3]

[1] Irkutsk National Research Technical University, Amur State University
[2] Amur State University
[3] Far East State Agricultural University

Abstract. One of the main operation modes that characterizes power quality in distribution networks is asymmetry of three-phase voltage system. Operation of an induction motor (IM) with disturbed voltage symmetry in the supply network can not be considered as a rated one. The system of voltages applied to the stator winding of IM under these conditions contains positive- and negative-sequence components. This worsens the performance characteristics of IM essentially. In order to balance the 0.38 kV network operation and enhance the efficiency of the three-phase electric motor operation it is suggested to use a special balancing unit (BU) that minimizes the negative-sequence components of current and voltage. The operation modes of the obtained system "supply source – induction motor – balancing unit" are simulated within the MATLAB software package of applied programs, which allows one to assess the impact of low quality of power on the operating characteristics of the electric motor and the efficiency of the balancing unit to increase the "durability" of the motor under the asymmetrical power consumption.

1 Characteristic of asymmetric conditions

The coefficient of negative-sequence voltages should be considered as the main index of power quality that characterizes the level of voltage asymmetry in three-phase three-wire networks:

$$K_{2U} = \frac{U_2}{U_1},$$

where U_2, U_1 - the negative- and positive sequence voltages.

Besides, the operation of induction motors under voltage asymmetry is complicated by the emergence of symmetrical components of the negative-sequence current. This causes additional power losses in the windings of rotor and stator which result in local overheating and affect greatly the motor lifetime and operation stability. Such losses for three-phase three-wire (without zero wire) electric networks can be assessed using a coefficient of additional power losses [1]:

$$K_P = \frac{\Delta P_{NES.}}{\Delta P_{SIM.}} = \frac{3I_1^2 R_1 + 3I_2^2 R_2}{3I_1^2} = 1 + K_{2i}^2,$$

where $K_{2i} = \frac{I_2}{I_1}$ - the coefficient of negative-sequence current; I_2, I_1 - the negative- and positive-sequence currents; R_2, R_1 - the negative- and positive- sequence resistances.

Numerous researches in the field of creation of imitating models of operation of the electric motor practically did not consider its work in the asymmetrical mode [2 - 11].

An aggregate impact of asymmetrical currents and voltages at motor operation in an asymmetric electric network causes a considerable change in its operating characteristics.

2 A modeling tool

In order to estimate real processes taking place in the motor during its operation under asymmetrical voltages the need arises to develop a mathematical model of these processes.

For modeling dynamic systems and devices we compose and solve the systems of differential equations which are most often of nonlinear character. The MATLAB system (MATrix LABoratory) with Simulink extension, in our opinion, is an ideal tool for implementation of the

* Corresponding author: professornaumov@list.ru

 https://doi.org/10.1051/e3sconf/20185803016

modeling. The said extension makes it possible to implement visually-oriented programming of the problems of automatic construction of a graphical model of the system or device, derive and solve the equations of state, present visually the results of modeling [12].

The MATLAB version 6.5 with Simulink 5.0 add-in and library of the block of Sim Power Systems extensions, section Machines, were used to construct the models of the three-phase induction motor operation under voltage asymmetry in the supply network. The component Asynchronous Machine SI Units simulates an asynchronous electric machine in either motor or generator modes (determined by the sign of electromagnetic torque of the machine).

In the studied simulation model (Fig.1) ports A, B and C are the terminals of the machine stator winding. Port Tm is intended to supply drag torque; at terminal port m a vector signal forms. The signal includes 21 elements: currents, fluxes and voltages of rotor and stator in fixed and rotating coordinates, electromagnetic torque, number of shaft revolutions and its angular position. For convenience of extracting variables from the vector the SimPowerSystems library has the block Machines Measurement Demux (MMD).

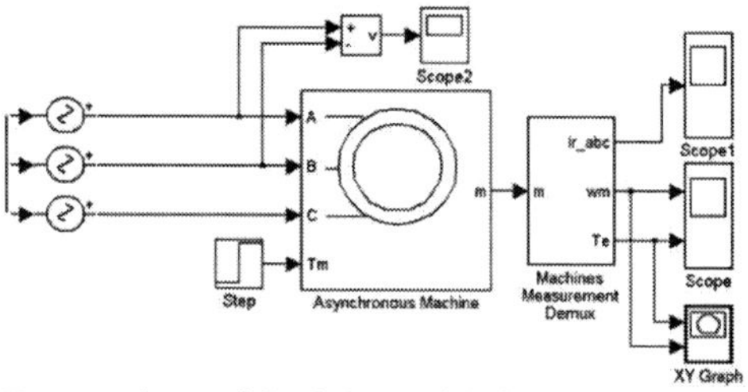

Fig.1. A scheme of simulation model of induction motor in the MATLAB system

The model of induction machine consists of a model of electric part represented by the fourth order state space and mechanical part being a second order system. All electric variables and parameters of the machine are reduced to the parameters of stator.

3 Computational example

The capacity of a simulated motor is 4 kW, rated rotation frequency - 950 rpm, rated voltage of the three-phase network is 0.38 kV with the frequency of 50 Hz. The parameters of the studied motor were set in the block for parameter input. The performance characteristics of the motor were measured since the time of the motor startup at a rated load. Based on the data from ports *ir, abc, wm,* Te of the MMD block we obtained the diagrams of currents in rotor, shaft angular velocity and electromagnetic torque as well as dynamic mechanical characteristics of the motor (Fig.2).

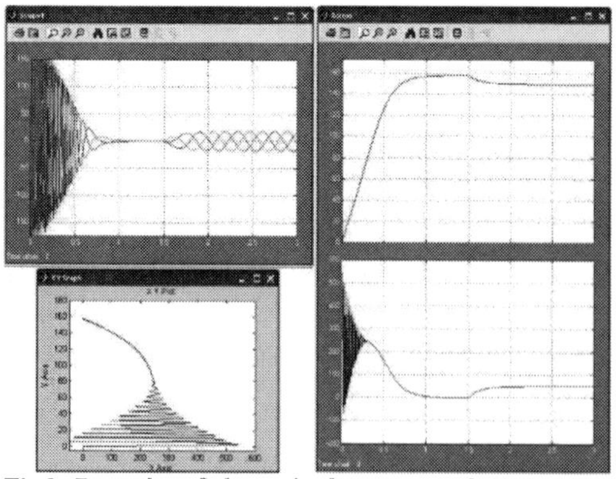

Fig.2. Dynamics of change in the motor performances at symmetrical voltage system

The symmetrizing device considerably reducing currents of the inverse sequence in the engine can serve as the most effective remedy of decrease in asymmetry of tension [13].

The parameters of balancing unit were calculated according to the largest set voltage asymmetry in the studied network [14]. According to the given initial data we conducted a series of experiments at different voltages of the power supply block of the induction motor. Here the change in the motor performances was studied both at connected and disconnected balancing unit.

The diagrams of the shaft angular velocity were analyzed as well as electromagnetic torque and dynamic mechanical characteristic of the motor.

The first experiment was conducted with symmetrical voltage system with rated parameters without balancing unit. The electromagnetic torque variations in the time range 0….0.35 sec (Fig.3) characterize the moment of motor start up. Then, after the speed is picked up and the operation mode is achieved, the operation of the motor stabilizes which is proved by the smooth section of the characteristics diagram. Electromagnetic field of the motor, which is generated by symmetrical voltage, has a regular radial form; the operation of the motor at a rated load is smooth without failures.

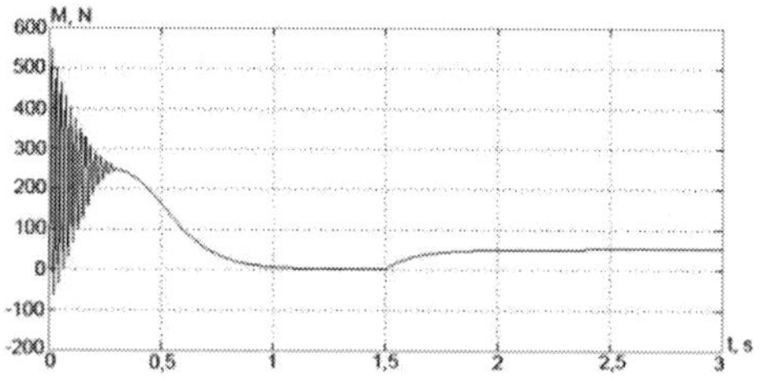

Fig.3. Dynamics of change in electromagnetic torque at a symmetrical voltage system in the supply network

The second series of experiments simulated different voltage asymmetry operation modes in supply network without balancing unit.

With increase in the asymmetry level the performance of the induction motor grows worse. In the operation mode with maximum asymmetry the increase in time of angular velocity acceleration was about 1%, the capacity developed on the shaft of the motor decreased by 5%. At the same time the operation of the motor itself becomes unstable which manifests itself through pulsation of velocity and electromagnetic torque (Fig.4). In order to determine more accurate relationship between the processes that occur during motor operation and coefficient of negative sequence voltage it is necessary to conduct additional research.

When the voltage asymmetry level (Ku2=3.9%) was simulated we observed an essentially unstable electromagnetic torque and rotation speed of the motor rotor (Fig.4).

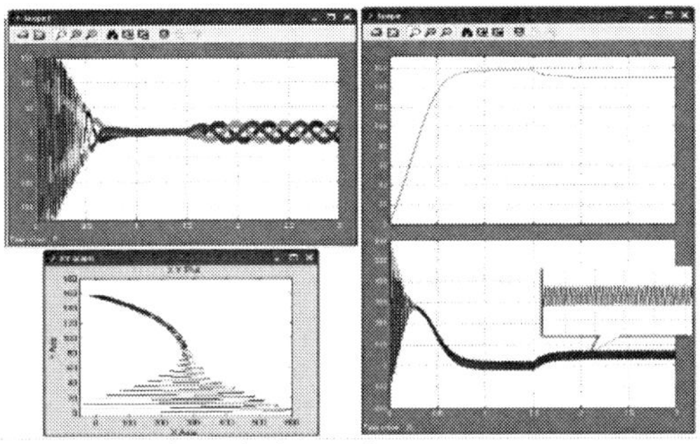

Fig.4. Dynamics of change in motor parameters at asymmetric voltage system in the network

This is caused by the change in the shape of stator and rotor magnetic fields from radial to elliptic under the influence of the negative sequence currents. This leads to emergence of a counteracting torque and decrease in available power of the motor.

The next series of experiments were conducted under the same level of voltage asymmetry in the supply network but with the balancing unit connected before the motor (Fig.5).

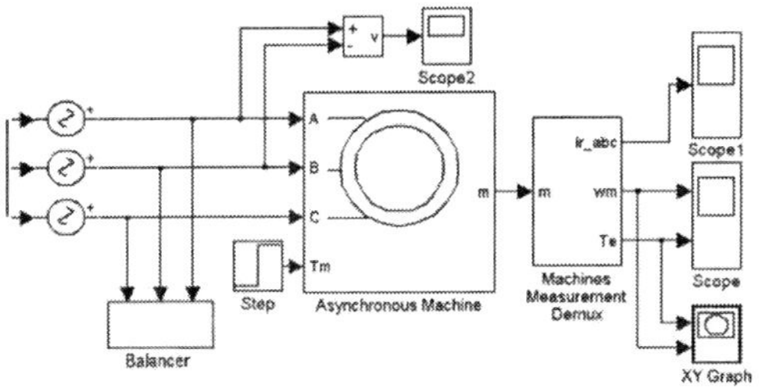

Fig.5. Flow chart of the simulation model of induction motor with the balancing unit in the MATLAB system

The balancing unit made it possible to compensate the voltage asymmetry in the supply network which affected positively the performance of the motor. For example when simulating the network asymmetry characterized by the negative sequence coefficient K2u=3.9% with balancing unit before the motor the performances of the motor became better (Fig.6). The amplitude of electromagnetic torque pulsation decreased approximately 3 times, the value of the negative sequence voltage coefficient at the motor supply input decreased up to 1.1%.

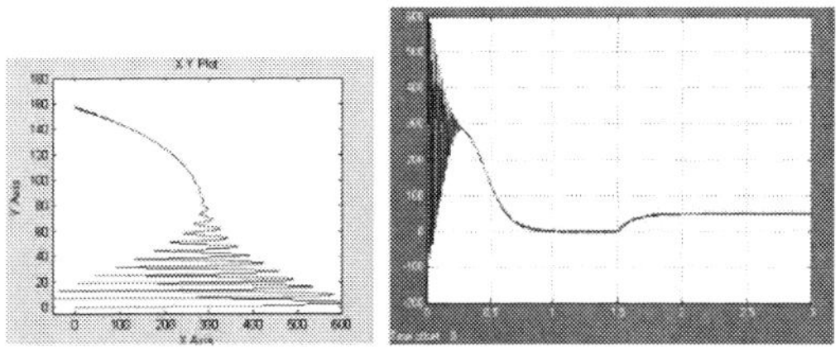

Fig.6. Dynamic mechanical characteristic and electromagnetic torque characteristic of the motor at asymmetry in the network with connected balancing unit.

4 Conclusion

Thus, the use of the simulation model in the MATLAB system allowed one in a rather short period to analyze the induction motor operation under different operating conditions of the supply network, reveal the extent of influence of the network voltage asymmetry, determine the effect of balancing unit connection. Hence, the use of balancing units for normalization of power quality and decrease of power losses enhances considerably the efficiency of three-phase electric motor operation in the public distribution networks.

The symmetrizing device is the most effective remedy of upgrading of operation of the three-phase motor in the conditions of asymmetry of tension

References

1. Kosoukhov F.D., I.V.Naumov Asymmetry of voltages and currents in rural distribution networks. Irkutsk: IS A A,2003. 257 p.
2. Shi K.L., Chan T.F., Wong Y.K., Ho S.L.. Modelling of the three-phase induction motor using SIMULINK/ Hong Kong Polytechnic University, Department of Electrical Engineerin// Int. J. Elect. Enging. Educ., Vol. 36, pp. 163–172. Manchester, UK, 1999.
3. Chani S.N. Digital computer simulation of three-phase induction machine-dynamics – a ganaralizd approach // IEEE Trans Industry Apple., No.1, pp. 106-114, 1988.
4. Waid S., Dunnigan M.V., Williams B.W. Modeling and simulation of induction machine vector control and rotor resistance indentificftion/ IEEE Trans. Power Electronicse, Vol. 12, No. 3, pp. 495-505, 1997.
5. Rrause P.C., Wasynezuk O., Sudholff S.D. Analysis of Electrik Mashinery, IEEE, 1995.
6. Krishnan R., Electric Motors Drives Modeling Analysis and Control, Publication Prentice Hall, India, 2002.
7. Bimal. K. Bose, Modern Power Electronics & AC Drives, Publication Prentice Hall, India, 2003
8. Tang L., Rahman M.F., A new direct torque control strategy for flux and torque ripple reduction for induction motors drive – a Matlab/Simulink model // IEEE International

Electric Machines and Drives Conference, 2001, pp. 884–890.
9. Leedy Aleck W. Simulink / MATLAB Dynamic Induction Motor Model for Use as a Teaching and Research Tool // International Journal of Soft Computing and Engineering (IJSCE), ISSN: 2231-2307, Volume-3, Issue-4, September, 2013.
10. Le-Huy H. Modeling and simulation of electrical drives using Matlab/Simulink and Power System Blockset // The 27th Annual Conference of the IEEE Industrial Electronics Society (IECON'01), Denver/Colorado, pp. 1603–1611.
11. Lakhya Jyoti Phukon, Neelanjana Baruah. A Generalized Matlab Simulink Model of a Three Phase Induction Motor // International Journal of Innovative Research in Science, Engineering and Technology. Vol. 4, Issue 5, May 2015.
12. MATLAB 6/6.1/6.5+Simulink 4/5. The Fundamentals of application. M.L SOLON-Press. 2004. 768 p.
13. Naumov I.V. The symmetrizing device for three-phase four-wire network with adjustable parameters / Naumov I.V., Ivanov D.A., Podiyachikh S.V., Shpak D. //Patent for the useful model No. 61063 – It is published 10.02.2007 Bulletin No. 4, (in Russian).
14. Naumov I.V. Choice of parameters of balancing devices in the 0.38 kV distribution networks / Naumov I.V., Belousova E.A. // Bulletin of KrasSAU, issue 1, 2017, No.1. –P.99-107. (in Russian).

AUTHOR INDEX

AUTHOR INDEX

EDP Sciences – Web of Conferences
17, Avenue du Hoggar
Parc d'Activité de Courtabouf
BP 112
F-91944 Les Ulis Cedex A
France

ISBN 978-1-5108-7296-7

CURRAN ASSOCIATES INC.
proceedings.com